Perfluoroalkyl Substances in the Environment

Theory, Practice, and Innovation

ENVIRONMENTAL AND OCCUPATIONAL HEALTH SERIES
Series Editors:
LeeAnn Racz
Bioenvironmental Engineer, US Air Force, Brandon, Suffolk, UK
Adedeji B. Badiru
Professor and Dean, Graduate School of Engineering and Management, AFIT, Ohio

Handbook of Respiratory Protection: Safeguarding Against Current and Emerging Hazards
LeeAnn Racz, Robert M. Eninger, and Dirk Yamamoto

Perfluoroalkyl Substances in the Environment: Theory, Practice, and Innovation
David M. Kempisty, Yun Xing, and LeeAnn Racz

For more information about this series, please visit: https://www.crcpress.com/Environmental-and-Occupational-Health-Series/book-series/CRCENVOCCHEASER

Perfluoroalkyl Substances in the Environment
Theory, Practice, and Innovation

Edited by
David M. Kempisty
Yun Xing
LeeAnn Racz

CRC Press
Taylor & Francis Group
Boca Raton London New York

CRC Press is an imprint of the
Taylor & Francis Group, an **informa** business

CRC Press
Taylor & Francis Group
6000 Broken Sound Parkway NW, Suite 300
Boca Raton, FL 33487-2742

© 2019 by Taylor & Francis Group, LLC
CRC Press is an imprint of Taylor & Francis Group, an Informa business

No claim to original US Government works

Printed on acid-free paper

International Standard Book Number-13: 978-1-4987-6418-6 (Hardback)

This book contains information obtained from authentic and highly regarded sources. Reasonable efforts have been made to publish reliable data and information, but the author and publisher cannot assume responsibility for the validity of all materials or the consequences of their use. The authors and publishers have attempted to trace the copyright holders of all material reproduced in this publication and apologize to copyright holders if permission to publish in this form has not been obtained. If any copyright material has not been acknowledged please write and let us know so we may rectify in any future reprint.

Except as permitted under US Copyright Law, no part of this book may be reprinted, reproduced, transmitted, or utilized in any form by any electronic, mechanical, or other means, now known or hereafter invented, including photocopying, microfilming, and recording, or in any information storage or retrieval system, without written permission from the publishers.

For permission to photocopy or use material electronically from this work, please access www.copyright.com (http://www.copyright.com/) or contact the Copyright Clearance Center, Inc. (CCC), 222 Rosewood Drive, Danvers, MA 01923, 978-750-8400. CCC is a not-for-profit organization that provides licenses and registration for a variety of users. For organizations that have been granted a photocopy license by the CCC, a separate system of payment has been arranged.

Trademark Notice: Product or corporate names may be trademarks or registered trademarks, and are used only for identification and explanation without intent to infringe.

Library of Congress Cataloging-in-Publication Data

Names: Kempisty, David M., editor. | Xing, Yun (Writer on chemistry) | Racz, LeeAnn, editor.
Title: Perfluoroalkyl substances in the environment : theory, practice, and innovation / editors, David M. Kempisty, Yun Xing, LeeAnn Racz.
Description: Boca Raton : CRC Press, Taylor & Francis Group, 2018. | Series: Environmental and occupational health series
Identifiers: LCCN 2018011967| ISBN 9781498764186 (hardback : alk. paper) | ISBN 9780429487125 (ebook)
Subjects: LCSH: Surface active agents--Environmental aspects. | Fluorocarbons--Environmental aspects.
Classification: LCC TD427.D4 P3945 2018 | DDC 628.5/2--dc23
LC record available at https://lccn.loc.gov/2018011967

Visit the Taylor & Francis Web site at
http://www.taylorandfrancis.com

and the CRC Press Web site at
http://www.crcpress.com

Dedicated to those who endeavor to ensure abundant, clean water for future generations.

Contents

Foreword xi
Editors xiii
Contributors xv
Introduction xix

SECTION I   Introduction and Analysis

Chapter 1   Fluorosurfactants in Firefighting Foams: Past and Present 3

Stephen H. Korzeniowski, Robert C. Buck, David M. Kempisty, and Martial Pabon

Chapter 2   Per- and Polyfluoroalkyl Substance Analysis to Support Site Characterization, Exposure, and Risk Management 35

Kavitha Dasu, David M. Kempisty, and Marc A. Mills

Chapter 3   Understanding Precursor Contributions: The Total Oxidizable Precursor Assay 67

Bharat Chandramouli

SECTION II   Regulations

Chapter 4   Managing Risk from Perfluorinated Compounds in Drinking Water 83

Steve Via

Chapter 5   Emerging Contaminant Monitoring as a Host Nation Guest: An Environmental Health Professional's Perspective 91

Andrew J. Wagner

Chapter 6   Uncharted Waters: Challenges for Public Water Systems Addressing Drinking Water Health Advisory Levels for PFOS and PFOA 103

Miranda Brannon

Chapter 7 Challenges of Managing Emerging Contaminants: Historical Per- and Polyfluorinated Alkyl Substance Use in the US Air Force 113

R. Hunter Anderson and David M. Kempisty

SECTION III Toxicology

Chapter 8 Human Health Risk Assessment of Perfluorinated Chemicals 123

Penelope Rice

Chapter 9 Perfluoroalkyl Substance Toxicity from Early-Life Exposure 171

David Klein and Joseph M. Braun

Chapter 10 PFAS Isomers: Characterization, Profiling, and Toxicity 203

Yun Xing, Gabriel Cantu, and David M. Kempisty

SECTION IV Remediation

Chapter 11 Water Treatment Technologies for Targeting the Removal of Poly- and Perfluoroalkyl Substances 241

Eric R. V. Dickenson and Edgard M. Verdugo

Chapter 12 Oxidation and Reduction Approaches for Treatment of Perfluoroalkyl Substances 255

Blossom N. Nzeribe, Selma M. Thagard, Thomas M. Holsen, Gunnar Stratton, and Michelle Crimi

Chapter 13 Reactivation of Spent Activated Carbon Used for PFAS Adsorption 303

John Matthis and Stephanie Carr

Chapter 14 Ion Exchange for PFAS Removal 325

Steven Woodard, Michael G. Nickelsen, and Marilyn M. Sinnett

Contents

Chapter 15 Occurrence of Select Perfluoroalkyl Substances at US Air Force Aqueous Film-Forming Foam Release Sites Other than Fire Training Areas: Field Validation of Critical Fate and Transport Properties 353

R. Hunter Anderson, G. Cornell Long, Ronald C. Porter, and Janet K. Anderson

Chapter 16 A Preliminary Treatment Train Study: Removal of Perfluorinated Compounds from Postemergency Wastewater by Advanced Oxidation Process and Granular Activated Carbon Adsorption 373

Sean M. Dyson, Christopher Schmidt, and John E. Stubbs

Chapter 17 Remediation of PFAS-Contaminated Soil 383

Konstantin Volchek, Yuan Yao, and Carl E. Brown

Chapter 18 Soil and Groundwater PFAS Remediation: Two Technology Examples 405

David F. Alden, John Archibald, Gary M. Birk, and Richard J. Stewart

Chapter 19 Per- and Polyfluoroalkyl Substances in AFFF-Impacted Soil and Groundwater and Their Treatment Technologies 417

Bo Wang, Abinash Agrawal, and Marc A. Mills

Chapter 20 Short-Chain PFAS: Their Sources, Properties, Toxicity, Environmental Fate, and Treatment 447

Yuan Yao, Justin Burgess, Konstantin Volchek, and Carl E. Brown

Chapter 21 Case Study: Pilot Testing Synthetic Media and Granular Activated Carbon for Treatment of Poly- and Perfluorinated Alkyl Substances in Groundwater 467

Brandon Newman and John Berry

Author Index 485

Subject Index 487

Foreword

Addressing the issues that poly- and perfluoroalkyl substances (PFAS) raise will be an enduring challenge for our generation of environmental engineers and chemists. Other classes of trace organic chemicals, such as pharmaceuticals and personal care products, may raise concerns wherever and whenever wastewater impacts our aquatic environment; however, to quote Calvin Coolidge, "Nothing in this world can take the place of persistence." In this manner, PFAS seem to have cornered the market.

It is not just the extreme persistence of their multiple carbon-fluorine bonds that have captured our attention. The unique surfactant behavior of many PFAS, particularly those associated with the use of aqueous film-forming foams (AFFFs), which are the focus of this book, is leading to many complications as we consider the environmental fate and transport of these otherwise relatively mobile substances. Moreover, as much as one might want to simplify the PFAS world into just perfluorooctanoate (PFOA) and perfluorooctane sulfonate (PFOS), the chemistry, and everything that goes along with it, is much more complex than I think many of us imagined when we first started to study these chemicals. It is not just about the anions—there are neutrals and cations too! Indeed, developing multiple lines of analytical evidence to demonstrate that one has accurately accounted for the diverse chemistry of PFAS present at AFFF-impacted sites is a current focus of many research groups. Though I am a reluctant analytical chemist, I seem to be more often suggesting that we simply embrace the complexity of the PFAS chemistry (as best we can) in the environment if we want any hope of making sense of the complicated web of precursors and products.

Of course, it is not just the chemistry of the PFAS themselves and what happens to them in the environment that is so complex: the multitude of uses of PFAS has also led to considerable confusion, or at least complexity, related to sources of exposure. Folks can get confused pretty quickly when you explain that the chemicals that they have used to keep stains off of their carpets and clothing are kissing cousins of those that were used to put out fuel fires and are now contaminating their drinking water supply. Never mind the fact that we are really only starting to scratch the surface with respect to understanding their toxicity. Nevertheless, the exposures that are happening and have happened are real, and the public is left wanting at least an explanation, if not a solution, to this complicated problem.

This leads me to why *this* book and why *now*. On the whole, this collection of information is just what is needed at this point in time. In November 2016 the US Environmental Protection Agency issued its long-anticipated health advisories for PFOS and PFOA in US drinking water supplies. Since then, there has been an onslaught of media coverage of the issue, with local press outlets reporting on every new groundwater plume. As a result, many nonacademicians are having to quickly come up to speed on a variety of topics and issues specific to PFAS. For those who have not been directly engaged with the scientific discussions on PFAS for the past 15 years, this could seem a very daunting task. What this book provides is information not just on the complicated chemistry and nomenclature, but also on the toxicology, regulatory approaches, analytical chemistry, and perhaps most importantly, the diverse ways in which environmental engineers are trying to remediate PFAS. Though it is downright impossible to capture every PFAS issue, particularly in this rapidly expanding research environment, Drs. Kempisty, Racz, and Xing have included a broad array of perspectives and cover an even broader array of PFAS topics. At the very minimum, this book will serve as an excellent gateway to the wild world of PFAS. Whether you are a federal, state, or local government official being thrust into the PFAS world, a student wanting to learn what all the fuss is about, or simply a member of the public trying to understand just why these chemicals are so difficult to treat, I would encourage you to engage with this book as much as you can. Take notes. Look up references. Follow your passion.

Good luck!

Christopher Higgins
Department of Civil and Environmental Engineering
Colorado School of Mines, Golden, Colorado

Editors

David M. Kempisty is a bioenvironmental engineer in the US Air Force with more than 20 years of active duty experience. He is currently serving in the grade of lieutenant colonel. In his present assignment, Dave is the chief of medical chemical, biological, radiological, and nuclear operations in the Surgeon's Directorate at the North American Aerospace Defense Command and US Northern Command at Peterson Air Force Base, Colorado. Previously, Dave was an assistant professor and the director for the Graduate Environmental Engineering and Science Program in the Department of Systems Engineering and Management at the Air Force Institute of Technology (AFIT). In this capacity, he and his students conducted research into perfluorinated alkyl substances (PFAS), to include both toxicological and remedial aspects of PFAS. He's been recognized for this research with the 2017 US Environmental Protection Agency's Office of Research and Development bronze medal for water infrastructure and technologies research. Previous to AFIT, Dave earned his PhD from the University of Colorado–Boulder in civil engineering. His dissertation involved investigated improved PFAS adsorption to granular activated carbon. Dave is a licensed professional engineer in the state of Michigan. He earned his BS degree from Michigan Technological University and his MS degree from AFIT. In addition to research and policy associated with PFAS, other areas of interest include environmental health issues, sustainability initiatives, and improving defense support to civil authorities after natural and man-made disasters. Dave is active in multiple professional societies, has led and presented at international and domestic conferences, and has authored many peer-reviewed and trade journal articles.

Yun Xing is a postdoctoral research scholar at the Air Force Institute of Technology (AFIT). She earned a BS in biochemical engineering from Tianjin University, China, and a PhD in bioengineering from the Georgia Institute of Technology. Her previous work experiences include cancer nanotechnology research at Stanford University and research on polyfluorinated alkyl substance remediation and isomer cytotoxicity at AFIT. Her areas of interest include bionanotechnology, toxicity of emerging materials, pollutants of emerging concern, and characterization of the effects and fate of biocontaminants in the environment. She has authored 19 peer-reviewed journal articles and 5 book chapters with more than 1500 citations. She is a member of several professional associations and has served as a reviewer for a number of journals, including the *Analyst*, *ACS Nano*, and *Applied and Environmental Microbiology*.

LeeAnn Racz is a bioenvironmental engineer in the US Air Force, having served at bases across the globe. She currently serves as commander of the 1st Special Operations Aerospace Medicine Squadron at Hurlburt Field, Florida. Previous assignments have included chief of consultative services at the US Air Force School of Aerospace Medicine, as well as assistant professor of environmental engineering and director of the Graduate Environmental Engineering and Science Program in the

Department of Systems Engineering and Management at the Air Force Institute of Technology. She presently holds the rank of lieutenant colonel. She is a licensed professional engineer, certified industrial hygienist, and board-certified environmental engineer. She earned a BS in environmental engineering from California Polytechnic State University, San Luis Obispo; an MS in biological and agricultural engineering from the University of Idaho; and a PhD in civil and environmental engineering from the University of Utah. Her areas of interest include characterizing the fate of chemical warfare agents and pollutants of emerging concern in the natural and engineered environments, as well as environmental health issues, and using biological reactors to treat industrial waste. She has authored dozens of refereed journal articles, conference proceedings, magazine articles, and presentations, and edited four handbooks. She is a member of several professional associations and honor societies and has received numerous prestigious teaching and research awards.

Contributors

Abinash Agrawal
Wright-State University
Dayton, Ohio

David F. Alden
Tersus Environmental
Wake Forest, North Carolina

Janet K. Anderson
Integral Consulting Inc.
San Antonio, Texas

R. Hunter Anderson
Air Force Civil Engineer Center
Joint Base San Antonio–Lackland
San Antonio, Texas

John Archibald
Tersus Environmental
Grimsby, Ontario, Canada

John Berry
Emerging Compounds Treatment
  Technologies, Inc.
Bedford, New Hampshire

Gary M. Birk
Tersus Environmental
Wake Forest, North Carolina

Miranda Brannon
US Air Force
Office of the Surgeon General
Falls Church, Virginia

Joseph M. Braun
Department of Epidemiology
Brown University
Providence, Rhode Island

Carl E. Brown
Environment and Climate Change
  Canada
Ottawa, Ontario, Canada

Robert C. Buck
The Chemours Company
Wilmington, Delaware

Justin Burgess
Environment and Climate Change
  Canada
Ottawa, Ontario, Canada

Gabriel Cantu
US Air Force School of Aerospace
  Medicine
Wright-Patterson Air Force Base
Dayton, Ohio

Stephanie Carr
Calgon Carbon Corporation - A
  Kuraray Company
Moon Township, Pennsylvania
Retired

Bharat Chandramouli
SGS AXYS
Sidney, British Columbia, Canada

Michelle Crimi
Clarkson University
Potsdam, New York

Kavitha Dasu
Battelle
Norwell, Massachusetts

Eric R. V. Dickenson
Southern Nevada Water Authority
Henderson, Nevada

Sean M. Dyson
Air Force Institute of Technology
Wright-Patterson Air Force Base
Dayton, Ohio

Thomas M. Holsen
Clarkson University
Potsdam, New York

David M. Kempisty
US Northern Command
Peterson Air Force Base
Colorado Springs, Colorado

David Klein
Department of Pathology and
 Laboratory Medicine
Brown University
Providence, Rhode Island

Stephen H. Korzeniowski
BeachEdge Consulting, LLC
Media, Pennsylvania

G. Cornell Long
Air Force Civil Engineer Center
Joint Base San Antonio–Lackland
San Antonio, Texas

John Matthis
Calgon Carbon Corporation - A
 Kuraray Company
Moon Township, Pennsylvania

Marc A. Mills
US Environmental Protection Agency
Cincinnati, Ohio

Brandon Newman
Amec Foster Wheeler Environment &
 Infrastructure, Inc.
Portland, Maine

Michael G. Nickelsen
Emerging Compounds Treatment
 Technologies, Inc.
Rochester, New York

Blossom N. Nzeribe
Clarkson University
Potsdam, New York

Martial Pabon
Dynax Corporation
Pound Ridge, New York

Ronald C. Porter
Noblis
San Antonio, Texas

Penelope Rice
Food and Drug Administration
College Park, Maryland

Christopher T. Schmidt
US Air Force
Little Rock Air Force Base
Little Rock, Arkansas

Marilyn M. Sinnett
Emerging Compounds Treatment
 Technologies, Inc.
Portland, Maine

Richard J. Stewart
Ziltek Pty Ltd.
Thebarton, South Australia, Australia

Gunnar Stratton
Clarkson University
Potsdam, New York

John E. Stubbs
Air Force Institute of Technology
Wright-Patterson Air Force Base
Dayton, Ohio

Contributors

Selma M. Thagard
Clarkson University
Potsdam, New York

Edgard M. Verdugo
Southern Nevada Water Authority
Henderson, Nevada

Steve Via
American Water Works Association
Washington, DC

Konstantin Volchek
Environment and Climate Change Canada
Ottawa, Ontario, Canada

Andrew J. Wagner
US Air Force
Joint Base Andrews
Camp Springs, Maryland

Bo Wang
Wright-State University
Dayton, Ohio

Steven Woodard
Emerging Compounds Treatment Technologies, Inc.
Portland, Maine

Yun Xing
Air Force Institute of Technology
Dayton, Ohio

Yuan Yao
Environment and Climate Change Canada
Ottawa, Ontario, Canada

Introduction

Per- and polyfluorinated alkyl substances (PFAS), often referred to as per- (and poly-) fluorinated compounds, have been used for years, providing desirable properties to many everyday products we use. From nonstick cookware to stain-resistant coatings on clothing and upholstery and even in our personal care products, perfluorinated compounds are common in our everyday environment. Particularly in the aviation and firefighting communities, perfluorinated chemicals have been incorporated into the agents used to fight hydrocarbon-based fires. A pivotal event highlighting the need for the best firefighting agent possible was a shipboard fire aboard the USS *Forrestal* in 1967. In the end, 134 sailors lost their lives, countless more were injured, and the loss of aircraft and ship damage was substantial. Lack of standardization, inadequate firefighting procedures, and a lack of capabilities for the firefighting operations were identified as major lessons learned from the incident. Improvements to the firefighting agents were made as a result of the tragic fire, and the use of aqueous film-forming foam (AFFF) containing perfluorinated compounds gained popularity over similar protein-based foams. AFFF use has continued, and although effective in extinguishing Class B hydrocarbon-based fires, perfluorinated compounds also have negative aspects associated with them; most notably, exposure has been linked to adverse health effects in humans. Other concerns include the long biological half-life in humans (on the order of years) and their persistence in the environment for periods of years and tens of years after application. Soluble in water, PFAS are now being identified in drinking water sources. Research continues to determine adverse health effect contributions due to various exposure routes, but nonetheless, a remedy for drinking water sources is needed.

This book discusses the various challenges of PFAS in our environment today. First, the diverse number of per- and polyfluorinated compounds is vast. In order to establish some kind of boundary condition for the scope of this book, discussion is primarily centered around PFAS contained in the firefighting and aviation community. Some discussion is involved outside of this scope, but it is limited. Section I, "Introduction and Analysis," provides background of the issues we face today. Chapter 1 sheds light on the history, nomenclature, and chemistry associated with the compounds in the PFAS suite. Discussion on the different types of firefighting foams and the military specification requirements is also covered in this chapter.

There are analytical challenges with reaching statistically reproducible detection limits. In addition to the detection of the trace concentrations, the ubiquitous presence of potentially cross-contaminating Teflon and other fluorinated compounds in an analytical laboratory make the baseline analytical work center requirements more stringent than those for other more common environmental contaminants. Chapters 2 and 3 provide an overview of the various detection methods available to determine PFAS concentrations in various environmental media.

Regulatory perspectives, discussed in Section II, are another dynamic landscape, and navigating them presents another challenge. The US Environmental Protection

Agency released a health advisory in 2016 stating that water with a combined perfluorooctane sulfonate (PFOS) and perfluorooctanoate (PFOA) concentration under 70 parts per trillion would result in no adverse health effects. The language went on to state that the advisory was not to be used for regulatory purposes, but with the lack of a maximum contaminant level for any perfluorinated substance, the health advisory has become a de facto regulatory limit. Further complicating things, individual states have established different allowable limits of PFAS to finished drinking water. Although not yet passed, New Jersey's Drinking Water Quality Institutes has recommended setting a limit of 14 parts per trillion for PFOA and 13 parts per trillion for another perfluorinated substance, the nine-carbon perfluorononanoic acid. Different concentrations and different regulatory language (maximum contaminant limits, health advisories, and summed concentrations) complicate remedial goals, drinking water standards, and risk communication language for agencies working in multiple states, such as often occurs in the aviation and defense communities. Chapters 4 through 7 explore these regulatory perspectives and give practical recommendations for operating with these changing requirements.

Section III discusses toxicological profiles, which are still being developed and advanced. Chapter 8 is a comprehensive review of the human health risks associated with exposure to PFAS. Toxicity with early life exposures is discussed further in Chapter 9. Isomers are another suite of PFAS that deserve toxicological investigation; depending on the manufacturing process, up to 30% of the finished PFAS product can be in an isomer form. The toxicological profile for these isomers is different than that of their straight-chained counterparts and is discussed in Chapter 10.

Determining the true toxicity, reacting to the changing regulatory landscape, and being aware of analytical and sampling challenges all present challenges to the community responsible for developing a PFAS management strategy. Perhaps the most challenging aspect of dealing with legacy PFAS contamination is the remediation of impacted soil and water. Remedial treatment objectives influenced by changing regulatory standards and advisories can make identifying efficient and effective remediation methods difficult. A considerable portion of this book, Section IV, discusses methods employed with varying degrees of success. The remediation of soil media and short-chain PFAS compounds are covered in Chapters 17 and 21. Conventional approaches for treating PFAS-impacted water, such as activated carbon and its reactivation of spent media, are discussed in Chapters 11 and 13. More novel approaches employing redox processes are covered in Chapter 12. Additionally, combinations of the treatment technologies, such as an advanced oxidization process with activated carbon and ion exchange and activated carbon polishing, are covered in Chapters 14 and 16.

Although perfluorinated compounds are a daunting challenge across multiple fronts, there is a plethora of ongoing work to resolve these challenges. This book explores the challenges across the topical areas of regulation and management, toxicology, environmental remediation, and analytical sampling and analysis. The text also provides insight into the great amount of work progressing to address the challenges. We hope the readers find this text helpful in understanding the complexities

Introduction

to a greater level. Our goal for this text is to have the analytical chemist and others practicing in the environmental engineering field appreciate PFAS regulatory issues and to increase understanding of PFAS toxicology. At the same time, an equal motivation for assembling this effort is to leave readers with an optimistic view of the meritorious efforts underway and the promising advances being seen across the PFAS spectrum.

David M. Kempisty
Yun Xing
LeeAnn Racz

Section I

Introduction and Analysis

1 Fluorosurfactants in Firefighting Foams
Past and Present

Stephen H. Korzeniowski, Robert C. Buck, David M. Kempisty, and Martial Pabon

CONTENTS

1.1 Introduction 3
1.2 Fluorosurfactant Chemistry 4
    1.2.1 Electrochemical Fluorination (ECF) Process 4
    1.2.2 Fluorotelomer Process 6
1.3 Long Chain and Short Chain: Terminology and Structures 6
1.4 Firefighting Foam before Fluorosurfactants 8
1.5 Fluorosurfactants in AFFF 9
    1.5.1 ECF-Based Fluorosurfactants Used in AFFF 9
    1.5.2 Fluorotelomer-Based Fluorosurfactants Used in AFFF 11
1.6 Firefighting Foams: Types and Composition 12
    1.6.1 Types of Firefighting Foam 13
    1.6.2 AFFF Ingredients and Composition 14
1.7 Qualified Products List (QPL) 15
1.8 Firefighting Foam Selection and Use 20
1.9 AFFF and F3: Field Test Performance and Efficacy 23
1.10 AFFF Use: Best Practices 25
1.11 Summary and Conclusions 27
Acknowledgments 29
References 29

1.1 INTRODUCTION

Fluorosurfactants are a unique class of surfactants whose properties originate from the substitution of hydrogen with fluorine along the carbon backbone that makes up the hydrophobic part of a surfactant (Taylor 1999; Kissa 2001; Pabon and Corpart 2002; Buck et al. 2012). Fluorinated surfactants have been commercially available since the 1950s. The most common commercial fluorosurfactants contain a perfluoroalkyl moiety, $F(CF_2)_n^-$, bound to a spacer group that is connected to a cationic, anionic, amphoteric, or nonionic functional group (Figure 1.1).

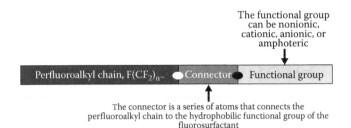

FIGURE 1.1 Fluorosurfactant schematic.

The carbon-fluorine bond is very strong. The atomic volume and ionic radius of fluorine shield carbon from a chemical attack. Thus, the perfluoroalkyl moiety is very rigid compared with the analogous hydrocarbon equivalent. The rigidity, ionic character, and large molecular volume of the perfluoroalkyl moiety drive the unique adsorption and aggregation behavior of fluorosurfactants. Moreover, the perfluoroalkyl functional moiety has very low surface energy and is both hydrophobic and oleophobic. Therefore, in both aqueous and hydrocarbon solutions, fluorosurfactants have a driving force to move to the air–liquid interface. The high chemical and thermal stability, surface tension lowering, and efficacy at low concentration (e.g., 100 parts per million [ppm]) have led to the widespread use of fluorosurfactants where traditional hydrocarbon or silicone surfactants cannot achieve the required performance. In parallel to a high number of other uses, these and other properties have made fluorosurfactants useful and valuable in aqueous film-forming foam (AFFF), which is used to extinguish high hazard hydrocarbon and polar solvent fires.

1.2 FLUOROSURFACTANT CHEMISTRY

The two primary commercial production processes employed to synthesize the perfluoroalkyl moiety that is used to make fluorosurfactants are electrochemical fluorination (ECF) and the fluorotelomer process (Figure 1.2) (Taylor 1999; Kissa 2001; Pabon and Corpart 2002; Buck et al. 2011, 2012). Among the first commercially available fluorosurfactants were perfluoroalkyl sulfonates (PFSAs) (e.g., perfluorooctane sulfonate [PFOS], $C_8F_{17}SO_3^-$) and perfluoroalkyl carboxylic acids (PFCAs) (e.g., perfluorooctanoic acid [PFOA], $C_7F_{15}COOH$), manufactured using the ECF process (Taylor 1999; Kissa 2001; Pabon and Corpart 2002; Buck et al. 2012). Subsequently, fluorosurfactants derived from the fluorotelomer process were introduced some time later (10–15 years after ECF products). A wide array of fluorosurfactants containing typical surfactant terminal functionalities (e.g., anionic, cationic, amphoteric, and nonionic) have been made and used (Taylor 1999; Pabon and Corpart 2002; Buck et al. 2012).

1.2.1 Electrochemical Fluorination (ECF) Process

A general synthetic scheme for synthesis of ECF-based products is shown in Figure 1.2(a). ECF accomplishes the substitution of hydrogen with fluorine by electrolysis of

Fluorosurfactants in Firefighting Foams

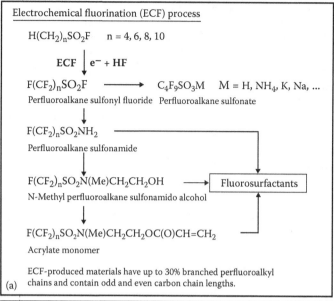

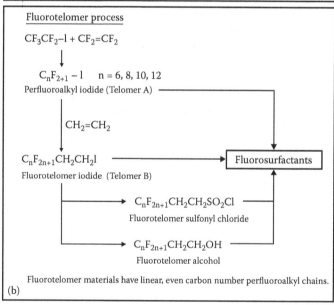

FIGURE 1.2 The two primary commercial processes to synthesize the perfluoroalkyl moiety: (a) the ECF process and (b) the telomerization process.

alkane sulfonyl chlorides or fluorides of the type $H(CH_2)_{n+1}SO_2X$, where X is Cl or F in a hydrogen fluoride (HF) solution (Alsmeyer et al. 1994). ECF is a free radical process yielding perfluoroalkane sulfonyl fluoride molecules with both linear and branched perfluoroalkyl chains and chain length homologues. For example, an eight-carbon starting material (e.g., octane sulfonyl fluoride) would yield a perfluoroalkane sulfonyl fluoride mixture composed of roughly 80% linear and 20% branched perfluoroalkyl chains and odd and even three- to eight-carbon-chain-length homologues. It is also worth noting that the electrolysis of alkane sulfonyl fluoride may also yield a homologous series of perfluorocarboxylic acids as impurities in percentage amounts. The perfluoroalkane sulfonyl fluoride is then reacted to create raw materials that can be used to make fluorosurfactants. Note that the "connector" in the ECF chemistry is a sulfonamidoalkyl group, $-SO_2N(R)CH_2CH_2-$, as shown in Figure 1.2(a). The historic production using ECF manufactured long-chain perfluorohexanesulfonyl (C6), perfluorooctane sulfonyl (C8), and perfluorodecane sulfonyl (C10) products, with the majority being C8, of which PFOS is the most well-known product. When 3M, the major global ECF manufacturer, withdrew from PFOS, PFOA, and related chemistry in 2000–2002, it transitioned to short-chain, C4, perfluorobutane sulfonyl chemistry. The reader is reminded that the short-chain (C4, perfluorobutane sulfonyl) ECF-based fluorosurfactants are not used in AFFF because to date, no AFFF formulation containing them has met required firefighting specifications.

1.2.2 Fluorotelomer Process

The second route most widely practiced for the synthesis of a perfluoroalkyl moiety is the fluorotelomer process (Taylor 1999; Pabon and Corpart 2002; Buck 2011, 2012). In the fluorotelomer process, tetrafluoroethylene (TFE) is oligomerized with perfluoroethyl iodide (PFEI) (CF_3CF_2I), yielding a mixture of exclusively linear, even-carbon-number perfluoroalkyl iodides that are commonly called Telomer A. The fluorotelomer chain length could range up to n = 20, but was primarily n = 6, 8, 10, and 12, with 6 and 8 predominant in commercial products. Ethylene is then inserted into Telomer A to make Telomer B, perfluoroalkyl ethyl iodide raw material, which may then be reacted to make fluorotelomer alcohol raw material and/or acrylate/methacrylate monomers. These raw materials or other types are then used to manufacture fluorosurfactants, as shown in Figure 1.2(b). Historic fluorotelomer-based fluorosurfactants contained mainly n = 6 (C6) and n = 8 (C8) perfluoroalkyl moieties ($C_nF_{2n+1}-$), with the majority being C6. As the major global fluorotelomer manufacturers have fulfilled their commitment to no longer make long-chain C8 chemistry (USEPA 2010), the short-chain C6 fluorotelomer-based fluorosurfactants are presently made and used.

1.3 LONG CHAIN AND SHORT CHAIN: TERMINOLOGY AND STRUCTURES

The historic use of fluorosurfactants, including their use in firefighting foams, has partially contributed to the widespread presence of PFOS and PFOA, aka long-chain perfluoroalkyl substances, in the environment. The discovery of environmental contamination began with the work by Moody and Field (1999) and continues today (Anderson et al. 2016). In 2000, 3M announced that it would exit from C8 manufacture, including PFOS and PFOA and all commercial products related to them (3M 2000). In 2006, eight major fluorochemical global manufacturers committed to exiting the manufacture and use of PFOA, longer-chain homologues, and related substances by the end of 2015 (USEPA 2010). Concurrent with these actions, the Stockholm Convention concluded that PFOS and PFOA meet the criteria to be classified as persistent, bioaccumulative, and toxic (PBT) substances (Stockholm Convention 2009; Matthies et al. 2016).

The PBT properties of the long-chain perfluoroalkyl substances, PFOS and PFOA, and the related long-chain raw materials and commercial products were the driving force behind manufacturers ceasing manufacture of these long-chain chemistries and transitioning to short-chain chemistry. To continue to deliver the unique and valuable properties of fluorosurfactants in uses where they are necessary for performance and to eliminate long-chain products, manufacturers have developed additional short-chain alternatives with fewer fluorinated carbons. The data on these alternatives are admittedly less than the data on their long-chain counterparts, but still exist in a considerable measure. Available data on short-chain fluorotelomer-based raw materials, products, and degradation products show that the short-chain alternatives have a more favorable environmental and human health profile (Environ 2014; Environ 2016; Buck 2015). Numerous short-chain products have been approved by regulators, and significant data are being required before the commercialization of new short-chain products. Therefore, today, short-chain ECF-based fluorosurfactants contain the perfluorobutane sulfonyl moiety, $C_4F_9SO_2-$, while fluorotelomer-based fluorosurfactants contain the perfluorohexylethyl moiety, $C_6F_{13}CH_2CH_2-$. Readers should note that there is a clear and distinct difference in what is considered long chain and short chain based on the origin chemistry and its connection to the corresponding perfluoroalkyl acid (PFAA) (Figure 1.3). Within the perfluoroalkane sulfonate (PFSA) family, all substances with a carbon chain length greater than or equal to six are considered long chain. In contrast, within the perfluorocarboxylic acid (PFCA) family, all substances with a carbon chain length greater than or equal to eight are considered long chain. This comes about because of the significant difference in bioaccumulation and toxicity properties between perfluorohexane sulfonate (PFHxS) and perfluorohexanoic acid (PFHxA), both called C6s, but with dramatically different properties (Gannon et al. 2011).

Long-chain		Short-chain
	Perfluoroalkane sulfonate	
	PFSA	
$n \geq 6$		$n \leq 5$
PFDS, PFOS, PFHxS	$C_nF_{2n+1}SO_3(H)$	PFBS
	Perfluoroalkyl carboxylate	
$x \geq 8$	PFCA	$x \leq 7$
PFDA, PFNA, PFOA	$C_{x-1}F_{2x-1}COOH$	PFHxA, PFPeA, PFBA

FIGURE 1.3 Long-chain versus short-chain PFAAs. (Adapted from Organisation for Economic Co-operation and Development, Portal on per and poly fluorinated chemicals, 2017. http://www.oecd.org/chemicalsafety/portal-perfluorinated-chemicals/aboutpfass/.)

Foam, fluorosurfactants, and firefighting have been inextricably linked for the past 50 years following the invention of AFFF in the 1960s. With the unique film-forming and surface tension properties provided by fluorosurfactants (Kissa 2001; Pabon and Corpart 2002), the category of AFFF was created approximately 50 years ago. This invention and subsequent application of fluorosurfactants for use in Class B foam concentrates forever changed the capability of firefighters worldwide to extinguish and rapidly control large-scale hydrocarbon liquid and polar solvent fires. This introductory chapter provides an overview of fluorosurfactant use in AFFF foams and concentrates. Subsequent chapters discuss challenges arising with the historic use of AFFF and the discovery of per- and poly-fluoroalkyl substances, PFAS, in the environment. Chapters explore issues associated with the remediation of impacted environmental media, particularly soil and water. The current regulatory landscape is also surveyed. The lack of published regulatory limits, and instead the presence of advisory limits, drives new decision logic regarding remedial options. Sampling and analysis considerations ensuring the integrity of analytical data at the low parts per trillion (ppt) range are discussed. Also, the ongoing development of the toxicological picture of PFAS is discussed in detail, as are efforts to understand the ecological impact of PFAS in the environment. The reader is also directed to a summary documenting the history and issues facing the use of AFFF prepared by the Fire Protection Association of Australia (FPAA 2017).

1.4 FIREFIGHTING FOAM BEFORE FLUOROSURFACTANTS

The need for and use of firefighting foams followed the introduction and wide-scale use of liquid hydrocarbons and later polar solvent fuels. Historically, water was used as the primary firefighting agent. However, water is more dense than liquid hydrocarbons and sinks, rendering it an ineffective bottom layer below a burning hydrocarbon surface. The ability to dilute or "lighten" water with air in concert with a foaming surfactant created a firefighting foam that floated on the burning liquid and prevented flammable vapor evolution. This higher-surface-area aqueous foam also provided a heat sink for the absorption of radiant heat and thus helped cool the burning surface and surrounding areas.

The early history of foam development was documented by Ratzer (1956). The earliest efforts to extinguish hydrocarbon liquid and/or petroleum fires were accomplished with aqueous mixtures of sodium bicarbonate, saponin (a glycoside with foaming characteristics), and acidic aluminum sulfate to extinguish a naphtha fire (Johnson 1877; Ratzer 1956). Concurrently, foam mechanically created from a solution of ammonium soap with dissolved borax (or ammonium sulfate) in a vessel that was connected to a gas cylinder containing either ammonia, nitrogen, or carbon dioxide under pressure was also developed (Gates 1903). The gas was used to force the liquid out of the vessel into another container. Additional cylinder gas was used to further enhance the foam generated in this mechanical process. By 1912, this equipment was broadly used in England (Ratzer 1956). Efforts ensued to find suitable materials to make mechanical foams more viable and effective. In 1937, water-soluble protein products were created by digestion of hoof or horn meal that were superior for mechanical foam generation (Weissenborn 1939). Consequently, other groups developed protein-based air foam concentrates that provided favorable firefighting characteristics (Friedrich 1940; Katzer 1943; Ratzer 1944).

1.5 FLUOROSURFACTANTS IN AFFF

The invention of fluorosurfactants provided revolutionary new technology to advance firefighting foam. The initial discovery and innovation were catalyzed because the US Navy was simultaneously spraying a protein foam and a Class A powder (also called "dry chemical") to fight Class B fires aboard ships. The Class A powder was helping to control the fire but was also partially breaking down the foam blanket, like any solid powder can do on classical aqueous foams. In 1963 (Gipe and Peterson 1972), it was shown that the addition of fluorosurfactants to a protein foam allowed that foam not to be significantly affected by a simultaneous projection of a Class A powder. After that first discovery, work continued on film-forming fluoroprotein (FFFP) and AFFF, where the foam was "lightened" further using fluorosurfactants. Incorporating fluorosurfactants dramatically lowered the foam's aqueous surface tension, delivering unprecedented rapid wetting and spreading and establishing a new standard for effectively extinguishing hydrocarbon fuel fires. The full-scale implementation of that discovery was accelerated by the July 1967 USS *Forrestal* tragedy, during which more efficient firefighting foams could have limited the propagation of the jet fuel fire that took place on the flight deck. Additionally, specific fluorosurfactants were also used to generate foam (e.g., amphoteric fluorosurfactants having a positive and a negative charge on their hydrophilic part).

The first work on AFFF containing fluorosurfactants was carried out by the Naval Research Laboratory and 3M in the early 1960s and led to the development of 3M's "Lightwater" AFFF products based on ECF chemistry (Gipe and Peterson 1972). As discussed above, there are two principal chemical processes for the synthesis of the perfluoroalkyl moiety to make fluorosurfactants used in AFFF: ECF and fluorotelomerization. Figure 1.4 provides structural examples of ECF-based and fluorotelomer-based fluorosurfactants that have been developed and used by the firefighting industry over four decades (Pabon and Corpart 2002; Buck et al. 2012; Place and Field 2012; Backe et al. 2013; Barzen-Hanson et al. 2017). Figure 1.4 shows structures of various fluorosurfactants, to include different functional types (e.g., betaine, sulfonate, carboxylate, and n-oxide), historic versus present-day products, and long-chain and short-chain fluorosurfactants.

1.5.1 ECF-Based Fluorosurfactants Used in AFFF

ECF-based fluorosurfactants used in AFFF are shown in Figure 1.4. As mentioned earlier, the ECF process was the first implemented to produce fluorosurfactants. The ECF process is also called the Simons process (Simons 1949). The two types of key fluorinated intermediates used to produce the ECF fluorosurfactants of Figure 1.4 are represented in Figure 1.5. They are obtained by the full fluorination of their corresponding hydrocarbon acid chloride and sulfochloride homologues. The full fluorination of alkyl acid chlorides also generates by-products that consist of fluorinated alkyl chains terminated by a fluorinated furan and pyran group (Banks et al. 1994).

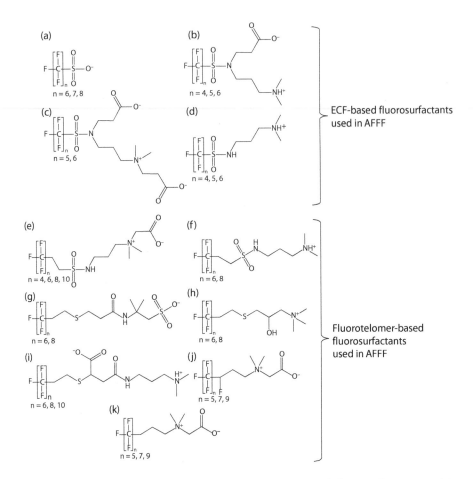

FIGURE 1.4 (a–d) ECF-based and (e–k) fluorotelomer-based fluorosurfactants used in AFFF. (Adapted from Place, B.J., and Field, J.A., *Environmental Science & Technology*, 46, 7120–7127, 2012.)

FIGURE 1.5 Key intermediates used for the production of the ECF fluorosurfactants presented in Figure 1.4.

These by-products were mainly used as fluids for the electronic industry. Increasing the number of carbon of the alkyl chain of the starting acid chloride (n value on Figure 1.4) increases the fraction of by-products that are produced.

Four major types of surfactants are used in ECF-based firefighting foams. Perfluoralkane sulfonates and perfluorocarboxylates are anionic surfactants, perfluoroalkane sulfamido amines are cationic surfactants, and perfluoroalkane sulfonamide amino carboxylates are amphoteric surfactants. Most firefighting foams are formulated with a mixture of anionic and amphoteric fluorosurfactants. Interestingly, typical amphoteric ECF fluorosurfactants, such as the one presented in Figure 1.4 (perfluoroalkane sulfonamido carboxylate), have their positive and negative charges on two distinct groups of the tertiary amine. For ECF-based fluorosurfactants, the amphoteric structure is designed in such a way that the isoelectric point of the surfactant is at the typical pH of the firefighting foam concentrate (in the range of pH 7–8.5). Cationic surfactants can be added as a minor fraction of the fluorosurfactant mixture for specific formulations.

The ECF process induces the formation of both linear and branched perfluoroalkyl moieties. Fluorotelomers are only linear for all processes starting with PFEI. In terms of performance in firefighting foams, branched fluorosurfactants are not as efficient as equivalent linear fluorosurfactants (Taylor 1999; Kissa 2001). This implies that, on average, to obtain a similar performance level, a firefighting foam based on an ECF-produced surfactant requires a higher fluorine content than a fluorotelomer-based firefighting foam. Historically, ECF-based fluorosurfactant firefighting foams used mainly long-chain perfluorooctane (C8) and perfluorohexane (C6) sulfonyl surfactants. In the early 2000s, attempts had been made to use short-chain ECF-based perfluorobutane sulfonyl fluorosurfactants. However, these did not meet the AFFF performance requirements stipulated in military specifications.

1.5.2 Fluorotelomer-Based Fluorosurfactant Used in AFFF

Fluorotelomer-based surfactants used in AFFF are shown in Figure 1.4 and are nominally derived from three structural families: (1) a thiol (–S–) in the spacer (e.g., $-CH_2CH_2-S-CH_2-$), (2) a sulfonyl (–SO$_2$–) in the spacer (e.g., $-CH_2CH_2-SO_2-N-$), or (3) an alkyl spacer (e.g., $-CH_2CH_2CH_2-$ or $-CFHCH_2CH_2-$) between the perfluoroalkyl ($C_nF_{2n+1}-$) moiety and the surfactant functionality.

Figure 1.4 shows that the fluorotelomer-based surfactants used in firefighting foams are either amphoteric (fluorotelomer thioamido sulfonates or fluorotelomer sulfamido betaines, fluorotelomer betaines), cationic (fluorotelomer sulfamido amines), or anionic (fluorotelomer sulfonates).

Historically, as of the early 1970s, fluorotelomer-based surfactants used in firefighting foams contained perfluoroalkyl moieties, $F(CF_2)_n-$, with n = 6 or a mixture of different perfluoroalkyl chains with n = 6, 8, and 10 carbons, with a majority (>50%) comprised of n = 6. Today, all currently produced fluorotelomer-based surfactants (by Voluntary Stewardship Program signees; USEPA 2010) used in firefighting foams are based on C6 raw materials that contain a six-carbon perfluoroalkyl moiety, $C_6F_{13}-$. Historical accounts (Cortina and Korzeniowski 2008; Place and Field 2012; Backe et al. 2013; Kleiner 2016; Barzen-Hanson et al. 2017) note that the fluorotelomer chemistry employed in the AFFF products used either a perfluoroalkyl moiety with n = 6 or a mixture of n = 6 (C6) and n = 8 (C8). This clearly means (not implies) that the fluorotelomer-based fluorosurfactants used in the early Ansul/Ciba-Geigy and Eau et Feu (Atochem) consisted of either short-chain or both short- and long-chain fluorosurfactants, from the mid-1970s onward. Back in those days, as surprising as it can be for nonspecialists, some producers where able to pass the performance requirements of the most demanding standard in the world (MIL-F-24385 1969) by using fluorosurfactants made with a perfluoroalkyl moiety, $C_nF_{2n+1}-$, where n = 6 only.

Fluorotelomer chemistry (Figure 1.4 lower part) was used to produce both the Ciba-Geigy Lodyne™ and Atochem Forafac™ branded surfactants. Atochem's fluorosurfactant business was acquired by DuPont in 2002. This business was spun out into Chemours in 2015 and is now part of the Capstone™ line of products. Likewise, the Ciba-Geigy Lodyne firefighting fluorosurfactant and foam stabilizer business was acquired by the Chemguard Specialty Chemicals Division from Ciba Specialty Chemicals in 2003–2004. The Chemguard Chemicals Division became part of the Tyco group in 2011, and Tyco was acquired by Johnson Controls in 2016. Each of these chemistries used the basic fluorotelomer scheme. The Ciba-Geigy/Chemguard fluorotelomer-based surfactants were generally characterized by the $-CH_2CH_2-S-CH_2-$ linkage. Likewise the Atochem/DuPont/Chemours fluorotelomer-based fluorosurfactants contained the $-CH_2CH_2-SO_2-N-$ linkage. The Qualified Products List (QPL) discussed below has all short-chain fluorotelomer products on the current 2017 QPL list for military specification MIL-F-24385F, the current acceptance specification used in the US Department of Defense (USDOD) (MIL-F-24385 1969).

1.6 FIREFIGHTING FOAMS: TYPES AND COMPOSITION

Fires are classified by the proper extinguishing agent. Class A fires consist of ordinary combustibles, such as wood, paper, fabric, and most kinds of trash. Water is usually used to extinguish Class A fires. For Class A, high-expansion and fluorine-free foams (F3, nonfluorinated hydrocarbon surfactants) are used. A Class B fire arises from flammable liquids or flammable gases, petroleum greases, tars, oils, oil-based paints, solvents, lacquers, or alcohols. Using water on a Class B fire (e.g., an oil-based liquid fire) is extremely dangerous. Petroleum-based fires burn significantly hotter (≥350°F) than the boiling point of water (212°F). When water is placed on a petroleum-based fire, it instantly vaporizes to form steam, which expands rapidly, causing the liquid hydrocarbon liquid to splatter, cause burns, and spread the fire. Hence, AFFF is preferred for extinguishing high-hazard Class B fires.

In use, AFFF concentrate is mixed with either freshwater or seawater by a proportioning nozzle (Figure 1.6). The foam forms spontaneously upon ejection from the nozzle. A thin layer of AFFF foam rapidly spreads over the hydrocarbon fuel surface, acting as a thermal and evaporation barrier to inhibit and eventually extinguish combustion. The "film-forming" characteristic refers to the fact that both in application and even after the foam has dissipated, the aqueous layer formed from the AFFF concentrate mixture remains on the flammable liquid hydrocarbon surface, thereby preventing reignition. This occurs because the fluorosurfactant makes the surface tension of the water lower than the surface tension of the hydrocarbon, enabling it to float on the hydrocarbon surface at the air interface.

Fluorosurfactants in Firefighting Foams

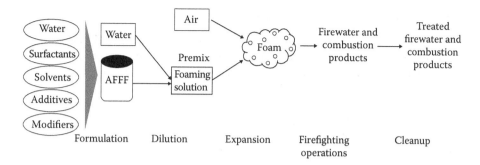

FIGURE 1.6 Schematic of AFFF concentrate and its use. It shows in pictorial form the ingredients used to concentrate foam to deploy to firewater.

1.6.1 Types of Firefighting Foam

There are a number of ways to classify firefighting foams. One way is by chemical type, and another is by expansion ratio. In general, expansion ratios are classified as (1) low-expansion foams (with an expansion ratio of ≤20), (2) medium-expansion foams (having an expansion ratio from 20 to 200), and (3) high-expansion foams (with expansion ratios of >200). We will focus on the two chemical types of firefighting foams: (1) F3 that are fluorine-free and (2) AFFF containing fluorosurfactants, with a focus on AFFF. Historically, nonfluorinated mechanical protein-type foams are derived from either hydrolyzed protein-rich animal or vegetable materials using alkali/lime. Vegetable materials were developed later and used in a "synthetic" formulation whose origin is from hydrocarbon-derived materials. Beginning in the 1960s, various fluorosurfactants were developed and incorporated into both types of synthetic and protein-based firefighting foam agents (Briggs 1996). Further, it is also well known that water-miscible liquids, such as polar solvents (e.g., alcohols and ketones), tend to destroy conventional foams. To overcome this characteristic, formulations were created that contained polymeric/gel-type materials, which help prevent foam destruction. These types of foam agents are called alcohol resistant (AR). For current-generation fluorine-free foams, there is basic F3, as well as alcohol-resistant fluorine-free foam (AR-FFF). It should be noted that protein, high-expansion, Class A, and most training foams have historically been fluorine-free. The types of firefighting foam are shown in Figure 1.7.

Fluorine-free foam (F3)	Fluorinated foams
· Basic fluorine-free foam (FFF)	· Fluoroprotein (FP)
· Protein	· Film-forming fluoroprotein (FFFP)
· Alcohol-resistant protein (AR-P)	· Alcohol-resistant film-forming fluoroprotein (AR-FFFP)
· Synthetic	· Aqueous film-forming foam (AFFF)
	· Alcohol-resistant aqueous film-forming foam (AR-AFFF)

FIGURE 1.7 Types of firefighting foam.

1.6.2 AFFF INGREDIENTS AND COMPOSITION

The precise chemical composition of AFFF formulations is generally proprietary confidential business information as the ingredients are viewed by AFFF manufacturers and formulators as providing unique competitive benefit to their products. A firefighting foam is obtained by the dilution of a foam concentrate in water (generally a 3% or 6% foam concentrate, v/v). This water–foam concentrate mixture is called a foaming solution or premix (Figure 1.6).

The foaming base of a foam concentrate consists of either a mixture of hydrocarbon surfactants or a hydrolyzed protein. Synthetic foam concentrates can be distinguished from protein foam concentrates according to the foaming base. To simplify matters, our discussion on ingredients and composition only involves synthetic foam concentrates made from surfactants. The types of firefighting foams are listed in Figure 1.7.

In general, AFFF concentrates contain four main ingredients: (1) water, (2) surfactants (to provide the foam and film-forming properties), (3) solvents, and (4) various additives and modifiers. As previously stated, water-soluble polymers are added to obtain AR-AFFF. Typical ingredients are shown in Figure 1.8. Water is the major ingredient and generally is more than 60% of the composition.

Typically, a synthetic foam concentrate designed to be diluted at 3% into water contains from 5% to 10% by weight of hydrocarbon surfactants. High-performance foam concentrates also contain one or several fluorosurfactants. Fluorosurfactants, given their relatively high expense and effectiveness at low concentration, are a minor but critical ingredient in AFFF. The fluorosurfactant plays a role in the formation of the foam as well as the formation of a water film at the surface of the solvent. In active matter, they account for 0.6–1.5 wt% of the total weight of a foam concentrate. Fluorosurfactants also provide oil repellency to firefighting foams, and this property is key when foams are used in larger-scale and real conditions.

The formation of the water film at the surface of the hydrocarbon has the result that the emission of solvent vapors is stopped or highly reduced. As those vapors are

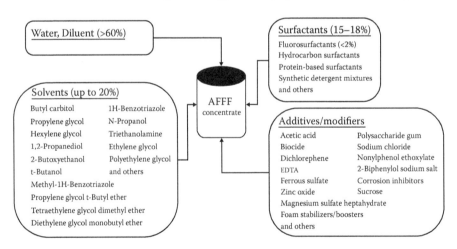

FIGURE 1.8 Type of ingredients for a 3% foam concentrate.

the fire source, it is very important to stop their emission. This water film formation occurs when the spreading coefficient (SC) of the foaming solution is positive. This coefficient is defined in Equation 1.1:

$$SC = \gamma_s - (\gamma_{HC} + \gamma_L) \tag{1.1}$$

where γ_s is the surface tension of the solvent, γ_{HC} is the interfacial tension between the solvent and the foaming solution, and γ_L is the surface tension of the aqueous foaming solution.

The use of hydrocarbon surfactants reduces the solvent–foaming solution interfacial tension (γ_{HC}), and fluorinated surfactants provide the foaming solution with a particularly low surface tension (γ_L). Without a suitable choice of these two types of components, the spreading coefficient of a foaming solution on a solvent would be negative and extinction impossible.

There are debates in the industry about the existence of an aqueous film with AFFF at elevated temperature conditions, but foams having the capacity to form an aqueous film at regular temperature demonstrate, in most cases, higher performances than F3 foams. A foam concentrate also contains polar solvents (up to 20 wt%) that contribute to the foam stability. 1,2-Propanediol and similar types of solvents are used to reduce the freezing point of foam concentrates stored in refineries, civil airports, or drilling platforms.

In addition to the classic foam concentrates, AR foam concentrates are polyvalent and can be used to extinguish polar solvent fires, such as ethers, ketones, or alcohols. Polar solvents are miscible in water and can solubilize most hydrocarbon and fluorinated surfactants. This results in the almost instantaneous disappearance of the firefighting foam when it is placed at the surface of the solvent. The most common solution to obtain a stable foam in contact with the polar solvent involves the addition of water-soluble polymers such as polysaccharide to the foam concentrate. In contact with the polar solvent, the polymer precipitates to form a protective layer at the interface that isolates the foam from the polar solvent and prevents its destruction. Modern AR-AFFF also contains water-soluble polymers modified with fluorotelomers. Additives providing anticorrosion, antimicrobial activity, and other properties are all part of typical AFFF formulations.

A generalized view of a typical AFFF Type 3 (3%) concentrate composition based on data taken from company websites, safety data sheet documents, company fact sheets, and various reverse engineering efforts and analytical testing is shown in Figure 1.9 (Cortina 2012). In Figure 1.9, a representative composition of the foam that is actually applied to a fire, a 3% AFFF concentrate diluted with 97 parts water, is shown. In this example, the combination of surfactants and solvents is approximately 1% of the final solution as employed in firefighting.

1.7 QUALIFIED PRODUCTS LIST (QPL)

AFFF is a highly efficient firefighting agent for extinguishing flammable liquid (Class B) fires. AFFF concentrates used by the US military must meet the requirements set forth in military specification MIL-F-24385 (1969). The specification is

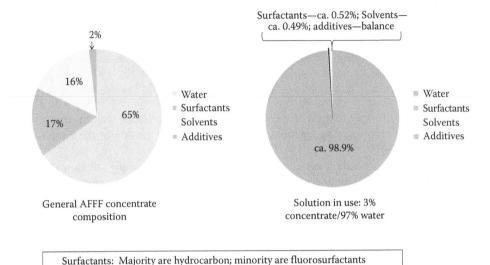

FIGURE 1.9 Pie chart showing the composition of a Type 3 (3%) AFFF concentrate.

under the control of the Naval Sea Systems Command, and the Naval Research Laboratory is the designated institution for certification evaluation for the USDOD's AFFF QPL. AFFF is used globally by the US military, by the militaries of other countries, and in the majority of civilian applications worldwide as either a 3% (Type 3) or a 6% (Type 6) concentrate. The military specification AFFF concentrates are effective against a wide variety of Class B liquid hydrocarbon fire threats. Special formulations exist for use against alcohol and other water-miscible fuels.

The early use of fluorosurfactants in firefighting foams (AFFF) is well documented in a Naval Research Laboratory report (Gipe and Peterson 1972). The first fluorosurfactant work was a collaborative effort between the US Navy and 3M and began around 1961. This effort resulted in the 1964 introduction of the first AFFF product into Navy firefighting service inventory, the ECF-based 3M product FC-183. FC-194, which replaced FC-183 about 3 years later, was less viscous and could be used at a 6% concentrate strength in freshwater with conventional foam-proportioning equipment. However, FC-194 had a significant drawback. FC-194 could not be used with seawater, essential for deployment on naval vessels. Subsequently, a seawater-compatible AFFF, FC-195, was developed, but it was never submitted for qualification under what became the well-known QPL and military specification MIL-F-24385, issued on November 21, 1969. The first AFFF on the QPL was FC-196 from 3M in 1969. Improvements and modifications to the AFFF agents continued. The next-generation AFFF, FC-200, had a pH of >7 and corrosion inhibitors added to limit any agent attack in seawater on the stainless steel storage vessels. FC-200 was QPL approved in January 1972. Of note, the FC-200 fluorine content (2.1% w/v) was significantly reduced in this formulation compared with its FC-196 predecessor (3.2% w/v).

For the first 4 years the QPL existed (1969–1973), 3M AFFFs containing ECF-based fluorosurfactants were the only products qualified under MIL-F-24385. 3M's development partnership with the Navy lasted nearly 10 years (~1963–1973) before the next product and company became listed. AFFF containing fluorotelomer-based fluorosurfactants from National Foam was qualified on the QPL in late 1973, followed by Ansul using a Ciba-Geigy fluorotelomer-based fluorosurfactant in 1974–1975. These products had between 0.7% and 0.83% w/v fluorine, or less than half of fluorine of the ECF product FC-200. In 1981, two more products were added to the QPL: 3M's FC-206C and Ansul's AFC-5, which contained 0.94% w/v and 0.61% w/v fluorine, respectively (Barzen-Hanson et al. 2017). In 1993, 3M's Lightwater™ FC-203CF contained 1.8 wt% fluorine while National Foam's Aer-O-Water 3EM contained 0.85 wt% fluorine; both were listed on the QPL (Kleiner and Jho 2009). Each of these products had passed the US military's QPL specification MIL-F-24385. The data indicate that the AFFF containing fluorotelomer-based fluorosurfactants provided firefighting performance equal to that of the ECF-based AFFF products with substantially less fluorosurfactant content.

A chronology of the QPL is shown in Figure 1.10, with AFFF product descriptions given in Table 1.1. Even though ECF-based long-chain products (e.g., PFOS containing) are still in inventory at various sites globally (Darwin 2011), none are on the QPL today, in 2017. The reader is reminded that the short-chain (C4, perfluorobutane sulfonyl) ECF-based fluorosurfactants are not used in AFFF because, to date, no AFFF formulation containing them has been qualified. There has been significant changes in the QPL over the past 6 years (2011–2017) as the transition to AFFF containing short-chain C6 fluorotelomer-based surfactants has occurred.

With 3M's exit (Type 6 withdrawn January 3, 2007, and Type 3 withdrawn August 8, 2010) from long-chain ECF-based fluorosurfactant manufacture for AFFF in the last decade, National Foam is the longest-standing company on the list, having come onto the QPL in 1973, followed close behind by Ansul (today Johnson Controls). Both Angus and Buckeye came off the QPL within the past 2 years, while at the same time ICL/Auxquimia (2015; now Perimeter Solutions), Amerex/Solberg (2016), and Dafo Fomtec AB are recent additions. Chemguard (since 1997) and Fire Service Plus (Type 6 added May 2011) are also on the May 2017 QPL (Figure 1.10). It is also noteworthy

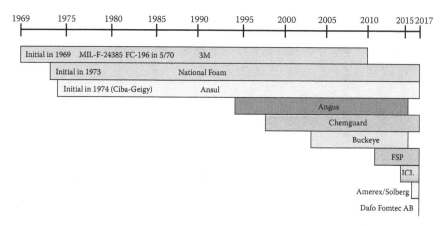

FIGURE 1.10 QPL chronology: 1969–2017 (MIL-F-24385) as of May 12, 2017.

TABLE 1.1
Qualified Products Database Comparison 2004–2017

Company	Chemistry Type	Test Date[a]		April 5, 2004		Test Date[a]		May 12, 2017	
		Type 3	Type 6	Type 3	Type 6	Type 3/6		Type 3	Type 6
3M Co.	E	NR	NR	FC-203CF	FC-206 CF				
Angus Fire	FT	February 18, 1994	—	Tridol M 3%	—				
Ansul, Inc.	FT	NR	NR	Ansulite 3%	Ansulite 6%				
Buckeye Fire Equipment	FT	November 12, 2003	—	Buckeye 3%	—				
Chemguard, Inc.	FT	January 30, 1997	July 25, 2001	Chemguard 3%	Chemguard 6%				
Kidde/National Foam	FT	NR	NR	Aer-O-Water 3-EM	Aer-O-Water 6-EM				
Fire Service Plus	FT					November 2, 2010		—	Fireade 2000-MIL 6 AFFF
National Foam	FT					May 2, 2016		Aer-O-Water 3-EM-C6 AFFF	Aer-O-Water 6-EM-C6 AFFF
	FT					May 2, 2016		Tridol-C6 M3 AFFF	Tridol-C6 M6 AFFF

(*Continued*)

TABLE 1.1 (CONTINUED)
Qualified Products Database Comparison 2004–2017

Company	Chemistry Type	Test Date[a] April 5, 2004			Test Date[a] May 12, 2017		
		Type 3	Type 6	Type 3	Type 3/6	Type 3	Type 6
Tyco Fire Procucts/Ansul	FT				November 13, 2015	Ansulite AFC-3MS	Ansulite AFC-6MS
Amerex Corp./Solberg Corp.	FT				February 29, 2016	Arctic 3% MIL-SPEC AFFF	Arctic 6% MIL-SPEC AFFF
Tyco Fire Procucts/Chemguard	FT				November 13, 2015	Chemguard C306-MS	Chemguard C606-MS
ICL/Auxquimia	FT				November 3, 2015	Phos-Chek 3% AFFF	
Dafo Tomtec AB	FT				January 19, 2017	Fomtec AFFF 3%M	

Note: FT = fluorotelomer; NR = tests were conducted but no date was noted; E = ECF.

[a] Test dates/Naval Research Laboratory report listed when available from QPL list. Listing on actual QPL followed at a later date.

TABLE 1.2
QPL List (March 2018) MIL Spec MIL-F 24385

Type 3, 3% AFFF		Type 6, 6% AFFF	
Product Name	Supplier	Product Name	Supplier
AER-O-WATER 3E-C6 AFFF	National Foam, Inc.	AER-O-WATER 6EM-C6 AFFF	National Foam, Inc.
ANSULITE AFC-3MS 3% AFFF	Tyco Fire Protection Products (JCI)	ANSULITE AFC-6MS 6% AFFF	TYCO Fire Protection Products (JCI)
ARCTIC 3% MIL-SPEC AFFF	Amerex Corp., The Solberg Company	ARCTIC 6% MILSPEC AFFF	Amerex Corp., The Solberg Company
CHEMGUARD C306-MS 3% AFFF	Tyco Fire Protection Products (JCI)	CHEMGUARD C606MS 6% AFFF	TYCO Fire Protection Products (JCI)
PHOS-CHEK 3% AFFF MS	ICL Performance Products (now Perimeter Solutions)	FIREADE MILSPEC 6 (Jan 2018)	Fire Service Plus, Inc.
TRIDOL-C6 M3 AFFF	National Foam, Inc.	TRIDOLC6 M6 AFFF	National Foam, Inc.
FOMTEC AFFF 3% M: SWE & USA	Dafo Foamtec AB	PHOS-CHEK 6% Milspec AFFF (12/17)	ICL Performance Products (now Perimeter Solutions)
FIREADE Milspec 3	Fire Service Plus, Inc.		

Source: http://qpldocs.dla.mil/search/parts.aspx?qpl=1910

that while three of the companies (National Foam, Ansul, and Chemguard) have been on the QPL for many years, there are an equal number of new entrants to qualify for arguably the world's most difficult fire standard. These new QPL entrants are Fire Service Plus (FSP); Solberg Corporation; ICL, who purchased Auxquimia's Class B foam business in 2014; Now Perimeter Solutions; and Dafo Fomtec AB. Current (2018) QPL AFFF products for Class B fires are shown in Table 1.2. The short-chain fluorosurfactants in all these products are C6 fluorotelomer-based products.

1.8 FIREFIGHTING FOAM SELECTION AND USE

The selection of a firefighting agent to extinguish a Class B fire is front and center today more so than it has been over the past 15 years due to the widespread environmental presence from historic use of AFFF (Willson 2016a). Numerous fire journal articles (and other media) have appeared debating the pros and cons of both general types of firefighting agents, F3 and AFFF containing fluorosurfactants. The selection decision requires a thoughtful consideration of multiple criteria, such as fire performance, life safety benefits, property protection, and environmental impact, including whether control of spent firewater is possible or practical. Balancing these criteria presents a set of difficult choices for regulators, firefighters, and those who routinely handle flammable liquids and require rapid fire extinguishment and control to protect human life and property. A recent firefighting foam seminar in Singapore (SAA-IAFPA 2016) provided an open forum for discussion of these decision-making criteria, including numerous presentations highlighting critical insights into why AFFF is often chosen (Castro 2016; Jho 2016; Plant 2016; Willson 2016b). To illustrate the issues, performance, and fundamental property parameters, this section discusses foam properties (Table 1.3) and environmental properties (Table 1.4).

TABLE 1.3
Foam Properties: Comparison of AFFF and F3

Foam Property	Advantage	AFFF[a]	F3
Fuel repellency	Yes	Yes	No
Film formation	Yes	Yes	No
Foam spreading on fuel	Yes	Yes	No
Fuel spreading on foam	No	No	Yes
Fuel shedding	High	High	Low
Fuel pickup	Low	Low	High
Fuel emulsification	Low	Low	High
Flammability of contaminated foam	Low	Low	High
Degradation of contaminated foam	Low	Low	High
Heat resistance of foam	High	High	High

[a] Includes short-chain AFFF.

TABLE 1.4
Environmental Properties: Comparison of Short-Chain Formulations of AFFF and F3

Environmental Property	Advantage	AFFF[a]	F3
Aquatic toxicity	Low	Yes	No
Persistence	No	Yes	No
Reduced foam, water resources used, and wastewater generation	Yes	Yes	No
Reduced smoke and breakdown products generated	Yes	Yes	No
Risk to life safety	Low	Low	Higher
Escalation potential	Low	Low	High
Bioaccumulation	No	No[b]	No
Disposal through PTOW or WWTP	Yes	Yes[c]	Yes

Note: PTOW = publicly owned treatment works; WWTP = wastewater treatment plant.
[a] Includes short-chain AFFF.
[b] Short-chain AFFF.
[c] With pretreatment.

Critical foam properties required to fight a Class B fire that are evaluated in selecting a firefighting agent are presented in Table 1.3, comparing AFFF (containing short-chain fluorosurfactants) and F3. For example, in hydrocarbon-type Class B fires, a firefighting agent that has fuel repellency, sheds the fuel, forms a film, is not contaminated by the fuel, and rapidly spreads across the burning fuel surface is highly advantaged versus an agent that does not have these critical properties. These properties are clear advantages of AFFF containing fluorosurfactants that fundamentally differentiate AFFF from F3. Both AFFF and F3, in general, appear to have a high resistance to heat, although one can argue that the mechanism for this heat resistance is different. For AFFF, the fluorosurfactants used in the formulation have better chemical and heat resistance than hydrocarbon surfactants. F3 is given a high rating as it believed that the current generation of fluorine-free foams have high loadings of polysaccharides (sugars and gums) to boost drain times and thus trap or keep water longer in the foam matrix. The presence of the trapped water should provide greater heat resistance.

The environmental properties of short-chain AFFF and F3 are shown in Table 1.4. Whereas the foam properties show many clear advantages for AFFF (Table 1.3), the environmental properties require discussion and explanation. Aquatic toxicity data for firefighting foam products have been discussed (FFFC 2006; Plant 2016). The available data show that AFFF products have significantly lower acute aquatic toxicity than the F3 used in comparative fire testing. The basis is also shown in Table 1.5 (FFFC 2006). Both indicate very clearly that each of the three AFFF products have significantly lower toxicity in both the rainbow trout and the fathead minnows. A lower LC_{50} indicates the agent is more toxic.

Another important environmental property is persistence. The stability of the carbon–fluorine bonds in fluorosurfactants makes them and/or their terminal biodegradation products stable and therefore persistent in the environment (see Section 1.11 for a discussion). In contrast, F3s are made with hydrocarbon ingredients that are generally expected to biodegrade under normal forseeable environmental conditions and thus are not considered persistent.

The next five properties in Table 1.4, from reduced foam and water resources used, deserve some discussion. AFFF has been selected as the product with the more

TABLE 1.5
Aquatic Toxicity Comparison

96 h LC_{50} Test Fingerling Rainbow Trout		96 h LC_{50} Flow-Through Test Fathead Minnow	
Agent	LC_{50} (mg/L)	Agent	LC_{50} (mg/L)
Wetting agent	1.06	Wetting agent	0.887
F3-A	63	F3-A	171
F3-B	71	F3-B	171
MIL-SPEC AFFF	2176	MIL-SPEC AFFF	884
AR-AFFF	3536	AR-AFFF	1487
UL AFFF	5657	UL AFFF	1726

Note: UL = Underwriters Laboratories.

desirable firefighting characteristics when compared with F3. The advantage originates from the demonstrated performance of AFFF versus F3 when considering time to control the fire to overall extinguishment time. It is generally accepted that AFFF will knock down, control, and extinguish Class B fires faster than F3. Some argue that there are some fire test data that show equal performance in putting out a fire. However, the majority of fire test data show AFFF perform superiorly to F3 (see Section 1.9). That AFFF extinguishes a fire faster leads to lower foam usage, reduced water usage, and reduced smoke and breakdown products due the shorter control and extinguishment times. With reduced foam and water resources used, it follows that firewater runoff and wastewater generation are also reduced. In light of faster control and extinguishment times, the edge for life safety risk is also given to AFFF. Life safety risk includes firefighters and any other personnel in the fire's vicinity. In considering these properties as a whole, faster fire control in the view of many leads to a lower fire escalation potential, thus making AFFF the selection for use in a large-scale high-hazard Class B fire event.

The present-day short-chain C6 fluorotelomer-based products, raw material and degradation products, have been shown to have a more favorable toxicological and environmental profile than historic long-chain equivalents (FluoroCouncil 2017). Further, fluorotelomer-based C6 short-chain alternatives have been shown through various evaluations not to be bioaccumulative under current regulatory criteria (Conder et al. 2008). F3 is generally considered biodegradable and not bioaccumulative. The last environmental property, disposal, is very important. It is essential that all firefighting foams be properly disposed. Both the Fire Fighting Foam Coalition (FFFC) (2016a, 2016b) and FPAA (2017) have provided industry guidance for proper foam disposal. Foam, spent foam, and firewater must be disposed of according to local and state (and other government) authorities' requirements.

In summary, selection of a firefighting foam for Class B fires involves consideration of multiple criteria, as shown in Tables 1.3 and 1.4. Over the years, many have invoked the phrase "fit for purpose" as guidance for selection. If the firefighter needs rapid knockdown and control, rapid extinguishment, extended burnback times, and the best life safety performance and asset protection, the choice in 2017 is clearly AFFF containing fluorosurfactants. F3 performance has improved dramatically over the past few years and has been selected by some users instead of AFFF where performance meets their requirements for life safety and ultimate property and asset protection. F3s have achieved many internationally recognized certifications, but to date not MIL-F-24385. A research initiative funded by the Strategic Environmental Research and Development Program (SERDP) titled "Fluorine-Free Aqueous Film Forming Foam," dated October 29, 2015, seeks research proposals to identify a fluorine-free foam that meets or exceeds the performance criteria required in MIL-F-24385 (SERDP 2015).

1.9 AFFF AND F3: FIELD TEST PERFORMANCE AND EFFICACY

In addition to foam and environmental properties, actual efficacy comparisons in the field are essential to assess fire performance and select firefighting foam. The FFFC has published two studies comparing field test performance and the efficacy of

AFFF and F3 (FFFC 2014, 2017). In 2011, the Naval Research Laboratory presented field test results of AFFF agents and F3 (Williams et al. 2011). AFFF extinguishment times on 28 ft² pool fires tested at full strength were on average 77% faster for gasoline, 88% faster for methylcyclohexane (MCH), and 70% faster for heptane than those of F3. For isooctane, the tested AFFF was unable to form a film and the F3 extinguished the fire 10% faster, highlighting the importance of film formation to foam performance. The AFFF tests extinguished the heptane and gasoline fires in 21–28 seconds. The time required to pass the MIL-F-24385 is 30 seconds. The F3 was unable to extinguish any gasoline or heptane fire in less than 30 seconds. To be on the QPL, AFFF products must meet these requirements to be used for military applications. The Federal Aviation Administration (FAA 2011) requires all US airports to carry AFFF agents that meet MIL-F-24385 and are listed on the QPL. In another field study reported at the 2013 Reebok Foam Conference, VS Focum (2013) summarized the company's F3 development. The presentation contained side-by-side field test data done at the same facility under the same conditions comparing the fire performance of AFFF and F3. The results showed that AFFF performed significantly better than F3 in spray extinction tests (0.785 m²) using heptane, gasoline, and Kerosene Jet A-1, and in Spray Pan Fire Out tests in sizes 0.25, 0.785, and 7.06 m², while the values were similar in the 4.52 m² test. The Naval Research Laboratory has conducted field tests comparing AFFF and F3 (Hinnant et al. 2017a, 2017b). In pool fire tests, AFFF achieved extinguishment in 18 seconds, compared with 40 seconds for the F3. In foam degradation tests, F3 degraded after 1–2 minutes, while AFFF lasted 35 minutes before degrading (longer is better). Similar results from a series of foam degradation tests on AFFF and F3 were published earlier in *International Fire Fighter* in 2012 (Jho 2012). At a 2016 foam seminar (SAA-IAFPA 2016), field fire test results conducted on five commercially available short-chain fluorotelomer-based AFFFs and five commercially available F3 were presented (Castro 2016). The tests were run on four fuels: gasoline, heptane, Jet A-1, and diesel. The results showed that AFFF performed significantly better than F3 on all fuels except diesel, where they both put out the fire. None of the F3s were able to extinguish Jet A-1.

Benefits to Firefighters	Key Firefighting Foam Properties
· Reduced risk to life safety of the firefighters and those in the adjacent community	· Suppression of flammable vapor release
· Reduced release of toxic products of combustion	· Cooling the fuel surface
	· Protection from reignition and flashback
· Rapid fire extinguishment	· Resist degradation due to radiant heat
· Reduced potential for the fire to spread	
· Reduced use of foam and water leading to reduced firewater effluent, including both foam and products of combustion	· Resist mixing with the fuel
	· Resist spreading over the fuel surface
	· Resist breakdown by the fuel (e.g., polar solvents)

FIGURE 1.11 FPAA information bulletin: "Selection and Use of Firefighting Foam" (From FPAA, Selection and use of firefighting foams, Information bulletin, Version 2.0, January 2017. http://www.fpaa.com.au/.)

Jet A-1 is the fuel used in International Civil Aviation Organization fire tests, which determine the acceptability of foams for airport use in many countries.

Lastly, the reader is directed to a recently published information bulletin titled "Selection and Use of Firefighting Foams," authored by the FPAA (2017). The FPAA has neatly articulated what a high level of firefighting performance provides firefighters, regardless of whether the foam contains fluorine (Figure 1.11). Additionally, the bulletin describes key foam properties that align with those presented in Table 1.3.

1.10 AFFF USE: BEST PRACTICES

It is imperative that the firefighting foam be fit for purpose. AFFFs should not be used for typical Class A fires (such as wood), forest fires and wildfires, residential fires, electrical fires (Class C), kitchens—commercial and residential, and computer, server, and any telecommunications equipment fires. In addition, AFFF products should not be used on a small car or truck fire where the potential for a large fuel spill is small and where water will suffice to extinguish the fire. AFFF is the foam of choice for fires where there is the potential for a significant flammable liquid hazard or where the rapid extinguishment of hydrocarbon fuel was necessary to avoid harm to people and/or property (e.g., armaments). This is why typical historic use of AFFFs has been in military operations, marine operations, and petroleum terminal, processing, and operations, as well as chemical and other industrial facilities. Considerations when determining the use of AFFF are many. The conversation appears to have shifted from performance to presence, and persistence is what really matters. Having perspective is critical in this discussion. Users of PFAS-containing products must consider the toxicological profile; the potential for exposure; the prevention of emissions; the protection of people, property, and the environment; and the use of risk-based analyses versus solely the precautionary principle (Figure 1.12) (Korzeniowski 2017).

AFFF safe handling, proper use, and disposal are essential. When it is appropriate to use AFFF and how a Class B fire is fought and extinguished are important. A universal consideration is minimizing human exposure and environmental release while protecting people and property. The potential environmental impact of Class B foam usage is critically important. There are three primary sources of potential environmental impact: (1) the fire's combustion products, (2) the AFFF foam itself, and (3) resulting firewater. Foam and firewater must be disposed of in compliance with local, state, and/or country regulations. All foams have the potential to contaminate surrounding areas where foam has been used and the potential to be harmful to the environment if the fire is not fought properly with the right products and if the firewater effluent is not managed appropriately.

Fluorosurfactants are highly stable and do not mineralize by biodegradation (see Section 1.11). As a result, AFFF emissions to the environment from historic use in fire training and firefighting have left a measurable environmental footprint, even decades after use according to recently published research (Anderson et al. 2016). The importance of proper use, handling, and disposal of AFFF has thus become a subject of intense discussion (Seow 2013; DEHP 2016a, 2016b; USEPA 2016). In

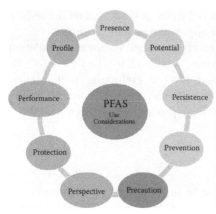

FIGURE 1.12 AFFF use considerations and best practices. (Adapted from FFFC, Best practice guidance for use of Class B firefighting foams, 2016. https://docs.wixstatic.com/ugd/331 cad_188bf72c523c46adac082278ac019a7b.pdf. FFFC, Guidance for fluorinated firefighting foams, 2016. www.fffc.org.)

response, guidance has been developed and promulgated by manufacturers and regulators to AFFF users to cease the use of AFFF for training and only use it in active firefighting to protect people and property. The Umwelt Bundes Amt (UBA) (the German Environment Agency), in collaboration with firefighting foam manufacturers and firefighters, published guidance for the proper use, handling, and disposal of AFFF in 2013 (UBA 2013). The two industry groups, FFFC and FPAA, have done a significant amount of work to raise awareness for best practices and proper use and disposal of foam. The FFFC published both a best practice guidance document and a flyer detailing critical best practices in handling AFFF to fight Class B fires (FFFC 2016a, 2016b). As described earlier, the FPAA has also recently provided guidance on foam use (FPAA 2017). The FFFC, FPAA, and UBA documents should be consulted before use of Class B foam. In brief, everyone using AFFF should diligently work to eliminate foam discharge to the environment. A first way to ensure this is to be sure that all training foams are nonfluorinated. AFFF and firewater should be contained followed by treatment and approved disposal to ensure minimal or no releases to the environment. Advanced planning, preparation, and practice are ways to identify and prevent uncontrolled foam or firewater releases and thereby minimize the environmental impact from firefighting events. Lastly, active monitoring of foam generation and delivery systems for leaks and proper preventative maintenance of foam systems should be a requirement.

AFFF products containing fluorosurfactants provide the firefighter with the most effective agents on the market today. As set forth in the FFFC guidance document (FFFC 2016a, 2016b), not only is AFFF the product of choice for Class B fires, but also the use of the multipurpose AR-AFFF or AR-FFFP foam concentrates allows the firefighter to stock one product that can be used to extinguish either a hydrocarbon or a polar solvent fire. Not using the correct foam agent would result in ineffective firefighting and could result in increased environmental contamination and slower fire extinguishment with potential catastrophic results.

Finally, while the FFFC and FPAA industry associations have done a significant amount of work to raise awareness and promote AFFF best practices, there are those that are opposed to the use of any fluorine-containing foam (Vierke 2016). It is the opinion of this chapter's authors that the reader—and the regulatory community, nongovernmental organizations, and firefighting community—should consider and carefully weigh the positives and negatives of fluorochemicals in protecting people and property when deciding if the use of AFFF is warranted.

1.11 SUMMARY AND CONCLUSIONS

This chapter has highlighted the historical aspects of fluorosurfactants used in the firefighting industry and their evolution until present day. From protein-based foams and long-chain fluorochemical-containing foams to today's short-chain fluorosurfactant and fluorine-free foams, a wide range of firefighting products were covered. The nomenclature, performance, and attributes of each of the above were discussed. The principle of operation as it pertains to extinguishing fire, the chemistry of the fluorosurfactants, and the differences between the two main manufacturing processes were covered. The QPL and the required firefighting specifications of the USDOD were explained. Although a lot of material was covered in this opening chapter, numerous other factors remain that need consideration and merit further exploration in order to fully understand the complexities assocatiated with PFAS decisions. This summary section provides a touch point on these issues, and subsequent chapters explore these challenges further.

The historic use of long-chain-containing AFFF has resulted in legacy contamination to soil, sediment, and waters, including groundwater, surface water, and drinking water. Being persistent, the environmental footprint of the fluorosurfactants in AFFF remains for extended periods of time, even decades after use (Place and Field 2012; Backe et al. 2013; Anderson et al. 2016; Barzen-Hanson et al. 2017). In addition, the fluorosurfactants in AFFF and their degradation products are often water soluble and migrate beyond the point of use, in some cases over substantial distances, and have impacted drinking water sources for adjacent communities (Hu et al. 2016). Much work continues in this area of fate and transport; numerous studies have been conducted and published that illuminate our understanding of what happens to these surfactants in the environment, to include biodegradation pathways (Liu and Mejia Avendaño 2013). Field monitoring studies continue to identify the fluorosurfactants and degradation products present at contaminated sites. These degradation products are of great interest to the AFFF community. The formation of stable end products, such as PFOS and PFOA, has been shown to occur with both ECF-based perfluoroalkane sulfonyl–based fluorosurfactants (Rhoads et al. 2008; Liu and Mejia Avendaño 2016; Mejia Avendaño et al. 2016; Zhang et al. 2017) and fluorotelomer-based fluorosurfactants (Moe et al. 2012; Liu and Mejia Avendaño 2013; Butt et al. 2014; Zhang et al. 2016; Buck 2017). Data on the physical-chemical properties (e.g., water solubility and adsorption) of the identified substances are being compiled and efforts to fill data gaps initiated. Environmental monitoring and scoping is ongoing and expanding. The presence of legacy contamination at sites historically associated with fire training activities and accident sites is generally understood (Moody and Field 1999, 2000; Moody et al. 2002, 2003), but a complete picture of the contamination footprint considering accidental releases, equipment testing and calibration areas, and undocumented disposal activities has not yet been realized. Determining the scope of the contamination is essential to understanding the magnitude of cleanup required to protect environmental and human health.

Work completed thus far to delineate the extent of the environmental affected has revealed widespread contamination and spurred intense work to identify effective remediation technologies. A recent review summarized many of the approaches (Merino et al. 2016). Additionally, some research has identified means for capturing fluorosurfactants in foam and firewater at the site of use as a best practice that can allow continued use of AFFF with minimal environmental impact (Pabon and Corpart 2002; Baudequin et al. 2011, 2014; Scholz 2014). Many chapters follow describing some the methods being investigated to remedy PFAS contamination in various environmental matrices in timely and cost-effective manners.

Another challenging aspect of PFAS is associated with the analytical methodology. The analytical methods and sample preparation procedures required to isolate, identify (aka structure speciation), and quantify fluorosurfactants and their degradation products have been the subject of numerous reviews (Martin et al. 2004; Powley et al. 2005; Larsen and Kaiser 2007; Place and Field 2012; Backe et al. 2013; Barzen-Hanson and Field 2015; Barzen-Hanson et al. 2017; Dauchy et al. 2017; Munoz et al. 2017). The importance of sample preparation and the use of controls, method blanks, isotopic standards, and rigorous laboratory hygiene practices, in order to be able to achieve reliable and reproducible identification and quantification, cannot be overstated. Fluorosurfactants used in AFFF are present, along with numerous other substances, including the products of combustion. Fluorosurfactants themselves can be especially difficult to isolate, identify, and quantify because they are often bound to environmental matrices, such as soil or sediment. Challenges remain when samples are prepared in aqueous solutions as the fluorosurfactants often stick to surfaces and/or rapidly concentrate at air–liquid interfaces. An additional challenge is ensuring that sample preparation does not induce chemical changes to the fluorosurfactant analyte. For ionic species with terminal sulfonate and carboxylate groups, liquid chromatography followed by tandem mass spectrometry (LC-MS/MS) is preferred. For neutral, nonionic species, such as fluorotelomer alcohols, gas chromatography followed by MS is the preferred analytical instrumentation. An extensive discussion on the nuances of the analytical chemistry is presented in Chapter 2.

As discussed earlier, short-chain C6 fluorotelomer-based products have been shown to have a more favorable toxicology than historic long-chain equivalents (Environ 2014 and 2016; Buck 2015; Jacobs 2016). Further, fluorotelomer-based C6 short-chain alternatives have been shown through various evaluations not to be bioaccumulative under current regulatory criteria (Conder et al. 2008; Gannon et al. 2011; Environ 2014; Buck 2015; Hoke et al. 2015). Work in this area continues and is extensive. Readers are directed to later chapters in the book describing more of these efforts and their evolving findings.

Understanding the scope of the environmental contamination is one important front that is advancing, but another question remains: How clean is clean? The current lifetime health advisory issued by the Environmental Protection Agency (EPA) is set at 70 ppt for the sum total of PFOS and PFOA. However, this is only an advisory and, as explicitly stated in the EPA language, is not to be used as a regulatory standard. If and when a maximum contaminant level under the Safe Drinking Water Act is established, will it be the same value? The toxicological data gaps being filled and the advancing analytical techniques with improved limits of detection will influence the federal regulatory landscape. Additionally, states can (and have) set their own limits on allowable levels of PFAS. The dynamic regulatory framework further contributes to the complexities of managing perfluorinated chemical use associated with firefighting activities.

ACKNOWLEDGMENTS

S.H.K. would like to thank C. Jho, B. Rambo, M. Willson, and T. Cortina for their invaluable help, insights, critique, and encouragement. The authors would like to thank our colleagues for their helpful review and comments during the preparation of this chapter.

REFERENCES

3M. 2000. Re: Phase-out plan for POSF-based products. USEPA Administrative Record AR226-0600. www.regulations.gov (accessed 3 October 2017) as document EPA-HQ-OPPT-2002-0051-0006.

Alsmeyer, Y., Childs, W., Flynn, R., Moore, C., Smeltzer, J. 1994. Electrochemical fluorination and its applications. In Banks, R.E., Smart, B.E., and Tatlow, J.C. (Ed.), *Organofluorine Chemistry—Principles and Commercial Applications*. Plenum Press, New York, pp. 121–144.

Anderson, R.H., Long, G.C., Porter, R.C., Anderson, J.K. 2016. Occurrence of select perfluoroalkyl substances at U.S. Air Force aqueous film-forming foam release sites other than fire-training areas: Field-validation of critical fate and transport properties. *Chemosphere*, 150:678–685.

Backe, W.J., Day, T.C., Field, J.A. 2013. Zwitterionic, cationic, and anionic fluorinated chemicals in aqueous film forming foam formulations and groundwater from U.S. military bases by nonaqueous large-volume injection HPLC-MS/MS. *Environmental Science & Technology*, 47:5226–5234.

Banks, R.E., Smart, B.E., Tatlow, J.C. 1994. Organofluorine chemistry: Principles and commercial applications. Plenum, New York.

Barzen-Hanson, K.A., Field, J.A. 2015. Discovery and implications of C2 and C3 perfluoroalkyl sulfonates in aqueous film-forming foams and groundwater. *Environmental Science & Technology Letters*, 2:95–99.

Barzen-Hanson, K.A., Roberts, S.C., Choyke, S., Oetjen, K., McAlees, A., Riddell, N., McCrindle, R., Ferguson, P.L., Higgins, C.P., Field, J.A. 2017. Discovery of 40 classes of per- and polyfluoroalkyl substances in historical aqueous film-forming foams (AFFFs) and AFFF-impacted groundwater. *Environmental Science & Technology*, 51:2047–2057.

Baudequin, C., Couallier, E., Rakib, M., Deguerry, I., Severac, R., Pabon, M. 2011. Purification of firefighting water containing a fluorinated surfactant by reverse osmosis coupled to electrocoagulation-filtration. *Separation and Purification Technology*, 76:275–282.

Baudequin, C., Mai, Z., Rakib, M., Deguerry, I., Severac, R., Pabon, M., Couallier, E. 2014. Removal of fluorinated surfactants by reverse osmosis—Role of surfactants in membrane fouling. *Journal of Membrane Science*, 458:111–119.

Briggs, T. 1996. Foams for firefighting. In *Foams: Theory, Measurements, and Applications*. Surfactant Science Series, Vol. 57. Marcel Dekker, Inc., New York, NY, pp. 465–509.

Buck, R.C. 2015. Toxicology data for alternative "short-chain" fluorinated substances. In DeWitt, J.C. (Ed.), *Toxicological Effects of Perfluoroalkyl and Polyfluoroalkyl Substances*. Humana Press, New York, pp. 451–477.

Buck, R.C. 2017. Short-chain fluorotelomer-based substances—Common biodegradation pathways. ENVR-882. American Chemical Society National Meeting, San Francisco, CA.

Buck, R.C., Franklin, J., Berger, U., Conder, J.M., Cousins, I.T., de Voogt, P., Jensen, A.A., Kannan, K., Mabury, S.A., van Leeuwen, S.P.J. 2011. Perfluoroalkyl and polyfluoroalkyl substances in the environment: Terminology, classification, and origins. *Integrated Environmental Assessment and Management*, 7:513–541.

Buck R.C., Murphy, P.M., Pabon, M. 2012. Chemistry, properties and uses of commercial fluorinated surfactants. In Knepper, T.P., Lange, F.T. (Eds.), *Handbook of Environmental Chemistry*, Vol. 17, *Polyfluorinated Chemicals and Transformation Products*. Springer, New York, pp. 1–24.

Butt, C.M., Muir, D.C.G., Mabury, S.A. 2014. Biotransformation pathways of fluorotelomer-based polyfluoroalkyl substances: A review. *Environmental Toxicology and Chemistry*, 33:243–267.

Castro, J. 2016. Fluorine free foams. Where is the limit? Presented at SAA-IAFPA Firefighting Foam Seminar, Singapore, 20–22 July.

Conder, J.M., Hoke, R.A., de Wolf, W., Russell, M.H., Buck, R.C. 2008. Are PFCAs bioaccumulative? A critical review and comparison with regulatory criteria and persistent lipophilic compounds. *Environmental Science & Technology*, 42:995–1003.

Cortina, T. 2012. Environmental and regulatory update on firefighting foams. Presented at NPFA Conference, Las Vegas, June 11.

Cortina, T., Korzeniowski, S. 2008. Firefighting foams—Reebok redux. *Industrial Fire Journal*, April 18–20.

Darwin, R.L. 2011. Estimated inventory of PFOS-based aqueous film forming foam (AFFF). Fire Fighting Foam Coalition, Arlington, VA.

Dauchy, X., Boiteux, V., Bach, C., Rosin, C., Munoz, J.F. 2017. Per- and polyfluoroalkyl substances in firefighting foam concentrates and water samples collected near sites impacted by the use of these foams. *Chemosphere*, 183:53–61.

DEHP (Queensland Department of Environment and Heritage Protection). 2016a. Management of firefighting foam policy. July. https://www.ehp.qld.gov.au/assets/documents/regulation/firefighting-foam-policy.pdf (accessed 3 October 2017).

DEHP (Queensland Department of Environment and Heritage Protection). 2016b. Management of firefighting foam policy's explanatory notes. July. https://www.ehp.qld.gov.au/assets/documents/regulation/firefighting-foam-policy-notes.pdf (accessed 3 October 2017).

Environ. 2014. Assessment of POP criteria for specific short-chain perfluorinated alkyl substances. Environ International Corporation, Arlington, VA. http://chm.pops.int/TheConvention/POPsReviewCommittee/Meetings/POPRC9/POPRC9Followup/PFOSSubmission/tabid/3565/Default.aspx.

Environ. 2016. Assessment of POP criteria for specific short-chain perfluorinated alkyl substances. Ramboll Environ, Arlington, VA. www.fluorocouncil.org/resources/research.

FAA (Federal Aviation Administration). 2011. Identifying MIL-SPEC aqueous film forming foam (AFFF). National Part 139 CertAlert No. 11-02. 15 February.

FFFC (Fire Fighting Foam Coalition). 2006. Aquatic toxicity of firefighting foams. Fire Fighting Foam Coalition newsletter, August.

FFFC (Fire Fighting Foam Coalition). 2014. Fact sheet on AFFF firefighting agents. www.fffc.org.
FFFC (Fire Fighting Foam Coalition). 2016a. Best practice guidance for use of Class B firefighting foams. https://docs.wixstatic.com/ugd/331cad_188bf72c523c46adac082278ac019a7b.pdf (accessed 3 October 2017).
FFFC (Fire Fighting Foam Coalition). 2016b. Guidance for fluorinated firefighting foams. www.fffc.org (accessed 3 October 2017).
FFFC (Fire Fighting Foam Coalition). 2017. Fact sheet on AFFF firefighting agents. www.fffc.org (accessed 3 October 2017).
FluoroCouncil. 2017. Web database. www.fluorocouncil.org/resources/research (accessed 3 October 2017).
FPAA (Fire Protection Association Australia). 2017. Selection and use of firefighting foams. Information bulletin, Version 2.0, January. http://www.fpaa.com.au/.
Friedrich, K. 1940. Fire extinguishing compound. U.S. Patent 2,212,470.
Gannon, S.A., Johnson, T., Nabb, D.L., Serex, T.L., Buck, R.C., Loveless, S.E. 2011. Absorption, distribution, metabolism, and excretion of [1–14C]-perfluorohexanoate ([14C]-PFHx) in rats and mice. *Toxicology*, 283:55–62.
Gates, E. 1903. Method of extinguishing fires. U.S. Patent 749,374.
Gipe, R.L., Peterson, H.B. 1972. Proportioning characteristics of aqueous film-forming foam concentrates. NRL Report 7437. Naval Research Laboratory, Washington, DC.
Hinnant, K.M., Conroy, M.W., Ananth, R. 2017a. Influence of fuel on foam degradation for fluorinated and fluorine-free foams. *Colloids and Surfaces A*, 522:1–17.
Hinnant, K.M., Giles, S.L., Ananth, R. 2017b. Measuring fuel transport through fluorocarbon and fluorine-free firefighting foams. *Fire Safety Journal*, 19:653–66. doi: 10.1016/j.firesaf.2017.03.077.
Hoke, R.A., Ferrell, B.D., Ryan, T., Sloman, T.L., Green, J.W., Nabb, D.L., Mingoia, R., Buck, R.C., Korzeniowski, S.H. 2015. Aquatic hazard, bioaccumulation and screening risk assessment for 6:2 fluorotelomer sulfonate. *Chemosphere*, 128:258–265.
Hu, X.C., Andrews, D.Q., Lindstrom, A.B., Bruton, T.A., Schaider, L.A., Grandjean, P., Lohmann, R., Carignan, C.C., Blum, A., Balan, S.A., Higgins, C.P., Sunderland, E.M. 2016. Detection of poly- and perfluoroalkyl substances (PFASs) in U.S. drinking water linked to industrial sites, military fire training areas, and wastewater treatment plants. *Environmental Science & Technology Letters*, 3(10):344–350.
Jacobs, L. 2016. Fluorine-free firefighting foam—The path to where we are today. Presented at SAA-IAFPA Firefighting Foam Seminar, Singapore, 20–22 July.
Jho, C. 2012. Flammability and degradation of fuel—Contaminated fluorine free foams. *International Fire Fighter*, 41(36).
Jho, C. 2016. Interactions of firefighting foam with hydrocarbon fuel—Some fundamental concepts. Presented at SAA-IAFPA Firefighting Foam Seminar, Singapore, 20–22 July.
Johnson, J.H. 1877. British Patent 560. July 20.
Katzer, A.F. 1943. Foam stabilizing composition. U.S. Patent 2,324,951.
Kissa, E. 2001. *Fluorinated Surfactants and Repellents*. Surfactant Science Series, Vol. 97. Marcel Dekker, New York.
Kleiner, E. 2016. 1976–2016: Forty years of saving lives, C6 fluorotelomer surfactants and their use in firefighting foams. American Chemical Society, Washington, DC, p. ENVR-130.
Kleiner, E., Jho, C. 2009. Recent developments in 6:2 fluorotelomer surfactants and stabilizers. Presented at 4th Reebok Foam Seminar, Bolton, UK, 6–9 July.
Korzeniowski, S.H. 2017. Presented at Fourth International Symposium on Bioremediation and Sustainable Environmental Technologies, Miami, 22–25 May, Poster Session F1, no. 110.
Larsen, B.S., Kaiser, M.A. 2007. Challenges in perfluorocarboxylic acid measurements. *Analytical Chemistry*, 79:3966–3973.

Liu, J., Mejia Avendaño, S. 2013. Microbial degradation of polyfluoroalkyl chemicals in the environment: A review. *Environment International*, 61:98–114.

Martin, J.W., Kannan, K., Berger, U., De Voogt, P., Field, J., Franklin, J., Giesy, J.P. et al. 2004. Analytical challenges hamper perfluoroalkyl research. *Environmental Science & Technology*, 38:248A–255A.

Matthies, M., Solomon, K., Vighi, M., Gilman, A., Tarazona, J.V. 2016. The origin and evolution of assessment criteria for persistent, bioaccumulative and toxic (PBT) chemicals and persistent organic pollutants (POPs). *Environmental Science: Processes & Impacts*, 18:1114–1128.

Mejia Avendaño, S., Duy, S.V., Sauvé, S., Liu, J. 2016. Generation of perfluoroalkyl acids from aerobic biotransformation of quaternary ammonium polyfluoroalkyl surfactants. *Environmental Science & Technology*, 50:9923–9932.

Merino, N., Qu, Y., Deeb, R.A., Hawley, E.L., Hoffmann, M.R., Mahendra, S. 2016. Degradation and removal methods for perfluoroalkyl and polyfluoroalkyl substances in water. *Environmental Engineering Science*, 33:615–649.

MIL-F-24385. 1969. Military specification: Fire extinguishing agent, aqueous film forming foam (AFFF) liquid concentrate for fresh and sea water. November 21. http://qpldocs.dla.mil/search/parts.aspx?qpl=1910. For Type 3 history, see http://www.dcfpnavymil.org/Systems/AFFF/QPL%2024385%20HISTORY%20-%20TYPE%203.pdf. For Type 6 history, see http://www.dcfpnavymil.org/Systems/AFFF/QPL%2024385%20HISTORY%20-%20TYPE%206.pdf (accessed 3 October 2017).

Moe, M.K., Huber, S., Svenson, J., Hagenaars, A., Pabon, M., Trümper, M., Berger, U., Knapen, D., Herzke, D. 2012. The structure of the fire fighting foam surfactant Forafac®1157 and its biological and photolytic transformation products. *Chemosphere*, 89:869–875.

Moody, C.A., Field, J.A. 1999. Determination of perfluorocarboxylates in groundwater impacted by Fire-fighting activity. *Environmental Science & Technology*, 33(16):2800–2806.

Moody, C.A., Field, J.A. 2000. Perfluorinated surfactants and the environmental implications of their use in fire-fighting foams. *Environmental Science & Technology*, 34:3864–3870.

Moody, C.A., Hebert, G.N., Strauss, S.H., Field, J.A. 2003. Occurrence and persistence of perfluorooctanesulfonate and other perfluorinated surfactants in groundwater at a fire-training area at Wurtsmith Air Force Base, Michigan, USA. *Journal of Environmental Monitoring*, 5:341–345.

Moody, C.A., Martin, J.W., Kwan, W.C., Muir, D.C.G., Mabury, S.A. 2002. Monitoring perfluorinated surfactants in biota and surface water samples following an accidental release of fire-fighting foam into Etobicoke Creek. *Environmental Science & Technology*, 36:545–551.

Munoz, G., Desrosiers, M., Duy, S.V., Labadie, P., Budzinski, H., Liu, J., Sauve, S. 2017. Environmental occurrence of perfluoroalkyl acids and novel fluorotelomer surfactants in the freshwater fish *Catostomus commersonii* and sediments following firefighting foam deployment at the Lac-Megantic Railway accident. *Environmental Science & Technology*, 51:1231–1240.

Organisation for Economic Co-operation and Development. 2017. Portal on per and poly fluorinated chemicals. http://www.oecd.org/chemicalsafety/portal-perfluorinated-chemicals/aboutpfass/ (accessed 3 October 2017).

Pabon, M., Corpart, J.M. 2002. Fluorinated surfactants: Synthesis, properties, effluent treatment. *Journal of Fluorine Chemistry*, 114:149–156.

Place, B.J., Field, J.A. 2012. Identification of novel fluorochemicals in aqueous film-forming foams used by the US military. *Environmental Science & Technology*, 46:7120–7127.

Plant, D. 2016. Firefighting foam: The real question of sustainability. Presented at SAA-IAFPA Firefighting Foam Seminar, Singapore, 20–22 July.

Powley, C.R., George, S.W., Ryan, T.W., Buck, R.C. 2005. Matrix effect-free analytical methods for determination of perfluorinated carboxylic acids in environmental matrixes. *Analytical Chemistry*, 77:6353–6358.

Ratzer, A.F. 1944. Method of making fire extinguishing foam. U.S. Patent 2,361,057.

Ratzer, A.F. 1956. History and development of foam as a fire extinguishing medium. *Industrial and Engineering Chemistry*, 48(11):2013–2016.

Rhoads, K.R., Janssen, E.M.L., Luthy, R.G., Criddle, C.S. 2008. Aerobic biotransformation and fate of N-ethyl perfluorooctane sulfonamidoethanol (N-EtFOSE) in activated sludge. *Environmental Science & Technology*, 42:2873–2878.

SAA-IAFPA (Singapore Aviation Academy–International Aviation Fire Protection Association). 2016. Firefighting foam seminar, Singapore, 20–22 July.

Scholz, M. 2014. Firewater storage, treatment, recycling and management: New perspectives based on experiences from the United Kingdom. *Water*, 6:367–380.

Seow, J. 2013. Firefighting foams with perfluorochemicals—Environmental review. Department of Environment and Conservation, Western Australia, June.

SERDP (Strategic Environmental Research and Development Program). 2015. Fluorine-free aqueous film forming foam. https://serdp-estcp.org/content/download/35979/344522/file/WPSON-17-01_Fluorine-Free+AFFF.pdf (accessed 3 October 2017).

Simons, J.H. 1949. Electrochemical process for the production of fluorocarbons. *Journal of the Electrochemical Society*, 95:47–59.

Stockholm Convention. 2009. At COP-4 in 2009, the Conference of the Parties decided to list perfluorooctane sulfonic acid (PFOS), its salts and perfluorooctane sulfonyl fluoride (PFOSF) in Annex B to the Stockholm Convention (decision SC-4/17). http://chm.pops.int/Implementation/PFOS/Overview/tabid/5221/Default.aspx (accessed 3 October 2017).

Taylor, C.K. 1999. Fluorinated surfactants in practice. In *Annual Surfactants Review*. Vol. 2. by D.R. Karsa, CRC Press LLC, Boca Raton, FL, pp. 271–316.

UBA (Umwelt Bundes Amt). 2013. Environmentally responsible use of fire-fighting foams. https://www.umweltbundesamt.de/sites/default/files/medien/378/publikationen/fluorinated_fire-fighting_foams_schaumloeschmittel_engl._version_25.6.2013.pdf (accessed 3 October 2017).

USEPA (U.S. Environmental Protection Agency). 2010/2015. PFOA Stewardship Program. https://www.epa.gov/assessing-and-managing-chemicals-under-tsca/fact-sheet-20102015-pfoa-stewardship-program (accessed 3 October 2017).

USEPA (U.S. Environmental Protection Agency). 2016. PFOS and PFOA health advisory [May 2016]. https://www.epa.gov/ground-water-and-drinking-water/drinking-water-health-advisories-pfoa-and-pfos (accessed 3 October 2017).

Vierke, L. 2016. Short-chain PFASs—No alternatives to PFASs—Are short-chain PFASs an alternative to long-chain PFASs? A regulatory perspective. Presented at OECD webinar, December. http://www.oecd.org/chemicalsafety/portal-perfluorinated-chemicals/webinars/ (accessed 3 October 2017).

VS Focum. 2013. A new high performance newtonian fluorine-free foam. Manual Acuna. Presented at 5th Reebok International Foam Conference, Bolton, UK, 19 March.

Weissenborn, A. 1939. Method of producing air foam. U.S. Patent 2,151,398.

Williams, B., Murray, T., Butterworth, C., Burger, Z., Sheinson, R., Fleming, J., Whitehurst, C., Farley, F. 2011. Extinguishment and burnback tests of fluorinated and fluorine-free firefighting foams with and without film formation. Presented at Suppression, Detection and Signaling Research and Applications—A Technical Working Conference (SUPDET 2011), Orlando, FL, 22–25 March.

Willson, M. 2016a. Can F3 agents take the fire security heat? *International Airport Review*, 20(6):3–6.

Willson, M. 2016b. Reducing environmental impacts of Class B firefighting foams. Presented at SAA-IAFPA Firefighting Foam Seminar, Singapore, 20–22 July.

Zhang, L., Lee, L.S., Niu, J., Liu, J. 2017. Kinetic analysis of aerobic biotransformation pathways of a perfluorooctane sulfonate (PFOS) precursor in distinctly different soils. *Environmental Pollution*, 229:159–167.

Zhang, S., Lu, X., Wang, N., Buck, R.C., 2016. Biotransformation potential of 6:2 fluorotelomer sulfonate (6:2 FTSA) in aerobic and anaerobic sediment. *Chemosphere* 154:224–230.

2 Per- and Polyfluoroalkyl Substance Analysis to Support Site Characterization, Exposure, and Risk Management

Kavitha Dasu, David M. Kempisty, and Marc A. Mills

CONTENTS

2.1	Introduction	36
2.2	General Extraction Methods	38
	2.2.1 Solid-Phase Extraction	38
	2.2.2 Liquid–Liquid Extraction	40
2.3	Overview of Analytical Methods	40
	2.3.1 Liquid Chromatography	40
	2.3.1.1 Orthogonal Chromatography	41
	2.3.1.2 Comparison of Analytical Instrumentation	42
	2.3.2 Gas Chromatography	42
	2.3.3 Emerging Technologies	43
	2.3.3.1 Total Fluorine Analysis	43
	2.3.3.2 Total Oxidizable Precursor Assay	48
	2.3.3.3 Particle-Induced Gamma-Ray Emission	48
	2.3.3.4 Fluorine-19 Nuclear Magnetic Resonance	48
2.4	Interlaboratory Comparisons	49
2.5	Standard Methods	50
2.6	Cross-Contamination Issues	51
	2.6.1 Sample Collection	51
	2.6.2 Sample Storage	52
	2.6.3 Sample Preparation	53
	2.6.4 Sample Analysis	53
2.7	Quality Assurance and Quality Control	54
2.8	Quantitation Methods	55
	2.8.1 Detection and Quantitation Limits	56

2.9   Quantitation Challenges 57
2.10 Conclusions 59
References 60

2.1 INTRODUCTION

Reliable, robust, and quantitative analysis of per- and polyfluoroalkyl substances (PFAS) is challenging. The widespread prevalence of these fluorinated organics in our world is a significant challenge to analytical accuracy and precision. One only needs to look at the 99% of the US populace with positive PFAS concentrations in blood serum (Calafat et al. 2006, 2007; Brede et al. 2010; Wu et al. 2015). There are many industrial and consumer sources (Liu et al. 2014; Allred et al. 2015; Anderson et al. 2016) for this anthropogenic class of chemicals to the environment in which we are constantly interacting (Lau et al. 2007; Fromme et al. 2010). PFAS have been found in our food (Senthilkumar et al. 2007; Jogsten et al. 2009; Schecter et al. 2010), water (Brede et al. 2010; Möller et al. 2010; Ahrens 2011), groundwater (Enevoldsen and Juhler 2010; Guelfo and Higgins 2013; Houtz et al. 2013), drinking water (Hölzer et al. 2008; Ericson et al. 2009; Boiteux et al. 2012), soils (Higgins and Luthy 2006; Gellrich and Knepper 2012; Dasu and Lee 2015), sediments (Ahrens et al. 2010; Naile et al. 2010; Gómez et al. 2011), dust (Huber et al. 2011; Shoeib et al. 2011; Jogsten et al. 2012), and air (Boulanger et al. 2005; Weinberg et al. 2011a, 2011b). PFAS are contained in the commercial and consumer products in our workplace and homes and the products that we interface with daily, such clothes, textiles, papers, and food contact surfaces. For example, the office environment from which I type this paragraph contains a myriad of fluorinated organics, such as the flooring, the furniture fabrics, and coated paper products.

To accurately determine the concentration of PFAS in an environmental media, it is important to minimize fluorinated chemicals from outside the media being tested (i.e., lower the background contamination). Strict good laboratory practices (GLPs), such as sample handling procedures and screening of laboratory and field chemicals and supplies, have to be abided by when determining environmental concentrations of PFAS. The opportunities for cross-contamination from both field activities and the analytical laboratory procedures are many. The residue from a gas station receipt or a food wrapping can alter the PFAS concentration in a water sample if proper GLP techniques are not followed. With limits of detection (LODs) in the single nanograms per liter (more commonly referred to as parts per trillion (ppt), the risk of background contamination is high. In addition, with recent regulatory limits at extremely low levels, such as the US Environmental Protection Agency's (USEPA) lifetime health advisory at 70 ng/L for the sum of perfluorooctanoic acid (PFOA) and perfluorooctane sulfonate (PFOS) (USEPA 2016a), the impacts of even very low levels of contamination can be significant. As a point of reference for "per trillion" as a very small number, the world population is 7.5 billion people. If we increase this number, almost double it to 14 billion people, the lifetime health advisory is the equivalent to finding a single person in the 14 billion. With such low detection limits, cross-contamination by even trace levels of PFAS from other sources will have a large impact on the accuracy and validity of the analytical results.

Now consider a typical, clean environmental analytical laboratory. Teflon™ is a fluorinated organic polymer, or fluoropolymer, that uses PFAS during the manufacturing process and contains PFAS by design in the final product. Fluoropolymers are more commonly used as instrumental parts and common laboratory supplies in an environmental laboratory. Stir bars, sample container caps, pipette tips, tubing, tape, seals, septa, chemical containers and caps, Viton™ elastomer materials, and much more all contain various levels and formulations of PFAS. Additionally, if you consider what you do not see, the plumbing and operational components of most analytical equipment and instruments contain fluoropolymers in various seals and linings that contact samples. Without proper precautions, these components may result in high background levels of PFAS for analysis. As stated above, accurate analysis of PFAS can be challenging.

For ease of explanation, PFAS analytes are broadly divided into two groups in this chapter, perfluoroalkylated acids and precursors/intermediates (Figure 2.1). Some of the commonly measured PFAS analytes are listed in Tables 2.1 and 2.2. This chapter describes in greater detail the challenges associated with PFAS analysis. Practices to avoid cross-contamination during sample collection and in the analytical laboratory are discussed. The various methodologies used to quantify PFAS from different environmental media are reviewed, including preparatory methods, quality assurance (QA) and quality control (QC) measures, and detection. More attention is given to aqueous media for the methodology, but soil, sediment, and air methods are briefly discussed. Often, the instrumentation for final separation and detection or quantification is the same for various media with sample extraction and preparation procedures appropriate to the matrices. Finally, challenges associated with analytical measurement, including the availability of common reference standards, variance in results between laboratories, and issues associated with PFAS precursor materials, are discussed.

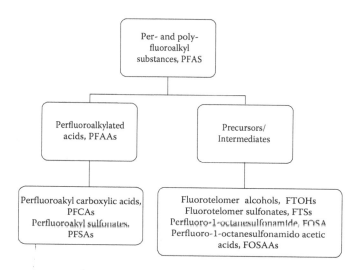

FIGURE 2.1 Broad classification of PFAS.

TABLE 2.1
List of Commonly Measured PFAS, Their Acronyms, and CAS Numbers

Perfluoroalkyl Acids (PFAAs)	Acronym	CAS
Perfluorocarboxylic Acids (PFCAs)		
Perfluoro-n-butanoic acid	PFBA	375-22-4
Perfluoro-n-pentanoic acid	PFPA	2706-90-3
Perfluoro-n-hexanoic acid	PFHxA	307-24-4
Perfluoro-n-heptanoic acid	PFHpA	375-85-9
Perfluoro-n-octanoic acid	PFOA	335-67-1
Perfluoro-n-nonanoic acid	PFNA	375-95-1
Perfluoro-n-decanoic acid	PFDA	335-76-2
Perfluoro-n-undecanoic acid	PFUnDA	2058-94-8
Perfluoro-n-dodecanoic acid	PFDoDA	307-55-1
Perfluoro-n-tridecanoic acid	PFTrDA	72629-94-8
Perfluoro-n-tetradecanoic acid	PFTeDA	376-06-7
Perfluoro-n-hexadecanoic acid	PFHxDA	67905-19-5
Perfluoro-n-octadecanoic acid	PFODA	16517-11-6
Perfluorosulfonates (PFSAs)		
Potassium perfluoro-1-butanesulfonate	L-PFBS	374-73-5
Sodium perfluoro-1-pentanesulfonate	L-PFPeS	N/A
Sodium perfluoro-1-hexanesulfonate	L-PFHxS	29420-49-3
Sodium perfluoro-1-heptanesulfonate	L-PFHpS	375-92-8
Sodium perfluoro-1-octanesulfonate	L-PFOS	1763-23-1
Sodium perfluoro-1-nonanesulfonate	L-PFNS	N/A
Sodium perfluoro-1-decanesulfonate	L-PFDS	335-77-3

Note: N/A = Not available.

2.2 GENERAL EXTRACTION METHODS

Analysis of environmental samples for PFAS target compounds requires methodic and meticulous sample preparation procedures to separate the target analytes from the matrices while avoiding background contamination. In general, extraction of aqueous samples involves concentration of the sample, and hence during the process, interfering nontargeted compounds in the matrix, such as biomolecules and organic matter, may also be concentrated. Therefore, the extraction technique should be designed to minimize the matrix interferences and provide reproducible recoveries of the target analytes. Common extraction techniques reported in the literature are discussed in Sections 2.2.1 and 2.2.2.

2.2.1 SOLID-PHASE EXTRACTION

For aqueous samples, such as surface water, groundwater, drinking water, and wastewater, solid-phase extraction (SPE) is commonly used for separating and

TABLE 2.2
List of Commonly Measured Precursors or Intermediates, Their Acronyms, and CAS Numbers

Precursors or Intermediates	Acronym	CAS
Sodium 1H,1H,2H,2H-perfluorohexane sulfonate	4:2 FTS	757124-72-4
Sodium 1H,1H,2H,2H-perfluorooctane sulfonate	6:2 FTS	27619-97-2
Sodium 1H,1H,2H,2H-perfluorodecane sulfonate	8:2 FTS	39108-34-4
Perfluoro-1-octanesulfonamidoacetic acid	FOSAA	N/A
N-Methylperfluoro-1-octanesulfonamidoacetic acid	N-MeFOSAA	2355-31-9
N-Ethylperfluoro-1-octanesulfonamidoacetic acid	N-EtFOSAA	2991-50-6
Perfluoro-1-octanesulfonamide	FOSA	754-91-6
N-Methylperfluoro-1-octanesulfonamide	N-MeFOSA	31506-32-8
N-Ethylperfluoro-1-octanesulfonamide	N-EtFOSA	4151-50-2
N,N-Dimethylperfluoro-1-octanesulfonamide	N,N-Me$_2$FOSA	N/A
2-Perfluorobutyl ethanol	4:2 FTOH	2043-47-2
2-Perfluorohexyl ethanol	6:2 FTOH	647-42-7
2-Perfluorooctyl ethanol	8:2 FTOH	678-39-7
2-Perfluorodecyl ethanol	10:2 FTOH	865-86-1
3-Perfluoropentyl propanoic acid	5:3 FTCA	N/A
3-Perfluoroheptyl propanoic acid	7:3 FTCA	812-70-4
2H-Perfluoro-2-octenoic acid	6:2 FTUCA	N/A
2H-Perfluoro-2-decenoic acid	8:2 FTUCA	70887-84-2
2H-Perfluoro-2-dodecenoic acid	10:2 FTUCA	N/A
2-Perfluorohexyl ethanoic acid	6:2 FTCA	53826-12-3
2-Perfluorooctyl ethanoic acid	8:2 FTCA	27854-31-5
2-Perfluorodecyl ethanoic acid	10:2 FTCA	53826-13-4

Note: N/A = Not available.

concentrating the target compounds from the sample. In general, the sample volume required for analysis varies from 5 to 1000 mL and depends on the SPE media and the sensitivity of the analytical instrument (ISO 2009; Quiñones and Snyder 2009; ASTM 2016). Common SPE phases used for sample concentration and cleanup are weak anion exchange (WAX), hydrophilic–lipophilic balance (HLB), hydrophobic C8, and C18 (Jin et al. 2009; Quiñones and Snyder 2009; Eschauzier et al. 2012). Typically, once the target analytes are bound to the SPE media, matrix interferences are washed off the media using a solvent (e.g., methanol). Further, the target analytes are subsequently eluted using various solvents or solvent mixtures (e.g., ammonium hydroxide in methanol, methyl tert-butyl ether, and methanol solvent mixtures). As PFAS analytes include different functionalities (e.g., sulfonates, carboxylates, and sulfonamides) and diverse chain lengths (e.g., C4–C14), it has been reported that the pH and polarity of the conditioning and elution solvents dramatically effect the analyte recoveries and reproducibility of the method (Taniyasu et al. 2005).

2.2.2 LIQUID–LIQUID EXTRACTION

Liquid–liquid extraction (LLE) techniques have been used for different matrices and reported low nanogram per liter quantitation limits. Flores et al. (2013) used LLE for the concentration of a 200 or 1000 mL volume of surface or drinking water, respectively, using (90:10 v/v) dichloromethane/isopropyl alcohol solvent mixture and Whatman filter paper for cleanup of the extracts. Backe et al. (2013) reported a micro-LLE of 3 mL of groundwater samples using 10% (v/v) 2,2,2-trifluoroethanol in ethyl acetate, followed by a large-volume (900 µL) injection method on a high-performance liquid chromatography (HPLC)–MS/MS for quantification of the PFAS.

Finally, a form of LLE involves large-volume injection techniques that use minimal sample preparation. This direct-injection technique usually involves sample dilution and direct analysis. This approach has less sample contamination potential from sample processing steps or other laboratory ware.

2.3 OVERVIEW OF ANALYTICAL METHODS

2.3.1 LIQUID CHROMATOGRAPHY

PFAS formulations used in commercial and consumer products are commonly complex mixtures of many PFAS compounds that provide specific performance characteristics (e.g., water–oil repellence, foaming characteristics, and thermal stability). As a result, though the analyst may be targeting specific compounds for quantification, there are many impurities, transformation products, and so forth, that are also present. Additionally, typical environmental samples are a mixture of anthropogenic and natural compounds. Due to these complex mixtures of target and nontarget compounds in an environmental sample, separating the intended target compound from nontargeted compounds and interfering matrix is typically done using chromatographic separation. Considering the physical-chemical properties of the PFAS, instrument analysis is routinely performed on liquid chromatographs coupled to a mass spectrometer.

Methods may use single-quadrupole mass spectrometry (MS) or triple-quadrupole tandem mass spectrometry (MS/MS) technique. However, due to the poor selectivity in complex matrices, moderate sensitivity, and matrix interferences, single-quadrupole MS is rarely used for PFAS analysis. Triple-quadrupole MS/MS consists of two quadrupole (Q_1 and Q_3) mass analyzers in series. When both Q_1 and Q_3 are set at a specific mass, allowing only a specific fragment ion from a certain parent molecular ion to be detected, it is called selected reaction monitoring (SRM), and if Q_1 and Q_3 are set to more than a single mass, it is called multiple reaction monitoring (MRM) mode. With the significant increase in selectivity and sensitivity, MS/MS techniques are preferred for PFAS analysis in complex environmental matrices.

The triple-quadrupole MS/MS technique offers high sensitivity and selectivity for PFAS, allowing very low LODs, as low as the picogram range on a column (Yamashita et al. 2004; Taniyasu et al. 2008; Ullah et al. 2011). For analysis of the anionic PFAS, such as carboxylates and sulfonates, MS/MS is most commonly operated in electrospray (ES) negative ionization mode, in which molecular ions are selected as parent ions. For perfluorocarboxylic acids (PFCAs), [M–H]⁻ ions (e.g., mass-to-charge ratio [m/z] of 413 for PFOA) are formed, and for perfluorosulfonates (PFSAs), [M–K/Na]⁻ ions (e.g., m/z 499 for PFOS) are formed, depending on the potassium or sodium salts of sulfonates. These molecular ions undergo further transition to form fragment MS^2 ions. Some of the neutral PFAS, such as perfluorooctyl sulfonamides, are also analyzed on the negative ES mode. In most cases, each target analyte is detected using two SRM transitions. One of the SRM transitions is used for the quantitation of the analyte, and the second transition is used for the confirmation of the analyte and to avoid any false-positive detections. To accomplish this, ion ratios of the two SRMs are monitored. Thus, the triple-quadrupole MS/MS technique offers excellent sensitivity and specificity.

The use of hybrid mass spectrometers is becoming more popular as they include a combination of two MS techniques, offering better selectivity and sensitivity. These instruments can perform a full scan in addition to the MS^n experiments and provide high mass resolution, which is the ability to distinguish between ions slightly different in m/z and provide accurate mass measurements. These techniques can be used for the quantitation of the known and complete characterization of the unknown analytes. This is very important for the PFAS analysis in complex matrices and the determination of unknown analytes with high mass accuracy. For example, aqueous film-forming foam (AFFF) contains many of unknown PFAS precursors that can be degraded or transformed to more persistent PFAS. As the commercial AFFF formulations are proprietary, the ability to conduct unknown analysis is crucial. For the remediation of AFFF-contaminated sites, it is important to know the extent of contamination by both the known and unknown PFAS and their precursors, which can potentially transform to stable PFAS over time under environmental conditions. The unknown PFAS may be present with different functionalities, chain lengths, degrees of fluorination, linear or branched chains, and so forth. Recently, a column-switching HPLC-MS/MS method with an online preconcentration method using a WAX column was applied for the determination of 14 perfluorinated compounds (PFCs) in smaller volumes of drinking water samples with reporting limits in the range 0.59–3.4 ng/L (Dasu et al. 2017). Hybrid instruments, such as ion trap mass spectrometry coupled to time-of-flight/quadrupole time-of-flight high-resolution mass spectrometry (IT-TOF/QTOF-HRMS), are also being used for the trace-level analysis of PFAS.

A few studies have reported the use of the LC–atmospheric pressure photoionization technique (LC-APPI/MS) (Takino et al. 2003; Chu and Letcher 2008). Takino et al. (2003) reported an online extraction LC-APPI/MS method for the determination of PFOS in river water and achieved a LOD of 5.35 ng/L. The method offered optimal sensitivity and selectivity and was less prone to matrix interferences. For neutral PFAS precursors, such as fluorotelomer alcohols (6:2 FTOH, 8:2 FTOH, and 10:2 FTOH) and polyfluorinated sulfonamides, Chu and Letcher (2008) reported an LC-APPI/MS method operated in negative mode in biotic samples with minimal matrix effects.

2.3.1.1 Orthogonal Chromatography

Backe et al. (2013) reported a nonaqueous large-volume (900 μL) direct-injection method using HPLC-MS/MS with minimal pretreatment for the quantification of

26 newly identified and 21 perfluoroalkyl acids (PFAAs) in groundwater and AFFF formulations. Unknown zwitterionic, cationic, and anionic fluorinated chemicals in AFFF were separated by orthogonal chromatography using cation exchange (silica) and anion exchange (propylamine) guard columns connected in series to a reverse-phase (C18) analytical column. The LOD for PFAS using this method in groundwater ranged from 0.71 to 67 ng/L for the analytes for which authentic standards were available.

2.3.1.2 Comparison of Analytical Instrumentation

Berger et al. (2004) reported a comparison of IT-MS/MS, TOF-HRMS, and triple-quadrupole MS techniques for PFAS analysis. Among these techniques, TOF-HRMS showed high selectivity with optimal sensitivity. The main drawback of the triple-quadrupole MS is the elimination of the matrix background by the instrument, whereas IT-TOF/QTOF-HRMS provides an estimation of matrix remaining in the sample extract, which could interfere with the ionization performance (Berger et al. 2004). Further, the data can also be used for the retrospective analysis of the nontargeted analytes in the future. The instrumental LODs of TOF-MS are approximately a factor of 10 lower than those for triple-quadrupole MS and cover a smaller linear range (Berger et al. 2004). With these considerations, the triple-quadrupole MS remains the most commonly used technique for highly sensitive routine analysis.

In another study, Llorca et al. (2010) showed a comparison of the analytical suitability of three techniques, triple-quadrupole MS, conventional three-dimensional ion trap (3D IT), and hybrid quadrupole linear ion trap–mass spectrometry (QLIT-MS) for the trace analysis of PFAS in fish and shellfish matrices. Among these techniques, QLIT-MS provided at least 20-fold higher sensitivity than triple-quadrupole MS. For the analysis of PFAS in complex matrices, QLIT-MS also offered the possibility to use enhanced product ion MS^3 modes in combination with the SRM mode for confirmation of the target analytes. Onghena et al. (2012) compared three chromatographic methods, HPLC-MS/MS, ultra-high-performance liquid chromatography–tandem mass spectrometry (UHPLC-MS/MS), and capillary liquid chromatography–tandem mass spectrometry (CLC-MS), for the analysis of 18 PFAS in surface waters. Using miniaturized chromatographic methods, such as UHPLC and CLC, decreased the chromatographic dilution and hence showed improved sensitivity compared with HPLC.

2.3.2 Gas Chromatography

Alternatively, the ionic and neutral PFAS may be analyzed using gas chromatography–mass spectrometry (GC-MS) techniques. The MS may be subdivided by the type of ionization and polarity of ions detected: electron impact ionization (EI-MS), positive chemical ionization (PCI/MS), and negative chemical ionization (NCI/MS) (Erickson 1997). For GC analysis, ionic PFAS need to be derivatized to convert them to volatile species. For PFCAs, Dufková et al. (2009) performed the derivatization using isobutyl chloroformate (IBCF) to convert the acids into the more volatile isobutyl esters, under catalysis by pyridine in phosphate-buffered medium and analyzed with GC-EI-MS. The LODs were relatively high and ranged from 0.030 to 0.314 μg/mL. Langlois et al. (2007) reported a qualitative high-resolution GC-MS method for the separation and identification of isomers of perfluoroalkyl sulfonates and carboxylates by derivatizing with isopropanol under acidic conditions. Recently, Shiwaku et al. (2016) derivatized the SPE-WAX extracts of well, surface, and tap water by benzylation and analyzed them using GC-NCI/MS. They reported method detection limits (MDLs) of 0.1–14 ng/L, which are comparable to the range reported using LC-MS/MS instruments.

Neutral PFAS precursors can be directly analyzed on GC due to their semivolatile nature. GC-MS methods have been used to monitor for PFAS precursors in air and soil matrices using GC-EI-MS, GC-PCI/MS, and GC-NCI/MS (Jahnke et al. 2007; Shoeib et al. 2008; Ellington et al. 2009). The PCI/MS mode is used for the quantitation, as it offers higher sensitivity and selectivity resulting from an abundant protonated molecular ion (i.e., $[M+H]^+$). Ellington et al. (2009) further confirmed the peaks using a derivatization step with trimethylsilyl agent and achieved a limit of quantitation (LOQ) of 100 ng/L for 8:2 FTOH. Precursor analysis was also reported using LC-MS/MS under the positive ES ionization mode (Wang et al. 2009; Dasu et al. 2012).

2.3.3 EMERGING TECHNOLOGIES

Table 2.3 and this section summarize emerging technologies for PFAS analytical methods in water matrices.

2.3.3.1 Total Fluorine Analysis

Miyake et al. (2007) reported a combustion ion chromatography (CIC) method for determining trace levels of total fluorine in water. The method was developed to determine low micrograms per liter of total fluorine in seawater. The extraction of total fluorine was done using organic solvents in two separate fractions of extractable organic fluorine and inorganic fluorine. Recently, Wagner et al. (2013) also reported a CIC method for the determination of trace levels of adsorbable organic fluorine from aqueous samples, such as tap water, wastewater effluent, and surface water and groundwater samples. The material used for the adsorption was synthetic polystyrene-divinylbenzene-based activated carbon with 0.3 μg/L LOQ for organic fluorine. In both studies, a mass balance analysis was done by analyzing the known PFAS by LC-MS/MS and comparing them with the extractable organic fluorine levels analyzed by CIC. They found that only 5%–40% of the PFAS were accounted for out of the total organic fluorine, and a major percentage (60%–95%) of the fluorine remains unknown. Hence, there is a need for a versatile technique to measure the complete suite of PFAS and their precursors, which can be used for the treatment and remediation of these chemicals.

TABLE 2.3
Summary of PFAS Analytical Methods for Water Matrices

Compounds	Sample Matrix	Extraction Method	Analytical Method	Sample Volume (mL)	LOD/LOQ (ng/L)	Reference
PFHxA, PFOA, PFHxS, PFOS	Surface water	SPE-C18	LC-MS/MS	200	9.17	Moody et al. (2001)
PFAS	Surface water	SPE-C18	19F NRMR	2–100	10,000	Moody et al. (2001)
PFOA, PFNA, PFBS, PFHxS, PFOS, FOSA	Surface water	Online extraction	LC-APPI-MS	1	5.35 (LOD)	Takino et al. (2003)
PFOA, PFOS, FOSA, FOSAA, N-EtFOSA, N-EtFOSAA, N-EtFOSE	Surface water	SPE-C18	LC-MS/MS	50,000	0.2–13 (LOQ)	Boulanger et al. (2004)
PFOA, PFNA, PFDA, PFUnDA, PFDoDA, PFTeDA, PFHxDA, PFOcDA, L-PFBS, L-PFHxS, L-PFOS, FOSA, 8:2 FTCA, 8:2 FTUCA	Wastewater influent and effluent	SPE (HLB)	LC-MS/MS	200	2.5–10	Sinclair and Kannan (2006)
PFPeA, PFHxA, PFHpA, PFOA, PFNA, PFDA, PFUnDA, PFODA, L-PFBS, L-PFHxS, L-PFOS, FOSA	Seawater, brackish water	SPE-WAX; different organic solvent elutions are fractionated	LC-MS/MS		0.01–0.1 (LOQ)	Miyake et al. (2007)
Total fluorine, extractable organic fluorine, inorganic fluorine	Seawater, brackish water	SPE-WAX; different organic solvent elutions are fractionated	CIC	1,000		Miyake et al. (2007)

(*Continued*)

TABLE 2.3 (CONTINUED)
Summary of PFAS Analytical Methods for Water Matrices

Compounds	Sample Matrix	Extraction Method	Analytical Method	Sample Volume (mL)	LOD/LOQ (ng/L)	Reference
TFA (C2), PFPrA, PFBA, PFPeA, PFHxA, PFHpA, PFOA, PFNA, PFDA, PFUnDA, PFDoDA, PFTrDA, PFTeDA, L-PFBS, L-PFHxS, L-PFOS, L-PFDS, FOSA, N-EtFOSA, N-EtFOSAA, 8:2 FTCA, 7:3 FTCA, 6:2 FTUCA, 8:2 FTUCA, 10:2 FTUCA	Precipitation	SPE (WAX)	LC-MS/MS (ion exchange column)	100	0.02–0.5 (LOQ)	Taniyasu et al. (2008)
PFOA, PFOS	Surface water, groundwater, and tap water	SPE (PreSep C Agri)	LC-MS/MS	1,000	0.03–0.05 (LOQ)	Jin et al. (2009)
PFHxA, PFOA, PFNA, PFDA, PFUnDA, PFDoDA, L-PFHxS, L-PFOS	Drinking water and its source water, including surface water, and secondary and tertiary treated effluent of wastewater	SPE (HLB)	LC-MS/MS	1,000	1–5	Quiñones and Snyder (2009)
PFPeA, PFHpA, PFOA, PFODA, PFTeDA, L-PFBS, L-PFHxS, L-PFOS, FOSA	Surface water	Online SPE, LC-MS/MS		1,000	9.0–49.0	Gosetti et al. (2010)

(Continued)

TABLE 2.3 (CONTINUED)
Summary of PFAS Analytical Methods for Water Matrices

Compounds	Sample Matrix	Extraction Method	Analytical Method	Sample Volume (mL)	LOD/LOQ (ng/L)	Reference
PFPeA, PFHxA, PFHpA, PFOA, L-PFBS, L-PFHxS, L-PFOS	Drinking water	SPE (WAX)	LC-MS/MS	1,000	0.5–0.92 (LOQ)	Thompson et al. (2011)
PFPeA, PFHxA, PFHpA, L-PFOA, Br-PFOA, PFNA, PFDA, PFUnDA, PFDoDA, L-PFBS, L-PFHxS, L-PFOS, Br-PFOS, L-PFDS, PFHxPA, PFOPA, PFDPA	Drinking water	SPE (C8) + quaternary amine	LC-QTOF-HRMS	500	0.014–0.17 (MDL)	Ullah et al. (2011)
PFBA, PFPeA, PFHxA, PFHpA, L-PFOA, Br-PFOA, PFNA, PFDA, PFBS, PFHxS, L-PFOS, Br-PFOS	Surface water, groundwater, and drinking water	SPE (WAX)	LC-MS/MS	250	0.1–9.5 (LOQ)	Eschauzier et al. (2012)
PFOS, PFOA	Drinking water	LLE	LC-MS/MS	200/1,000	1.1–4.2	Flores et al. (2013)
PFBA, PFPeA, PFHxA, PFHpA, PFOA, PFNA, PFDA, PFUnDA, PFDoDA, PFTrDA, PFTeDA, PFBS, PFHxS, PFHpS, PFOS, PFDS, 4:2 FTS, 6:2 FTS, 8:2 FTS, 26 newly identified PFAS	Groundwater	Micro-LLE; large volume, direct injection (900 μL)	HPLC-MS/MS (orthogonal chromatography—cation exchange and anion exchange guard columns connected in series to C18 column)	3	2.4–14 (LOQ)	Backe et al. (2013)

(Continued)

TABLE 2.3 (CONTINUED)
Summary of PFAS Analytical Methods for Water Matrices

Compounds	Sample Matrix	Extraction Method	Analytical Method	Sample Volume (mL)	LOD/LOQ (ng/L)	Reference
PFBA, PFPeA, PFHxA, PFHpA, PFOA, PFNA, PFDA, PFUnDA, PFDoDA, PFBS, PFHxS, PFOS, PFDS, 4:2 FTS, 6:2 FTS, 8:2 FTS, N-MeFOSAA, N-EtFOSAA, PFHxPA, PFOPA	Wastewater effluent	Precursor oxidation using basic persulfate followed by SPE-WAX	LC-MS/MS	50	0.04–5.7 unoxidized samples; 0.1–4.8 oxidized samples	Houtz et al. (2016)
PFBA, PFPeA, PFHxA, PFHpA, PFOA, PFNA, PFDA, PFUnDA, PFDoDA, PFTrDA, PFTeDA	Well water, surface water, and tap water	SPE-WAX, eluate derivatized by benzylation	GC-NCI/MS	500	0.2–14 (MDL)	Shiwaku et al. (2016)
PFPeA, PFHxA, PFHpA, PFOA, PFNA, PFDA, PFUnDA, PFDoDA, PFTrDA, PFTeDA, L-PFBS, L-PFHxS, L-PFOS, L-PFDS	Drinking water	SPE (HLB) + (WAX)	LC-IT-TOF	10	0.43–3.4 (LCMRL)	Dasu et al. (2017)

Note: PFHxPA = perfluorohexylphosphonic acid; PFOPA = perfluorooctylphosphonic acid; PFPrA = perfluoropropionic acid; TFA = Trifluoroacetic acid.

2.3.3.2 Total Oxidizable Precursor Assay

Another emerging indirect technique that is used for the quantitation of the fluorotelomer precursors is total oxidizable precursor (TOP) analysis (Houtz and Sedlak 2012). In this technique, the fluorotelomer precursor chemicals present in the PFAS-contaminated samples are oxidized using persulfate in the presence of base to produce persistent perfluorinated alkyl acids, such as PFOA. The PFAS composition in AFFF formulations or the AFFF-contaminated sites is largely unknown, and only a limited number of analytes can be quantitated using the targeted analytical procedures. Therefore, using this method, total precursors are calculated based on the difference between the PFAAs generated postoxidation compared with the initial PFAAs. However, individual precursors undergoing oxidation to produce PFAS cannot be identified using this technique, and the approach is applicable only to precursors with an oxidizable CH_2 carbon capable of undergoing oxidation. Chandramouli provides more extensive discussion on the TOP assay in Chapter 3.

2.3.3.3 Particle-Induced Gamma-Ray Emission

Particle-induced gamma-ray emission (PIGE) is a spectroscopic technique used for direct surface analysis to quantify total elemental fluorine (Srivastava et al. 2014). The method was applied to measuring elemental fluorine on papers and textiles (Robel et al. 2017; Schaider et al. 2017) with the LOQ ranging from 42 to 150 nmol F/ cm^2 for papers and fabrics of varying thickness. This surface analysis method is also applied to liquid samples by extracting the samples onto a surface for fluorine measurement. The method can be used as the prescreening tool for PFAS-contaminated site assessments and groundwater investigations. Therefore, the PIGE method seems promising, with the detection limit of 60 µg/L, to quantitate the total fluorine for the site characterization and to estimate the level of contamination. As the PIGE method is a surface analysis, it may underestimate solid material thicker than the penetration depth of 0.1 mm. The PIGE technique uses irradiation; therefore, the accessibility of the instrumentation and operation of the instrument might be limited. Chandramouli provides more extensive discussion on PIGE analysis in Chapter 3.

2.3.3.4 Fluorine-19 Nuclear Magnetic Resonance

Additionally, fluorine-19 nuclear magnetic resonance (^{19}F NMR) was used to investigate total PFAS concentrations in surface water samples. However, the MDLs were relatively high (10 µg/L) (Moody et al. 2002). Arsenault et al. (2008) reported isolation and structural elucidation of 11 PFOS isomers in a technical mixture using ^{19}F NMR spectroscopy. The technique allowed for the quantification of individual CF_3-branched isomers in a technical mixture of PFOS. However, for the structural elucidation, using this technique requires a pure isomer, and therefore due to the complex nature of the environmental matrices, the applicability of this technique to environmental samples is limited. Chandramouli provides more extensive discussion on ^{19}F NMR in Chapter 3.

2.4 INTERLABORATORY COMPARISONS

For more than a decade, much attention has been given to the quantitation of PFAS in different environmental and biological matrices. Therefore, many laboratories around the world have developed various methods using multiple classes of instruments for PFAS analysis. To coordinate the evaluation of method performance, the first worldwide interlaboratory study (ILS) was conducted on PFCs in human and environmental matrices (Martin et al. 2004). In this study, 13 PFAS were analyzed by 38 worldwide participating laboratories reporting the results following their in-house methodologies. This study showed significant variability between the results obtained in different laboratories analyzing the same samples. Some of the problems in this study could be attributed to the limited availability of high-quality standards and mass-labeled standards, matrix interferences, and background contamination from laboratory ware and instrumentation.

A second ILS was organized by Van Leeuwen et al. (2009) to assess the performance of 21 participating laboratories from North America and Europe for analyzing PFAS in water and fish matrices. In this study, care was taken to address some of the drawbacks of the first ILS, such as (1) determination of the precision of individual laboratories and (2) utilization of the shared native and mass-labeled standards by the participating laboratories. In this second study, two quantitation methods, the mass-labeled internal standard method and the standard addition method, were evaluated by the participating laboratories. However, the participants used their own in-house extraction methods. The mass-labeled internal standard–based quantitation method showed good performance. This study showed improved agreement of results from different participating laboratories, which can be attributed to the use of quality native standards, the use of internal standards, and better cleanup of the samples to reduce matrix effects. The study recognized the need for the standard reference materials to improve reproducibility between the laboratories.

A multi-laboratory-validated method, USEPA 537 Rev 1 (USEPA 2012), was reported for the analysis of 14 PFAS analytes in drinking water samples. PFOA and PFOS are included in the USEPA Contaminant Candidate Lists 3 and 4 of chemicals under consideration for future drinking water regulation in the United States (USEPA 2009, USEPA 2016b). The agency also included six PFAS in Unregulated Contaminant Monitoring Rule 3 (UCMR 3)—PFOS, PFOA, perfluorononanoic acid (PFNA), perfluorohexanesulfonic acid (PFHxS), perfluoroheptanoic acid (PFHpA), and perfluorobutanesulfonic acid (PFBS) (USEPA 2012, 2015)—and collected nationwide occurrence data (USEPA 2017) using the EPA 537 method. It should be noted that the EPA 537 standard is strictly a drinking water method with the listed number of analytes and their respective reporting limits. Often, a "modified EPA 537" method is seen in the literature being used for different matrices or involving a different analyte list. There is no standardized method available as a modified EPA 537 method regardless of the many laboratories reporting their in-house methods as such. Therefore, caution needs to be exercised while adapting such nonstandard methods. Recently, the Department of Defense issued Table B-15 in Quality Systems

Manual 5.1 (DOD 2017), which provides a comprehensive set of quality standards for PFAS analysis. Although not a USEPA standard method, these standards outline specific quality processes for sample preparation, instrument calibration, and analysis when working with PFAS.

Another interlaboratory comparison study conducted with five participating laboratories to measure the PFAS in sludge standard reference material (SRM 2781) (Reiner et al. 2015) was produced by the National Institute of Standards and Technology (NIST) (Gaithersburg, Maryland). This study showed high variability for some of the PFAS measured in SRM 2781 using different methods by the participating laboratories. Therefore, it is imperative to identify sources of the differences between analytical results and minimize them in the future. There is a need for reproducible extraction and cleanup methods for measuring these chemicals in different matrices. Currently, the USEPA is undertaking a multilaboratory validation study for the method development and validation of internal USEPA methods for the analysis of PFAS in nonpotable water and solid matrices.

2.5 STANDARD METHODS

Very few standard methods are reported in the literature, and they all have certain limitations. A standard ISO 25101:2029(E) method (ISO 2009) is reported by Yamashita et al. (2008) for the analysis of PFOS and PFOA in water using LC-ES-MS/MS. Although the method performance was good with <32% relative standard deviation for both analytes, it was limited to only PFOA and PFOS. In this standard method, recommendations for method validation and qualification of results if using a single-quadrupole MS instrument were given. As described above, a multi-laboratory-validated method, USEPA method 537 (USEPA 2012), reported using an SPE extraction of a 250 mL sample and is limited to drinking water samples with 14 PFAS analytes included. The American Society for Testing and Materials (ASTM) International recently published ASTM D7979-16 (ASTM 2016). This method is a direct-injection method with a reported sensitivity of 10–8000 ng/L for a wide number of PFAS analytes in liquids without much sample preparation. This method can be applied for a wide range of liquid environmental samples, such as drinking water, surface water, groundwater, and wastewater influent and effluents. However, a highly sensitive instrument is needed for such analysis and hence limits the wide applicability of the method. Another ASTM D7968-14 (ASTM 2014) method was developed for analyzing PFAS in soil matrices. The sample extraction method involved (1:1) MeOH:H_2O solvent extraction, followed by filtration using polypropylene (PP) filters. The reporting limits ranged from 25 to 1000 ng/kg. Table 2.4 briefly summarizes the standard methods and criteria.

TABLE 2.4
Comparison of Standard Methods for PFAS Analysis

	EPA 537	ISO 25101:2029(E)	ASTM D7979-16	ASTM D7968-14
Sample volume	250 mL	500 mL	5 mL	2 g
Sample matrix	Drinking water	Drinking water, groundwater, surface water, seawater	Water; wastewater sludge, influent, and effluent	Soil, biosolid
Analytes	PFAS, FOSAAs 14 PFAS	PFOS, PFOA	PFAS, FOSAAs, FTSs, n:2 FTUCAs, FTCAs	PFAS, FOSAAs, FTSs, n:2 FTUCAs, FTCAs
Preservation	Trizma for buffering and removal of free chlorine	Sodium thiosulfate pentahydrate for removal of free chlorine	None	None
Holding time	Before extraction: 14 days refrigerated at ≤6°C Postextraction: 28 days at room temperature	14 days at 4°C ± 2°C	28 days at 0°C–6°C	28 days at 0°C–6°C
Extraction method	SPE-WAX	SPE	Direct injection	Solvent extraction followed by filtration using PP filters
Analytical instrument	LC-MS/MS	LC-MS/MS and LC-MS	LC-MS/MS	LC-MS/MS
Reporting limits	2.9–14 ng/L	2–10,000 ng/L	10–400 ng/L	25–1,000 ng/kg

2.6 CROSS-CONTAMINATION ISSUES

2.6.1 Sample Collection

PFAS have a wide range of commercial applications as they are used as coatings on many consumer products, such as water- and stain-resistant coatings on textiles, paper, and leather (Prevedouros et al. 2006; Buck et al. 2011). They are also used in the fluoropolymer industry in the creation of surfactants and as polymerization aids (Prevedouros et al. 2006; Buck et al. 2011). Due to the myriad applications, care needs to be exercised to avoid any contamination while collecting samples for PFAS analysis. The following precautions should be taken during sample collection:

1. Avoid sampling equipment or tubing made with polytetrafluoroethylene (PTFE) (e.g., Tygon®) or Teflon. If possible, fill sample containers directly; however, if needed, PFAS-free high-density polyethylene (HDPE) and silicone are acceptable tubing materials for sampling (requires the user to verify that they are PFAS-free).
2. Use sample containers made of PP (resin code PP 5) or HDPE (resin code HDPE 2). Avoid glass, as some studies have shown that some PFAS sorb to glass. Pay attention to the material of the closure and closure liner (e.g., Teflon septa) as this may be a different material than the bottle.
3. In between the sample collection, drilling and sampling equipment needs to be rinsed thoroughly with tap water followed by Milli-Q water. A detergent may be required. Alconox® and Liquinox® soaps do not contain fluorinated ingredients and therefore are acceptable. A sample of the final rinse water can be collected to verify that the washing process is not a source of cross-contamination as an equipment blank or rinse.
4. Collect samples from the locations that are anticipated to have low concentrations first and progress the sample collection to areas with greater expected PFAS concentrations.
5. Disposable sampling equipment is preferred to reduce chances for cross-contamination.
6. Avoid handling sample containers with bare hands, as PFAS may be present in personal care products, sunscreens, and insect repellants. If these materials are used, select products that do not contain PFAS. The use of nitrile or appropriate gloves is suggested. Personnel should wash their hands prior to donning gloves.
7. Avoid wearing stain-resistant and water-resistant-treated clothing and shoes. This includes a wide variety of "performance" clothing made from synthetic materials. Rain gear from polyurethane or wax-coated materials is acceptable. Previously washed (as opposed to new and unwashed) garments should be worn. Avoid Gore-Tex® clothing and footwear as the active membrane has PFAS-containing components. Tyvek® materials are also known to contain fluorinated materials.
8. Avoid waterproof or treated field notebooks and paper.
9. Avoid handling prepackaged and fast food, as PFAS are used in oil-resistant coatings on food packaging wraps.

2.6.2 Sample Storage

The most common materials for sample containers for PFAS are HDPE and PP. It is recommended to thoroughly rinse all sampling and sample processing containers with methanol, followed by PFAS-free water prior to use. Glass containers are not recommended as studies have shown that there is a potential loss of PFAS due to sorption to the glass. However, it was observed that longer-chain PFCAs, such as perfluoroundecanoic acid (PFUnDA) and perfluorododecanoic acid (PFDoDA), are adsorbed onto the PP containers and showed lower holding time recoveries (<50%) at days 60 and 90 (Berger et al. 2011). Therefore, it is recommended to decant the sample and rinse the emptied sample bottle with methanol to remove sorbed components. The rinsate should be added to the sample extract to increase the recoveries of longer-chain acids. Samples are stored in the refrigerator below 6°C until extraction up to 14 days (ISO 2009; Shoemaker 2009). Although PFAS are persistent and do not degrade at typical laboratory temperatures, keeping the samples refrigerated until extraction is important to prevent the biodegradation of precursors and other matrix interferences, which might result in low recoveries of the analytes. Sample preservatives commonly used are Trizma® and sodium thiosulfate pentahydrate, which are often used for the removal of free chlorine (ISO 2009; Shoemaker 2009). Trizma also functions as the buffer for the extraction of PFAAs from water samples.

2.6.3 Sample Preparation

For sample preparation and analysis of PFAS in environmental samples, care needs to be practiced, avoiding use of any Teflon-coated or fluoropolymer materials, such as sample containers used for sample collection, pipette tips, and other laboratory ware used for sample processing. It was also observed that glass droppers used to transfer samples often contain low levels of PFAS and hence might result in cross-contamination during sample preparation for sensitive ultra-low-level detection analytical methods. It is always advisable to periodically test and screen all the laboratory ware for any background levels of PFAS. Apart from the laboratory ware, high-purity water (e.g., Milli-Q water) used as the laboratory blank and the field blank should be analyzed to ensure that it is free of any background PFAS.

Analysis of environmental samples for PFAS requires methodic and meticulous sample preparation protocols to separate the target analytes from the matrices. For aqueous samples, such as surface water, groundwater, drinking water, and wastewater, SPE is most commonly used. Some of the SPE media frequently used are WAX, HLB, C8, and C18 (Jin et al. 2009; Quiñones and Snyder 2009; Eschauzier et al. 2012). The sample volume required for analysis varies from 5 to 1000 mL and depends on the extraction method and sensitivity of the analytical instrument (ISO 2009; Quiñones and Snyder 2009; Ullah et al. 2011; Eschauzier et al. 2012; ASTM 2016).

2.6.4 Sample Analysis

Many of the analytical components (e.g., HPLC), such as solvent lines, tubing, seals, mobile phase filters, and parts inside the degassing unit, contain fluoropolymer materials. These components contribute to significant contamination for PFAS analysis, resulting in high background levels. Most of these parts can be replaced with either stainless steel or polyether ether ketone (PEEK) materials to minimize the background PFAS contamination from the instrumentation.

On the column separation, the PFAS coming from the background contamination might also elute at the same retention time as the target analytes, producing similar mass transitions and hence peak enhancement. Therefore, to differentiate the peaks coming from the background, a small isolator or delay column can be installed between the mixing chamber and the analytical column, which results in completely resolved background peaks from the target analyte peaks. As most of the PFAS analytes are acidic, the analytical column used for the chromatographic separation should have a wide range of pH (1–10) stability.

For HPLC analysis using autosampler injection needles, PTFE-sealed vials are most widely used. It is a common practice to vortex the samples before analysis while they are capped with these PTFE septa; however, there is potential for cross-contamination using this practice. To avoid any contamination from sample coming in contact with the PTFE septa, PP or polyethylene vial caps free of fluoropolymers should be used.

2.7 QUALITY ASSURANCE AND QUALITY CONTROL

QA and QC at each step of the sample analysis process are imperative to ensure the reliability and accuracy of the data generated. A brief overview of the QC considerations for PFAS analysis is provided in Table 2.5. The following are some of the vital components of a proper QA/QC protocol: field blanks, equipment blanks, the trip blank, procedural blanks, laboratory-fortified blanks, matrix-fortified samples, and replicates. A brief description of each is included below.

Field blanks: To assess potential field contamination, field blanks are exposed to the same field conditions as the samples. A field blank bottle is empty upon arrival at the field site and then filled with the PFAS-free laboratory water in the field. Any contamination from the field environment (e.g., volatile PFAS at the location or PFAS contributions from contaminated hands due to improper hygiene practices) will be detected in this sample and be indicative of possible additional cross-contamination to all other samples collected at the same time and location.

Equipment blanks: To assess potential contamination from the equipment used to collect the sample, PFAS-free laboratory water is run through the equipment, similar to the approach used for field samples. Any positive PFAS detected is indicative of PFAS contributions from the sampling equipment, and all samples collected with the equipment set are likely cross-contaminated. Wash water can also be collected after washing equipment between sampling events; PFAS-free results are an indication of an adequate cleaning methodology.

Travel blank: A travel blank is used to assess potential contamination from the overall field trip. The travel blank is not opened on site but is merely along for the trip. Sometimes this blank is referred to as a "trip blank." Any contamination present is indicative of the presence of an unwanted PFAS source, such as cooler or synthetic ice products and sample bottles.

Procedural blanks: Laboratory PFAS-free water is extracted or processed following the same procedure as that for the samples to assess any contamination from the sample processing method.

Laboratory-fortified blanks: Laboratory PFAS-free water is fortified with a fixed amount of target analyte(s) and processed in a similar manner as the samples to ensure the accuracy and reliability of the sample processing method.

Matrix-fortified samples: Samples in the environmental matrix are fortified with a fixed amount of target analytes and processed to assess interferences that may result in ionization suppression or enhancement, depending on the background matrix of the sample.

Replicates: The precision of the entire method is evaluated by measuring the reproducibility of the replicate samples. Therefore, to assess the reproducibility of the sample preparation and analytical method, replicates are included. It is common practice to do replication for at least 10% of the samples.

2.8 QUANTITATION METHODS

The most typically used quantitation method is the internal standard approach. Other quantitation methods are the external standard approach, isotope dilution method, and standard addition method.

In the internal standard method, a known quantity of structurally similar compound or mass-labeled compound is added to all the calibration standards and samples. The calibration curves are prepared by plotting the area ratio of analyte to the assigned internal standard versus the concentration. This method of quantitation helps correct for instrument drifts, or matrix effects. For example, the EPA 537.1 (USEPA 2012) drinking water method requires the internal standard quantitation method to quantify the concentrations of PFAS analytes.

An external standard quantitation method is commonly used in the absence of mass-labeled analyte compounds. The calibration curves are prepared by plotting the response area of analyte versus the known concentration of the analyte. This method of quantitation does not account for analyte loss due to sample preparation, instrument drift, or matrix effects, and hence is most suitable for clean matrices with little to no interferences. The ASTM D7979-16 (ASTM 2016) method uses external standard quantitation for PFAS analysis in liquid matrices using a direct-injection method, but importantly, it includes mass-labeled standards as surrogates to report extraction recoveries.

Isotope dilution methods are accurate methods that may reduce the matrix interferences caused by complex sample matrices. This approach requires two different mass-labeled compounds for each target analyte. Obtaining multiple mass-labeled compounds for all analytes may be difficult and costly to obtain. To implement this method, a known amount of analyte-matched mass-labeled compound is added to the sample and equilibrated followed by the addition of differently labeled internal standard before analysis. This method of quantitation accounts for any analyte loss due to sample preparation, matrix effects, and instrument drifts. The criteria in Table B-15 in Quality Systems Manual 5.1 (DOD 2017) currently require isotope dilution quantitation for PFAS analysis. This method may perform well in complex matrices and accounts for the matrix interferences.

Standard addition methods are used for the quantitation of analytes in complex matrices where signal suppression or enhancement matrix effects impact the analyte signal. In this method, different concentrations of standard are added to a series of samples with the same sample matrix, with one sample receiving no standard addition. Quantitation of analytes in the sample matrix is performed, and the suppression or enhancement of the instrument signal can be determined. This method assumes that instrument response is operating in the linear region and can be applied where analyte-matched mass-labeled internal standards are not available (Furdui et al. 2008).

2.8.1 Detection and Quantitation Limits

Mass spectrometers are the most essential tools for the trace analysis of environmental contaminants. These advanced modern-day instruments can be used to detect trace levels of single ion transitions with very minimal background noise or with multiple transitions for more complex matrices. Therefore, to properly quantitate and report the valuable data, certain detection and quantitation limits need to be considered. Understanding these limits and how they are determined is very important for proper data analysis. Commonly, the proper use of these terminologies can be confusing. Below are the brief definitions of the terminology used.

Instrument detection limit (IDL): The IDL is the minimum analyte concentration required to produce a signal that is distinguishable from the noise level within a particular statistical confidence limit. The IDL is determined by measuring the response variability in standards of known composition and concentration. The IDL does not account for measurement response variability due to the sample preparation procedure, or bias due to the sample matrix (USEPA 2007).

Method detection limit (MDL): In 2016, the USEPA Office of Water updated the MDL from 40 CFR 136 Appendix B as Revision 2 (USEPA 2016c). According to the revised version, the MDL is defined as the minimum measured concentration of a substance that can be reported with 99% confidence that the measured concentration is distinguishable from method blank results. The revised MDL procedure accounts for the background contamination in determining the MDLs and is representative of the laboratory performance overtime. The method also allows the creation of a single MDL if multiple instruments are used for analysis in a laboratory. The MDL procedure is designed to be a straightforward technique for estimation of the detection limit for a broad variety of physical and chemical methods. The procedure requires a complete, specific, and well-defined analytical method (e.g., a standard operating procedure [SOP] or standard method). It is essential that all sample processing steps used by the laboratory be included in the determination of the MDL.

To determine MDLs, a laboratory processes a minimum of seven spiked samples and seven blank samples that go through all steps of the method. The spiking concentrations used to determine an MDL should be >10 times the estimated MDL and should be reevaluated annually. The MDL of the spiked sample is calculated by using the standard deviation of the results and the appropriate Student t-value and blank sample response.

Minimum reporting limit (MRL): The MRL is defined as 10 times the standard deviation obtained from the MDL study (approximately 3 times the MDL). This is most commonly reported for each analyte, along with the data, as a more conservative reporting value for an analyte.

Lowest-concentration minimum reporting level (LCMRL): The LCMRL is a relatively new approach to reporting limits for drinking water data. The LCMRL is the lowest true concentration for which future analyte recovery is predicted (with at least 99% confidence) to fall between 50% and 150% (Winslow et al. 2006). The LCMRL allows for the simultaneous incorporation of precision and accuracy in analytical measurements. The LCMRL is a laboratory- and analyte-specific value (Martin et al. 2007). The LCMRL values can be determined using the LCMRL calculator or manually using the procedures described by Winslow et al. (2006) and a USEPA statistical protocol.

Accurately determining and reporting PFAS sample concentrations is dependent on a wide variety of factors. Proper sampling procedures and sound QA and QC all require appropriate and well-thought-out planning. Table 2.5 provides a list of considerations to avoid the introduction of error and inaccurate analytical results.

2.9 QUANTITATION CHALLENGES

PFAS are manufactured by two methods, a telomerization process and electrochemical fluorination (ECF) (Prevedouros et al. 2006). These processes result in widely different isomeric chemicals. The telomerization process results in even-chain-length PFAS chemicals, mainly linear isomers. ECF yields even- and odd-chain-length chemicals, mainly PFSAs, such as PFOS, PFHxS, and PFBS. The ECF process yields 30%–40% branched isomers. These branched isomers interact differently with SPE cleanup and chromatographic columns, and good chromatographic separation of these branched isomers requires a proper recovery and gradient mobile phase program. Although many of these branched isomers are unknown and with the limited availability of standards, it is becoming more important to have such isomer-specific data for more comprehensive toxicological understanding and remedial implications. Berger et al. (2011) discussed the isomer-specific analysis of PFAS in human blood and observed that branched PFOS isomers show a lower response to the transition m/z 499 > 99, used for linear PFOS quantification. Hence, using a linear PFOS standard for quantification of total PFOS could lead to the underestimation of PFOS concentration

TABLE 2.5
QC Considerations for PFAS Analysis

Study Design
    Prepare a project-specific QA project plan
    Follow the SOPs, if available
    Decide on data reporting procedures

Sampling
    Use a well-designed sampling scheme to fit the study objectives
    Include field controls to assess losses (approximately 10% of samples)
    Include field blanks to assess contamination (approximately 10% of samples)
    Use a validated sample preservation and storage procedure
    Avoid sampling equipment, sampling containers, closures, and liners made with PTFE and related material
    Avoid handling sampling equipment or containers with bare hands to prevent contamination
    Avoid wearing water-resistant- or stain-resistant-treated clothes and shoes

Method Validation
    Analyze blanks
    Analyze replicates (precision)
    Analyze spikes
    Assess potential interferences
    Conduct a ruggedness test
    Check recoveries (accuracy)
    Establish instrumental performance criteria (chromatographic resolution, sensitivity)
    Determine IDLs
    Determine MDLs
    Determine MRLs or LCMRLs
    Determine range of quantitation
    Decide on data reporting
    Conduct a collaborative study or interlaboratory comparison

Method Execution
    Monitor instrumental performance (sensitivity, resolution)
    Calibrate instrumentation
    Check the de-ionized water used as the blanks for extraction and field blanks for any potential background levels
    Periodically check the sampling containers and other laboratory ware used for extraction for any background levels, especially when changing lots or vendors
    Periodically calibrate apparatus (e.g., pipettes)

Sample Preparation and Analysis
    Include laboratory blanks to assess contamination (approximately 10% of samples with each extraction batch)
    Include replicates (approximately 10% of samples with each extraction batch)
    Include laboratory controls to assess losses (approximately 10% of samples with each extraction batch)
    Include matrix-fortified samples (approximately 10% of samples with each extraction batch)
    Use mass-labeled surrogates to measure method recovery for each sample

(Continued)

TABLE 2.5 (CONTINUED)
QC Considerations for PFAS Analysis

    If possible, monitor two MS/MS transitions for each analyte, one for quantitation and the other for qualification

    Use internal standards for quantitation

    Separate quantitation of linear and branched isomers can be done if proper chromatographic resolution is observed

Data Reporting

    In the analysis report, include sample number, date of sampling, date of extraction, date of analysis, notebook reference, and other relevant information to trace the data

    Report QC data and the acceptable criteria

    Report MDL and MRL

    Repeat questionable analyses

Source: Adapted from Erickson, M. D., *Analytical Chemistry of PCBs*, Lewis Publishers, Boca Raton, FL, 1997.

(up to 30%–40%). Due to the lack of authentic branched isomer standards, the quantification of the total branched isomers is often done using the calibration curve of the linear isomers but reported as the total branched isomer concentration. Therefore, there is a need for high-purity and well-characterized isomeric standards for accurate measurements of branched and linear PFAS isomers as more fate, transport, and toxicity data are being developed to evaluate these various PFAS isomers.

2.10 CONCLUSIONS

This chapter provides an overview of the sampling and analytical challenges and a review of the recent developments in the analysis of PFAS in environmental matrices focused on water samples. As PFAS are ubiquitously used for a wide range of applications, extreme care is required to avoid cross-contamination during sampling, sample processing, and analysis. To produce reliable data, it is important to include many QA/QC samples, such as replicates, laboratory and field blanks (free of any PFAS contamination), travel blanks, and laboratory and matrix-fortified samples.

Recently, significant improvements in the trace-level analysis of PFAS in environmental and biological matrices can be attributed to the availability of good chemical standards and mass-labeled standards and avoiding the instrument background contamination by using PEEK tubing for solvent lines and a precolumn to delay the solvent peaks. However, quantitation of unknown fluorinated chemicals and branched isomers of some of the PFAS is a significant challenge without the availability of pure isomeric standards. Although many of these branched isomers are unknown and with the limited availability of standards, it is becoming more important to have isomer-specific data in the treatment and remediation processes. Analyzing these structural isomers separately also provides information about their different manufacturing sources (ECF and telomerization processes). Due to different physical and chemical properties, water solubility, sorption, and bioaccumulation potentials of the branched isomers compared with their linear counterparts, different partitioning and transport behavior in the environment is expected. Therefore, there is a need for pure and well-characterized isomeric standards for the accurate measurement of branched and linear PFAS isomers to improve understanding of the environmental behavior of PFAS.

Another major challenge for the analysis of PFAS chemicals in AFFF formulations or at AFFF-contaminated sites is the large number of unknown PFAS, and the limited number of analytes can be quantitated using the known analytical procedures. The known quantifiable PFAS account for a very small fraction of the total fluorinated organic chemicals. Therefore, there is a need for emerging methods like total fluorine analysis, TOP assay, and PIGE techniques, which can be used as pre-screening tools for PFAS-contaminated site assessments and groundwater investigations. Although these techniques do not provide any information on the exact chemical structure of the precursors, they can be used for site characterization and remediation efforts to estimate the mass balance of the total PFAS contamination. It should also be noted that hybrid instruments, such as QTOF, IT-TOF, QLIT, and Orbitrap, can achieve accurate mass measurements and acquire qualitative information of the unknown chemicals; however, these instruments are not commonly used by typical contract environmental laboratories due to their high cost.

Robust, high-quality data are imperative to understand the environmental fate of PFAS. To obtain such data, there is a need for more standardized, multi-laboratory-validated methods for analysis of a broad list of PFAS in different matrices. Also, there is a need for more ILSs, analytical standards, and standard reference materials in different matrices for PFAS and their precursors to validate the comparability of analytical methods between different laboratories.

REFERENCES

Ahrens, L.; Taniyasu, S.; Yeung, L. W.; Yamashita, N.; Lam, P. K.; Ebinghaus, R. Distribution of polyfluoroalkyl compounds in water, suspended particulate matter and sediment from Tokyo Bay, Japan. *Chemosphere* **2010**, *79* (3), 266–272.

Ahrens, L. Polyfluoroalkyl compounds in the aquatic environment: A review of their occurrence and fate. *J Environ Monit* **2011**, *13* (1), 20–31.

Allred, B. M.; Lang, J. R.; Barlaz, M. A.; Field, J. A. Physical and biological release of poly- and perfluoroalkyl substances (PFAS) from municipal solid waste in anaerobic model landfill reactors. *Environ Sci Technol* **2015**, *49* (13), 7648–7656.

Anderson, R. H.; Long, G. C.; Porter, R. C.; Anderson, J. K. Occurrence of select perfluoroalkyl substances at US Air Force aqueous film-forming foam release sites other than fire-training areas: Field-validation of critical fate and transport properties. *Chemosphere* **2016**, *150*, 678–685.

Arsenault, G.; Chittim, B.; Gu, J.; McAlees, A.; McCrindle, R.; Robertson, V. Separation and fluorine nuclear magnetic resonance spectroscopic (19F NMR) analysis of individual branched isomers present in technical perfluorooctanesulfonic acid (PFOS). *Chemosphere* **2008**, *73* (1 Suppl.), S53–S59.

ASTM (American Society for Testing and Materials). Standard test method for determination of perfluorinated compounds in soil by liquid chromatography tandem mass spectrometry (LC/MS/MS). D7968-14. ASTM, West Conshohocken, PA, **2014**.

ASTM (American Society for Testing and Materials). Standard test method for determination of perfluorinated compounds in water, sludge, influent, effluent and wastewater by liquid chromatography tandem mass spectrometry (LC/MS/MS). D7979-16. ASTM, West Conshohocken, PA, **2016**.

Backe, W. J.; Day, T. C.; Field, J. A. Zwitterionic, cationic, and anionic fluorinated chemicals in aqueous film forming foam formulations and groundwater from U.S. military bases by nonaqueous large-volume injection HPLC-MS/MS. *Environ Sci Technol* **2013**, *47* (10), 5226–5234.

Berger, U.; Kaiser, M. A.; Kärrman, A.; Barber, J. L.; van Leeuwen, S. P. Recent developments in trace analysis of poly- and perfluoroalkyl substances. *Anal Bioanal Chem* **2011**, *400* (6), 1625–1635.

Berger, U.; Langlois, I.; Oehme, M.; Kallenborn, R. Comparison of three types of mass spectrometers for HPLC/MS analysis of perfluoroalkylated substances and fluorotelomer alcohols. *Eur Mass Spectrom* **2004**, *10* (5), 579–588.

Boiteux, V.; Dauchy, X.; Rosin, C.; Munoz, J.-F. National screening study on 10 perfluorinated compounds in raw and treated tap water in France. *Arch Environ Contam Toxicol* **2012**, *63* (1), 1–12.

Boulanger, B.; Peck, A. M.; Schnoor, J. L.; Hornbuckle, K. C. Mass budget of perfluorooctane surfactants in Lake Ontario. *Environ Sci Technol* **2005**, *39* (1), 74–79.

Boulanger, B.; Vargo, J.; Schnoor, J. L.; Hornbuckle, K. C. Detection of perfluorooctane surfactants in Great Lakes water. *Environ Sci Technol* **2004**, *38* (15), 4064–4070.

Brede, E.; Wilhelm, M.; Göen, T.; Müller, J.; Rauchfuss, K.; Kraft, M.; Hölzer, J. Two-year follow-up biomonitoring pilot study of residents' and controls' PFC plasma levels after PFOA reduction in public water system in Arnsberg, Germany. *Int J Hyg Environ Health* **2010**, *213*, 217–223.

Buck, R. C.; Franklin, J.; Berger, U.; Conder, J. M.; Cousins, I. T.; de Voogt, P.; Jensen, A. A.; Kannan, K.; Mabury, S. A.; van Leeuwen, S. P. Perfluoroalkyl and polyfluoroalkyl substances in the environment: Terminology, classification, and origins. *Integr Environ Assess Manag* **2011**, *7* (4), 513–541.

Calafat, A. M.; Needham, L. L.; Kuklenyik, Z.; Reidy, J. A.; Tully, J. S.; Aguilar-Villalobos, M.; Naeher, L. P. Perfluorinated chemicals in selected residents of the American continent. *Chemosphere* **2006**, *63* (3), 490–496.

Calafat, A. M.; Wong, L. Y.; Kuklenyik, Z.; Reidy, J. A.; Needham, L. L. Polyfluoroalkyl chemicals in the U.S. population: Data from the National Health and Nutrition Examination Survey (NHANES) 2003–2004 and comparisons with NHANES 1999–2000. *Environ Health Perspect* **2007**, *115* (11), 1596–1602.

Chu, S.; Letcher, R. J. Analysis of fluorotelomer alcohols and perfluorinated sulfonamides in biotic samples by liquid chromatography-atmospheric pressure photoionization mass spectrometry. *J Chromatogr A* **2008**, *1215* (1), 92–99.

Dasu, K.; Lee, L. S. Aerobic biodegradation of toluene-2,4-di(8:2 fluorotelomer urethane) and hexamethylene-1,6-di(8:2 fluorotelomer urethane) monomers in soils. *Chemosphere* **2015**, *144*, 2482–2488.

Dasu, K.; Liu, J.; Lee, L. S. Aerobic soil biodegradation of 8:2 fluorotelomer stearate monoester. *Environ Sci Technol* **2012**, *46* (7), 3831–3836.

Dasu, K.; Nakayama, S. F.; Yoshikane, M.; Mills, M. A.; Wright, J. M.; Ehrlich, S. An ultrasensitive method for the analysis of perfluorinated alkyl acids in drinking water using a column switching high-performance liquid chromatography tandem mass spectrometry. *J Chromatogr A* **2017**, *1494*, 45–54.

DOD (Department of Defense). Per- and polyfluoroalkyl substances (PFAS) using liquid chromatography tandem mass spectrometry (LC/MS/MS) with isotope dilution or internal standard quantification in matrices other than drinking water. Quality Systems Manual 5.1. DOD, Washington, DC, **2017**, Table B-15.

Dufková, V.; Čabala, R.; Maradová, D.; Štícha, M. A fast derivatization procedure for gas chromatographic analysis of perfluorinated organic acids. *J Chromatogr A* **2009**, *1216* (49), 8659–8664.

Ellington, J. J.; Washington, J. W.; Evans, J. J.; Jenkins, T. M.; Hafner, S. C.; Neill, M. P. Analysis of fluorotelomer alcohols in soils: Optimization of extraction and chromatography. *J Chromatogr A* **2009**, *1216* (28), 5347–5354.

Enevoldsen, R.; Juhler, R. K. Perfluorinated compounds (PFCs) in groundwater and aqueous soil extracts: Using inline SPE-LC-MS/MS for screening and sorption characterisation of perfluorooctane sulphonate and related compounds. *Anal Bioanal Chem* **2010**, *398* (3), 1161–1172.

Erickson, M. D. *Analytical Chemistry of PCBs*. Lewis Publishers, Boca Raton, FL, **1997**.

Ericson, I.; Domingo, J. L.; Nadal, M.; Bigas, E.; Llebaria, X.; van Bavel, B.; Lindström, G. Levels of perfluorinated chemicals in municipal drinking water from Catalonia, Spain: Public health implications. *Arch Environ Contam Toxicol* **2009**, *57* (4), 631–638.

Eschauzier, C.; Beerendonk, E.; Scholte-Veenendaal, P.; De Voogt, P. Impact of treatment processes on the removal of perfluoroalkyl acids from the drinking water production chain. *Environ Sci Technol* **2012**, *46* (3), 1708–1715.

Flores, C.; Ventura, F.; Martin-Alonso, J.; Caixach, J. Occurrence of perfluorooctane sulfonate (PFOS) and perfluorooctanoate (PFOA) in NE Spanish surface waters and their removal in a drinking water treatment plant that combines conventional and advanced treatments in parallel lines. *Sci Total Environ* **2013**, *461*, 618–626.

Fromme, H.; Mosch, C.; Morovitz, M.; Alba-Alejandre, I.; Boehmer, S.; Kiranoglu, M.; Faber, F.; Hannibal, I.; Genzel-Boroviczény, O.; Koletzko, B. Pre-and postnatal exposure to perfluorinated compounds (PFCs). *Environ Sci Technol* **2010**, *44* (18), 7123–7129.

Furdui, V. I.; Crozier, P. W.; Reiner, E. J.; Mabury, S. A. Trace level determination of perfluorinated compounds in water by direct injection. *Chemosphere* **2008**, *73* (1), S24–S30.

Gellrich, V.; Knepper, T. P. Sorption and leaching behavior of perfluorinated compounds in soil. In: Knepper, T.; Lange, F. (eds.). *Polyfluorinated Chemicals and Transformation Products. The Handbook of Environmental Chemistry*, vol. 17. Springer, Berlin, **2012**, pp. 63–72.

Gómez, C.; Vicente, J.; Echavarri-Erasun, B.; Porte, C.; Lacorte, S. Occurrence of perfluorinated compounds in water, sediment and mussels from the Cantabrian Sea (North Spain). *Mar Pollut Bullet* **2011**, *62* (5), 948–955.

Gosetti, F.; Chiuminatto, U.; Zampieri, D.; Mazzucco, E.; Robotti, E.; Calabrese, G.; Gennaro, M. C.; Marengo, E. Determination of perfluorochemicals in biological, environmental and food samples by an automated on-line solid phase extraction ultra high performance liquid chromatography tandem mass spectrometry method. *J Chromatogr A* **2010**, *1217* (50), 7864–7872.

Guelfo, J. L.; Higgins, C. P. Subsurface transport potential of perfluoroalkyl acids at aqueous film-forming foam (AFFF)-impacted sites. *Environ Sci Technol* **2013**, *47* (9), 4164–4171.

Higgins, C. P.; Luthy, R. G. Sorption of perfluorinated surfactants on sediments. *Environ Sci Technol* **2006**, *40* (23), 7251–7256.

Hölzer, J.; Midasch, O.; Rauchfuss, K.; Kraft, M.; Reupert, R.; Angerer, J.; Kleeschulte, P.; Marschall, N.; Wilhelm, M. Biomonitoring of perfluorinated compounds in children and adults exposed to perfluorooctanoate-contaminated drinking water. *Environ Health Perspect* **2008**, *116* (5), 651.

Houtz, E. F.; Higgins, C. P.; Field, J. A.; Sedlak, D. L. Persistence of perfluoroalkyl acid precursors in AFFF-impacted groundwater and soil. *Environ Sci Technol* **2013**, *47* (15), 8187–8195.

Houtz, E. F.; Sedlak, D. L. Oxidative conversion as a means of detecting precursors to perfluoroalkyl acids in urban runoff. *Environ Sci Technol* **2012**, *46* (17), 9342–9349.

Houtz, E. F.; Sutton, R.; Park, J. S.; Sedlak, M. Poly- and perfluoroalkyl substances in wastewater: Significance of unknown precursors, manufacturing shifts, and likely AFFF impacts. *Water Res* **2016**, *95*, 142–149.

Huber, S.; Haug, L. S.; Schlabach, M. Per- and polyfluorinated compounds in house dust and indoor air from northern Norway—A pilot study. *Chemosphere* **2011**, *84* (11), 1686–1693.

ISO (International Organization for Standardization). 25101:2009(E). Water quality—Determination of perfluorooctane sulfonate (PFOS) and perfluorooctanoate (PFOA)—Method for unfiltered samples using solid phase extraction and liquid chromatography/mass spectrometry. ISO, Geneva, **2009**.

Jahnke, A.; Ahrens, L.; Ebinghaus, R.; Temme, C. Urban versus remote air concentrations of fluorotelomer alcohols and other polyfluorinated alkyl substances in Germany. *Environ Sci Technol* **2007**, *41* (3), 745–752.

Jin, Y. H.; Liu, W.; Sato, I.; Nakayama, S. F.; Sasaki, K.; Saito, N.; Tsuda, S. PFOS and PFOA in environmental and tap water in China. *Chemosphere* **2009**, *77* (5), 605–611.

Jogsten, I. E.; Nadal, M.; van Bavel, B.; Lindström, G.; Domingo, J. L. Per- and polyfluorinated compounds (PFCs) in house dust and indoor air in Catalonia, Spain: Implications for human exposure. *Environ Int* **2012**, *39* (1), 172–180.

Jogsten, I. E.; Perelló, G.; Llebaria, X.; Bigas, E.; Martí-Cid, R.; Kärrman, A.; Domingo, J. L. Exposure to perfluorinated compounds in Catalonia, Spain, through consumption of various raw and cooked foodstuffs, including packaged food. *Food Chem Toxicol* **2009**, *47* (7), 1577–1583.

Langlois, I.; Berger, U.; Zencak, Z.; Oehme, M. Mass spectral studies of perfluorooctane sulfonate derivatives separated by high-resolution gas chromatography. *Rapid Commun Mass Spectrom* **2007**, *21* (22), 3547–3553.

Lau, C.; Anitole, K.; Hodes, C.; Lai, D.; Pfahles-Hutchens, A.; Seed, J. Perfluoroalkyl acids: A review of monitoring and toxicological findings. *Toxicol Sci* **2007**, *99*, 366–394.

Liu, X.; Guo, Z.; Krebs, K. A.; Pope, R. H.; Roache, N. F. Concentrations and trends of perfluorinated chemicals in potential indoor sources from 2007 through 2011 in the US. *Chemosphere* **2014**, *98*, 51–57.

Llorca, M.; Farre, M.; Pico, Y.; Barcelo, D. Study of the performance of three LC-MS/MS platforms for analysis of perfluorinated compounds. *Anal Bioanal Chem* **2010**, *398* (3), 1145–1159.

Martin, J. J.; Winslow, S. D.; Munch, D. J. A new approach to drinking-water-quality data: Lowest-concentration minimum reporting level. *Environ Sci Technol* **2007**, *41* (3), 677–681.

Martin, J. W.; Kannan, K.; Berger, U.; Voogt, P. D.; Field, J.; Franklin, J.; Giesy, J. P.; et al. Peer reviewed: Analytical challenges hamper perfluoroalkyl research. *Environ Sci Technol* **2004**, *38* (13), 248A–255A.

Miyake, Y.; Yamashita, N.; Rostkowski, P.; So, M. K.; Taniyasu, S.; Lam, P. K.; Kannan, K. Determination of trace levels of total fluorine in water using combustion ion chromatography for fluorine: A mass balance approach to determine individual perfluorinated chemicals in water. *J Chromatogr A* **2007**, *1143* (1), 98–104.

Möller, A.; Ahrens, L.; Surm, R.; Westerveld, J.; van der Wielen, F.; Ebinghaus, R.; de Voogt, P. Distribution and sources of polyfluoroalkyl substances (PFAS) in the River Rhine watershed. *Environ Pollut* **2010**, *158* (10), 3243–3250.

Moody, C. A.; Kwan, W. C.; Martin, J. W.; Muir, D. C.; Mabury, S. A. Determination of perfluorinated surfactants in surface water samples by two independent analytical techniques: Liquid chromatography/tandem mass spectrometry and 19F NMR. *Anal Chem* **2001**, *73* (10), 2200–2206.

Moody, C. A.; Martin, J. W.; Kwan, W. C.; Muir, D. C.; Mabury, S. A. Monitoring perfluorinated surfactants in biota and surface water samples following an accidental release of fire-fighting foam into Etobicoke Creek. *Environ Sci Technol* **2002**, *36* (4), 545–551.

Naile, J. E.; Khim, J. S.; Wang, T.; Chen, C.; Luo, W.; Kwon, B.-O.; Park, J.; Koh, C.-H.; Jones, P. D.; Lu, Y. Perfluorinated compounds in water, sediment, soil and biota from estuarine and coastal areas of Korea. *Environ Pollut* **2010**, *158* (5), 1237–1244.

Onghena, M.; Moliner-Martinez, Y.; Pico, Y.; Campins-Falco, P.; Barcelo, D. Analysis of 18 perfluorinated compounds in river waters: Comparison of high performance liquid chromatography-tandem mass spectrometry, ultra-high-performance liquid chromatography-tandem mass spectrometry and capillary liquid chromatography-mass spectrometry. *J. Chromatogr. A* **2012**, *1244*, 88–97.

Prevedouros, K.; Cousins, I. T.; Buck, R. C.; Korzeniowski, S. H. Sources, fate and transport of perfluorocarboxylates. *Environ Sci Technol* **2006**, *40* (1), 32–44.

Quiñones, O.; Snyder, S. A. Occurrence of perfluoroalkyl carboxylates and sulfonates in drinking water utilities and related waters from the United States. *Environ Sci Technol* **2009**, *43* (24), 9089–9095.

Reiner, J. L.; Blaine, A. C.; Higgins, C. P.; Huset, C.; Jenkins, T. M.; Kwadijk, C. J.; Lange, C. C.; et al. Polyfluorinated substances in abiotic standard reference materials. *Anal Bioanal Chem* **2015**, *407* (11), 2975–2983.

Robel, A. E.; Marshall, K.; Dickinson, M.; Lunderberg, D.; Butt, C.; Peaslee, G.; Stapleton, H. M.; Field, J. A. Closing the mass balance on fluorine on papers and textiles. *Environ Sci Technol* **2017**, *51* (16), 9022–9032.

Schaider, L. A.; Balan, S. A.; Blum, A.; Andrews, D. Q.; Strynar, M. J.; Dickinson, M. E.; Lunderberg, D. M.; Lang, J. R.; Peaslee, G. F. Fluorinated compounds in U.S. fast food packaging. *Environ Sci Technol Lett* **2017**.

Schecter, A.; Colacino, J.; Haffner, D.; Patel, K.; Opel, M.; Päpke, O.; Birnbaum, L. Perfluorinated compounds, polychlorinated biphenyls, and organochlorine pesticide contamination in composite food samples from Dallas, Texas, USA. *Environ Health Perspect* **2010**, *118* (6), 796.

Senthilkumar, K.; Ohi, E.; Sajwan, K.; Takasuga, T.; Kannan, K. Perfluorinated compounds in river water, river sediment, market fish, and wildlife samples from Japan. *Bull Environ Contam Toxicol* **2007**, *79* (4), 427–431.

Shiwaku, Y.; Lee, P.; Thepaksorn, P.; Zheng, B.; Koizumi, A.; Harada, K. H. Spatial and temporal trends in perfluorooctanoic and perfluorohexanoic acid in well, surface, and tap water around a fluoropolymer plant in Osaka, Japan. *Chemosphere* **2016**, *164*, 603–610.

Shoeib, M.; Harner, T.; Lee, S. C.; Lane, D.; Zhu, J. Sorbent-impregnated polyurethane foam disk for passive air sampling of volatile fluorinated chemicals. *Anal Chem* **2008**, *80* (3), 675–682.

Shoeib, M.; Harner, T.; Webster, G. M.; Lee, S. C. Indoor sources of poly- and perfluorinated compounds (PFCS) in Vancouver, Canada: Implications for human exposure. *Environ Sci Technol* **2011**, *45* (19), 7999–8005.

Shoemaker, J. Determination of selected perfluorinated alkyl acids in drinking water by solid phase extraction and liquid chromatography/tandem mass spectrometry (LC/MS/MS). Method 537. U.S. Environmental Protection Agency, Washington, DC, **2009**.

Sinclair, E.; Kannan, K. Mass loading and fate of perfluoroalkyl surfactants in wastewater treatment plants. *Environ Sci Technol* **2006**, *40* (5), 1408–1414.

Srivastava, A.; Chhillar, S.; Singh, D.; Acharya, R.; Pujari, P. Determination of fluorine concentrations in soil samples using proton induced gamma-ray emission. *J Radioanal Nucl Chem* **2014**, *302* (3), 1461–1464.

Takino, M.; Daishima, S.; Nakahara, T. Determination of perfluorooctane sulfonate in river water by liquid chromatography/atmospheric pressure photoionization mass spectrometry by automated on-line extraction using turbulent flow chromatography. *Rapid Commun Mass Spectrom* **2003**, *17* (5), 383–390.

Taniyasu, S.; Kannan, K.; So, M. K.; Gulkowska, A.; Sinclair, E.; Okazawa, T.; Yamashita, N. Analysis of fluorotelomer alcohols, fluorotelomer acids, and short- and long-chain perfluorinated acids in water and biota. *J Chromatogr A* **2005**, *1093* (1–2), 89–97.

Taniyasu, S.; Kannan, K.; Yeung, L. W. Y.; Kwok, K. Y.; Lam, P. K. S.; Yamashita, N. Analysis of trifluoroacetic acid and other short-chain perfluorinated acids (C2–C4) in precipitation by liquid chromatography–tandem mass spectrometry: Comparison to patterns of long-chain perfluorinated acids (C5–C18). *Anal Chim Acta* **2008**, *619* (2), 221–230.

Thompson, J.; Eaglesham, G.; Mueller, J. Concentrations of PFOS, PFOA and other perfluorinated alkyl acids in Australian drinking water. *Chemosphere* **2011**, *83* (10), 1320–1325.

Ullah, S.; Alsberg, T.; Berger, U. Simultaneous determination of perfluoroalkyl phosphonates, carboxylates, and sulfonates in drinking water. *J Chromatogr A* **2011**, *1218* (37), 6388–6395.

USEPA (U.S. Environmental Protection Agency). Statistical protocol for the determination of the single-laboratory lowest concentration minimum reporting level (LCMRL) and validation of laboratory performance at or below the minimum reporting level (MRL). EPA Document 815-R-05-006. **2004**. https://nepis.epa.gov (accessed 29 September 2017).

USEPA (U.S. Environmental Protection Agency). Report of the Federal Advisory Committee on Detection and Quantitation Approaches and Uses in Clean Water Act Programs. **2007**. https://www.epa.gov/sites/production/files/2015-10/documents/detection-quant-faca _final-report_2012.pdf (accessed 8 December 2017).

USEPA (U.S. Environmental Protection Agency). Contaminant Candidate List 3—CCL 3. **2009**. https://www.epa.gov/ccl/contaminant-candidate-list-3-ccl-3 (accessed 29 September 2017).

USEPA (U.S. Environmental Protection Agency). Lowest concentration minimum reporting level (LCMRL) calculator. **2010**. http://water.epa.gov/scitech/drinkingwater/labcert /analyticalmethods_ogwdw.cfm (accessed 29 September 2017).

USEPA (U.S. Environmental Protection Agency). Determination of selected perfluorinated alkyl acids in drinking water by solid phase extraction and liquid chromatography/ tandem mass spectrometry (LC/MS/MS). Method 537 Rev 1. **2012**. https://cfpub.epa .gov/si/si_public_record_report.cfm?dirEntryId=198984&simpleSearch=1&searchAll =EPA%2F600%2FR-08%2F092 (accessed 29 September 2017).

USEPA (U.S. Environmental Protection Agency). Third Unregulated Contaminant Monitoring Rule. **2015**. https://www.epa.gov/dwucmr/third-unregulated-contaminant -monitoring-rule (accessed 29 September 2017).

USEPA (U.S. Environmental Protection Agency). Fact sheet—PFOA & PFOS drinking water health advisories. **2016a**. https://www.epa.gov/sites/production/files/2016-06/documents /drinkingwaterhealthadvisories_pfoa_pfos_updated_5.31.16.pdf, p. 5 (accessed 29 September 2017).

USEPA (U.S. Environmental Protection Agency). Contaminant Candidate List 4—CCL 4. **2016b**. https://www.epa.gov/ccl/contaminant-candidate-list-4-ccl-4-0 (accessed 29 September 2017).

USEPA (U.S. Environmental Protection Agency). Definition and procedure for the determination of the method detection limit. Revision 2. **2016c**. https://www.epa.gov/sites /production/files/2016-12/documents/mdl-procedure_rev2_12-13-2016.pdf (accessed 8 December 2017).

USEPA (U.S. Environmental Protection Agency). Data summary of the Third Unregulated Contaminant Monitoring Rule. **2017**. https://www.epa.gov/dwucmr/data-summary-third-unregulated-contaminant-monitoring-rule (accessed 29 September 2017).

Van Leeuwen, S.; Swart, C.; Van der Veen, I.; De Boer, J. Significant improvements in the analysis of perfluorinated compounds in water and fish: Results from an interlaboratory method evaluation study. *J Chromatogr A* **2009**, *1216* (3), 401–409.

Wagner, A.; Raue, B.; Brauch, H. J.; Worch, E.; Lange, F. T. Determination of adsorbable organic fluorine from aqueous environmental samples by adsorption to polystyrene-divinylbenzene based activated carbon and combustion ion chromatography. *J Chromatogr A* **2013**, *1295*, 82–89.

Wang, N.; Szostek, B.; Buck, R. C.; Folsom, P. W.; Sulecki, L. M.; Gannon, J. T. 8-2 fluorotelomer alcohol aerobic soil biodegradation: Pathways, metabolites, and metabolite yields. *Chemosphere* **2009**, *75* (8), 1089–1096.

Weinberg, I.; Dreyer, A.; Ebinghaus, R. Waste water treatment plants as sources of polyfluorinated compounds, polybrominated diphenyl ethers and musk fragrances to ambient air. *Environ Pollut* **2011a**, *159* (1), 125–132.

Weinberg, I.; Dreyer, A.; Ebinghaus, R. Landfills as sources of polyfluorinated compounds, polybrominated diphenyl ethers and musk fragrances to ambient air. *Atmos Environ* **2011b**, *45* (4), 935–941.

Winslow, S. D.; Pepich, B. V.; Martin, J. J.; Hallberg, G. R.; Munch, D. J.; Frebis, C. P.; Hedrick, E. J.; Krop, R. A. Statistical procedures for determination and verification of minimum reporting levels for drinking water methods. *Environ Sci Technol* **2006**, *40* (1), 281–288.

Wu, X.; Bennett, D. H.; Calafat, A. M.; Kato, K.; Strynar, M.; Andersen, E.; Moran, R. E.; Tancredi, D. J.; Tulve, N. S.; Hertz-Picciotto, I. Serum concentrations of perfluorinated compounds (PFC) among selected populations of children and adults in California. *Environ Res* **2015**, *136*, 264–273.

Yamashita, N.; Kannan, K.; Taniyasu, S.; Horii, Y.; Okazawa, T.; Petrick, G.; Gamo, T. Analysis of perfluorinated acids at parts-per-quadrillion levels in seawater using liquid chromatography-tandem mass spectrometry. *Environ Sci Technol* **2004**, *38*, 5522–5528.

Yamashita, N.; Taniyasu, S.; Petrick, G.; Wei, S.; Gamo, T.; Lam, P. K.; Kannan, K. Perfluorinated acids as novel chemical tracers of global circulation of ocean waters. *Chemosphere* **2008**, *70* (7), 1247–1255.

3 Understanding Precursor Contributions
The Total Oxidizable Precursor Assay

Bharat Chandramouli

CONTENTS

3.1 Background 67
3.2 Approaches to Fluorine Content Estimation 69
    3.2.1 Particle-Induced Gamma Emission 69
    3.2.2 Adsorbable Organic Fluorine 69
3.3 Total Oxidizable Precursor Assay 70
    3.3.1 TOP Procedure 70
    3.3.2 TOP Applications 72
    3.3.3 Regulatory Applications 74
3.4 Analytical Issues, Limitations, and Data Usability Considerations 75
    3.4.1 Quality of Underlying Analytical Procedure 75
    3.4.2 Understanding and Monitoring for Reaction Completeness 76
    3.4.3 Precursor Volatility 76
    3.4.4 Ultra-Short-Chain Products 76
    3.4.5 Completeness of Extraction 77
3.5 Summary 77
References 78

3.1 BACKGROUND

Perfluorinated carboxylates (PFCAs), including perfluorooctanoic acid (PFOA), and perfluorinated sulfonates (PFSAs), including perfluorooctane sulfonate (PFOS) (collectively known as perfluorinated alkyl acids [PFAAs]), have been the most extensively characterized of the poly- and perfluoroalkyl substances (PFAS) (Rankin et al., 2016). This is in part due to their persistence in the environment and, in the case of the longer-chain acids, bioaccumulation and long half-lives in humans (Olsen et al., 2007). However recent work (Place and Field, 2012; Barzen-Hanson et al., 2017), and as detailed in several reviews (Buck et al., 2011; Wang et al., 2017; Xiao, 2017), has shown that the traditionally measured PFSAs and PFCAs represent a small portion of the PFAS mass balance. This measurement gap complicates the understanding of PFAS sources, sinks, fate, and transport (Wang et al., 2015). In addition, given the phaseout of most direct sources of PFCAs and PFSAs, there is increasing attention on "precursors," or PFAS that can transform into perfluorinated acids, either in the environment or in vivo in humans and wildlife (Paul et al., 2009; Martin et al., 2010). Modern formulations of aqueous film-forming foam (AFFF) products are, for example, designed to only have residual or negligible amounts of PFAAs. The main active ingredients are a variety of fluorotelomer substances consisting of continuous C–F chains of six carbons or less, a hydrocarbon portion, and a variety of functional moieties. They can be cationic, anionic, and in some cases, neutral (Place and Field, 2012). The Xiao (2017) review has identified 455 new PFAS between 2009 and 2017, in addition to the already identified PFAS. This is but a small fraction of the 3000 or so PFAS estimated to be in the global market (KEMI: Swedish Chemicals Agency, 2015; Wang et al., 2017). This proliferation of PFAS presents a serious prioritization problem for regulators and risk assessors alike.

Why do we need to get a better understanding of PFAS, especially precursors to medium- and long-chain PFAS?

- Each of them is a potential new xenobiotic of environmental concern to be studied for fate, transport, and health effects.
- We must understand which of these can transform (precursors) into C6 and longer PFAAs, which are of concern due to their persistence and bioaccumulation potential.
- We wish to know how much time it takes for them to transform in the environment.
- We need to know how they behave in a spill or use site or in a plume, and whether monitoring a select set of PFAAs is representative of this entire set of PFAS.
- We need to understand how they affect the design, maintenance, and effectiveness of remediation systems, such as granular activated carbon (GAC) and ion exchange.

The number and complexity of PFAS in the environment means that targeted analysis focusing on a select set (<50 at the most) of PFAS will not provide a complete feature. While untargeted analysis utilizing high-resolution mass spectrometry is of growing interest (Strynar et al., 2015; Newton et al., 2017), the availability of instrumentation, for example, and complicated workflows limit their use for the next 5–10 years.

There is thus clearly a need for proxy measurements of PFAS that attempt to estimate total loadings. This chapter describes a variety of tools for the indirect estimation of total fluorine content and focuses on one of them, the total oxidizable precursor (TOP) assay, which, if used appropriately, can provide estimates of the potential of unknown precursors in a sample to transform into stable and persistent PFAAs in the environment.

3.2 APPROACHES TO FLUORINE CONTENT ESTIMATION

There are multiple approaches to assessing total fluorine in a sample, each with different and complementary applications.

3.2.1 PARTICLE-INDUCED GAMMA EMISSION

Particle-induced gamma emission (PIGE) uses protons to stimulate the formation of gamma rays from molecules. ^{19}F emits gamma rays at characteristic energies of 110 and 197 keV. This signal can be converted into a total elemental fluorine measurement using a PFAS standard dried onto filter paper (Ritter et al., 2017). The current detection limit for this technique is 16 nmol F/cm^2. Recent publications (Ritter et al., 2017; Schaider et al., 2017) showcase applications of PIGE as a screening tool to understand the occurrence of fluorinated content in paper and textiles (Ritter et al., 2017) and in fast food packaging (Schaider et al., 2017). This is an application PIGE is uniquely suited for due to its ability to quickly test surfaces for fluorine. Schaider et al. (2017) reported that 46% of food contact papers and 20% of paperboard samples contained fluorine. This application of PIGE does not distinguish between small-molecule PFAS, the compounds of most concern, and fluoropolymers that may not be bioavailable, or inorganic fluorine. PIGE is also limited to surface measurement as the ion beam can only penetrate up to 200 μm. Research on the applications of PIGE in environmental samples by capture of PFAS onto a solid-phase extraction cartridge or disk is ongoing.

3.2.2 ADSORBABLE ORGANIC FLUORINE

The adsorbable organic fluorine (AOF) technique (Wagner et al., 2013) uses the relatively uncomplicated technique of hydropyrolysis combustion ion chromatography to measure organofluorine compounds adsorbed on a polystyrene-divinylbenzene-based activated carbon. A limit of quantitation of 0.3 μg/L F (100 mL sample) was achieved in river water by Wagner et al. (2013). AOF can provide an estimate of organic fluorine in water and solid samples. Recent applications of AOF in peer-reviewed literature include its use in understanding PFAS flows and fate in a wastewater treatment plant (WWTP) treating fluorochemical industry waste (Dauchy et al., 2017), where it was used in conjunction with TOP, and its use to determine PFAS levels in German rivers and groundwater (Willach et al. 2016). The latter study showed that only 5% of total fluorine levels as estimated by AOF was attributable to the 17 individual PFAS measured using liquid chromatography–tandem mass spectrometry (LC-MS/MS). While AOF is a promising tool for the estimation of total fluorine, it does not provide any information on which of these PFAS are precursors, or information on the chain length of the PFAS, which may limit its applications in fingerprinting, source characterization, and precursor relevance.

3.3 TOTAL OXIDIZABLE PRECURSOR ASSAY

The TOP assay, developed by Houtz and Sedlak (2012), takes a different approach to estimating unknown PFAS contributions. The rest of this chapter details the TOP procedure, applications, and limitations.

3.3.1 TOP Procedure

The TOP procedure (Houtz and Sedlak, 2012) uses persulfate at basic pH and elevated temperature to enable the formation of the hydroxyl radical (Equations 3.1 and 3.2).

$$S_2O_8^{2-} + \text{heat} \rightarrow 2SO_4^{-} \tag{3.1}$$

$$SO_4^{-} + OH^{-} \rightarrow SO_4^{2-} + OH^{\cdot} \tag{3.2}$$

The hydroxyl radical can transform PFAS precursors containing C–F chains and C–H linkages in a way that can cleave the nonfluorinated parts off and convert the C–F chains into PFCAs. For example, with perfluorooctanesulfonamides, or acids, this conversion is an almost 100% molar conversion to the corresponding perfluorocarboxylic acid. With fluorotelomers, 6:2 fluorotelomer sulfonate (FTS), for example, there can be further cleaving of the C–F chain to C4 (and below), and C5 acids in addition to the C6 acid. Table 3.1 shows typical molar conversion rates for a set of precursor compounds studied. The developers of the TOP assay studied the effects of varying persulfate concentrations and precursor concentrations and settled on using 2 g (60 mM) of potassium persulfate and 1.9 mL of 10 N NaOH (150 mM) amended to the sample, which was then maintained at 85°C for 6 hours. The hydroxyl radical can be consumed by a variety of other organic compounds present in the sample; therefore, while this procedure has become the standard for TOP analysis, it is important to note that sample size and sample type have a large influence on the completion of the reaction for the fluorinated precursors present. The consumption of hydroxyl ion from the KOH reduces the pH of the sample, and once the pH drops below neutral, the reaction will slow down and eventually stop as production of the OH radical stops.

Independent evaluations of these yields have been carried out in other laboratories and have generally shown results similar to the ones published in Houtz and Sedlak (2012). An example of replication from the SGS AXYS laboratory (Sidney, British Columbia, Canada; author's data, not yet published) is shown in Figure 3.1. There is excellent agreement, mostly within experimental error and analytical protocol differences for both the PFCA chain length breakdown and individual molar yields. There are a few differences; for example, the perfluorinated sulfonamidoacetic acids showed reproducible formation of lower-carbon-chain compounds in our study compared with Houtz and Sedlak (2012). The reason for this is yet unclear, perhaps linked to the analytical standard used for the reaction, or differences in the analytical measurement protocols. However, overall molar yields were indistinguishable within experimental error.

TABLE 3.1
Published Molar Yields of Select Precursors after the TOP Assay

Precursor	Molar Yields as a Fraction of Precursor Concentration						
	Perfluorobutanoate (PFBA)	Perfluoropentanoate (PFPeA)	Perfluorohexanoate (PFHxA)	Perfluoroheptanoate (PFHpA)	Perfluorooctanoate (PFOA)	Perfluorononanoate (PFNA)
n-Ethyl sulfonamidoacetic acid (N-EtFoSAA)					92% ± 4%	
n-Methyl sulfonamidoacetic acid (N-MeFOSAA)					110% ± 8%	
Perfluorooctane-sulfonamide (FOSA)					97% ± 3%	
6:2 fluorotelomer sulfonate (6:2 FTS)	22% ± 5%	27% ± 2%	22% ± 2%	2% ± 1%		
8:2 fluorotelomer sulfonate (8:2 FTS)	11% ± 4%	12% ± 4%	19% ± 3%	27% ± 3%	21% ± 2%	3% ± 0.1%
6:2 polyfluoroalkyl phosphoric acid diester (6:2 DiPAP)	27% ± 3%	47% ± 3%	33% ± 2%	15% ± 3%		
8:2 polyfluoroalkyl phosphoric acid diester (8:2 DiPAP)	10% ± 2%	17% ± 1%	24% ± 1%	43% ± 2%	38% ± 2%	13% ± 1%

Source: Adapted with permission from Houtz, E.F., and Sedlak, D.L., *Environ. Sci. Technol.*, 46, 9342–9349, 2012.

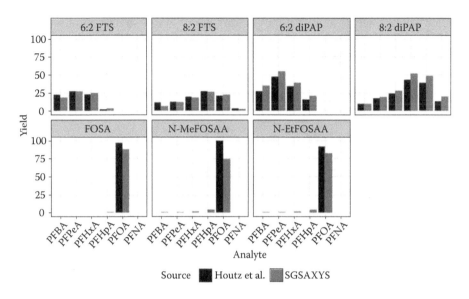

FIGURE 3.1 Verification of published molar yields. (Reprinted with permission from Houtz, E.F., and Sedlak, D.L., *Environ. Sci. Technol.*, 46, 9342–9349, 2012. SGS AXYS's work is internal and unpublished.)

3.3.2 TOP Applications

As of the time of this writing, the paper by Houtz and Sedlak (2012) has been cited 74 times in peer-reviewed literature, representing a rapid uptake in the use of this technique in research publications (not all citations used TOP in the work). The original paper describing TOP (Houtz and Sedlak, 2012) used the assay to understand the contributions of precursors to urban runoff in the San Francisco Bay Area. They found that

> Following oxidative treatment, the total concentrations of PFCAs with 5–12 membered perfluoroalkyl chains increased by a median of 69%, or between 2.8 and 56 ng/L. Precursors that produced PFHxA [perfluorohexanoic acid] and PFPeA [perfluoropentanoic acid] upon oxidation were more prevalent in runoff samples than those that produced PFOA, despite lower concentrations of their corresponding perfluorinated acids prior to oxidation. Direct measurements of several common precursors to PFOS and PFOA (e.g., perfluorooctanesulfonamide and 8:2 fluorotelomer sulfonate) accounted for less than 25% of the observed increase in PFOA, which increased by a median value of 37%.

In follow-up work, Houtz et al. (2013) used the assay in conjunction with the direct measurement of 22 precursors in AFFF-impacted groundwater, soil, and aquifer solids and in archived AFFF formulations. Figure 3.2 shows the results of the TOP assay performed on the product formulations. The results show that most formulations other than the C8 PFOS-containing ones showed less than detectable concentrations of perfluorinated acids before TOP, but 5.9 and 10.8 g/L of PFCAs post-TOP. The occurrence and concentration of precursors measured provide a good indication

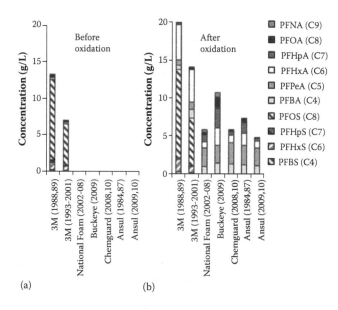

FIGURE 3.2 Results of TOP assay performed on AFFF product formulations (a) before and (b) after oxidation. (b) Shows large increases in known PFAAs measured after TOP oxidation, with PFAA concentration profiles providing insight on the type of AFFF product tested. (Reprinted with permission from Houtz, E.F. et al., *Environ. Sci. Technol.*, 47, 8187–8195, 2013.)

of the type of formulation in use, with transformation of the fluorotelomer precursors into a mix of $C_{n+1}:C_n:C_{n-1}:C_{n-2}$ PFCA products at approximately 0.25:1.0:1.3:1.0, and the transformation of C_n-sulfonamide-based AFFF products into equimolar quantities of the C_n carboxylates. The authors used this information to link the post-TOP measurements in AFFF-contaminated groundwater and soil samples to potential sources. In the aquifer samples from the firefighter sample training area, measurement of a set of precursors only accounted for 40% of the precursor mass attributable after TOP. Additionally, the PFCA profiles suggested that the unidentified precursors were primarily C6 compounds, indicating a distinct C_6 sulfonamide source in addition to the expected 6:2 telomer products.

This application linking formulation results with environmental sample results showcases the utility of the TOP assay as a potential fingerprinting tool. The data show the potential fate of these precursors over a period of decades as a reservoir of persistent PFAS. Further use of this technique requires a more detailed study of formulations using the TOP assay and an understanding of reaction uncertainties in these complex matrices.

Further work from Houtz et al. (2016) used the TOP assay in conjunction with an expanded PFAS analysis to understand the role of AFFF product shifts on concentrations of PFAS and precursor contributions in WWTP effluent in the San Francisco Bay Area. They found PFAS precursors, estimated by TOP, to be responsible for 33%–63% of the total molar concentration of PFAS, with much of the TOP results showing short-chain acids, reflecting manufacturing shifts toward shorter-chained PFAS.

Eberle et al. (2017) used TOP to assess the efficacy of *in situ* chemical oxidation (ISCO) for PFAS remediation in a former firefighter training area. They first identified the presence of precursors forming C4–C6 acids in samples prior to remediation and then used the concentrations to confirm the historic use of a C6 electrochemical-based AFFF formulation containing perfluorohexane sulfonamide amine and perfluorohexane sulfonamide amino carboxylate. In the postremediation samples, only PFHxA was observed at concentrations more than an order of magnitude lower than in preremediation samples. This use of TOP served to demonstrate remediation effectiveness for not just the target compounds monitored but also for unknown precursors.

Robel et al. (2017) used TOP as part of a diverse set of analyses, including LC-MS/MS, gas chromatography–mass spectrometry (GC-MS), liquid chromatography–time-of-flight spectrometry (LC-TOF), and PIGE, to understand the fluorine mass balance in food packaging before and after extraction. Their work showed that the nonextractable portion of the papers and textiles comprised the bulk of the total fluorine (64% ± 28% to 110% ± 30% of the original nmol F/cm^2), as determined by PIGE. In addition, unknown precursors measured by TOP were the overwhelming fraction of the extractable portion (0.021%–14%) compared with volatile PFAS by GC-MS (0%–2.2%) and ionic PFAS by LC-MS/MS (0.0.41%). These results show the utility of TOP in understanding the mass balance in extractable (small-molecule) PFAS, and the different applications between PIGE (all fluorine) and TOP (extracted precursors only).

Weber et al. (2017) used the TOP assay to understand the movement of precursors in a contaminated site plume. For this particular site in Cape Cod, Massachusetts, TOP results indicated that the precursors were cotransporting with the main PFAS plume originating from the firefighter training site and showed a similar pattern as PFOS (Figure 3.3). This cotransport could be a site-specific feature, as it has been reported previously (Xiao et al., 2015) that certain precursors are less mobile than PFCAs or PFSAs and that interactions of PFAS with other plume components are dependent on chemical structure.

Unpublished data from SGS-AXYS have shown 20- to 250-fold increases in PFCAs post-TOP in WWTP biosolids, indicating that the transformation of precursors into PFCAs or PFSAs that is known to occur in WWTPs is far from complete.

3.3.3 REGULATORY APPLICATIONS

The application of TOP is still in its early stages, with only one jurisdiction (at the time of writing), the Department of Environment and Heritage Protection (Queensland, Australia), having included the use of TOP analysis in setting limits for the presence of PFAS in AFFF (50 mg/kg F sum of C7–C14 products) and in contaminated water (interim level of 1 µg/L sum of C4–C14 PFCA and C4–C8 PFSA) (Department of Environment and Heritage Protection, 2016). At the time of writing, the state of Victoria in Australia has included TOP in its draft PFAS management plan (PFAS National Environmental Management Plan, 2017).

Understanding Precursor Contributions

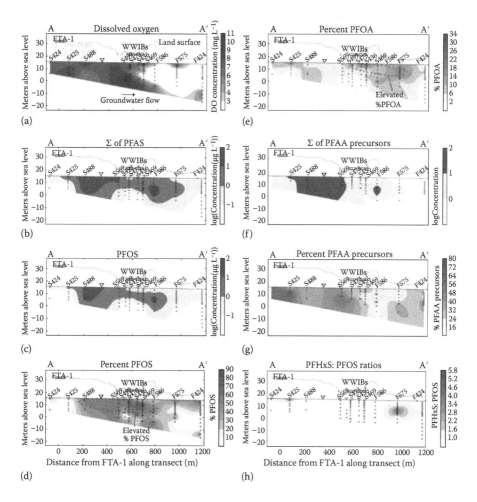

FIGURE 3.3 Using TOP to understand the movement of precursors in a groundwater plume. Note the comparison of TOP vertical and horizontal profiles (f) with ΣPFAS as measured individually (b); the unknown precursors (f) show a similar transport profile to the known and quantitated PFAS. (Reprinted with permission from Weber, A.K. et al., *Environ. Sci. Technol.*, 51, 4269–4279, 2017.)

3.4 ANALYTICAL ISSUES, LIMITATIONS, AND DATA USABILITY CONSIDERATIONS

3.4.1 Quality of Underlying Analytical Procedure

While sometimes overlooked in discussions around the quality of TOP data, the underlying analytical procedure is a significant contributor to data quality. The TOP assay simplifies measurement of PFAS greatly by enabling focus on a select set of PFCAs (and sulfonates for quality assessment). This means that the use of isotope dilution, multipoint calibrations, multiple transitions to assess interferences, weak anion exchange (WAX) cleanup methodologies to reduce interference in the extracts, the monitoring of internal standard recoveries, the use of control standards to monitor for analyte losses during the TOP reaction, and more can be implemented without too much difficulty. Table B15 in the US Department of Defense (DOD) Quality Systems Manual (QSM 5.1) is a good reference point to evaluate the quality of the LC-MS/MS analytical parts of the TOP assay (DOD/DOE, 2017).

3.4.2 UNDERSTANDING AND MONITORING FOR REACTION COMPLETENESS

Because a fixed amount of persulfate and KOH is added to the sample, it is quite possible that due to the presence of other organics in the sample, or high precursor levels, the reaction does not go to completion. There is currently no procedure for monitoring pH through the course of the reaction and adding more reactant as necessary, and this is likely too impractical to be implemented for large-scale TOP analysis. Careful management of sample size, adherence to standardized reaction conditions, and monitoring for reaction completeness thus become important. Internal work at SGS AXYS is currently monitoring the use of $^{13}C_8$-PFOSA as a measure of reaction completion, and potentially formation as well, as it converts to $^{13}C_8$-PFOA upon reaction, which can be monitored. However, the relative volatility of this neutral compound, especially in comparison with ionic precursors, makes it an imperfect choice. At the time of writing, there are no other suitable standards that could be used for this purpose that will not interfere with the analysis.

Monitoring of certain precursors may also be important in understanding data quality. 6:2 FTS, for example, is a known intermediate in the decomposition of 6:2 telomer-based products (Harding-Marjanovic et al., 2015) and, as a strong acid, can consume hydroxyl ions as well. In a TOP sample where the reaction has gone to completion, there should be no 6:2 FTS present. While monitoring for a longer, more exhaustive list of precursors can be time consuming, select measurements, such as the telomer sulfonates, can act as a good indicator for reaction completion.

3.4.3 PRECURSOR VOLATILITY

Methanol, the most commonly used extraction solvent in PFAS analysis, and the solvent for analytical standards as well, consumes the hydroxyl radical. Therefore, removal of all methanol is critical to the performance of the assay. This issue imposes procedures that take extracts, especially solids, to dryness prior to TOP reagent addition, and hence can cause significant losses of the more volatile precursors, especially neutral, relatively hydrophobic precursors, such as fluorotelomer alcohols. This effect is chain length dependent, with the PFAS of eight or greater chain lengths less affected than C4 or C6.

3.4.4 ULTRA-SHORT-CHAIN PRODUCTS

It is likely that there is significant formation of C2 and C3 PFCAs from the TOP reaction, as evidenced by the 25% "missing mass" in the reaction of 6:2 FTS.

While there are analytical approaches to measure these ultrashort acids (Barzen-Hanson and Field, 2015), they are not routinely analyzed in most laboratories, and there are analytical challenges, including the potential of high laboratory background levels of C2 (trifluoroacetic acid), a commonly used additive in LC-MS/MS.

3.4.5 COMPLETENESS OF EXTRACTION

Most extraction procedures developed for PFAS were targeted at and optimized for the anionic perfluorinated acids. While this is not an issue in TOP analysis of aqueous samples, solid extraction procedures optimized for anionic PFAS may show lower recoveries of cationic precursors, while not necessarily affecting the neutral precursors. Standardization of TOP methods needs to take this issue into account.

3.5 SUMMARY

The use of TOP is still in its beginning stages. Given that the overwhelming mass of PFAS currently going into the receiving environment are not PFCAs or PFSAs monitored in routine analysis, it becomes increasingly important to elucidate the PFAS "dark matter," and specifically flag the sources that can act as ongoing reservoirs of persistent PFAS over time. The applications detailed in Section 3.3.2 are examples of where the use of the TOP assay, especially in conjunction with relevant precursor measurement, can be useful. Applications in ambient monitoring are critical in identifying hitherto unknown sources, and potentially in understanding the historical contribution from known sources. Applications in sites impacted by industrial or AFFF product use can provide a better understanding of plume movement, source attribution, and scale of problem and provide much more comprehensive data than a limited targeted analysis. The use of TOP or total fluorine measurement in remediation applications is highly recommended as it provides insight into capacity, efficacy of treatment, maintenance schedules, and more. The use of TOP in product testing provides information on the soluble and hence potentially bioavailable PFAS loads. Given that WWTPs are some of the largest sources and concentrators of PFAS in the environment, tracking precursor oxidation through the treatment process and assessing TOP potential in the effluent and biosolids can help elucidate the PFAS mass balance and benchmark better treatment strategies. Given the known capacity limits on GAC in drinking water systems, assessing performance using TOP measurements can help us understand capacity better.

The use of TOP, however, will greatly benefit from standardization of method and reaction conditions, the use of standards to track method performance, and a clear understanding that TOP provides an estimate of extractable precursor potential, not a comprehensive mass balance. While some of the quality issues have been addressed in this chapter, further study may be required. It is recommended that end users and regulators drive this process of standardization and quality by appropriate specification of TOP-specific quality standards in quality assurance project plans (QAPPs) and other project planning procedures. If these issues are addressed, the TOP assay can provide a valuable set of data that provide more insight than the measurement of a short list of PFAS targets.

REFERENCES

Barzen-Hanson, K.A., Field, J.A. 2015. Discovery and implications of C2 and C3 perfluoroalkyl sulfonates in aqueous film-forming foams and groundwater. *Environ. Sci. Technol. Lett.* 2, 95–99. https://doi.org/10.1021/acs.estlett.5b00049.

Barzen-Hanson, K.A., Roberts, S.C., Choyke, S., Oetjen, K., McAlees, A., Riddell, N., McCrindle, R., Ferguson, P.L., Higgins, C.P., Field, J.A. 2017. Discovery of 40 classes of per- and polyfluoroalkyl substances in historical aqueous film-forming foams (AFFFs) and AFFF-impacted groundwater. *Environ. Sci. Technol.* 51, 2047–2057. https://doi.org/10.1021/acs.est.6b05843.

Buck, R.C., Franklin, J., Berger, U., Conder, J.M., Cousins, I.T., de Voogt, P., Jensen, A.A., Kannan, K., Mabury, S.A., van Leeuwen, S.P. 2011. Perfluoroalkyl and polyfluoroalkyl substances in the environment: Terminology, classification, and origins. *Integr. Environ. Assess. Manag.* 7, 513–541. https://doi.org/10.1002/ieam.258.

Dauchy, X., Boiteux, V., Bach, C., Colin, A., Hemard, J., Rosin, C., Munoz, J.-F. 2017. Mass flows and fate of per- and polyfluoroalkyl substances (PFASs) in the wastewater treatment plant of a fluorochemical manufacturing facility. *Sci. Total Environ.* 576, 549–558. https://doi.org/10.1016/j.scitotenv.2016.10.130.

Department of Environment and Heritage Protection. 2016. Environmental management of firefighting foam. https://www.ehp.qld.gov.au/assets/documents/regulation/firefighting-foam-policy.pdf (accessed 21 November 2017).

DOD/DOE (Department of Defense/Department of Energy). 2017. Department of Defense (DoD) Department of Energy (DOE) consolidated quality systems manual (QSM) for environmental laboratories. http://www.denix.osd.mil/edqw/documents/documents/qsm-version-5-1-final/ (accessed 21 November 2017).

Eberle, D., Ball, R., Boving, T.B. 2017. Impact of ISCO treatment on PFAA co-contaminants at a former fire training area. *Environ. Sci. Technol.* 51, 5127–5136. https://doi.org/10.1021/acs.est.6b06591.

Harding-Marjanovic, K.C., Houtz, E.F., Yi, S., Field, J.A., Sedlak, D.L., Alvarez-Cohen, L. 2015. Aerobic biotransformation of fluorotelomer thioether amido sulfonate (Lodyne) in AFFF-amended microcosms. *Environ. Sci. Technol.* 49, 7666–7674. https://doi.org/10.1021/acs.est.5b01219.

Houtz, E.F., Higgins, C.P., Field, J.A., Sedlak, D.L. 2013. Persistence of perfluoroalkyl acid precursors in AFFF-impacted groundwater and soil. *Environ. Sci. Technol.* 47, 8187–8195. https://doi.org/10.1021/es4018877.

Houtz, E.F., Sedlak, D.L. 2012. Oxidative conversion as a means of detecting precursors to perfluoroalkyl acids in urban runoff. *Environ. Sci. Technol.* 46, 9342–9349. https://doi.org/10.1021/es302274g.

Houtz, E.F., Sutton, R., Park, J.-S., Sedlak, M. 2016. Poly- and perfluoroalkyl substances in wastewater: Significance of unknown precursors, manufacturing shifts, and likely AFFF impacts. *Water Res.* 95, 142–149. https://doi.org/10.1016/j.watres.2016.02.055.

KEMI: Swedish Chemicals Agency. 2015. Occurrence and use of highly fluorinated substances and alternatives. KEMI: Swedish Chemicals Agency, Stockholm. https://www.kemi.se/global/rapporter/2015/report-7-15-occurrence-and-use-of-highly-fluorinated-substances-and-alternatives.pdf (accessed 21 November 2017).

Martin, J.W., Asher, B.J., Beesoon, S., Benskin, J.P., Ross, M.S. 2010. PFOS or PreFOS? Are perfluorooctane sulfonate precursors (PreFOS) important determinants of human and environmental perfluorooctane sulfonate (PFOS) exposure? *J. Environ. Monit.* 12, 1979–2004. https://doi.org/10.1039/C0EM00295J.

Newton, S., McMahen, R., Stoeckel, J.A., Chislock, M., Lindstrom, A., Strynar, M. 2017. Novel polyfluorinated compounds identified using high resolution mass spectrometry downstream of manufacturing facilities near Decatur, Alabama. *Environ. Sci. Technol.* https://doi.org/10.1021/acs.est.6b05330.

Olsen, G.W., Burris, J.M., Ehresman, D.J., Froehlich, J.W., Seacat, A.M., Butenhoff, J.L., Zobel, L.R. 2007. Half-life of serum elimination of perfluorooctanesulfonate, perfluorohexanesulfonate, and perfluorooctanoate in retired fluorochemical production workers. *Environ. Health Perspect.* 115, 1298–1305. https://doi.org/10.1289/ehp.10009.

Paul, A.G., Jones, K.C., Sweetman, A.J. 2009. A first global production, emission, and environmental inventory for perfluorooctane sulfonate. *Environ. Sci. Technol.* 43, 386–392. https://doi.org/10.1021/es802216n.

PFAS National Environmental Management Plan. 2017. http://www.epa.vic.gov.au/PFAS_NMP (accessed 20 October 2017).

Place, B.J., Field, J.A. 2012. Identification of novel fluorochemicals in aqueous film-forming foams used by the US military. *Environ. Sci. Technol.* 46, 7120–7127. https://doi.org/10.1021/es301465n.

Rankin, K., Mabury, S.A., Jenkins, T.M., Washington, J.W. 2016. A North American and global survey of perfluoroalkyl substances in surface soils: Distribution patterns and mode of occurrence. *Chemosphere* 161, 333–341. https://doi.org/10.1016/j.chemosphere.2016.06.109.

Ritter, E.E., Dickinson, M.E., Harron, J.P., Lunderberg, D.M., DeYoung, P.A., Robel, A.E., Field, J.A., Peaslee, G.F. 2017. PIGE as a screening tool for per- and polyfluorinated substances in papers and textiles. *Nucl. Instrum. Methods Phys. Res. B* 407, 47–54. https://doi.org/10.1016/j.nimb.2017.05.052.

Robel, A.E., Marshall, K., Dickinson, M., Lunderberg, D., Butt, C., Peaslee, G., Stapleton, H.M., Field, J.A. 2017. Closing the mass balance on fluorine on papers and textiles. *Environ. Sci. Technol.* 51, 9022–9032. https://doi.org/10.1021/acs.est.7b02080.

Schaider, L.A., Balan, S.A., Blum, A., Andrews, D.Q., Strynar, M.J., Dickinson, M.E., Lunderberg, D.M., Lang, J.R., Peaslee, G.F. 2017. Fluorinated compounds in U.S. fast food packaging. *Environ. Sci. Technol. Lett.* 4, 105–111. https://doi.org/10.1021/acs.estlett.6b00435.

Strynar, M., Dagnino, S., McMahen, R., Liang, S., Lindstrom, A., Andersen, E., McMillan, L., Thurman, M., Ferrer, I., Ball, C. 2015. Identification of novel perfluoroalkyl ether carboxylic acids (PFECAs) and sulfonic acids (PFESAs) in natural waters using accurate mass time-of-flight mass spectrometry (TOFMS). *Environ. Sci. Technol.* 49, 11622–11630. https://doi.org/10.1021/acs.est.5b01215.

Wagner, A., Raue, B., Brauch, H.-J., Worch, E., Lange, F.T. 2013. Determination of adsorbable organic fluorine from aqueous environmental samples by adsorption to polystyrene-divinylbenzene based activated carbon and combustion ion chromatography. *J. Chromatogr. A* 1295, 82–89. https://doi.org/10.1016/j.chroma.2013.04.051.

Wang, Z., Cousins, I.T., Scheringer, M., Hungerbuehler, K. 2015. Hazard assessment of fluorinated alternatives to long-chain perfluoroalkyl acids (PFAAs) and their precursors: Status quo, ongoing challenges and possible solutions. *Environ. Int.* 75, 172–179. https://doi.org/10.1016/j.envint.2014.11.013.

Wang, Z., DeWitt, J.C., Higgins, C.P., Cousins, I.T. 2017. A never-ending story of per- and polyfluoroalkyl substances (PFASs)? *Environ. Sci. Technol.* https://doi.org/10.1021/acs.est.6b04806.

Weber, A.K., Barber, L.B., LeBlanc, D.R., Sunderland, E.M., Vecitis, C.D. 2017. Geochemical and hydrologic factors controlling subsurface transport of poly- and perfluoroalkyl substances, Cape Cod, Massachusetts. *Environ. Sci. Technol.* 51, 4269–4279. https://doi.org/10.1021/acs.est.6b05573.

Willach, S., Brauch, H.-J., Lange, F.T. 2016. Contribution of selected perfluoroalkyl and polyfluoroalkyl substances to the adsorbable organically bound fluorine in German rivers and in a highly contaminated groundwater. *Chemosphere* 145, 342–350. https://doi.org/10.1016/j.chemosphere.2015.11.113.

Xiao, F. 2017. Emerging poly- and perfluoroalkyl substances in the aquatic environment: A review of current literature. *Water Res.* 124, 482–495. https://doi.org/10.1016/j.watres.2017.07.024.

Xiao, F., Simcik, M.F., Halbach, T.R., Gulliver, J.S. 2015. Perfluorooctane sulfonate (PFOS) and perfluorooctanoate (PFOA) in soils and groundwater of a U.S. metropolitan area: Migration and implications for human exposure. *Water Res.* 72, 64–74. https://doi.org/10.1016/j.watres.2014.09.052.

Section II

Regulations

4 Managing Risk from Perfluorinated Compounds in Drinking Water

Steve Via

CONTENTS

4.1　Introduction 83
4.2　SDWA Risk Management Tools 84
4.3　Selecting a Contaminant and Deciding to Regulate 85
4.4　Health Advisories for PFOS and PFOA 85
4.5　Perfluorinated Compounds and the SDWA Risk Management Paradigm 87
References 89

4.1 INTRODUCTION

In the United States, the Safe Drinking Water Act (SDWA) provides the overarching federal construct for ensuring drinking water quality. The SDWA provides a process for identifying contaminants of potential health concern and mechanisms to manage exposure in drinking water (USEPA, 2009b). The standards developed under the SDWA are also a point of reference for other statutes, including quality standards for bottled water, as well as cleanup goals for contaminated sites.

The SDWA applies to public water systems (PWSs), which means "a system for the provision to the public of water for human consumption through pipes or other constructed conveyances, if the system has at least fifteen service connections or regularly serves at least twenty-five individuals" (42 US Code § 300f(4)). PWSs can take the form of communities, individual businesses, schools, and community facilities. The SDWA does not apply to water supplies (e.g., wells) for individual homes or smaller collections of homes.

The SDWA is a delegated federal program. That is, individual states take responsibility for implementing drinking water regulations that reflect federal regulatory requirements or are more stringent than the federal standards. At present, the US Environmental Protection Agency (USEPA) has responsibility for direct implementation, through USEPA regional offices, in the District of Columbia, Wyoming, the US territories, and most of the Native American nations.

4.2 SDWA RISK MANAGEMENT TOOLS

Under the SDWA, there are three primary risk management options: primary standards, which may be maximum contaminant levels (MCLs) or treatment technique requirements; health advisories; and secondary maximum contaminant levels (SMCLs). Only primary standards are enforceable under federal law, though some states treat SMCLs as enforceable standards.

There are two types of primary standards: MCLs and treatment techniques. MCLs are the maximum permissible level of a contaminant in drinking water that is delivered to users of a PWS. In setting the MCL, the USEPA administrator is required to consider what level is economically and technologically feasible. When an MCL is not feasible, for instance, if there is not an appropriate analytical method for the contaminant to allow adequate quality control, the administrator can require PWSs to apply criteria or procedures that reduce the contaminant of concern. An example is disinfection. All water systems relying on surface water must provide sufficient disinfection to kill *Giardia*, bacteria, and viruses, and most systems must use filtration. While it is possible to monitor control parameters for both disinfection and filtration, monitoring for hundreds of waterborne pathogens is not feasible, particularly in a manner that allows adequate process control.

In setting primary standards, the Agency must meet a number of statutory obligations. Those obligations include basing decisions on the best available science, being transparent with the public about its analysis, and considering both the quantifiable and nonquantifiable benefits and costs of potential standards. While setting an MCL does take cost and feasibility into account, the statute also calls for the MCL to be as close to a level that the USEPA believes will result in no known or anticipated adverse effects on public health. This level is known as the maximum contaminant level goal (MCLG). The MCLG, as a matter of policy, is set at a default value of zero for carcinogens. It is also worth noting that the Agency has a duty to consider sensitive subpopulations when determining the MCLG and calculating benefits associated with a regulation. What subgroup within the US population that is sensitive will vary with contaminant, but children are frequently viewed as a sensitive population, particularly with respect to reproductive and developmental health effects. The feasibility of treatment is an essential element of setting an MCL; the SDWA explicitly identifies granular activated carbon as a feasible best available technology for the control of synthetic organic chemicals.

While the SDWA outlines the requirements for setting primary standards, there is little direction given to the establishment of health advisories, other than to note that health advisories are not regulations under 42 US Code § 300g–1(b)(1)(f). In setting a health advisory level, USEPA's analytical process seeks to identify a concentration of a contaminant in water that is likely to be without adverse effects on health and aesthetics for the period it is derived (USEPA 2012). The underlying USEPA risk assessment guidance, policy, and basic steps used to calculate the MCLG and a health advisory are the same. Both MCLGs and health advisories are based on a sensitive subpopulation and particular health effects. MCLGs are a single benchmark value, but advisories can include a range of values representing both cancer and noncancer end points to facilitate situation-specific risk management decisions.

Managing Risk from Perfluorinated Compounds in Drinking Water

While there are currently 97 drinking water MCLGs with associated MCLs and treatment techniques, there are an additional 100 contaminants for which there are health advisories, as well as a small number of chemical contaminants for which the USEPA has selected a specific reference dose and calculated a drinking water equivalent level. In calculating MCLGs and health advisories, USEPA's process considers

1. Exposure from other sources
2. Sources of uncertainty in the underlying health effects data
3. Differences as a function of human life stage

Neither MCLGs nor health advisories reflect practical implementation issues, like the availability of analytical methods at the level of health concern, opportunities to reduce exposure, or the cost of drinking water treatment.

4.3 SELECTING A CONTAMINANT AND DECIDING TO REGULATE

The SDWA describes three clear questions for the Administrator to ask and answer when determining that a contaminant should be regulated:

1. Does the contaminant have an adverse health effect on people?
2. Is the contaminant likely to occur in drinking water at a frequency and concentration that make it a public health concern?
3. Is there a meaningful opportunity for health risk reduction by managing exposure through drinking water?

The act also includes a four-step process to both ensure that these questions are answered and meet the statutory obligations described previously. Those steps are

1. Prepare a list of contaminants, known as the Contaminant Candidate List (CCL), that warrant evaluation against these criteria every 5 years.
2. Make a determination for at least five contaminants every 5 years of whether regulation should be pursued.
3. Collect national occurrence data to support regulatory determinations.
4. Propose a standard within 18 months of making a positive regulatory determination and finalize that standard within 2 years of its proposal.

The USEPA has the authority to both issue emergency rules and develop drinking water standards outside this regulatory process, but the underlying expectations of the act are intrinsic to the Agency's deliberative process.

4.4 HEALTH ADVISORIES FOR PFOS AND PFOA

In January 2009, the USEPA published provisional drinking water health advisories for perfluorooctanoic acid (PFOA) and perfluorooctane sulfonic acid (PFOS) as individual contaminants at 0.4 and 0.2 µg/L, respectively (USEPA 2009a). After further evaluation, 7 years later, in May 2016, the Agency published a final drinking water health advisory level for both PFOA and PFOS individually or as a combined sum at a substantially lower value, 0.07 µg/L (USEPA 2016b, 2016c). There is not a body of epidemiology data sufficient to provide a basis for a health advisory; both the provisional and final health advisories are based on rodent studies (USEPA 2016c).

In both 2009 and 2016, the USEPA framed the advisory levels as protecting sensitive subpopulations from both cancer and noncancer health effects from short-term (i.e., weeks to months) and chronic (i.e., years) exposure. In both instances, the USEPA highlighted short-term exposure to elevated levels of PFOA and PFOS as constituting a risk warranting management. Sensitive populations of concern in the final health advisory include fetuses during pregnancy and breastfeeding infants.

While health advisories are not enforceable regulatory standards, risk management actions were mandated at individual water systems because of both the provisional and final health advisories. The final health advisory was accompanied by an explicit list of recommendations for action by individuals, water systems, and states. The USEPA recommended that consumers not drink or prepare food with water above the health advisory levels. The Agency recommended that water systems that observe elevated PFOA and/or PFOS:

1. Conduct additional sampling to assess the level and scope and localize the source of contamination
2. Notify their primacy agency
3. Provide consumers with information on levels, hazard, and risk reduction steps (e.g., use an alternative water source)
4. Take steps to reduce exposure (e.g., close wells, blend water sources to lower PFOA or PFOS concentrations, or initiate treatment, such as install granular activated carbon) (USEPA 2016d)

Individual states are utilizing the PFOA-PFOS health advisory in different ways. Since the advisory is not a federal regulation, states utilize it under state-specific authorities. State agencies with oversight of drinking water systems can require systems to follow the recommended actions. Some states, like Pennsylvania, utilize other authorities to evaluate private wells. In Pennsylvania, testing of private wells near sites known to be contaminated with PFOA or PFOS is recommended (PDEP 2017). In California, the Office of Environmental Health Hazard Assessment (OEHHA) has indicated it would list PFOA and PFOS as known to cause reproductive toxicity. Under state authorities that will require both notification of elected officials and the public about the presence of PFOA and PFOS, potentially at levels below the health advisories (California OEHHA 2016). In New Hampshire, the state set ambient groundwater quality standards at the advisory level based on the final health advisory (NHDES 2016). Both Maine and Connecticut have set guideline levels for drinking water (CDPH 2016; MCDC 2016) based on the USEPA health advisory; the guideline in Maine is for PFOA and PFOS, while in Connecticut the guideline is cumulative for PFOS, PFOA, perfluorononanoic acid (PFNA), perfluorohexanesulfonic acid (PFHxS), and perfluoroheptanoic acid (PFHpA). Minnesota also set drinking water guideline values for PFOA and PFOS, but developed its own values—35 ng/L PFOA and 27 ng/L (PFOS) (MDH 2017).

4.5 PERFLUORINATED COMPOUNDS AND THE SDWA RISK MANAGEMENT PARADIGM

Selected perfluorinated compounds (PFOS and PFOA) were included in the third CCL in 2009, setting the stage for additional data collection and potentially regulatory action (USEPA 2016b). In 2012, the USEPA promulgated the Third Unregulated Contaminant Monitoring Rule (UCMR3). The UCMR3 included monitoring for PFOS, PFOA, and four other perfluorinated compounds: PFNA, PFHxS, PFHpA, and perfluorobutanesulfonic acid (PFBS).

To ensure that decisions are made on nationally representative occurrence information, data are collected through a rulemaking where participation is compulsory. The resulting dataset reflects the following:

1. A census of PWSs that serve more than 10,000 persons
2. A statistically derived sample representing systems serving less than 10,000 persons
3. A sample representative of current finished water concentrations
4. A consistent analytical method used across all samples
5. Use of a high-quality laboratory analysis with rigorous quality control
6. Sampling to accomplish a snapshot of occurrence across times and seasons

In January 2017, the USEPA released the final compilation of observed occurrence from UCMR3 (USEPA 2017). There were 36,972 samples analyzed using USEPA Method 537. The method reporting level ranged from 0.01 to 0.09 μg/L, depending on the poly- and perfluorinated alkyl substances (PFAS), with the lowest MRL being for PFHpA and the highest being for PFBS. The MRLs for PFOS and PFOA were 0.04 and 0.02 μg/L, respectively (Table 4.1).

The USEPA had an opportunity to utilize available information to make a determination to regulate PFAS compounds in its January 2016 regulatory determinations

TABLE 4.1
Occurrence of PFASs Monitored in UCMR3

PFAS	MRL (μg/L)	Number of PWSs with Results ≥MRL
PFOS	0.04	95
PFOA	0.02	117
PFNA	0.02	14
PFHxS	0.03	55
PFHpA	0.01	86
PFBS	0.09	8

notice, but it did not. This was presumably because finalization of the Agency's health effects analysis and the UCMR3 data collection process was not complete at the time (USEPA 2016a).

Federal action to create a regulatory standard (MCL) appears unlikely given the limited scope of contamination nationally and the state-specific regulatory actions underway. In UCMR3, 0.9% of systems monitoring observed PFOS above 70 ng/L and only 0.3% of systems found PFOA above 70 ng/L. These percentages could increase slightly if one were to sum PFOA and PFOS or monitor and sum low-level occurrences of a large suite of compounds, but the occurrence of these well-documented compounds is limited. Analysis presented by Hartz (2017) illustrates more occurrence at very low levels (MRL of 2.5 ng/L). This data includes detections of PFOA and PFOS, as well as other UCMR-monitored perfluorinated compounds. In the Eurofins drinking water dataset, 20.5% of samples are positive for PFOS and 23.5% of samples are positive for PFOA (Hartz, 2017).

There are clearly countervailing views on the need for a national standard. A federal standard would provide national consistency and ensure that the regulatory process meets expectations for use of the best available science and consideration of cost–benefit analysis. Such assurances do not exist if, for example, state legislatures were to set state-specific standards. Absent federal regulation, under the SDWA individual states may develop and enforce their own standards and regulate drinking water quality more stringently than the USEPA. Historically, a limited number of states have pursued setting a small number of MCLs for chemicals other than those included in federal regulation. Currently, the combination of USEPA health advisories, the uncertainty in the human toxicology of the entire suite of PFAS compounds that can be found in the environment, and public concern that an anthropogenic compound is so widespread in the environment has made state-level decision making more difficult. Furthermore, having different state regulations complicates the regulatory burden for water systems. Where there are multiple states with different guidelines or standards of care, water systems are left to explain why they are adhering to one set of treatment objectives when the public is aware of states with different expectations. To enterprises that work in different states, such as the Department of Defense, there are also the practical considerations of understanding and planning for different sampling, analysis, and monitoring requirements, which adds to the complexity of the effective drinking water program management. For the foreseeable future, it is expected that the health advisory will continue in place as is. Doing such creates a challenging situation for purveyors of drinking water and state regulators. Although there is not a federal regulatory "requirement," systems where PFAS compounds are found must act in the absence of a clear standard of care and, in doing so, confront decisions with significant cost implications associated with the available treatment responses in order to maintain public confidence. From the public's perspective, equal concern is expressed if either an exceedance of an MCL, a health advisory, or a state-determined guidance value is found. Realizing this, purveyors are compelled to treat the health advisory as if it were regulatory (although it is explicitly stated that they are not regulatory in the health advisory definition), thereby creating a de facto regulatory standard.

REFERENCES

California OEHHA (Office of Environmental Health Hazard Assessment). 2016. Notice of intent to list perfluorooctanoic acid (PFOA) and perfluorooctane sulfonate (PFOS). https://oehha.ca.gov/proposition-65/crnr/notice-intent-list-perfluorooctanoic-acid-pfoa-and-perfluorooctane-sulfonate (accessed 22 November 2017).

CDPH (Connecticut Department of Public Health). 2016. Drinking water action level for perfluorinated alkyl substances (PFAS). http://www.ct.gov/dph/lib/dph/environmental_health/eoha/groundwater_well_contamination/052317_pfas_action_level_dec_2016.pdf (accessed 22 November 2017).

Hartz, M. 2017. PFAS monitoring in a post health advisory world—What should we be doing? AWWA Water Works Association New York Section. https://nysawwa.org/docs/presentations/2017/FINAL-PFAS%20Monitoring%20in%20Post%20health%20Advisory%20World-What%20Should%20We%20Be%20Doing-2017.pdf (accessed 22 November 2017).

MCDC (Maine Centers for Disease Control). 2016. Maximum exposure guidelines (MEGs) for drinking water. http://www.state.me.us/dhhs/mecdc/environmental-health/eohp/wells/documents/megtable2016.pdf (accessed 22 November 2017).

MDH (Minnesota Department of Health). 2017. Perfluorochemicals (PFCs) and health. http://www.health.state.mn.us/divs/eh/hazardous/topics/pfcshealth.html#levels (accessed 22 November 2017).

NHDES (New Hampshire Department of Environmental Services). 2016. Press release: NHDES establishes ambient groundwater quality standard for perfluorooctanoic acid (PFOA) and perfluorooctane sulfonate (PFOS). https://www.des.nh.gov/media/pr/2016/20160531-pfoa-standard.htm (accessed 22 November 2017).

PDEP (Pennsylvania Department of Environmental Protection). 2017. Private water wells. http://www.dep.pa.gov/Citizens/My-Water/PrivateWells/Pages/default.aspx#.V1cuU9krJMw (accessed 22 November 2017).

USEPA (U.S. Environmental Protection Agency). 2009a. Provisional health advisories for perfluorooctanoic acid (PFOA) and perfluorooctane sulfonate (PFOS). https://www.epa.gov/sites/production/files/2015-09/documents/pfoa-pfos-provisional.pdf (accessed 22 November 2017).

USEPA (U.S. Environmental Protection Agency). 2009b. Drinking water Contaminant Candidate List 3—Final. *Federal Register* 74(194):51850–51862. https://www.federalregister.gov/documents/2016/01/04/2015-32760/announcement-of-final-regulatory-determinations-for-contaminants-on-the-third-drinking-water (accessed 22 November 2017).

USEPA (U.S. Environmental Protection Agency). 2012. 2012 edition of the drinking water standards and health advisories. EPA 822-S-12-001. https://www.epa.gov/sites/production/files/2015-09/documents/dwstandards2012.pdf (accessed 22 November 2017).

U.S. Environmental Protection Agency. 2016a. Announcement of final regulatory determinations for contaminants on the Third Drinking Water Contaminant Candidate List. *Federal Register* 81(1):13–19. https://www.gpo.gov/fdsys/pkg/FR-2016-01-04/html/2015-32760.htm (accessed 22 November 2017).

USEPA (U.S. Environmental Protection Agency). 2016b. Lifetime health advisories and health effects support documents for perfluorooctanoic acid and perfluorooctane sulfonate. *Federal Register* 81(101):33250–33251. https://www.federalregister.gov/documents/2016/05/25/2016-12361/lifetime-health-advisories-and-health-effects-support-documents-for-perfluorooctanoic-acid-and (accessed 22 November 2017).

USEPA (U.S. Environmental Protection Agency). 2016c. Drinking water health advisory for perfluorooctanoic acid (PFOA). EPA 822-R-16-005. https://www.epa.gov/sites/production/files/2016-05/documents/pfoa_health_advisory_final_508.pdf (accessed 22 November 2017).

USEPA (U.S. Environmental Protection Agency). 2016d. Factsheet PFOA & PFOS drinking water health advisories. EPA 800-F-16-003. https://www.epa.gov/sites/production/files/2016-06/documents/drinkingwaterhealthadvisories_pfoa_pfos_updated_5.31.16.pdf (accessed 22 November 2017).

USEPA (U.S. Environmental Protection Agency). 2017. The Third Unregulated Contaminant Monitoring Rule (UCMR 3): Data summary. January. https://www.epa.gov/sites/production/files/2017-02/documents/ucmr3-data-summary-january-2017.pdf (accessed 22 November 2017).

5 Emerging Contaminant Monitoring as a Host Nation Guest
An Environmental Health Professional's Perspective*

Andrew J. Wagner

CONTENTS

5.1 Introduction 91
    5.1.1 Background on US Military and Neighboring Host Nation Communities 92
    5.1.2 Background on Drinking Water Sanitary Surveys: Role of a USAF Environmental Consultant 93
5.2 Common Sanitary Survey Trends 94
5.3 Working Groups and Their Stakeholders 95
5.4 Source Water Protection and Root Cause Analysis 96
5.5 Source Water Assessment Protection Programs 97
5.6 Avoiding Policy Gaps 98
5.7 Finding Positive Ways to Connect 99
5.8 Conclusions 100
References 100

5.1 INTRODUCTION

Past regulatory provisions created by the 1972 Clean Water Act (CWA) and the 1974 Safe Drinking Water Act (SDWA) have been very effective in improving water quality in the United States with respect to point source pollution (obvious pollution sources directly affecting large bodies of water). Under proper management, these efforts focus on the well-regulated 91 SDWA contaminants that have well-established and defined human health risks. Many developed counties have regulatory standards

---

* Nothing in this publication is intended to imply federal, Department of Defense, or Department of the Air Force endorsement. The Privacy Act of 1974, 5 USC 522 (a), and/or the Health Insurance Portability and Accountability Act (PL 104-191), 10 USC 1102, and its various implementing regulations protect the information in this package. Unauthorized release, use, or failure to maintain confidentiality subjects you to appropriate sanctions.

similar to, if not mirroring, standards used or developed by the United States. A new challenge we face today is that pollution has evolved from an easily traced point source for these 91 contaminants to a non–point source (not obvious pollution source meandering from many smaller tributaries to a body of water) of thousands of relatively unknown micropollutants and emerging contaminants (ECs) (Fishman 2012). With the advancement of technology, we can detect many of these new pollutants down to the parts per trillion or nanogram per liter level, but in recent years, the understanding of health effects of these new contaminants at these extremely minute concentrations has lagged behind (Salina 2010; Fishman 2012; Prud'homme 2012). With regards to poly- and perfluorinated alkyl substances (PFAS), the US Environmental Protection Agency (USEPA) has taken steps to bridge this knowledge gap and in 2016 provided a lifetime health advisory for PFAS at a concentration level of 70 parts per trillion (USEPA 2016). It is imperative to remember that other nations are likely paying attention to these changes and new "standards" and will take necessary actions to develop their policies to protect their people and environment.

This chapter develops a case for how to address ECs like PFAS and other environmental health concerns in nations where US partners may have various levels of influence or activity via the military, nongovernmental organizations (NGOs), or even in business. The chapter presents a different way of leveraging already established policies and practices in a proactive manner.

To simplify the context, these foreign or non-US nations that "host" organizations with US interests will be addressed as "host nations" (HNs). Ideas presented are based on observations, experiences, and lessons learned from a US Air Force (USAF) environmental health consultant working with other sister service military professionals, overseas partners, and various international and HN professionals in Japan, South Korea, and across the Pacific.

The consultant's role in this case was primarily to support preventive medicine professionals in protecting the health of US Armed Forces and to advise leadership and regional regulatory authorities (i.e., drinking water primacy agencies) on various environmental health concerns. Although the focus of this book is on PFAS, on occasion this chapter discusses other related environmental health challenges to help illustrate critical points. The intended purpose for this chapter is to give insight to others outside the military, including NGOs and US businesses, who may have goals to either help improve water quality or simply establish good environmental stewardship practices while doing business as a guest of an HN.

5.1.1 Background on US Military and Neighboring Host Nation Communities

In order to gain proper perspective, it is necessary to understand a military environmental health consultant's scope of work in the community for which they are assigned to advise. Military installations overseas are essentially industrialized towns ranging from a few hundred military members to tens of thousands of military, civilian workers, and their families living together as a community. These military communities are guests to an HN, and how they interact with their neighboring HN communities is critical to international relations and the overall mission for that area. There is a level of trust that must be established between these local and "outsider" military communities. To be effective, both of them must trust that the military in place will help protect everyone's overall best interests, but not at the perceived expense of the environment or local public health.

The stress of an overseas environment, along with increased emphasis on sustaining a large mission, can cause focus on environmental compliance and safety to lapse for various reasons, such as a high operations tempo, frequent changes in personnel, and challenges in communicating with HN workers, subject matter experts, and government regulators. In addition, there may be political sensitivities between local HN community governments and larger centralized HN government agencies that do not always agree.

Adding another layer of complexity is attempting to understand various negotiation styles and tactics used by different cultures. Japan and US relationships over the past 70 years provide an excellent example of striving for cultural balance. Cohen (1992) noted that through "supplicant" negotiation tactics, Japan has influenced the United States to locate most of its forces on the very small, yet strategically relevant, Japanese island Okinawa. This has created a strained relationship between mainland Japan and Okinawa as over the last six decades Okinawa's small prefecture has supported two-thirds of almost 50,000 US forces. Contentious relationships like these have the potential to help magnify environmental miscues or accidents by the US military and put them in the local spotlight when neighboring communities to our US installations feel they are bearing a larger share of the burden.

5.1.2 Background on Drinking Water Sanitary Surveys: Role of a USAF Environmental Consultant

The role of a USAF environmental health consultant in an HN is similar to what a local regulatory or public health authority in the United States may provide to a community or even an NGO assisting a foreign nation community. That role is to provide information and advice to key individuals in a region on how to maintain clean and reliable drinking water in these communities. This is primarily accomplished through observations made during assessments, such as drinking water sanitary surveys. In the United States, these surveys are generally provided by the state EPA or delegated to a regional drinking water authority.

Based on the USEPA definition and those found in US Department of Defense (DOD) regulations, a sanitary survey can be described as an on-site tour and review of a complete water system from source to consumer. Its purpose is to evaluate the adequacy of drinking water sources, facilities, equipment, operation, and maintenance for producing and distributing safe drinking water. Conducting sanitary surveys on a routine basis is an important element in preventing contamination of drinking water supplies and is an opportunity for collaboration among water system stakeholders in order to prevent unnecessary risk. These surveys provide a working knowledge of the operation, maintenance, management, and technology of water systems to identify sanitary risks that may interrupt a protection system and adversely affect the ability of a water system to provide safe water (USEPA 1999; DOD 2007; KEGS 2012; JEGS 2016).

In overseas locations, the DOD criteria for sanitary surveys is based on USEPA criteria as well as requirements found in various other DOD policies and directives. These requirements have their foundation in the USEPA and in the Overseas Environmental Baseline Guidance Document (OEBGD) (DOD 2007). The OEBGD serves as a foundation to create agreements between HNs and the US military on various environmental requirements. The OEBGD is refined and made specific to regions by further defined requirements for specific areas in final governing standards, for instance, the Japan Environmental Governing Standards (JEGS 2016) and the Korea Environmental Governing Standards (KEGS 2012). These environmental standards (OEBGD, JEGS, and KEGS) require sanitary surveys every 3 years for potable water systems utilizing surface water sources and every 5 years for systems operating solely on groundwater as a source.

A sanitary survey consists of an extensive review of these eight elements (USEPA 1999):

1. Operator compliance with regulations
2. Source protection, physical components, and condition
3. Treatment
4. Finished water storage
5. Distribution
6. Pumps and pump facility and controls
7. Monitoring, reporting, and data verification
8. Water system management and operations

5.2 COMMON SANITARY SURVEY TRENDS

Some of the more common observations seen during sanitary surveys include infrastructure maintenance needs, hydraulic modeling update needs, unmet requirements for unidirectional flushing, and inconsistent compliance with pipeline disinfection standards. Each installation has its own concerns related to increased risk of water contamination due to various reasons, but a significant challenge is managing aging and deteriorating infrastructure. Observations and survey findings may be easily corrected with actions such as replacing missing storage tank ventilation screens (designed to keep bugs and dirt from finished water) and honoring maintenance requirements or recurring work program requirements. Other observations may require more emphasis on risk assessment, planning and management for cost-effectiveness, and complex coordination, such as bringing submerged wellheads above ground level, extensive water main upgrades, and coordinating and executing unidirectional flushing plans. Of significant importance to regulated contaminants and new ECs like PFAS is the need to develop and execute effective programs similar to source water assessment protection programs (SWAPPs) for these drinking water systems (AFRL 2013).

Sanitary surveys often uncover a wide variety of concerns, and most will have challenges with timely resolution if there are limited resources, poor coordination, and competing priorities. Many have the unfortunate destiny of becoming unresolved repeat observations for the next sanitary survey to address. Unfortunately, some concerns that are not adequately prioritized will only be addressed once they develop into a larger issue. Management of ECs like PFAS at HN locations have only recently become a priority due to increased public concern from the US military and HN communities.

5.3 WORKING GROUPS AND THEIR STAKEHOLDERS

To avoid concerns like these from turning into larger issues, the USAF mandates installations to have a drinking water working group to

> address all local drinking water issues involving compliance, risk reduction and continuous improvement. The drinking water working group will establish procedures for interfacing with regulators on matters such as enforcement actions, reporting of compliance/non-compliance, and any other local drinking water concerns that may arise. (USAF 2014)

These groups provide an opportunity to communicate with stakeholders and prioritize and document progress made toward solving problems that threaten to interrupt the "multiple barrier" protection of a drinking water system. This group, in the opinion of the author, if utilized effectively, could become the primary tool to facilitate action and start rectifying contamination trends. These types of working groups are a USAF "best practice" that other US entities should consider when working with HNs. They are most effective when leveraged appropriately with current guidance and consistent direction with strong leadership.

Many concerns revealed during sanitary surveys at overseas locations could be addressed and managed with an effective drinking water working group. As seen by the author, the more effective groups employ a standardized agenda and metrics to steer efforts toward specific prioritized issues and concerns. HN involvement is critical, and their subject matter experts should be viewed as stakeholders and invited to attend or present at these working groups. If we share the same water sources, or have a potential impact on each other's water, their involvement would help bring credibility, diversity of ideas, and a different point of reference on concerns, issues, and potential solutions. The idea of a drinking water working group need not be limited to just US forces, but can be used by local community organizers and stakeholders. Work should be done to make these meetings effective and relevant to the concerns of a particular location.

In recent years, a number of preventive medicine leaders across the Pacific region from sister US military services (i.e., Air Force, Army, Navy, and Marines) have come together and chartered joint working groups. They often include HN subject matter experts and local government representation, which is key to the joint working groups' success. This type of representation increases the credibility and diversity of ideas. These groups were designed to address and solve environmental health concerns that may affect military members and their dependents in these regions. Developed in 2012 and formalized with a charter in 2013, the US Forces Japan Public Health Working Group (JPHWG) is a joint military working group set up to provide crisis management during public health emergencies (JPHWG 2013).

It has been instrumental in tackling various environmental health concerns in the area, including, but not limited to, Japan military housing mold and radon concerns (Slavin 2014); endemic disease updates, for example, leptospirosis (Nava 2014); dengue (Wingfield-Hayes 2014); aeolian dust public service announcements (Wagner 2016; USFJ 2017); and PFAS updates and risk communication (Kadena 2016). The Joint Public Epidemiological Action Center for Health (JPEACH) is a group that is similar to the JPHWG but specific to Okinawa, and it has a very active membership that includes HN public health experts. Representatives from these working groups often write policy, develop risk communication products, create subcommittees, and advise higher levels of military leadership on high-visibility environmental health issues. They often serve as the public presence for leadership that may not be engaged in day-to-day local issues. These teams fundamentally are used to gather more information to help formulate possible solutions in order to make better decisions. Maintaining current levels of HN subject matter experts in these groups is critical.

5.4 SOURCE WATER PROTECTION AND ROOT CAUSE ANALYSIS

When investigating an environmental health concern, such as PFAS water contamination, members of a drinking water working group or any environmental health team should pursue an understanding of the problem's root cause. A root cause, as defined by the American Society of Quality, is "a factor that causes a nonconformance and should be permanently eliminated through process improvement." Root cause analysis (RCA) is "a collective term that describes a wide range of approaches, tools and techniques used to uncover causes of problems" (ASQ 2017).

Originally developed by Taiichi Ohno, former executive vice president of the Toyota Motor Company, the "five-whys" method is a problem-solving technique that has been proven successful at identifying the root cause of a problem (Ohno 1988). This method is seen in various process improvement practices today, such as Kaizen and Six Sigma (Toyota Traditions 2006), and is the foundation of today's Lean Manufacturing (Devane 2004). To uncover the root of a problem using the five-whys RCA process, the question "why?" is essentially asked over and over again until the root cause of a problem is uncovered. It typically requires asking why at least five times to complete a RCA in complicated systems or problems (Ohno 1988).

A simple RCA for the case of an EC being found in HN drinking water is illustrated in Figure 5.1. It proposes that the root cause for EC contamination of drinking water is that current policies and procedures are based on already well-regulated contaminants and not on ECs. Expanding on this, policies are not adequate to address new ECs identified in a timely manner and procedures are not in place to monitor for high-risk ECs or to determine which ECs are high risk. Additionally, there is no set process to know when an EC will become enough of a threat to be assigned a health advisory, and polluters are not aware of how to handle past contamination events for ECs they may no longer use but could still be polluting water sources. Bottom line: Once a high-risk EC is established, predetermined policies and procedures on how to assess and take measures to prevent the ECs from entering drinking water sources are needed.

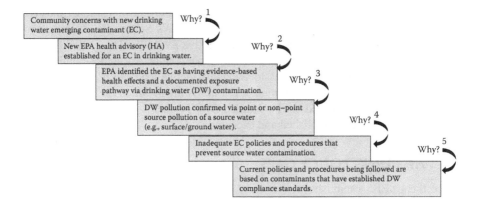

FIGURE 5.1 RCA (five whys) for PFAS exposure concern.

For locations in the United States, the 1996 amendment to the SDWA outlines steps for conducting source water assessments for public water systems and forming SWAPPs, previously known as wellhead protection plans prior to 1996 (USEPA 2017). At overseas military installations, this protection of water sources is implied in the OEBGD, but it is not clear whether it covers both regulated contaminants and unregulated contaminants, such as ECs. The OEBGD requires US forces to "protect all water supply aquifers (groundwater) and surface water sources from contamination by suitable placement and construction of wells, by suitable placing of the new intake (heading) to all water treatment facilities, by siting and maintaining septic systems and onsite treatment units, and by appropriate land use management on DoD installations" (DOD 2007). Similar versions of this statement are also found in final governing standards (KEGS 2012; JEGS 2016).

5.5 SOURCE WATER ASSESSMENT PROTECTION PROGRAMS

Organizations may look to the American Water Works Association (AWWA) and Water Research Foundation (WRF), two examples of organizations that have promoted the SDWA's concepts of SWAPPs. Products like AWWA's Source Water Protection (AWWA 2014) and WRF's Source Water Protection Vision and Roadmap (WRF 2012) have established evidence-based practices, provided practical implementation guidance, and promoted the need for a strategic vision on the use of SWAPPs to protect drinking water sources in the United States and around the world.

Established SWAPPs should not be limited to developed areas. NGOs that focus on clean water projects in developing countries could help add this strategy to the tool kit in areas they support. Protecting source waters against ECs may not be an initial priority in developing areas, but if SWAPPs are established and used to protect against contaminants that have acute health effects, such as biological or industrial point source pollution events, they may be viewed as valuable and later expanded to include additional contaminants.

An organization's leadership may consider directing preventive medicine and civil engineering partners to use guidance, such as AWWA's Source Water Protection guide (AWWA 2014), to plan and maintain effective SWAPPs. These SWAPPs can be installation or area specific and reach beyond local regulatory authority minimum requirements at home and overseas locations. Effective leadership and the use of established best practices may promote a cultural shift in which the use of a water source is not even considered until a detailed SWAPP that is flexible and capable of addressing new ECs is established and executed.

5.6 AVOIDING POLICY GAPS

Once the USEPA establishes the concept of a new or emerging contaminant and assigns a health advisory standard to it, the need to address that contaminant in policy with the HN is critical. While not a regulatory standard, these health advisories could result in certain expectations of water quality. Current US military and HN policy agreements are written to address specific contaminants of concern that have well-defined SDWA regulatory standards. More specifically, ECs were not identified in the latest version of the environmental policy agreement (DOD 2007).

If an HN's assumption is for the United States to uphold a nonregulatory standard, specifically the USEPA's health advisories for ECs, the 2007 OEBGD does not support this expectation. The concept of only addressing contaminants that have well-established US regulatory standards (e.g., for drinking water, a maximum contamination level) is, in the opinion of the author, contrary to what is provided in other guidance and directives, such as the Department of Defense Instruction (DODI) 6055.05, "Occupational and Environmental Health." This DODI expands on the need for exposure characterization to protect our military and DOD civilians (DOD 2008). The USAF captures its expectations and execution for this DODI in its Occupational Environmental Health Site Assessment (OEHSA) process, which is governed by various instructions and manuals. These directives were established to help fulfill the requirement of Presidential Review Directive 5, established by President Clinton in 1998 in response to lessons learned from the 1991 Gulf War and other recent military engagements, in order to protect military members, veterans, and their families by improving monitoring, exposure tracking, and risk communication (National Science and Technology Council 1998).

In the OEHSA process, preventive medicine professionals characterize the risk and potential for exposure to a contaminant of concern (e.g., PFAS or other EC) that may affect a population at risk (e.g., drinking water consumers) via a defined route of exposure (e.g., polluted drinking water source) (USAF 2010). Based on OEHSA alone, monitoring for ECs with an established USEPA health advisory would be logical if there was a population at risk and a possible exposure route. The established 2007 OEBGD alone does not account for ECs, and characterizing these exposures would not be within its scope.

This chapter proposes that the adoption and expansion of both the SDWA's SWAPP and the DOD's OEHSA initiatives to include ECs would allow them to complement each other and provide a viable direction to take. For example, if a US-owned company determines that it utilizes a material at one of its HN locations that currently has no regulatory requirements but is potentially hazardous to human health (e.g., EC with USEPA health advisory), and the material has the potential to be released into a drinking water source via storm runoff, what should it do? By utilizing these two concepts (OEHSA and SWAPP), that company would likely set up a process to periodically monitor for that contaminant in its storm water runoff before the water leaves the site. This would be a proactive approach to monitor and, if needed, eliminate that EC from entering the drinking water. In this example, there would be no waiting for the HN to realize that a potential environmental health hazard may exist. This type of approach would be a best practice for the US military, NGOs, corporations, and developed countries to incorporate into some of their processes or policies.

5.7 FINDING POSITIVE WAYS TO CONNECT

Environmental health and preventive medicine experts need to reach out to various types of HN partners in order to be effective. Creating positive avenues of communication with HN members with assigned roles in the visiting organization is a good place to start. It is also important to have an understanding of cultural differences. There may be anthropologic sensitivities that create potential barriers to establishing effective relationships. What may seem to be a minor thing may have significant implications and create barriers in communication even between HN members themselves. For instance, in Japan and Okinawa it can be important to some where an individual was raised and received their education. If an HN subject matter expert living in Okinawa was raised and attended school on mainland Japan, he might not always be received immediately by another HN subject matter expert that was raised and studied on Okinawa. It will take additional time and proactive interaction for them to gain trust and garner each other's professional respect. With this in mind, imagine the challenges that guests to this HN may face. Where there are communication barriers, a lack of trust and professional respect should be assumed. Therefore, breaking down communication and cultural barriers in imperative.

Guests would be wise to create opportunities to build relationships with HN partners. Finding reasons to meet and socialize prior to an environmental crisis can pay dividends. HN environmental health partners on Okinawa recall periods of time where local environmental professionals in the Okinawa Prefecture Government (OPG) were highly involved in social events with US military preventive medicine personnel, often participating together in league softball, foot races, and Friday afternoon socials. In 2013, US military preventive medicine experts embraced the retired director of the Okinawa Prefectural Institute of Public Health and Environment. He reached out in a desire to share his knowledge and present topics of mutual interest. His proactive leadership was integral in repairing a communication void between the US military and local HN professionals. Later, US military environmental health personnel built on his initial efforts by developing a seminar group with military environmental health experts from all branches of service and a number of HN members, including local OPG members. The interactions with OPG members and the relationships formed during the seminars were extraordinarily valuable during discussions on various topics, including ECs in drinking water.

5.8 CONCLUSIONS

Water crises are usually restricted to specific areas or regions. It is usually up to a nation, region, or community to solve its own problem (Fishman 2012). When US entities are invited as guests, the military expands its footprint onto another nation's soil, a corporation sets up a business, or an NGO makes a commitment to make lasting improvements to a region's drinking water, they all accept a certain level of responsibility. Water is an invaluable resource, and guests cannot risk contributing or causing an HN's water crisis. Instead, visiting organizations can look for ways to use these best practices established by laws, policies, and professional organizations to be better stewards of their resources.

REFERENCES

AFRL (Air Force Research Laboratory). WestPAC drinking water sanitary survey trends May 2013–December 2015. AFRL-SA-WP-CL-2016-0004, Consultative letter, 4 January 2016.

ASQ (American Society of Quality). Root cause analysis overview. 2017. http://asq.org/learn-about-quality/root-cause-analysis/overview/overview.html (accessed 23 August 2017).

AWWA (American Water Works Association). Source water protection. G300-14 utility management standard. Denver, CO: AWWA, 2014.

Cohen, R. *Negotiating Across Cultures: Communication Obstacles in International Diplomacy.* Washington, DC: U.S. Institute of Peace Press, 1992.

Devane, T. *Integrating Lean Six Sigma and High-Performance Organizations.* Hoboken, NJ: John Wiley & Sons, 2004.

DOD (Department of Defense). Overseas environmental baseline guidance document (OEBGD). DODI 4715.05-G. 1 May 2007. http://www.dtic.mil/dtic/tr/fulltext/u2/a637297.pdf (accessed 26 July 2017).

DOD (Department of Defense). Occupational and environmental health. DODI 6055.05. 11 November 2008. http://www.dtic.mil/whs/directives/corres/pdf/605505p.pdf (accessed 23 June 2017).

Fishman, C. *The Big Thirst: The Secret Life and Turbulent Future of Water.* New York: Free Press, 2012.

Kadena AB Public Service Announcement. Kadena web page. 24 January 2016. http://www.kadena.af.mil/portals/40/documents/AFD-160124-001.pdf (accessed 24 August 2017).

KEGS (Korean Environmental Governing Standards). Regulation 201-1. Headquarters, U.S. Forces Korea. June 2012.

JEGS (Japan Environmental Governing Standards). Headquarters, U.S. Forces Japan. April 2016.

JPHWG (U.S. Forces Japan Public Health Working Group). Charter, signed April 2013.

USFJ-PHWG milSuite web page. https://www.milsuite.mil/book/docs/DOC-112272 (accessed 22 August 2017).

National Science and Technology Council. Presidential Review Directive 5: A national obligation, planning for health preparedness for and readjustment of the military, veterans, and their families after future deployments. August 1998. https://fas.org/irp/offdocs/prd-5-report.htm (accessed 26 July 2017).

Nava, J. Commanders investigate source of illness. MCIPAC Public Affairs Office, Okinawa. Marines in the Asia-Pacific Region, October 2014. http://www.okinawa.marines.mil/News/News-Article-Display/Article/504888/commands-investigating-source-of-illness/ (accessed 24 August 2017).

Ohno, T. *Toyota Production System: Beyond Large-Scale Production*. New York: Productivity Press, 1988.

Prud'homme, A. The ripple effect: The fate of freshwater in the twenty-first century. New York: Scribner, 2012.

Salina, I. *Written in Water: Messages of Hope for Earth's Most Precious Resource*. Washington, DC: National Geographic Society, 2010.

Slavin, E. IG report finds health hazards in Japan base housing, questions Pentagon policy. *Stars and Stripes*, October 2014. https://www.stripes.com/news/pacific/japan/ig-report-finds-health-hazards-in-japan-base-housing-questions-pentagon-policy-1.306126 (accessed 24 August 2017).

Toyota Traditions. Ask 'why' five times about every matter. In newsletter published March 2006. http://www.toyota-global.com/company/toyota_traditions/quality/mar_apr_2006.html (accessed 26 March 2018).

USAF (U.S. Air Force). Occupational Environmental Health Site Assessment. Air Force Manual 48-145. Published 28 March 2007; certified current 22 March 2010. http://static.e-publishing.af.mil/production/1/af_ja/publication/afman48-154/afman48-154.pdf (accessed 26 July 2017).

USAF (U.S. Air Force). Drinking water surveillance program. Air Force Instruction 48-144. 21 October 2014. http://static.e-publishing.af.mil/production/1/af_sg/publication/afi48-144/afi48-144.pdf (accessed 26 July 2017).

USEPA (U.S. Environmental Protection Agency), Office of Water. Guidance manual for conducting sanitary surveys of public water systems; surface water and ground water under the direct influence (GWUDI) of surface water. EPA 815-R-99-016. Washington, DC: USEPA, April 1999.

USEPA (U.S. Environmental Protection Agency). Lifetime health advisories and health effects support documents for perfluorooctanoic acid and perfluorooctane sulfonate. *Federal Register* 81(101):2016.

USEPA (U.S. Environmental Protection Agency). Source water protection. 2017. https://www.epa.gov/sourcewaterprotection (accessed 23 June 2017).

USFJ (U.S. Forces Japan) air quality awareness. Public service announcement/web page resource, Kadena AB web page. http://www.kadena.af.mil/Portals/40/documents/AFD-160602-077.pdf?ver=2016-06-08-144505-697 (accessed 22 August 2017).

Wagner, A. Use of host nation tools to provide surveillance and risk communication: Aeolian dust in South Korea and Japan. Poster presentation at Aerospace Medical Association Annual Meeting, Atlantic City, New Jersey, March 2016.

Wingfield-Hayes Report. Japan tackles first dengue fever outbreak in 70 years. BBC News, September 2014. https://wwwnc.cdc.gov/eid/article/21/3/14-1662_article (accessed 22 August 2017).

WRF (Water Research Foundation). Source water protection vision and roadmap. WRF 4176b. 2012. http://www.waterrf.org/PublicReportLibrary/4176b.pdf (accessed 26 July 2017).

6 Uncharted Waters
Challenges for Public Water Systems Addressing Drinking Water Health Advisory Levels for PFOS and PFOA

Miranda Brannon

CONTENTS

6.1	Introduction 103
6.2	Background 104
6.3	Terminology 104
	6.3.1 Maximum Contaminant Level and Treatment Technique 105
	6.3.2 Health Advisory Level 105
	6.3.3 Drinking Water Advisories 106
6.4	Challenges for Dealing with PFOS and PFOA HALs 106
	6.4.1 Response Actions 106
	6.4.2 Risk Communication 107
	6.4.3 No Time to Prepare for Implementation and Lack of Treatment Technologies 107
6.5	Unintended Consequences for the Unregulated Contaminant Monitoring Rule Program 108
6.6	Regulatory Process 108
6.7	Future State 109
6.8	Recommendations for PWSs Addressing Drinking Water HALs 109
References 110

6.1 INTRODUCTION

Imagine you are the captain of a passenger ship. You take your job very seriously, because you know your passengers place their lives in your hands when boarding the ship. You are responsible for transporting your passengers to their destination safely. You depend on your navigation systems to show you the way, tell you how to correct mistakes if you veer off course, and tell you how close you are to reaching your destination. Now imagine your navigation system goes down en route to a new destination. How are you going to get there? How do you know where it is? How will you know if you are doing the right things to get you to the new place? What will you tell your passengers who have entrusted their lives to you on this trip?

In 2012, the US Air Force (USAF) drinking water program began a long, difficult, and still ongoing journey: understanding and responding to US Environmental Protection Agency (USEPA) health advisory levels (HALs) for perfluorooctane sulfonate (PFOS) and perfluorooctanoic acid (PFOA) in drinking water. For a long time, the USAF did not even know what the final destination was; we were working with provisional HALs that we knew could change at any time. We had no "navigation system"—no standardized monitoring framework, mature analytical techniques, or National Primary Drinking Water Regulations (NPDWRs) to guide us. Personnel who manage the USAF drinking water program have learned a lot during this journey, and we hope that by sharing our experiences and comments with others, we may help others be better prepared if they find themselves in similar situations.

6.2 BACKGROUND

For the USAF drinking water community, our first engagement with PFOS or PFOA was providing input to the Air Force Civil Engineer Center's (AFCEC) 2012 Interim Air Force Guidance on Sampling and Response Actions for Perfluorinated Compounds at Active and Base Realignment and Closure Installations. Scientists at AFCEC were monitoring developments with respect to emerging contaminants within the USEPA and across the country, and understood that even though PFOS and PFOA were not yet regulated, they would likely have an impact on the USAF environmental program in the near future. They developed the 2012 document to assess this impact and establish guidance for initiating an AF-wide strategy for approaching perfluorinated compounds (PFCs). The guidance focused mostly on environmental cleanup approaches, but it included guidance to sample installation drinking water "when knowledge gained from environmental sampling indicates the system may be impacted." Few in the drinking water community realized at the time what an immense issue this would become over the next 5 years. Many presumed that there may be a handful of sites that might need drinking water sampling, if the environmental cleanup investigations discovered a completed pathway to drinking water. As it turned out, the USAF sampled all USAF public water systems (PWSs) for PFOS and PFOA and ensured that any USAF PWS purchasing water from a non-USAF purveyor obtained PFOS and PFOA sample results from the purveyor or sampled on its own (Ballentine 2016).

6.3 TERMINOLOGY

This chapter discusses the differences between maximum contaminant levels (MCLs) and HALs, potential challenges encountered by PWSs when responding to USEPA drinking water HALs (specifically those for PFOS and PFOA), the SDWA regulatory process, and observations about the future of drinking water HALs.

6.3.1 MAXIMUM CONTAMINANT LEVEL AND TREATMENT TECHNIQUE

The Safe Drinking Water Act (SDWA) requires the USEPA to promulgate NPDWRs to limit the level of contaminants in drinking water. NPDWRs must specify for each contaminant either an MCL or, if it is not economically and technically feasible to determine the level of the contaminant, a treatment technique (TT). An MCL is the maximum permissible level of a contaminant in water that is delivered to any user of a PWS. Often, water professionals conversationally use the term *MCL* to mean the same thing as NPDWR. An example of an MCL is 10 parts per billion (ppb) for arsenic. This is the "measuring stick" that indicates whether drinking water has a safe level of arsenic (USEPA 2007). An example of a TT exists for lead and copper. Lead and copper do not have established MCLs; rather, the regulations dictate a series of actions that the PWS must undertake if drinking water exceeds a certain action level (USEPA 2008). These actions could include additional monitoring, changes to water treatment, conducting public education, or replacement of lead service lines.

Beyond establishing MCLs and TTs, the USEPA also prescribes a monitoring framework for each regulated contaminant. Going back to our earlier example, the USEPA's standardized monitoring framework outlines a monitoring strategy for arsenic. Depending on what type of water source and the amount of arsenic, you could sample anywhere from one sample every 9 years to quarterly sampling. For lead and copper, the USEPA prescribes specific frequencies, numbers of samples, and sample locations. The SDWA establishes clear-cut criteria to evaluate the safety of drinking water for regulated contaminants (Tiemann 2017).

6.3.2 HEALTH ADVISORY LEVEL

Like MCLs, HALs are also numeric concentrations of contaminants in drinking water. Unlike MCLs, HALs are nonenforceable and nonregulatory, and most of them do not provide monitoring frameworks or recommended response actions. HALs may be developed for a number of reasons, which can include providing technical information to state agencies and other public health officials on health effects, analytical methodologies, and treatment technologies associated with drinking water contamination, or to respond quickly to a localized situation where no guidance exists (USEPA 2016a). For example, the USEPA first established provisional HALs for PFOS and PFOA in 2009 to respond to a rapidly developing situation in Alabama, where some agricultural sites were found to be contaminated by sewage sludge containing PFOS and PFOA. These provisional HALs were defined as "reasonable, health-based hazard concentrations above which action should be taken to reduce exposure to unregulated contaminants in drinking water" (USEPA 2009). However, there were no recommended "actions" provided—no direction on implementing a sampling and monitoring program or information on how drinking water exceeding these provisional HALs could affect consumers. Later, in May 2016, the USEPA published lifetime HALs for PFOS and PFOA based on more robust health assessments than produced in 2009. This time, the HALs were accompanied by some general recommendations, including retesting, informing consumers, and taking steps to limit exposure (USEPA 2016b).

Besides the difference in regulatory status, MCLs and HALs have a number of very important differences. One of these is the process used to derive the actual numeric concentration, or level. On the way to establishing a final MCL, a maximum contaminant level goal (MCLG) is first established. This is defined as the maximum level of a contaminant in drinking water at which no known or anticipated adverse effect on the health of persons would occur, allowing an adequate margin of safety. MCLGs only consider health and do not consider feasibility. They are "goals" for what ideally should be in drinking water. However, to promulgate an enforceable MCL, the USEPA must consider other factors to ensure that the MCL is something achievable, such as the availability and efficacy of treatment technologies and laboratory analyses. The SDWA also requires that the USEPA perform a health risk reduction and cost analysis (HRRCA) to develop any MCL. It is understood that any public policy will have both benefits and costs. The HRRCA seeks to answer the question, "Will the public health benefits of implementing this MCL outweigh the costs?" HALs do not incorporate considerations for feasibility, availability, and efficacy of treatment technologies and laboratory analyses, nor are they based on any cost–benefit analysis (USEPA 2016a).

6.3.3 Drinking Water Advisories

There is yet another term that should be introduced here, if only to explain what it is not. A drinking water advisory is not the same thing as a HAL. A drinking water advisory is a notification made to drinking water consumers by PWSs when they believe or know water quality is or may be compromised. Advisories give information about the situation and what actions consumers should take, if any. Examples of drinking water advisory actions are boil water notices, do not drink notices, and do not use notices (USEPA 2012; CDC 2013).

6.4 CHALLENGES FOR DEALING WITH PFOS AND PFOA HALs

6.4.1 Response Actions

Under the SDWA, requirements for PWSs to monitor for regulated contaminants are straightforward. Based on its classification and source type, a PWS has a prescribed monitoring schedule that details when, where, and how often each contaminant will be monitored. In addition, if results exceed a specified level, there are prescribed follow-up actions to take.

For most unregulated contaminants with a drinking water health advisory, there is no prescribed monitoring framework and no recommended actions to take following an exceedance of the HAL. This leaves PWSs in a quandary. How often should they sample? How often is enough? How many locations? How close to the HAL is close enough to warrant continued monitoring? What do the results mean? Should we continue to allow consumers to drink the water? Are certain subpopulations more vulnerable? How do we communicate the risk to our consumers?

When dealing with PFOS or PFOA in drinking water, the USAF found that the lack of a USEPA-prescribed monitoring framework resulted in a more severe and conservative response than if dealing with a regulated contaminant. In the absence of specific monitoring and response guidance, and in the presence of significant attention in the press and pressure from consumers, we immediately took measures, to include offering alternative water, shutting down wells, and installing treatment systems, if even one sample exceeded the HAL for PFOS or PFOA. Contrast this with actions taken to respond to an exceedance of a regulated contaminant with an MCL. Using the example of arsenic, the action prescribed by the USEPA for an exceedance of the MCL is to commence a program of increased monitoring until it is deemed by the primacy agency to be "reliably and consistently less than or equal to the MCL." An exceedance of an MCL is, in most cases, not an acute concern (except for microbial contaminants such as *Escherichia coli*). An MCL represents the highest allowable concentration of a chemical in drinking water for a lifetime of exposure, with the underlying calculations based on a 70 kg adult consuming 2 L of water every day.

6.4.2 Risk Communication

Even though HALs are defined as nonregulatory and nonenforceable, and are not intended to be observed as definitive thresholds of what is considered safe and unsafe in drinking water, the general public does not appreciate this nuance. The majority of community members view any numerical limit set by the USEPA as a determination of what is safe and not safe. It is very difficult to explain to most drinking water consumers that exceedance of a HAL may not be a cause for immediate concern or a reason to discontinue drinking the water.

6.4.3 No Time to Prepare for Implementation and Lack of Treatment Technologies

Under the normal regulatory process for promulgating MCLs, the regulated community has time to prepare for implementation of a new rule. Typically, proposed MCLs are published in the *Federal Register* and subject to public comment. Approximately 1 year later, the final rule is published. Then, the rule is not considered "in effect" until 3 years past its publication date. This gives the regulated community time to prepare for implementation of such a rule. With a HAL, there is no preparation period. The HAL can be published one day and implementation expected immediately (again, even though it is not enforceable per se). For example, the USEPA published the lifetime HALs for PFOS and PFOA on May 19, 2016. On May 20, 2016, the Ohio Environmental Protection Agency issued a letter to Wright-Patterson Air Force Base to immediately shut down drinking water wells containing PFOS and PFOA concentrations above the HAL, issue a drinking water advisory to drinking water consumers, offer alternative sources of water for pregnant and lactating women and bottle-fed infants, and continue to sample affected wells monthly (Barber 2016).

As previously mentioned, MCL development requires the availability and feasibility of treatment technologies. Publishing a HAL does not require that a treatment technology be available or feasible. A PWS may exceed a HAL but have no ability or proven technology to mitigate the situation. This results in significant risk communication challenges and anxiety to consumers. The PWS may be forced to seek an alternative source of water without the benefit of a clearly understood risk level.

6.5 UNINTENDED CONSEQUENCES FOR THE UNREGULATED CONTAMINANT MONITORING RULE PROGRAM

The USEPA uses its Unregulated Contaminant Monitoring Rule (UCMR) to help determine what contaminants need regulating in drinking water. Every 5 years, the USEPA uses the UCMR to collect data on a list of contaminants that are suspected to be present in drinking water and do not have health-based standards set under the SDWA (USEPA 2017). All large PWSs (those that serve 10,000 or more consumers) and a randomly chosen number of small PWSs must collect UCMR data and report it to the USEPA. The USEPA uses these data to determine if those contaminants should be regulated. UCMR results are not intended to be used as fodder to enforce regulatory action on PWSs; however, that is exactly what has happened with PFOS and PFOA under UCMR3. Sixty-three PWSs nationwide were found to have levels of PFOS in their drinking water above the HAL, and in some cases these PWSs were pressured by regulatory agencies to take response actions. For example, Artesian Water Company's Wilmington Manor 3 Treatment Plant exceeded the provisional HAL for PFOS, discovered as a result of UCMR3 sampling. Artesian subsequently removed the affected treatment plant from service after consultation with the Delaware Division of Public Health (Artesian Water Company 2014). In another example, Liberty Utilities in Arizona shut down a drinking water well after discussions with the USEPA and Arizona Department of Environmental Quality. This was also due to sample results from UCMR3 sampling (City of Litchfield Park 2016). This type of reaction makes for a chilling effect on UCMR participants. What UCMR contaminants will be deemed unacceptable on the next round? This completely alters the intent of the UCMR program and turns it into a challenging prospect for PWSs.

6.6 REGULATORY PROCESS

The USEPA faces its own challenges when it comes to promulgating new drinking water standards. All federal agencies must follow the procedural rules of the Administrative Procedures Act (APA) to make sure that their rulemakings are transparent and open to the public. Although this is an extremely important part of rulemaking, it does make for a long and drawn-out process, making passage of new standards difficult and a years-long endeavor (Ferrey 2010). In addition, the USEPA needs robust science to inform the development of an MCL, and for many contaminants, that level of science is not yet achieved. For many contaminants, there are neither reliable analytical methods nor treatment technologies available. The USEPA would need to conduct additional research and studies to build the body of knowledge required, which is another long-term endeavor. One can see why the USEPA may begin to rely on other tools at their disposal to force action in the absence of regulations. Amid pressures from Congress and communities nationwide, establishing a health advisory is often one of the only ways the USEPA can respond quickly and substantively. For example, in response to the August 2014 harmful algal bloom (HAB) that shut down the entire city of Toledo's drinking water source, congressional members called hearings to examine the issues surrounding HABs. One item that was discussed in these hearings was the lack of federal standards for HABs and the acknowledgment that establishing such standards would take an extensive period of time. As a result, the USEPA developed HALs for two cyanotoxins in order to quickly provide some level of guidance to PWSs in the absence of a regulatory standard (US House of Representatives 2014).

6.7 FUTURE STATE

In light of the current climate of increased public awareness and regulatory scrutiny on drinking water issues, the United States is likely to increase its development of drinking water HALs. In their review of the USEPA's draft Fourth Contaminant Candidate List, the USEPA Science Advisory Board acknowledged the cumbersome process to promulgate an MCL, and even recommended that the USEPA develop more health advisories in the future (USEPA 2016c). The USEPA states in its Drinking Water Action Plan that one of its goals is to "better leverage non-regulatory tools." The plan also discusses the USEPA's desire to collaborate with stakeholders to develop better risk management and risk communication approaches, including a framework to assess risk to PWSs from unregulated contaminants (USEPA 2016d). PWSs across the country would welcome these developments so that they could increase their abilities to effectively manage drinking water contaminants with HALs.

6.8 RECOMMENDATIONS FOR PWSs ADDRESSING DRINKING WATER HALs

- Establish a collaborative relationship among health, environmental compliance, legal, public affairs, and infrastructure teams.
- Communicate often with state and local regulators with jurisdiction over drinking water.
- Have a proactive risk communication and public notification program.
- Become familiar with terminology and calculations or assumptions behind HALs versus MCLs.
- Understand what HALs are intended for and be able to explain the differences between HALs and MCLs.
- Connect with other drinking water systems that have experienced or are experiencing similar situations
- Scan news for drinking water–related articles to foster situational awareness of potential issues.

REFERENCES

AFCEC (Air Force Civil Engineer Center). 2012. Interim Air Force guidance on sampling and response actions for perfluorinated compounds at active and base realignment and closure installations. Internal Air Force guidance, pp. 6–14. http://alaskacollection.library.uaf.edu/eafbsc/cd2/AR901-1.pdf (accessed 26 September 17).

Artesian Water Company. 2014. Drinking water notice. http://www.artesianwater.com/wp-content/uploads/2014/06/06-12-2014-PFOS-DRINKING-WATER-NOTICE-6-12-14-DPH-clean-final1.pdf (accessed 26 May 2017).

Ballentine, M.A.A. 2016. Testing drinking water for perfluorooctane sulfonate (PFOS) and perfluorooctanoic acid (PFOA). Memorandum to the Air Force Surgeon General and the Air Force Deputy Chief of Staff, Logistics, Engineering, and Force Protection. 12 August.

Barber, B. 2016. Ohio EPA fears Dayton drinking water could be at risk. *Dayton Daily News*, 20 May. http://www.daytondailynews.com/news/wright-patt-ordered-shut-down-well/9EU5yUW7bbXyrFxTEPohVI/.

CDC (Center for Disease Control and Prevention). 2013. Drinking water advisory communication toolbox. https://www.cdc.gov/healthywater/pdf/emergency/drinking-water-advisory-communication-toolbox.pdf (accessed 17 February 2017).

City of Litchfield Park. 2016. Drinking water notice. http://www.litchfield-park.org/DocumentCenter/View/6009 (accessed 26 May 2017).

Ferrey, S. 2010. *Environmental Law: Examples & Explanations*. New York: Aspen Publishers.

Tiemann, M. 2017. Safe Drinking Water Act (SDWA): A summary of the act and its major requirements. Congressional Research Service, 1 March. https://fas.org/sgp/crs/misc/RL31243.pdf (accessed 26 July 2017).

USEPA (U.S. Environmental Protection Agency). 2007. Arsenic in your drinking water. https://nepis.epa.gov/Exe/ZyPDF.cgi?Dockey=60000E1E.txt (accessed 17 February 2017).

USEPA (U.S. Environmental Protection Agency). 2008. Lead and copper rule: A quick reference guide. https://nepis.epa.gov/Exe/ZyPDF.cgi?Dockey=60001N8P.txt (accessed 17 February 2017).

USEPA (U.S. Environmental Protection Agency). 2009. Provisional health advisories for perfluorooctanoic acid (PFOA) and perfluorooctane sulfonate (PFOS). https://www.epa.gov/sites/production/files/2015-09/documents/pfoa-pfos-provisional.pdf (accessed 21 February 2017).

USEPA (U.S. Environmental Protection Agency). 2012. Drinking water standards and health advisories. https://nepis.epa.gov/Exe/ZyPDF.cgi/P100N01H.PDF?Dockey=P100N01H.PDF (accessed 17 February 2017).

USEPA (U.S. Environmental Protection Agency). 2016a. Drinking water health advisories for PFOA and PFOS. https://www.epa.gov/ground-water-and-drinking-water/drinking-water-health-advisories-pfoa-and-pfos (accessed 21 February 2017).

USEPA (U.S. Environmental Protection Agency). 2016b. Fact sheet: PFOA & PFOS drinking water health advisories. https://www.epa.gov/sites/production/files/2016-06/documents/drinkingwaterhealthadvisories_pfoa_pfos_updated_5.31.16.pdf (accessed 21 February 2017).

USEPA (U.S. Environmental Protection Agency), Science Advisory Board Drinking Water Committee. 2016c. Review of the EPA's draft Fourth Contaminant Candidate List (CCL4). https://nepis.epa.gov/Exe/ZyNET.exe/P100ROIS.TXT?ZyActionD=ZyDocument&Client=EPA&Index=2011+Thru+2015&Docs=&Query=&Time=&EndTime=&SearchMethod=1&TocRestrict=n&Toc=&TocEntry=&QField=&QFieldYear=&QFieldMonth=&QFieldDay=&IntQFieldOp=0&ExtQFieldOp=0&XmlQuery=&File=D%3A%5CZyfiles%5CIndex%20Data%5C11thru15%5CTxt%5C00000025%5CP100R

OIS.txt&User=ANONYMOUS&Password=anonymous&SortMethod=h%7C-&MaximumDocuments=1&FuzzyDegree=0&ImageQuality=r75g8/r75g8/x150y150g16/i425&Display=hpfr&DefSeekPage=x&SearchBack=ZyActionL&Back=ZyActionS&BackDesc=Results%20page&MaximumPages=1&ZyEntry=1&SeekPage=x&ZyPURL (accessed 26 July 2017).

USEPA (U.S. Environmental Protection Agency). 2016d. Drinking water action plan: Priority area 4. https://www.epa.gov/ground-water-and-drinking-water/drinking-water-action-plan-priority-area-4 (accessed 17 February 2017).

USEPA (U.S. Environmental Protection Agency). 2017. Learn about the Unregulated Contaminant Monitoring Rule. https://www.epa.gov/dwucmr/learn-about-unregulated-contaminant-monitoring-rule (accessed 17 February 2017).

U.S. House of Representatives, Subcommittee on Environment and Economy, Committee on Energy and Commerce. 2014. Hearing: Cyanotoxins in drinking water. http://docs.house.gov/Committee/Calendar/ByEvent.aspx?EventID=102747 (accessed 17 February 2017).

7 Challenges of Managing Emerging Contaminants
Historical Per- and Polyfluorinated Alkyl Substance Use in the US Air Force

R. Hunter Anderson and David M. Kempisty

CONTENTS

7.1 Introduction 113
7.2 Regulatory Background 114
7.3 AFFF Use 116
References 119

7.1 INTRODUCTION

The number of chemicals in existence today is vast. The Chemical Abstract Services currently cites the number of chemicals registered in their database at 118×10^6 (www.CAS.org). This number will invariably continue to grow. These chemicals afford countless conveniences in our daily lives. Plastics and their multitude of uses, fertilizers and pesticides and their contribution to increased food supplies, pharmaceuticals and the prevention and treatment of disease that they enable are only a handful of examples of beneficial chemical use that affect our daily lives. "Better living through chemistry" was a popular slogan in advertising campaigns that touted the use of science for our benefit. However, many of the same chemicals that provide benefit can also be hazardous to the environment and to human health. Determining the balance between gain and harm is important; the perspective from which one looks can influence the outcome. Considering only individual benefits and consequences misses externalized costs borne by society. In Garrett Hardin's article "The Tragedy of the Commons", published in the journal *Science*, Hardin makes the statement that "freedom in a commons brings ruin to all" (Hardin 1968). By "commons", Hardin is referring to a shared resource for all. One example would be the air we breathe. Another example is an entity we formerly considered a commons but no longer today, the sea. Exploitation of fish stocks has resulted in low fish numbers discontinuing "without limit" withdrawals. The tragedy of the commons involves the sharing of externalized consequences among all users of the commons but having the benefits garnered by a singular individual. In this manner, each individual maximizes benefit at the cost of the collective group. The classic example is the shared common pasture among shepherds. With each additional sheep, the shepherd is better off in terms of profitability (having more sheep results in more wool and more sales at the market). This is true until the carrying capacity of the commons is met. Once this occurs, an additional sheep increases a shepherd's profitability by +1, but the consequence due to overexploitation of the commons (inadequate food supply for all sheep, excessive waste, increased runoff, etc.) is not −1. Since the commons is shared by the network of shepherds, the consequence to the shepherd bringing the additional sheep over the pasture's carrying capacity is only a fraction of −1. Unchecked, commons with finite resources will inevitably end in tragedy or failure.

Moving from shepherds to chemicals, how do we maximize benefit while minimizing harm? How do we avoid damages to the commons (soil, water, and air) and ensure safety and environmental health? Is it through legislation and regulation and management? These tenets also provide oversight to avoid, or at least limit, long-term adverse health effects to users of chemicals. Occupational health and environmental management are not trivial manners and can be onerous tasks, especially for a large organization dealing with hundreds of chemicals. Consider past chemical use presenting legacy contamination issues and the process becomes more difficult. This chapter focuses on poly- and perfluoroalkyl substances (PFAS) and describes challenges faced by personnel in the US Air Force (USAF) from a restoration perspective: identifying PFAS contamination from historical use of PFAS-containing aqueous film-forming foam (AFFF).

7.2 REGULATORY BACKGROUND

The Toxic Substances Control Act (TSCA) of 1976 is the regulatory tool used to manage chemicals at a federal level—but the regulation has its limitations and shortfalls. For instance, chemicals already in the marketplace prior to 1976, approximately 62,000 in number, were exempted from regulation (Schmidt 2016). This led to the creation of piecemeal regulation by the states and an intractable position for chemical manufacturers to understand and abide by different state regulations for chemicals and materials they wish to solicit across state boundaries to the entire country. Addressing some of the limitations of TSCA, the Frank Lautenberg Chemical Safety for the 21st Century Act was signed into law in June 2016. The Lautenberg Act provides more federal control of new and existing chemicals encountered in the environment but should not be considered without challenges.

From a Department of Defense (DOD) perspective, the overall effect of new and emerging contaminants (ECs), or new regulations associated with existing chemicals, is facilitated through a working group under the Office of the Assistant Secretary of Defense (Energy, Installations and Environment), Science and Technology Directorate's Chemical and Material Risk Management Program (OASD EI&E 2008). The group is known by the acronym MERIT, standing for Materials of Evolving Regulatory Interest Team.

An EC is defined by MERIT as chemicals and materials with

- Perceived or real threat to human health or the environment
- Either no peer-reviewed health standard or an evolving standard

ECs may have

- Insufficient human health data or science
- New detection limits
- Newly identified exposure pathways

ECs can affect the DOD and its members in many ways. The health and well-being of DOD employees is an obvious reason for concern. This rationale can be expanded to include members of the public whom the DOD ultimately serves. From an operational perspective, mission accomplishment is paramount in the military. ECs can provide a risk to successful execution of military goals and objectives. In the context of protecting the freedom of the United States, this can potentially have enormous implications. Preparedness and readiness are unequivocally linked to the military's overall mission of preservation of national security. ECs can affect training opportunities of the military and thereby affect military readiness. Logistically, resources and materiel can also be affected by ECs. Specific and unique chemicals and materials used on world-class weapon systems can be adversely affected by evolving regulation on ECs. Finally, in today's fiscally constrained environment, cleanup costs associated with remediation of legacy EC contamination can negatively affect the operational budget of today's DOD.

To address the issues with ECs and their potential effect on the DOD, MERIT has developed a "scan–watch–act" process. The initial activity, *scanning*, is an activity to gain awareness of ECs that may be coming to the forefront of regulatory interest. Heightened interest can be due to a variety of reasons, including newly discovered exposures and/or adverse toxicity, increasing public concern, advancements in analytical detection capabilities, or truly new chemicals introduced into the marketplace. Regardless of the reason, the scan activity looks to scholastic literature, periodicals, and communications with the regulatory community to identify ECs of concern and evaluate possible effects on specific DOD missions. The second phase, *watch*, involves a Phase I Impact Assessment of the EC. This includes an estimation of the likelihood of regulation and a qualitative assessment of the potential effects on five different DOD functional categories. Functional categories include environmental safety and health, readiness and training, acquisition, production and operation and maintenance and disposal, and cleanup. Depending on the outcome of the Phase I Impact Assessment, a decision is made whether to move the EC to the *act* (or action) category. If regulatory action on the EC is determined to likely affect the DOD, risk management (RM) options are developed in a quantitative Phase II Assessment. Potential RM options that could be explored involve the availability of product substitution or changes to the process involving the EC, research addressing identified knowledge gaps (e.g., incomplete toxicological profiles or lack of effective remedial technologies, or inadequate analytical methodologies), establishing communication with the regulatory community, and exploring exposures to personnel working with the EC and potential controls (engineering or personnel protective equipment or administrative controls). Additional RM alternatives may involve evaluating the acquisition process for weapon systems involving the EC and developing risk communication products to educate and train personnel associated with the EC. The RM options identified are then presented to the MERIT governance council for further action. The likelihood of occurrence and the severity of impact to the DOD mission will influence RM decisions. Additional actions performed for ECs on the action list involve assigning a materiel champion to coordinate further DOD actions (e.g., development of executive information sheets, resource management, and budget development).

7.3 AFFF USE

Recently, PFAS (one particular class of ECs) have gained interest to DOD stakeholders and regulators. PFAS have been associated with adverse environmental and human health effects. Contained as an active- and military-specified ingredient in AFFF used for extinguishing Class B hydrocarbon fires, the DOD has used PFAS-containing AFFF since approximately 1970.

The USAF Civil Engineer Center (AFCEC) first identified the use of AFFF as an environmental issue following MERIT's Phase I Impact Assessment of perfluorooctane sulfonate (PFOS) and perfluorooctanoic acid (PFOA) in 2008. While this assessment was focused on potential mission-related implications of future AFFF formulation regulations or restrictions, potential impacts to the DOD's Environmental Restoration Program (DERP) were noted as a result of historic AFFF use and operational practices (referred to herein as "the PFAS problem"). This conclusion was reinforced by several already-published studies reporting PFAS contamination in soil and groundwater at former fire training areas (FTAs) on military installations (Levine et al. 1997; Moody and Field 1999; Moody et al. 2003; Schultz et al. 2004). As a collective result, in 2009 the AFCEC initiated limited investigations at select ($n = 10$) FTAs with known historic AFFF use in order to more comprehensively gauge the likelihood and magnitude of soil and groundwater contamination at FTAs across the USAF portfolio. Select PFAS were detected in soil and groundwater, and concentrations of PFOS and PFOA exceeded the (then provisional) health advisory values from the US Environmental Protection Agency (USEPA) at every FTA investigated; groundwater results are presented in Figure 7.1. Given that large quantities of AFFF were episodically discharged in the same physical location during a period of time when FTAs were not lined and were not constructed to prevent infiltration or runoff of AFFF and combustion products, it is perhaps not surprising that all FTAs evaluated resulted in quantifiable contamination. Nevertheless, until these investigations, the number of PFAS-contaminated FTAs could only be speculated based on a few published studies.

Subsequent to the FTA investigations, AFCEC began addressing remaining data gaps to more fully address and respond to the PFAS problem. Several preliminary remediation-oriented projects were soon thereafter funded under both the AFCEC's Broad Agency Announcement (BAA) and the DOD's Strategic Environmental

Challenges of Managing Emerging Contaminants

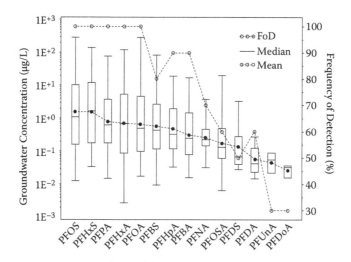

FIGURE 7.1 PFAS data initially collected by AFCEC exclusively from FTAs to characterize the frequency of occurrence and magnitude of concentrations in groundwater. FoD = frequency of detection, PFBA = perfluorobutanoate, PFBS = perfluorobutane sulfonate, PFDA = perfluorodecanoate, PFDoDA = perfluorododecanoate, PFDS = perfluorodecane sulfonate, PFHpA = perfluoroheptanoate, PFHxA = perfluorohexanoate, PFHxS = perfluoroheptane sulfonate, PFNA = Perfluorononanoate, PFOA = perfluorooctanoate, PFOS = perfluorooctane sulfonate, PFOSA = perfluorooctanesulfonamide, PFPA = perfluoropentanoate, PFUnA = Perfluoroundecanoate.

Research and Development Program (SERDP). Initial remediation projects focused on *in situ* chemical oxidation and reduction (e.g., Park et al. 2016), electrochemical oxidation (e.g., Schaefer et al. 2015), and enzyme-catalyzed oxidative humification reactions (e.g., Luo et al. 2015). Also of primary interest was the magnitude and frequency of PFAS contamination at other (i.e., non-FTA) AFFF-impacted sites where relatively smaller quantities of AFFF were discharged and at lesser frequencies, for example, emergency response sites with a one-time release and at hangars with automatic overhead suppression systems where either occasional testing or equipment failure resulted in few but sporadic releases. Consequently, investigation of AFFF-impacted soil and groundwater at these other AFFF-impacted sites was initiated in 2013; these results are presented in Anderson et al. (2016) and are reproduced in Chapter 15. Ultimately, the non-FTA investigation demonstrated that even relatively small historic AFFF releases have the potential to contaminate groundwater at levels of concern.

As a result of the collective weight of evidence regarding the scale of the PFAS problem, AFCEC initiated in 2014 a programmatic response to investigate all AFFF releases across the entire portfolio of installations within the continental United States (CONUS). Given DERP policy and congressional mandates on DERP funding expenditures to follow the Comprehensive Environmental Response, Compensation, and Liability Act (CERCLA) and associated USEPA guidance, the initial effort began with preliminary assessments (PAs) at all CONUS installations. Procedurally, PAs are document and record reviews and include interviews with applicable base personnel regarding the location and magnitude of all AFFF releases or spills. All relevant information was compiled in the publicly available PA reports for all active USAF installations, former (i.e., Base Realignment and Closure [BRAC]) installations, and Air National Guard (ANG) facilities. More than 1700 potential AFFF releases were identified, and they are categorized in Figure 7.2. It should be noted, however, that these are best estimates based on available information and should not be interpreted as definitive. Subsequently, and again in accordance with CERCLA procedure, site inspection (SI) investigations have been initiated and are currently ongoing whereby PA conclusions are being verified and soil and groundwater samples are being collected specifically to confirm environmental contamination resulting from all documented AFFF releases. Remedial investigations (RIs) to delineate the nature and extent of PFAS contamination at validated sites will be conducted in follow-on efforts.

Historical use of AFFF within the USAF was arguably warranted. Extinguishing fire in the most expeditious manner to save valuable resources, including human lives, is a judicious use of chemicals. However, at the time of application, policy and procedures did not fully address the environmental implications to today's level of concern. Therefore, reparations addressing legacy contamination and the risk to human health and the environment need to be considered today. In the case of the USAF, more than 1700 potential AFFF release areas have been identified and are currently being investigated to determine soil, water, and sediment contamination. Understanding the scope of the issues is one challenging aspect to effective management of areas where historical use of AFFF occurred. After the extent of contamination and site-specific risks are quantified, remedial alternatives may be considered. These investigations and the development of the various courses of action require considerable resources, including time and money. The situation is furthered challenged by the incomplete toxicological profile, the changing regulatory landscape, and limited remediation alternatives associated with PFAS contamination.

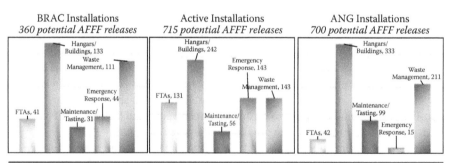

FIGURE 7.2 Potential AFFF-impacted areas identified during AFCEC's programmatic PA effort. Note that these are best estimates based on available information and should not be interpreted as definitive. The definitive site count will be determined once all ongoing SI results are compiled.

REFERENCES

Anderson, R. H., Long, G. C., Porter, R. C., & Anderson, J. K. (2016). Occurrence of select perfluoroalkyl substances at US Air Force aqueous film-forming foam release sites other than fire-training areas: Field-validation of critical fate and transport properties. *Chemosphere, 150*, 678–685.

Hardin, G. (1968). The Tragedy of the Commons. *Science, 162*, No. 3859, 1243–1248.

Levine, A. D., Libelo, E. L., Bugna, G., Shelley, T., Mayfield, H., & Stauffer, T. B. (1997). Biogeochemical assessment of natural attenuation of JP-4-contaminated ground water in the presence of fluorinated surfactants. *Science of the Total Environment, 208*(3), 179–195.

Luo, Q., Lu, J., Zhang, H., Wang, Z., Feng, M., Chiang, S. Y. D. et al. (2015). Laccase-catalyzed degradation of perfluorooctanoic acid. *Environmental Science & Technology Letters, 2*(7), 198–203.

Moody, C. A., & Field, J. A. (1999). Determination of perfluorocarboxylates in groundwater impacted by fire-fighting activity. *Environmental Science & Technology, 33*(16), 2800–2806.

Moody, C. A., Hebert, G. N., Strauss, S. H., & Field, J. A. (2003). Occurrence and persistence of perfluorooctanesulfonate and other perfluorinated surfactants in groundwater at a fire-training area at Wurtsmith Air Force Base, Michigan, USA. *Journal of Environmental Monitoring: JEM, 5*(2), 341.

OASD EI&E. (2008). Requests for this document shall be referred to the Office of the Assistant Secretary of Defense (Energy, Installations and Environment) (OASD EI&E), Chemical and Material Risk Management Program, 4800 Mark Center Dr., Suite 16G14, Alexandria, VA 22350-3600. osd.pentagon.ousd-atl.mbx.cmrmp@mail.mil.

Park, S., Lee, L. S., Medina, V. F., Zull, A., & Waisner, S. (2016). Heat-activated persulfate oxidation of PFOA, 6:2 fluorotelomer sulfonate, and PFOS under conditions suitable for in-situ groundwater remediation. *Chemosphere, 145*, 376–383.

Schaefer, C. E., Andaya, C., Urtiaga, A., McKenzie, E. R., & Higgins, C. P. (2015). Electrochemical treatment of perfluorooctanoic acid (PFOA) and perfluorooctane sulfonic acid (PFOS) in groundwater impacted by aqueous film forming foams (AFFFs). *Journal of Hazardous Materials, 295*, 170–175.

Schmidt, C. W. (2016). TSCA 2.0—A new era in chemical risk management. *Environmental Health Perspectives, 124*. doi.org/10.1289ehp.124-A182.

Schultz, M. M., Barofsky, D. F., & Field, J. A. (2004). Quantitative determination of fluorotelomer sulfonates in groundwater by LC MS/MS. *Environmental Science & Technology, 38*(6), 1828–1835.

Section III

Toxicology

8 Human Health Risk Assessment of Perfluorinated Chemicals

Penelope Rice

CONTENTS

8.1 Introduction 123
8.2 Pharmacokinetics 125
8.3 Health Effects and Toxicology 127
    8.3.1 Systemic Toxicity: Nonneoplastic Effects on Organ Systems 127
        8.3.1.1 Effects on the Liver 128
        8.3.1.2 Effects on the Immune System 139
        8.3.1.3 Effects on Male Reproductive Organs 141
        8.3.1.4 Effects on Female Reproductive Organs
                 and the Mammary Gland 143
        8.3.1.5 Effects on the Thyroid 145
        8.3.1.6 Other Effects 147
    8.3.2 Carcinogenicity and Genotoxicity 148
8.4 Common Mechanisms of Action and Class-Based Analysis 150
8.5 Regulatory Agency Risk Assessments for PFAS 152
8.6 PFAS "Dark Matter," Conclusions, and Future Prospects 157
References 158

8.1 INTRODUCTION

The broad category of poly- and perfluoroalkyl substances (PFAS) substances may be grouped into various subclasses of PFAS based on two factors: the number of carbons in the perfluoroalkyl chain and the functional group attached to this chain. Perfluorinated carboxylic acids (PFCAs) and perfluorosulfonic acids (PFSAs) are subsets of perfluorinated acids that consist of a nonpolar perfluorinated alkyl chain and a polar carboxylic acid (PFCA) or sulfonate (PFSA) end group (Buck et al., 2011). Long-chain PFAS (LC-PFAS) are defined as those PFAS that contain an extended perfluoroalkyl chain of eight carbons (C8) or longer for PFCAs and their precursors or six carbons (C6) for PFSAs and their precursors. Several concerns have been identified for LC-PFAS, including persistence in the environment and human tissues, potent reproductive and developmental toxicity, and carcinogenicity (USEPA, 2009). PFCAs and PFSAs with perfluorinated chain lengths shorter than those of LC-PFAS do not accumulate in mammalian tissues (Chengelis et al., 2009), although they are still persistent in the environment.

Aqueous film-forming foams (AFFFs) have been used extensively as fire suppressants for both civilian and military uses, and studies have found increased levels of PFAS in groundwater from watersheds containing military training sites or civilian airports within the watershed's hydrological unit. One published study found that each additional military training site within a given hydrological unit was associated with a 20% increase in groundwater perfluorohexane sulfonate (PFHxS), a 10% increase in groundwater perfluoroheptanoic acid (PFHpA) and perfluorooctanoic acid (PFOA), and a 35% increase in groundwater perfluorooctane sulfonate (PFOS) levels (Hu et al., 2016). In general, the main perfluorinated contaminants identified in groundwater deriving from AFFF-contaminated sites are primarily PFCAs and PFSAs of perfluorinated carbon-chain-lengths C2–C10 and 4:2, 6:2, and 8:2 fluorotelomer sulfonates (Barzen-Hanson and Field, 2015). Contamination of groundwater with PFAS may result in contamination of food and feed with these materials. A survey conducted by the European Food Safety Authority of PFAS in food during the period 2000–2009 found PFAS contamination in 68% of fish offal samples, 64% of game animal offal samples, 22% of game animal meat samples, 20% of mollusk samples, and 17% of crustacean samples (EFSA, 2011). Contamination levels as high as 216 ppb PFOS, 10.3 ppb perfluorononanoic acid (PFNA), 7.1 ppb PFOA, and 6 ppb perfluorodecanoic acid (PFDA) were found in samples of offal from game animals in this study. General population average and upper-bound exposure levels to PFAS from all routes of exposure (air, house dust, drinking water, and food) have been estimated to be 1.6 and 8.8 ng/kg bw/d for PFOS and 2.9 and 12.6 ng/kg bw/d for PFOA (Fromme et al., 2009), with exposure through the diet comprising the majority of the total daily exposure (Haug et al., 2011).

PFOS, PFOA, PFHxS, and PFNA are also detected in the sera of populations worldwide. National Health and Nutrition Examination Surveys (NHANESs) conducted during the years 1999–2008 have reported geometric mean serum levels of 13.2 ng/mL PFOS, 4.13 ng/mL PFOA, 1.49 ng/mL PFNA, and 1.96 ng/mL PFHxS in the general US population for the years 2007–2008. PFOS and PFOA serum levels showed a decreasing trend from the 1999–2000 survey to the 2007–2008 survey, while PFHxS and PFNA levels increased over the same time period (Kato et al., 2011). These serum levels represent the result of direct exposure to these four PFAS, as cited above, as well as exposure to perfluorinated precursors to these four compounds, such as fluorinated telomer products, that are metabolized *in vivo* or biotransformed in the environment to PFOS, PFNA, or PFOA (D'eon and Mabury, 2007). Populations exposed to heavily contaminated drinking water sources, such as the Washington Works cohort and the Arnsberg, Germany, cohort, have elevated levels of PFAS in their sera. PFOA levels in sera from the Washington Works cohort, whose groundwater was contaminated with PFOA from a fluorochemical production facility located nearby, had median serum PFOA levels that were 20-fold greater than levels in the general US population (Hoffman et al., 2011). Firefighters with occupational exposure to AFFF also have had elevated serum PFAS levels. One biomonitoring study conducted in Australian firefighters with more than 19 years' experience with handling AFFF measured serum levels of 92–343 ng PFOS/mL and

49–326 ng PFHxS/mL. This same study reported increased detection frequencies of perfluoroheptane sulfonate (PFHpS), perfluoropentane sulfonate (PFPeS), Cl-PFOS, ketone-PFOS, and Cl-PFHxS, as well as detection of several PFAS found only in firefighters, namely, perfluorononanesulfonate and ether-PFHxS (Rotander et al., 2015a, 2015b). Consumption of fish from marine sectors heavily contaminated by AFFF also resulted in increased serum PFAS concentrations in high- versus low-frequency fish consumers, with modeled serum concentration increases of 26 ng/mL PFOS, 0.8 ng/mL PFNA, and 0.56 ng/mL PFHxS based on biomonitoring data from a population located near Harstad/Narvik Airport in Norway (Hansen et al., 2016).

This chapter focuses on potential human health risks associated with oral exposure to PFAS present in groundwater contaminated with AFFF, emphasizing a class-based approach to understanding the toxic potential of this group of chemicals. Specifically, the available database for PFCAs and PFSAs is assessed with respect to adequate chemical space coverage and study quality, identification of common apical effects, comparison of potency values for these apical effects, and discussion of common modes of action. The pharmacokinetic properties of PFSAs and PFCAs are briefly summarized, followed by sections summarizing the effects of PFCAs and PFSAs in mammalian animal models noted in guideline and nonguideline single- or repeated-dose studies of varying duration conducted with adult animals. Studies conducted using the oral route are preferred, but studies conducted using inhalation or parenteral administration are also included since these PFAS are well absorbed by all routes of administration and are not metabolized. This chapter does not address pre- or postnatal toxicity or long-term effects of PFAS exposure during pre- or early postnatal development, as these effects are discussed in Chapter 9. Moreover, this chapter focuses primarily on data derived from studies conducted in mammalian animal models. *In vitro* and epidemiological data are discussed only in the context of establishing mode of action or human relevance of observed effects in animals.

8.2 PHARMACOKINETICS

PFCAs and PFSAs are readily absorbed after ingestion (USEPA, 2009) and are not metabolized. Accumulation occurs mainly in well-perfused tissues, particularly the liver, blood, and kidneys, with extremely low levels in other tissues (Benskin et al., 2009). PFSAs and PFCAs are largely bound to albumin in serum and other proteins (Ohmori et al., 2003) and excreted unchanged in the urine and/or bile; biliary excretion dominates at ≥C9 for PFCAs and at ≥C8 for PFSAs (Goecke et al., 1992; Ohmori et al., 2003). Urinary excretion is rapid for PFCAs of ≤C6 and PFSAs of ≤C4 in length in rats; urinary clearance rates in rats rapidly decrease for PFCAs of ≥C8 and PFSAs of ≥C6, leading to increased systemic half-lives (USEPA, 2009). Renal clearance rates for PFOA and PFNA are much lower in male rats than female rats due to reuptake of these PFCAs from the renal filtrate in males via renal organic anion transporters (OATs) whose expression is upregulated by testosterone (Weaver et al., 2010). In rats, systemic half-life values for perfluorobutane sulfonate (PFBS), perfluorohexanoic acid (PFHxA), n-PFOA, n-PFOS, PFNA, and PFDA have been estimated to be 2.4 hours, 2.3–2.6 hours, 2–4 hours, 73–95 days, 1.4–2.44 days, and 58.6 days in females and 3.1 hours, 2.2–2.8 hours, 6–7 days, 66–107 days,

29.5–30.6 days, and 39.9–40 days in males, respectively (Ohmori et al., 2003; Olsen et al., 2007; Chengelis et al., 2009; De Silva et al., 2009; Tatum-Gibbs et al., 2011). Systemic half-lives in mice have been estimated to be 5.22–16.25 hours for males and 2.79–3.08 hours for females for perfluorobutanoic acid (PFBA) (Chang et al., 2008b), <24 hours for both sexes for PFHxA (Gannon et al., 2011), 17–19 days for PFOA (USEPA, 2009), 27–30 days for PFHxS (Sundstrom et al., 2012), and 25.8–68.4 days in males and 34.3–68.9 days in females for PFNA (Tatum-Gibbs et al., 2011). Another study reported that PFBS reached systemic steady state in mice after 3 days, indicating that the systemic half-life for this compound is likely to be less than 24 hours, as it generally takes three to four half-lives to reach steady-state concentrations (Bogdanska et al., 2014). Half-life values for PFBA, PFBS, PFHxA, PFHxS, PFOA, and PFOS in monkeys have been estimated to be 40.32–41.04 hours (Chang et al., 2008b), 83.2–95.2 hours (Olsen et al., 2009), 24 hours (Gannon et al., 2011), 141 days in males and 87 days in females (Sundstrom et al., 2012), 14–42 days (Butenhoff et al., 2004a), and 45 days (Olsen et al., 2007).

In contrast to the relatively rapid rate of excretion of PFAS in rodents of hours to months, systemic half-lives of PFCAs and PFAS in humans are much longer. Estimates of systemic half-lives in humans differ based on the population samples. Estimates for PFBS, PFHxS, PFOS, and PFOA half-lives from biomonitoring data collected from occupationally exposed humans have been reported to be 25.8 days, 7.3 years, 4.8 years, and 3.5 years, respectively (Olsen et al., 2007, 2009). In contrast, PFHxS, PFOS, and PFOA geometric mean half-life values estimated from biomonitoring data collected from adults in the general population have been reported to be 7.1, 5.8, and 1.5 years in young adult females and 25, 18, and 1.2 years for all adult males and postmenopausal females, respectively (Zhang et al., 2013), indicating that significant sexual dimorphism in PFSA total body clearance exists in the general population. PFNA, PFDA, and perfluoroundecanoic acid (PFUnDA) geometric mean half-life values were estimated to be 1.7, 4, and 4 years for young female adults and 3.2, 7.1, and 7.4 years for all males and postmenopausal females, respectively, in the same study. Zhang et al. (2013) also estimated half-life values for PFHpA, which has not traditionally been considered a biopersistent compound, of 1 year in young females and 0.82 years in males and postmenopausal females. The reason for this apparent sexual dimorphism in the general population is, in addition to excretion in urine and bile, the ability of menstruating females to offload a portion of the PFAS body burden in menstrual blood, as well as the products of conception (fetus and placenta).

Systemic half-life may also be significantly affected by administered dose. Several studies conducted in animals have reported inverse relationships between calculated systemic half-life and dose. This effect may reflect differences in tissue partitioning, intracellular partitioning, and the percentage of the dose excreted in bile. One study that examined the tissue distribution and hepatic subcellular distribution of different doses of PFOA in male rats found that 52% of the dose was recovered in the liver after administration of 0.041 mg/kg bw (low dose) versus 27% of the dose in the livers of rats administered at the high dose of 16.56 mg/kg bw (Kudo et al., 2007). The intracellular distribution of the majority of the PFOA dose also shifted from the nuclear/mitochondrial fraction at the low dose to the cytoplasmic fraction

Human Health Risk Assessment of Perfluorinated Chemicals

at the high dose. At the low dose, the vast majority of the dose was recovered in the liver > serum = kidney > blood, while at the high dose, the order of recovery of the dose was serum > liver > kidney > blood. Further, while the percentage of the dose cleared via renal excretion was constant, the percentage of the dose excreted in bile increased 3.6-fold from the low to the high dose (Kudo et al., 2007). These data may explain the variance in calculated systemic half-life values for PFAS between those derived from biomonitoring studies in the general population and those derived from occupationally exposed populations. This finding also complicates interpolation of data from animal studies conducted at extremely high doses of PFAS to possible effects of low doses in the human population.

8.3 HEALTH EFFECTS AND TOXICOLOGY

8.3.1 Systemic Toxicity: Nonneoplastic Effects on Organ Systems

Toxicity data from studies conducted in mammalian animal models are available for PFBS, PFHxS, and PFOS for PFSAs, and PFBA, PFHxA, PFOA, PFNA, PFDA, perfluorododecanoic acid (PFDoDA), and perfluorooctadecanoic acid (PFOcDA). The majority of the studies have been conducted in adult rats and/or mice in repeated-dose study designs of between 7 days and 2 years in duration. Additional specialized studies assessing the effects of PFAS on immune system, hepatic, testicular, and/or thyroid end points in rodents have been conducted with PFOA, PFOS, PFNA, PFDA, and PFDoDA. There are no mammalian toxicity or pharmacokinetic data available for ≤C3 PFCAs and PFSAs, C5 and C7 PFCAs and PFSAs, PFSAs of ≥C9, PFCAs of C13–C17, or any of the fluorotelomer sulfonic acids (FTSAs). By far, PFOA and PFOS have the most complete toxicological profile associated with them, including guideline oral toxicity studies conducted in rats 14–90 days in duration; subchronic oral studies in monkeys; repeated-dose oral toxicity data in mice; 2-year chronic oral bioassays in rats; reproductive toxicity studies in rats and mice; teratology studies in multiple species; single- and repeated-dose pharmacokinetic studies in rats, mice, and monkeys; and epidemiological and biomonitoring data in humans. The other compounds for which data are available are much less data rich, often only having been assessed in a single species or only in nonguideline studies. PFOcDA (Hirata-Koizumi et al., 2012), PFHxS (Butenhoff et al., 2009), PFDoDA (Kato et al., 2015), and PFUnDA (Takahashi et al., 2014) have only been assessed in OECD 422 repeated-dose and reproductive toxicity screening assessment studies conducted in rats. The OECD 422 design, where adult animals are dosed with the test compound for 42–56 days, is not a particularly sensitive study design due to the short duration of dosing time, low power, and limited end points assessed during the study (OECD, 2015). As such, high-confidence potency comparisons between compounds can only be made across the same study design and animal model. Therefore, overall systemic potency comparisons can only be made for the 90-day studies in rats conducted with PFBA (Van Otterdijk, 2007), PFHxA (Chengelis et al., 2009), sodium perfluorohexanoate (NaPFHx) (Slezak, 2007), PFOA (Perkins et al., 2004), PFBS (Lieder et al., 2009a), and PFOS (Seacat et al., 2003). Potency after 2-year oral exposure can only be compared in rats for PFOS (Butenhoff et al., 2012b), PFHxA

(Klaunig et al., 2014), and PFOA (Butenhoff et al., 2012a). Potency for adverse effects on male and female fertility in standard reproductive toxicity assays (i.e., OECD Guideline 416 two-generation reproductive toxicity tests [OECD, 2001]) can only be assessed in rats for PFOA (Butenhoff et al., 2004b; York et al., 2010), PFOS (Luebker et al., 2005), and PFBS (Lieder et al., 2009b). OECD 416 studies typically dose parental males for 70 days prior to cohabitation through a 2-week cohabitation period, for a total of 70–84 days. Parental females are dosed through the same 70- to 84-day period, plus an additional 43 days comprising gestation and lactation through postnatal day 22, for a total of 113–127 days.

Overall, the no observed effect levels (NOELs) for systemic toxicity in rats from 90-day studies are 6 mg/kg/d in males (540 mg/kg total administered dose [TAD]) and 30 mg/kg/d (2700 mg/kg TAD) in females for PFBA (Van Otterdijk, 2007); 10 mg/kg/d in males (900 mg/kg TAD) and 50 mg/kg/d in females (4500 mg/kg TAD) for PFHxA (Chengelis et al., 2009); <20 mg/kg/d in males (<1800 mg/kg TAD) and 20 mg/kg/d in females (1800 mg/kg TAD) for NaPFHx (Slezak, 2007); 0.06 mg/kg/d in males (5.4 mg/kg TAD) for PFOA (Perkins et al., 2004); <60 mg/kg/d in males (<5400 mg/kg TAD) and 60 mg/kg/d (5400 mg/kg TAD) in females for PFBS (Lieder et al., 2009a); and 0.13 mg/kg/d in males (13.1 mg/kg TAD) and 0.4 mg/kg/d (39 mg/kg TAD) in females for PFOS (Seacat et al., 2003). NOEL values for systemic toxicity from 182-day repeated-dose studies in cynomolgus monkeys are <3 mg/kg/d (<1638 mg/kg TAD) for PFOA (Butenhoff et al., 2002) and 0.15 mg/kg/d (22.9 mg/kg TAD) for PFOS (Seacat et al., 2002). No effects on male or female fertility parameters were noted in any of the studies with any compound, except for PFDoDA, which had a reproductive toxicity NOEL value of 0.5 mg/kg/d (21–23.5 mg/kg TAD in females) for decreased gestation index, delivery index, and number of litters at 2.5 mg/kg/d (105–117.5 mg/kg TAD in females) and produced adverse effects on the parental male and female reproductive tracts at the same dose level (Kato et al., 2014). These data indicate that the order of potency for systemic toxicity in male rats in 90-day studies is PFOA > PFOS > PFBA > PFHxA ~ NaPFHx > PFBS. For female rats, the order of potency is PFOS > NaPFHx > PFBA > PFHxA > PFBS. The critical effects at the lowest observed effect levels (LOELs) for these studies differ for each compound. High doses often result in decreased body weight gain and feed efficiency or intake, with PFCA of ≥C10 producing frank body weight loss in rats (Seacat et al., 2002, 2003; Perkins et al., 2004; Shi et al., 2007; Zhang et al., 2008). Target organs common to several of the tested compounds are the liver, the immune system, and the testes. Various other targets of toxicity include the kidney, thyroid, hematopoetic system, and mammary gland. Effects on each target organ are discussed separately below. Overall NOEL and LOEL values and LOELs for effects on the liver, immune system, thyroid, and male and female reproductive systems are listed in Table 8.1.

8.3.1.1 Effects on the Liver

The liver is one of the most sensitive target organs for PFAS toxicity. Increased liver weight and hepatocellular hypertrophy have been noted with almost all of the PFAS assessed in rodent models, with male rats being more sensitive to these effects than female rats. LOELs from OECD guideline studies for increased liver weight

TABLE 8.1
NOEL and LOEL Values

Compound	Reference/Study Type	Overall NOEL/Effect at LOEL (TAD)	Liver Effects LOEL	Thyroid Effects LOEL	Immunotoxicity LOEL	Male Reproductive System Effects LOEL	Female Reproductive System LOEL
PFBA	Van Otterdijk (2007); rat 90-day study	>2700 mg/kg F; 540 mg/kg M; liver and thyroid effects	↑ weight parameters, hepatocellular hypertrophy, 2700 mg/kg M No effects F	↑ incidence and severity of follicular cell hypertrophy, 2700 mg/kg M No effects F	No effects	No effects	No effects
PFHxA	Sezak (2007); Na PFHx rat 90-day study	1800 mg/kg F; ↓ incidence degeneration/atrophy olfactory epithelium, turbinate adhesions, respiratory metaplasia None reached M, ↓ thymic weights	≥9000 mg/kg M; ↑ weight parameters, hepatocellular hypertrophy, microsomal β-oxidation, serum ALT, AST; ↓ bilirubin ≥9000 mg/kg F; ↓ bilirubin; ↑ liver weights, hepatocellular hypertrophy, microsomal β-oxidation, 45,000 mg/kg	↑ incidence minimal follicular cell hypertrophy; both, 45,000 mg/kg	↓ thymic weights, ≥1800 mg/kg M	No effects	No effects
	Chengelis et al. (2009); rat 90-day study PFHxA	900 mg/kg M; ↓ BW, serum Ca 4500 mg/kg F; ↓ RBC parameters, serum total protein, globulin	↓ serum total protein, globulin, 18,000 mg/kg F ↑ liver weight parameters, peroxisome β-oxidation, centrilobular hepatocyte necrosis, 18,000 mg/kg M	No effects	No effects	No effects	No effects

(Continued)

TABLE 8.1 (CONTINUED)
NOEL and LOEL Values

Compound	Reference/ Study Type	Overall NOEL/Effect at LOEL (TAD)	Liver Effects LOEL	Thyroid Effects LOEL	Immunotoxicity LOEL	Male Reproductive System Effects LOEL	Female Reproductive System LOEL
	Klaunig et al. (2014); rat 2-year study, PFHxA	3650 mg/kg F: hepatocellular necrosis, pulmonary hemorrhage and inflammation, stomach erosion/inflammation 10,950 mg/kg M: clinical signs toxicity; ↓ serum TG and FFA; ↑ InorgP; ↓ urine pH; lung hemorrhage/inflammation	≥21,900 mg/kg F: hepatocellular necrosis 73,000 mg/kg M: ↓ serum TG	No effects	No effects	No effects	No effects
PFOA	York et al. (2010), Butenhoff et al. (2004b); two-generation study in rats, APFO	M: none; ↑ liver and kidney weight parameters, ≥70–84 mg/kg F: 3210–3330 mg/kg: highest dose tested	M: ≥70–84 mg/kg; ↑ liver weight parameters F: >3210–3330 mg/kg; no effects	No effects	No effects	↑ BW—relative seminal vesicles + fluid, ≥70–84 mg/kg; ↑ BW—relative testes and accessory sex organs, ≥700–840 mg/kg; ↓ abs epididymes, seminal vesicles (± fluid), and prostate weights, 2100–2520 mg/kg	No effects
	Butenhoff et al. (2002); 26-week capsule study, APFO, male cynomolgous monkeys	None; at lowest dose of 1638 mg/kg, mortality; ↑ liver and kidney weights, Kupffer cell pigmentation, thymic involution; ↓ free T4, lymphohistiocytic infiltration	↑ liver weights, Kupffer cell pigmentation, ≥1638 mg/kg ↑ liver palmitoyl CoA oxidase activity, ≥1820 mg/kg	↓ free T4, ≥1638 mg/kg ↓ total T4, ≥1820 mg/kg ↓ free T3, total T3, 5460/3640 mg/kg	Thymic involution, ≥1638 mg/kg ↓ leukocyte, lymphocyte counts, ≥1820 mg/kg	Testicular atrophy and degeneration, one 5460 mg/kg male; nonstatistically significant; ↓ serum E1 and E2, ≥1820 mg/kg; ↓ T, 5460 mg/kg	N/A

(Continued)

TABLE 8.1 (CONTINUED)
NOEL and LOEL Values

Compound	Reference/Study Type	Overall NOEL/Effect at LOEL (TAD)	Liver Effects LOEL	Thyroid Effects LOEL	Immunotoxicity LOEL	Male Reproductive System Effects LOEL	Female Reproductive System LOEL
	IRDC (1978); 90-day study, rhesus monkeys	270 mg/kg for effects seen at that dose, occasional diarrhea; ↓ urine pH (not statistically significant); ↑ adrenal weight parameters, slight hyperkeratosis	↑ ALT, ≥900 mg/kg	No effects	≥2700 mg/kg; both lymphoid follicle atrophy, spleen and bone marrow hypocellularity	No effects	No effects
	Perkins et al. (2004); 90-day feeding study, male rats, APFO	5.4 mg/kg; at ≥57.6 mg/kg, ↑ liver weights and palmitoyl CoA oxidase activity, hepatocellular hypertrophy	≥57.6 mg/kg; ↑ liver weights and palmitoyl CoA oxidase activity, hepatocellular hypertrophy	No effects	No effects	↑ serum E, ≥57.6 mg/kg; ↑ BW—relative testes weights, 553.5 mg/kg; no changes serum T or LH, brain weight or abs testes weights, accessory sex organ weights; no histopathology	N/A
	Butenhoff et al. (2012a); 2-year feeding study, rats, APFO	None; 949–1175 mg/kg; ↑ liver weights (M); ↑ kidney weights (both), hepatocellular hypertrophy and cystic hepatic degeneration (both, M > F), vascular mineralization testes, ovarian tubular mineralization; ↑ ALT, AST, ALP, albumin (M)	949–1175 mg/kg; ↑ liver weights (M), hepatocellular hypertrophy and cystic hepatic degeneration (both, M > F); ↑ ALT, AST, ALP, albumin (M)	No effects	No effects	↑ Incidence vascular mineralization of testes, ≥949 mg/kg; Leydig cell adenomas, 10,366 mg/kg	↑ Ovarian tubular mineralization, ≥1175 mg/kg
PFUnDA	Takahashi et al. (2014); OECD 422 study, rats	4.2 mg/kg M, 4.1–4.6 mg/kg F; liver histology and weight changes	↑ liver weight parameters, ≥12.6 mg/kg M, 41–46 mg/kg F; ≥12.6–13.8 mg/kg; centrilobular hepatocyte hypertrophy, both	No effects	↓ spleen weight parameters, 42 mg/kg M only	No effects	No effects

(Continued)

TABLE 8.1 (CONTINUED)
NOEL and LOEL Values

Compound	Reference/Study Type	Overall NOEL/Effect at LOEL (TAD)	Liver Effects LOEL	Thyroid Effects LOEL	Immunotoxicity LOEL	Male Reproductive System Effects LOEL	Female Reproductive System LOEL
PFDoDA	Kato et al. (2014); OECD 422 study, rats	<4.2 mg/kg; ↓ cholesterol M at ≥4.2 mg/kg 21–23.5 mg/kg F: mortality; ↓ BW and feed consumption through gestation, pancreatic interstium edema, liver and thymus pathology, hemorrhage at uterine implant site	M: ↓ CHOL ≥4.2 mg/kg; ↑ BW—relative liver, ≥21 mg/kg: diffuse hepatocyte hypertrophy, 105 mg/kg F: hepatocyte single-cell necrosis and diffuse hepatocyte hypertrophy, 105–117.5 mg/kg	No effects	Thymic cortex atrophy, 105–117.5 mg/kg F 105 mg/kg M: ↓ leukocytes, lymphocytes, monocytes, eosinophils	105 mg/kg: glandular epithelium atrophy of prostate, seminal vesicle, and coagulating gland; ↓ spermatozoa, cell debris in seminiferous tubule; ↓ elongated spermatids, spermatic granulomas	Persistent diestrus, 105–117.5 mg/kg; recovery group only
PFOcDA	Hirata-Koizumi et al. (2012); OECD 422 study, rats	1680 mg/kg M, 1680–2240 mg/kg both sexes; ↑ liver weights, histopathology changes in liver, pancreas, thymus in M only	↑ liver weights, M ≥8400 mg/kg, F ≥8400–11,200 mg/kg M ≥8400 mg/kg, F 42,000–56,000 mg/kg: centrilobular hepatocellular hypertrophy; ↓ CHOL	No effects	Thymic cortex atrophy: M 42,000 mg/kg, F 42,000–56,000 mg/kg	No effects	No effects
PFBS	Lieder et al. (2009a); rats, 90-day gavage	None M, non-DR; ↓ spleen weight parameters, ≥5400 mg/kg F: 5400 mg/kg, multifocal necrosis/atrophy and inflammation olfactory epithelium with foci epithelial hyperplasia	54,000 mg/kg F: ↓ total protein and albumin	No effects	M non-DR; ↓ spleen weight parameters, all doses (≥5400 mg/kg)	No effects	No effects

(Continued)

TABLE 8.1 (CONTINUED)
NOEL and LOEL Values

Compound	Reference/Study Type	Overall NOEL/Effect at LOEL (TAD)	Liver Effects LOEL	Thyroid Effects LOEL	Immunotoxicity LOEL	Male Reproductive System Effects LOEL	Female Reproductive System LOEL
	Loeder et al. (2009b); two-generation study, rats	7100–8400 mg/kg M, 10,710–11,100 mg/kg F; renal tubule epithelial hyperplasia (both), renal papillary focal edema (F), hepatocellular hypertrophy (M); ↑ liver weights (M), ≥21,300–25,200 mg/kg M, 32,130–33,300 mg/kg F	≥21,300–25,200 mg/kg M: ↑ liver weights and hepatocellular hypertrophy	No effects	No effects	No effects	No effects
PFHxS	Butenhoff et al. (2009); rats, OECD 422 study	None, M: ↓ CHOL, prothrombin time, ≥13.2 mg/kg; >590–610 mg/kg F	↓ CHOL, all doses; ↑ liver weights and hepatocellular hypertrophy, ≥132 mg/kg M	Follicular cell hypertrophy/hyperplasia, ≥132 mg/kg M	No effects	No effects	No effects
PFOS	Seacat et al. (2003); rats, 90-day study	M: 13.1 mg/kg, hepatocellular hypertrophy and midzonal/centrilobular vacuolation, ≥33 mg/kg F: 39 mg/kg; at 148 mg/kg, ↑ liver weights, BUN; ↓ feed consumption; hepatocellular hypertrophy and midzonal/centrilobular vacuolation	Hepatocellular hypertrophy and midzonal/centrilobular vacuolation, ≥33 mg/kg M, 148 mg/kg F ↑ liver weights and BUN (130 mg/kg M, 148 mg/kg F), ALT (130 mg/kg M); ↓ CHOL (130 mg/kg M) No peroxisome proliferation detected	No effects	↑ PMN, 130 mg/kg M	No effects	No effects

(*Continued*)

TABLE 8.1 (CONTINUED)
NOEL and LOEL Values

Compound	Reference/ Study Type	Overall NOEL/Effect at LOEL (TAD)	Liver Effects LOEL	Thyroid Effects LOEL	Immunotoxicity LOEL	Male Reproductive System Effects LOEL	Female Reproductive System LOEL
	Seacat et al. (2002); 182-day cynomolgus monkey study, 1-year recovery	22.9, liver effects, thyroid effects; ↓ estradiol (M), ↓ RBC M at 114.7 mg/kg	114.7 mg/kg both: hepatocellular hypertrophy/ vacuolation/glycogen accumulation; ↓ CHOL, HDL 114.7 mg/kg F: ↑ ALP, ALT 114.7 mg/kg M: ↓ total bilirubin; ↑ SBA No peroxisome proliferation or PCNA antigen detected	↓ free T3, TT3; ↑ TSH (114.7 mg/kg M)	No effects	No effects	No effects
	Butenhoff et al. (2012b); rats, 2-year feeding study with 52-week *dosing* → 52-week recovery (recovery arm)	17.52 mg/kg M, 21.17 mg/kg F; serum biochemistry changes (F) and liver histopathology (M) at 71.54 mg/kg M and 87.6 mg/kg F	↓ CHOL, ≥87.6 mg/kg F ≥71.54 mg/kg M: hepatocellular hypertrophy and eosinophilic clear cell foci, cystic hepatocellular degeneration ↑ incidence hepatocellular adenoma: 718.32 mg/kg M, 913.23 mg/kg F	↑ incidence follicular cell adenoma, recovery group, 416.4 mg/kg M	No effects	No effects	No effects

Note: M = males; F = females; APFO = ammonium perfluorooctanoate; BUN = blood urea nitrogen; BW = bodyweight; DR = dose responsive; CHOL = cholesterol; FFA = free fatty acids; LH = luteinizing hormone; N/A = not assessed; TG = triglycerides; T = testosterone; PCNA = proliferating cell nuclear antigen; HDL = high density lipoprotein cholesterol; RBC = red blood cells; ALP = alkaline phosphatase; ALT = alanine aminotransferase; TSH = thyroid stimulating hormone; TT3 = total triiodothyronine; PMN = neutrophils; AST = aspartate aminotransferase; InorgP = inorganic phosphate; T4 = thyroxine; E1 = estrone; E2 = estradiol; SBA = serum bile acids; coA = coenzyme A.

and/or histopathological evidence for hepatocellular hypertrophy in male rats have been reported to be (in terms of TAD) 2700 mg/kg for PFBA (Van Otterdijk, 2007), 9000 mg/kg for NaPFHx (Slezak, 2007), 18,000 mg/kg for PFHxA (Chengelis et al., 2009), 57.6 mg/kg for PFOA (Perkins et al., 2004), 12.6 mg/kg for PFUnDA (Takahashi et al., 2014), 21 mg/kg for PFDoDA (Kato et al., 2015), and 8400 mg/kg for PFOcDA (Hirata-Koizumi et al., 2012) for the PFCAs. For the PFSAs, LOEL values for this effect in male rats are reportedly 12,600 mg/kg for PFBS (Lieder et al., 2009b), 132 mg/kg for PFHxS (Butenhoff et al., 2009), and 33 mg/kg for PFOS (Seacat et al., 2003). For studies conducted in cynomolgus monkeys, LOEL TAD values for increased liver weights and/or hepatocellular hypertrophy were 546 mg/kg for PFOA (Butenhoff et al., 2002) and 136.5 mg/kg PFOS (Seacat et al., 2002). Dose levels that produce liver histopathology may be higher (Nakagawa et al., 2012), lower (Seacat et al., 2003; Curran et al., 2008), or the same (Perkins et al., 2004; Van Otterdijk, 2007) as those that increase liver weight parameters.

Hepatocellular hypertrophy noted in these studies is often, but not always, accompanied by histopathological or biochemical evidence of peroxisomal proliferation and biochemical evidence of increased peroxisomal β-oxidation rates, commonly reported as increased cyanide-insensitive palmitoyl coenzyme A (CoA) oxidation rates in liver microsomal fractions. Increased liver weights and peroxisomal enzymes appear to be the critical effects for PFOA, occurring at serum levels as low as 14.1 μg PFOA/mL in mice (Wolf et al., 2008) and 70 μg PFOA/mL in rats (Perkins et al., 2004). The potency for induction of peroxisome proliferation increases with perfluorocarbon chain length and is proportional to the degree of hepatic accumulation of the compound (Kudo et al., 2006). PFOA and PFNA appear to be among the most potent peroxisome proliferator–activated receptor (PPAR)-α agonists (Wolf et al., 2012). PFDA may be the most potent peroxisomal enzyme inducer of the PFCAs, as a single injection of 50 mg/kg increased peroxisomal β-oxidation in the livers of male rats (Harrison et al., 1988). Peroxisomal enzyme induction may coexist with peroxisomal enzyme inhibition, such as the case with PFDA, PFDoDA, and PFOA (Singer et al., 1990; Borges et al., 1992). This increased peroxisomal β-oxidation is often accompanied by dose-related decreases in serum cholesterol, the ratio of high-density lipoprotein (HDL) to low-density lipoprotein (LDL), and triglycerides (Borges et al., 1992; Zhang et al., 2008; Minata et al., 2010; Bijland et al., 2011) after administration of PFCAs and PFSAs, with potency increasing with chain length. However, repeated-dose oral toxicity studies conducted with rats (Seacat et al., 2003) and monkeys (Seacat et al., 2002) with PFOS reported no signs of peroxisomal proliferation in the liver, even though hepatocellular hypertrophy and signs of liver toxicity, such as elevated serum liver enzymes and hepatocyte vacuolation, and decreased serum cholesterol were noted in both species. The reason for this discrepancy is unclear, as other studies conducted with PFOS in rats have reported clear induction of hepatocellular peroxisomal enzymes (Curran et al., 2008; Elcombe et al., 2012).

The effects of PFASs on lipid metabolism commonly manifest themselves as increased liver lipid and cholesterol content, especially in mice, and PFCAs and PFSAs both increased liver cholesterol and triglyceride content in rodents at chain lengths of ≥C8 and ≥C6, respectively (Harrison et al., 1988; Bijland et al., 2011; Fang et al., 2012a; Nakagawa et al., 2012). PFCAs of C6–C10 in length increase hepatic concentrations of unsaturated lipids and alter phosphatidyl choline acyl composition via induction of stearoyl CoA desaturase, 1-acylglycerophosphocholine acyltransferase, chain elongase, and delta-6 desaturase enzymes (Kawashima et al., 1989; Uy-Yu et al., 1990; Kudo et al., 2011) in mice and male rats, with the effects in male rats being testosterone dependent. Kudo and Kawashima (2003) demonstrated that hepatic peroxisomal β-oxidation and triglyceride accumulation in male and female rat hepatocytes are dependent on the total molar concentration of PFCA in hepatocytes, regardless of perfluorinated chain length. The relative potency of the PFCAs for inducing these liver effects was attributable to the differences in the toxicokinetics of these compounds in the liver. For example, PFHpA, which did not accumulate in the livers of rats of either sex, did not induce peroxisome β-oxidation or triglyceride accumulation. PFDA, which accumulates in the livers of both sexes to an equal extent, produced peroxisome β-oxidation or triglyceride accumulation in both sexes to a roughly equal extent. In contrast, PFOA, which accumulates in the livers of male but not female rats, induces hepatic peroxisome β-oxidation, but not triglyceride accumulation, in males only. PFNA also produces a gender-specific liver effect, inducing peroxisome β-oxidation in both sexes, but triglyceride accumulation in males only. The relative hepatocyte molar threshold for effect induction is lower for peroxisome β-oxidation than triglyceride accumulation. Dose-responsive, chain-length-independent increases in peroxisome β-oxidation occurred up to a saturating dose level of 500 nmol/g liver, which was the lower-bound threshold for induction of triglyceride accumulation (Kudo and Kawashima, 2003). This accumulation of lipid in the liver may be due to activation of PPAR-γ genes (Rosen et al., 2008), increased liver very low-density lipoprotein (VLDL) particle uptake (Bijland et al., 2011), and/or activation of PPAR-α (Rosen et al., 2008). In contrast to rodent liver cells, human liver cells are refractory to peroxisomal proliferation (Bjork and Wallace, 2009) even though PFAS are agonists at both the human and rodent PPAR-α receptor with apparently equal potency (Vanden Heuvel et al., 2006). This complicates the extrapolation of the results of rodent studies to humans.

Many PFAS induce xenobiotic-metabolizing enzymes in the liver and increase hepatocellular mitochondrial protein content. A comparative study of C2–C10 PFCA (Permadi et al., 1992) in mice reported that PFCAs of >C4 increased liver mitochondrial protein content by 5- to 9-fold, increased cytochrome P450 (CYP) protein content by 3- to 6-fold, increased liver DT-diaphorase content by 3- to 10-fold, doubled cytoplasmic epoxide hydrolase content, and increased mitochondrial thiobarbaturic acid (TBA) content by 60%, with PFOA being the most potent of the tested compounds. Changes in hepatic xenobiotic enzyme activity often occur at dose levels below those that cause increased liver weights and histopathological changes, making this effect one of the most sensitive markers of PFAS effects in the liver (Permadi et al., 1993; Curran et al., 2008). This effect may be due to activation of the constitutive androstane receptor (CAR) in hepatocytes. Abe et al. (2016) report that PFCAs of ≥C8 indirectly activate both mouse and human CAR, and PFOA-mediated induction of CYP 2B10 expression was abolished in CAR knockout (KO) mice; PFOA and PFNA, but not PFHpA or PFDA, both increased CYP 2B10 expression in wild-type (WT) mice.

Increased lipid content in the liver is frequently, although not always, associated with hepatotoxicity and consequent increases in serum alanine aminotransferase (ALT) and/or asparatate aminotransferase (AST) levels (Kawashima et al., 1995; Perkins et al., 2004; Son et al., 2009; Hirata-Koizumi et al., 2012). Increased serum enzyme levels in guideline studies were noted at dose levels (TAD) of 2700 mg/kg PFBA in male rats (Van Otterdijk, 2007), 9000 mg/kg NaPFHx in male rats (Slezak, 2007), 949 mg/kg PFOA in male rats (Butenhoff et al., 2012a), 900 mg/kg PFOA in rhesus monkeys (IRDC, 1978), 130 mg/kg PFOS in male rats (Seacat et al., 2003), and 114.7 mg/kg PFOS in female cynomolgus monkeys (Seacat et al., 2002). No other effects on serum liver enzyme levels were noted in guideline studies, except as noted above, even at doses where histopathological evidence of hepatocellular damage was present. In contrast to the data from studies in rats, mice are far more susceptible to PFAS-induced hepatotoxicity, with elevations in serum liver enzyme levels occurring at doses as low as 35 mg/kg PFOA TAD in male mice (Yang et al., 2014) and hepatocellular single-cell necrosis occurring at 210 mg/kg PFOA in WT, KO, and humanized PPAR-α mice (Nakagawa et al., 2012). This hepatotoxicity is PPAR-α independent, as PFOA caused marked accumulation of lipid in hepatocytes, hepatocellular damage, and apoptosis of hepatocytes and bile duct cells in both WT and PPAR-α KO mice, with the KO mice suffering a higher degree of damage, inflammatory cytokine production, and oxidative stress (Minata et al., 2010). Further, PFOA administration exacerbates hepatocellular toxicity associated with concanavalin A administration (Qazi et al., 2013) and a high-fat diet (Tan et al., 2013) in mice. The accumulation of hepatic lipids co-occurs with increased liver inflammatory cytokine content and induction of inflammatory cytokine gene expression in Kupffer cells (Fang et al., 2012b; Nakagawa et al., 2012; Qazi et al., 2013; Yang et al., 2014) and indicators of increased oxidative stress, such as increased isoprostane content (Kamendulis et al., 2014) and malondialdehyde (MDA) content (Yang et al., 2014). Increased inflammatory secretion by Kupffer cells may be the immediate causative agent of PFAS liver toxicity as elimination of Kupffer cells via pretreatment with $GdCl_2$ protected male rats from hepatotoxicity caused by PFNA administration (Fang et al., 2012b). Complement activation and deposition in the liver may also contribute to the observed cytotoxicity (Botelho et al., 2015) as liver deposition of C3 fragments and decreased classical and alternative pathway complement activity have been noted in mice administered with PFOA in the diet. Hepatotoxicity of PFCAs may also be mediated through indirect activation of CARs as CAR KO mice are protected from hepatic injury seen in CAR WT mice fed with a methionine- and choline-deficient diet (Yamazaki et al., 2007). The pathogenesis of the hepatic injury seen in mice fed with a methionine- and choline-deficient diet involves accumulation of lipid in the liver, followed by increased oxidative stress injury in hepatocytes induced by lipid peroxidation. Yamazaki et al. (2007) demonstrated that CAR activation mediates the increase in lipid peroxidation that constitutes the "second hit" producing injury; the mechanisms of toxicity for the methionine- and choline-deficient diet are also produced by PFCAs as MDA and isoprostanes are molecular signatures of lipid peroxidation. Further studies in CAR KO mice administered PFCAs will be needed to determine whether this proposed "two-hit" mechanism is indeed operative in PFAS-induced liver injury. CAR activation in the absence of concomitant PPAR-α activation may also explain the qualitatively distinct hepatic toxicity PFOA produces in PPAR-α KO mice (Wolf et al., 2008).

Epidemiological studies investigating the association between PFCA serum levels and serum liver enzymes have found positive associations in the general population (Lin et al., 2010; Gleason et al., 2015) and the highly exposed Washington Works population (Gallo et al., 2012) but no association in occupationally exposed works at three different sites (Gilliland and Mandel, 1996; Olsen et al., 2000, 2003; Olsen and Zobel, 2007). No association with liver disease was noted in an occupationally exposed cohort (Lundin et al., 2009) or in the Washington Works cohort (C8 Science Panel, 2012a). Several epidemiology studies have noted positive associations between serum PFAS levels and blood total cholesterol and/or triglyceride levels and increased odds ratios of high cholesterol in the general population, highly exposed general population cohorts, and some occupationally exposed cohorts (Olsen et al., 2003; Sakr et al., 2007a, 2007b; Steenland et al., 2009; Frisbee et al., 2010; Nelson et al., 2010; Eriksen et al., 2013; Fisher et al., 2013; Fu et al., 2014), with one study reporting a parallel 50% decrease in serum total cholesterol and LDL with halving of serum PFOA and PFOS levels in a repeated-measures study of the Washington Works cohort (Fitz-Simon et al., 2013). In contrast, some studies have noted only a negative association with serum HDL levels (Gilliland and Mandel, 1996; Olsen and Zobel, 2007; Wang et al., 2012). As decreased HDL cholesterol levels and increased total cholesterol and LDL cholesterol levels are associated with increased risk of cardiovascular disease, the Lin et al. (2013a) finding of a positive association between serum PFOA and carotid artery intima media thickness is congruent with the findings cited above, as is the finding of a positive association of serum PFOA levels with increased odds ratios for peripheral artery disease and cardiovascular disease in the general US population (Shankar et al., 2012). There is a striking difference between the findings of epidemiology studies and the decreased serum lipids reported from studies conducted in both rodents with PFCAs and PFSAs and monkeys with PFOS. However, studies conducted with PFOA in rhesus and cynomolgus monkeys reported increased serum lipids after PFOA administration (IRDC, 1978; Butenhoff et al., 2002). This effect was noted at dose levels that produced overt toxicity, such as severe immunotoxicity and mortality in both studies, making the relevance of these findings to humans exposed to environmentally relevant doses unclear.

Preferential retention of LC-PFSAs and PFCAs in the liver may partially explain the relative sensitivity of this organ to the toxic effects of these compounds. Retention occurs due to, in part, relatively high-affinity binding of these compounds to liver fatty acid–binding protein (L-FABP), which is highly expressed in the liver, kidneys, and intestines. The binding affinity of PFSAs and PFCAs to human L-FABP increases with perfluorinated carbon chain length, with relative binding affinity decreasing in the order of PFNA > PFOS > PFHxS/PFHxA; PFNA and PFOA binding affinity constants to human L-FABP were estimated to be 3.14 and 6.49 µM, respectively (Sheng et al., 2014). IC50 values for binding of PFOS and PFOA to rat L-FABP were calculated to be 4.9 and >10 µM, respectively, which are significantly greater than the IC50 value of 0.01 µM for oleate (Luebker et al., 2002), indicating that perfluorinated compounds have lower affinity for L-FABP than endogenous fatty acids.

In summary, the liver is a sensitive target organ for PFCA and PFSA toxicity, commonly producing hepatocellular hypertrophy or increased liver weight, induction of peroxisome proliferation and increased peroxisomal β-oxidation, induction of xenobiotic-metabolizing enzymes, alterations in hepatocyte lipid composition, and decreased serum cholesterol and triglycerides. The more potent LC-PFAS, such as PFOA and PFOS, induce triglyceride and cholesterol accumulation in hepatocytes, which in turn are associated with increased oxidative stress and lipid peroxidation, increased inflammatory cytokine production by Kupffer cells, and liver toxicity, manifesting as hepatocellular necrosis and/or elevations in serum liver enzymes. The effects of PFASs in the liver and the potency for producing these effects are directly proportional to the degree of accumulation of the PFAS compound in hepatocytes. Effects on peroxisomal and xenobiotic metabolism enzyme activity are mediated by PPAR-α and CAR activation, but hepatotoxicity itself appears to be PPAR-α independent. No effects of PFAS on serum liver enzymes have been seen in epidemiology studies, and the majority of the epidemiology studies suggest that PFAS exposure increases serum cholesterol in humans, in contrast to the decreases seen in rodents, even though PFAS are agonists for both mouse and human PPAR-α. Therefore, the relevance of the observed effects seen in rodent livers to human health risk assessment remains unclear.

8.3.1.2 Effects on the Immune System

Most of the specialized studies evaluating the immunotoxicity of perfluorinated compounds have focused on PFOS, PFOA, and PFNA, although splenic and thymus weights and histopathology and standard whole-blood differential cell counts have been assessed in guideline subchronic, chronic, or reproductive toxicity studies conducted in rats with PFBA, PFHxA, PFOS, PFOA, PFBS, PFHxS, PFUnDA, PFDoDA, and PFOcDA. Decreased spleen weights were reported at LOEL TAD values of 5400 mg/kg PFBS and 42 mg/kg PFUnDA in male rats (Lieder et al., 2009a; Takahashi et al., 2014). Decreased thymus weights were reported at 1800 mg/kg NaPFHx in rats (Slezak, 2007). Overt thymic atrophy was noted at 105 mg/kg in male rats and 105–117.5 mg/kg in female rats with PFDoDA (Kato et al., 2015) and 42,000 mg/kg in male rats and 42,000–56,000 mg/kg in female rats with PFOcDA (Hirata-Koizumi et al., 2012). PFOA produced lymphoid follicle atrophy in spleen and lymph nodes and bone marrow hypocellularity at 2700 mg/kg in rhesus monkeys (IRDC, 1978); PFOA induced thymic involution at 1638 mg/kg and decreased leukocyte and lymphocyte counts in male cynomolgus monkeys (Butenhoff et al., 2002).

Short-term (14-day) specialized studies conducted with PFOS, PFOA, or PFNA in mice have investigated functional parameters of the immune system as well as histopathology and organ weights. Splenic and thymic atrophy with inhibition of cell proliferation and/or induction of apoptosis (Yang et al., 2001; Fang et al., 2008, 2009) altered T-cell phenotypes (Son et al., 2009), and decreased cell-mediated and humoral immune responses (Yang et al., 2002a) were noted in mice at oral doses of ≥26 mg/kg/d (260 mg/kg TAD) of PFOA, 0.083 mg/kg/d (5 mg/kg TAD) of PFOS (Dong et al., 2009), or ≥3 mg/kg/d (42 mg/kg TAD) of PFNA. Decreased T-cell-dependent antibody responses (TDARs), such as plaque-forming cell responses to sheep red blood cells (RBCs), are the most sensitive effect of PFAS on the immune system, occurring at doses as low as 0.083 mg/kg/d of PFOS (Peden-Adams et al., 2008;

Dong et al., 2009, 2012) and 15 mg/kg/d (150 mg/kg TAD) of PFOA (DeWitt et al., 2009) in mice. A single dose of ≥20 mg/kg PFDA was sufficient to suppress antibody formation to keyhole limpet hemocyanin (KLH) and cell-mediated delayed-type hypersensitivity (DTH) responses in male rats (Nelson et al., 1992). Decreased TDARs were not secondary to increased serum corticosterone as TDAR suppression was also noted in adrenalectomized mice administered PFOA (DeWitt et al., 2009). Decreased immune organ weights and cellularity and thymic or splenic atrophy occur at higher doses than those that produce functional immune system changes and are most often associated with decreased body weight, increased liver weights, and increased serum corticosterone (Lefebvre et al., 2008; Zheng et al., 2009; Wang et al., 2014).

Published studies have also reported that PFOA and PFOS decreased (Qazi et al., 2009) numbers of circulating lymphocytes, neutrophils, and monocytes at doses (TAD) of 60 mg/kg PFOS and 52 mg/kg PFOA in mice. Decreased myeloid and B-cell progenitor numbers in the bone marrow of mice administered PFOS or PFOA at 60 and 52 mg/kg TAD, respectively, may explain the observed lymphopenia and neutropenia (Qazi et al., 2009, 2013) and possibly the effects of PFOA and PFOS administration on TDAR. Concomitantly, Corsini et al. (2012) reported that PFBS, PFOS, PFOA, and PFDA decreased human whole-blood and THP-1-cell lipopolysaccharide (LPS)-stimulated cytokine secretion via PPAR-α-independent inhibition of activation of the NF-κB pathway at concentrations ≥0.1 μg/mL, with PFOS being the most and PFOA the least potent compound tested. In contrast, PFOA and PFOS reportedly increased LPS-stimulated inflammatory cytokine production from peritoneal macrophages (Qazi et al., 2009; Mollenhauer et al., 2011) but not splenic macrophages. Increased anaphylactic and immunoglobulin E (IgE)–mediated responses, such as histamine release and airway eosinophilia in response to allergen challenge, were also reported in mice administered with PFOA (Fairley et al., 2007; Singh et al., 2012) prior to assessment in the passive cutaneous anaphylaxis assay or sensitization to ovalbumin, respectively. Taken together, these results imply that PFAS suppress adaptive immune system responses at the same time as they augment inflammatory cell responses in some compartments. This conclusion is consistent with the observed increase in pro-inflammatory cytokine content in the livers and sera of PFAS-treated mice (Fang et al., 2012a; Yang et al., 2014), but it is inconsistent with the known anti-inflammatory effect of PPAR-α activation *in vivo* (Straus and Glass, 2007), which is mediated by suppression of NF-κB signaling. Therefore, it is likely that PFAS are acting through non-PPAR-mediated mechanisms in addition to PPAR-mediated signaling to produce the observed effects. The effects on splenic and thymic cellularity may be mediated via PPAR-α activation, as administration of PFOA to PPAR-α KO mice had no effects on these parameters (Yang et al., 2002b). Other potential mechanisms include increased oxidative stress and activation of the apoptotic signaling pathway in immune cells and/or PPAR-γ activation (Fang et al., 2009, 2010). Increased oxidative stress as a mechanism of action may be specific to PFNA, as PFOA increased numbers of apoptotic cells in the spleens of mice administered with ≥5 mg PFOA/kg/d for 14 days (70 mg/kg TAD) without affecting splenic reactive oxygen species (ROS) levels (Wang et al., 2014).

Comparing the relative potency of PFSAs and PFCAs is difficult due to the paucity of studies and variation in study protocols used. However, when expressed in terms of TAD, PFOS is the most potent PFAS assessed in rodents, with a TAD of 0.05 mg/kg in mice for suppressed TDAR and a TAD of 0.1 mg/kg for changes in T-cell subsets in the spleens of male mice (Peden-Adams et al., 2008). PFNA is also extremely potent, decreasing thymic and splenic cellularity and altering T-cell subsets at doses as low as 1.856 mg/kg TAD in mice of both sexes (Rockwell et al., 2013). As Rockwell et al. did not assess TDAR, this study may underestimate the immunotoxic potency of PFNA. Notably, the TAD values cited in this section are far lower than those for liver effects, making the immune system one of the most sensitive target organs for PFAS toxicity.

Epidemiology studies suggest that the observed immunosuppressive effects of PFAS in animals may be relevant to humans. Serum PFOA was inversely related to serum antibody titer to tetanus and diphtheria in vaccinated children in the Faroe Islands (Grandjean et al., 2012), and serum PFDA, PFNA, PFDoDA, PFUnDA, and PFOS levels were negatively associated with the rate of increase in antibody responses to tetanus and/or diphtheria booster vaccines (Kielsen et al., 2016). Serum PFOA was positively associated with the incidence of osteoarthritis, especially in young, nonobese adults, while serum PFOS was negatively associated with osteoarthritis incidence (Innes et al., 2011) in the Washington Works population. Other epidemiology studies found no association between PFOA exposure and infectious disease morbidity and/or mortality in children from the general population (Okada et al., 2012), in the Washington Works cohort (C8 Science Panel, 2012b), or in occupationally exposed populations (Leonard et al., 2008).

In summary, relatively high doses of LC-PFAS induce thymic and splenic atrophy in rodents and monkeys, manifesting as histopathological changes, decreased weight parameters, and decreased cellularity. Inhibition of TDAR responses occurs at much lower doses of LC-PFAS; the TAD value of 0.05 mg PFOS/kg for suppression of TDAR in male mice is the lowest value calculated for any effect herein, making TDAR suppression the most sensitive marker for PFAS toxicity in adults. TDAR suppression is independent of PFAS effects on serum corticosterone and may be mediated in part by PPAR-α activation. Suppression of adaptive immune responses has also been seen in epidemiology studies, indicating that the effect seen in rodent studies is human health relevant. Increased inflammatory responses have also been seen in both rodent studies and some epidemiology studies. As most of the studies assessing functional immune system parameters have been conducted with PFNA, PFOA, and PFOS, it is unclear how prevalent TDAR suppression is for PFAS as a class.

8.3.1.3 Effects on Male Reproductive Organs

Effects of PFAS on male reproductive organ and fertility parameters, reproductive organ weights, and reproductive organ histopathology have been evaluated for PFBA, PFHxA, PFOA, PFNA, PFDA, PFUnDA, PFDoDA, PFOcDA, PFBS, PFHxS, and PFOS. No effects on male fertility or spermatogenesis were noted with any compound, except for PFDoDA, where decreased spermatid and spermatozoa counts were reported at 105 mg/kg TAD in male rats (Kato et al., 2014). Decreased testis and/or accessory sex organ weights were only noted at 70–350 mg/kg PFOA TAD in rats (Cook et al., 1992; Liu et al., 1996), 80 mg/kg PFDA TAD in rats (Bookstaff et al., 1990), and 140 mg/kg PFDoDA TAD in rats (Shi et al., 2007). Testicular histopathology was noted in any of the guideline studies, except for testicular atrophy noted in male cynomolgus monkeys at 5460 mg/kg PFOA (Butenhoff et al., 2002); Leydig cell adenomas and testicular vascular mineralization in male rats at 10,366 mg/kg PFOA; cell debris in seminiferous tubules, glandular epithelial atrophy in the prostate, seminal vesicle, and coagulating gland, and spermatic granulomas at 105 mg/kg PFDoDA TAD; and seminiferous tubule edema and detachment of spermatids in male rats at 28 mg/kg PFOS TAD (Lopez-Doval et al., 2014). TAD LOEL values for decreased serum testosterone in male rats were 70 mg/kg for PFNA (Feng et al., 2009), 40 mg/kg for PFDA (Bookstaff et al., 1990), 22 mg/kg for PFDoDA (Shi et al., 2009), and 14 mg/kg for PFOS (Lopez-Doval et al., 2014). PFOA administration decreased serum testosterone in male cynomolgus monkeys at a TAD of 5460 mg/kg (Butenhoff et al., 2002). Increased serum estradiol was noted in male rats administered with PFOA at TAD values as low as 28 mg/kg (Liu et al., 1996), which corresponded to dose levels in the study that also increased hepatic aromatase activity. This same study noted a linear correlation between serum estradiol levels and hepatic aromatase activity, indicating that enzyme induction in the liver was the proximate cause of the observed effect.

In mice, both PFOS and PFOA decreased serum or testes testosterone, decreased testes steroidogenesis and expression of steroidogenesis enzymes, and damaged the testes germinal epithelium and/or germ cells (Wan et al., 2011; Qui et al., 2013; Zhang et al., 2014). Disassembly of the blood–testis barrier was also noted in mice after PFOS administration at doses that also decreased sperm count and increased germ cell vacuolation (Qin et al., 2013). PFNA at a TAD of 42 mg/kg induced vacuolation between Sertoli cells, increased WT1 expression and decreased transferrin expression in Sertoli cells, and decreased serum inhibin B in rats; 70 mg/kg TAD induced vacuolation between spermatogonia and Sertoli cells, and germ cell degeneration (Feng et al., 2010). Morphological changes included degenerative and/or apoptotic changes in seminiferous tubules or tubular atrophy and/or apoptosis and sloughing of germ cells. Decreased expression of steroidogenic enzymes in the testes appeared to be the most sensitive effect of PFOA administration, occurring in mice at doses as low as 35 mg/kg TAD versus the LOEL of 140 mg/kg TAD for histopathological effects and decreased sperm counts (Zhang et al., 2014). PFOA decreased serum testosterone and caused testicular degeneration and increased sperm abnormalities in mice (Li et al., 2011) but not in rats (York et al., 2010), indicating that mice are more sensitive to the adverse effects of PFOA on the testes than rats.

The adverse effects of PFOA in the testes of mice were noted in both WT and humanized PPAR-α mice, but not in KO mice, indicating these effects were mediated by PPAR-α activation (Li et al., 2011). Direct inhibition of steroidogenesis in Leydig cells may also play a significant role in the observed toxicity, as PFOA, PFDA, and PFDoDA inhibited steroidogenesis in primary Leydig cells derived from adult rats and in Leydig cell tumor lines (Biegel et al., 1995; Boujrad et al., 2000; Shi et al., 2010) via competitive inhibition of steroidogenic enzymes and also possibly downregulation of steroidogenic gene expression. PFAS reportedly inhibited human

17β-hydroxysteroid dehydrogenase 3 (17βHSD3) and rat 3β-hydroxysteroid dehydrogenase (3βHSD3), two enzymes critical to the conversion of cholesterol to testosterone in Leydig cells; the IC50 values for PFOS inhibition of rat 3βHSD3 and human 17βHSD3 were 1.35 and 6.02 µM, respectively (Zhao et al., 2010b). This study also reported that the IC50 value for inhibition of 17βHSD3 by PFOA was 127.6 µM; PFAS preferentially inhibited rat 3βHSD3, while human 17βHSD3 was more sensitive to PFAS inhibition than human 3βHSD3. Antagonist activity at the androgen receptor may also play a role in the observed toxicity, as PFOS, PFOA, PFNA, PFDA, and PFHxS inhibited dihydrotestosterone activity at the androgen receptor at a concentration range of 4.7–52 µM, which is 100- to 1000-fold less potent than flutamide (Kjeldsen and Bonefeld-Jorgensen, 2013). Finally, indicators of cell damage, such as lipid accumulation and ROS accumulation in mitochondria, were also noted in Leydig cells treated with PFAS (Boujrad et al., 2000; Shi et al., 2010), indicating that direct PFAS induction of oxidative stress may act synergistically to exacerbate the effects of hormone synthesis disruption on the testes.

In contrast to the clear evidence of adverse effects of PFAS on male reproductive organs in animal studies, evidence from epidemiological studies is far less clear-cut. Impaired semen quality was associated with high levels of PFAS in nonoccupationally exposed males in Denmark (Joensen et al., 2009) but not in other studies (Raymer et al., 2012; Specht et al., 2012; Toft et al., 2012). Positive associations between decreased percentages of normal sperm and combined serum concentrations of PFOS and PFOA (Joensen et al., 2009) and serum concentrations of PFOA and PFHxS (Toft et al., 2012) have also been reported. No associations between serum PFOA and reproductive hormone levels were noted in occupationally exposed men (Olsen et al., 1998) or males in the general population (Specht et al., 2012; Lewis et al., 2015).

In summary, C8–C12 LC-PFAS are toxic to the male reproductive system in rodents. Although direct adverse effects on male fertility are not observed in rats with any PFAS except for PFDoDA, decreased serum concentrations of male hormones and testicular biomarkers, testicular damage and/or atrophy, sperm abnormalities, and decreased testicular steroidogenesis have been noted in both rats and mice. Rats are more sensitive to the effects of PFSAs than mice, while the opposite is true for the effects of PFCAs. The observed effects may be mediated by PPAR-α activation, androgen receptor antagonism, inhibition of steroidogenic enzymes in Leydig cells, and increased oxidative stress in Leydig cells. Increased metabolism of serum androgens via induction of aromatase in the liver may play a role in producing the Leydig cell adenomas noted in PFOA-exposed rats (see the discussion on carcinogenicity in Section 8.3.2). No alterations in serum androgens have been reported in epidemiology studies, and reported relationships between serum PFAS and semen quality or male fertility are conflicting. Therefore, the relevance to humans of the observed adverse effects of PFAS on the male reproductive system is uncertain at the present time.

8.3.1.4 Effects on Female Reproductive Organs and the Mammary Gland

No adverse effects on female fertility or reproductive organ parameters were noted in female rats exposed as adults to PFCAs or PFSAs, except for continuous diestrus noted at doses of 102.5–117.5 mg/kg of PFDoDA TAD in recovery group females only (Kato et al., 2014). Non-dose-responsive increases in the incidence and severity of ovarian tubular hyperplasia were also noted in females administered PFOA at a dose of 1175 mg/kg TAD (Butenhoff et al., 2012a). In contrast to the lack of effect of PFOS in guideline studies, one study reported that female rats injected with 10 mg/kg/d of PFOS for 14 days (140 mg/kg TAD) exhibited persistent diestrus, and decreased numbers of normal estrous cycles were noted at ≥1 mg/kg/d (14 mg/kg TAD) (Austin et al., 2003). These effects were associated with increased norepinephrine concentrations in the paraventricular nucleus of the hypothalamus, implying that the estrous cycle effects were mediated by altered hypothalamic–pituitary–gonadal (HPG) axis signaling. The highest dose tested in the OECD 416 study conducted with PFOS was 3.2 mg/kg/d (361.6–406.4 mg/kg TAD), so it is not surprising that persistent diestrus was not noted in this study (Luebker et al., 2005). However, the effect on numbers of normal estrous cycles should have been apparent in the OECD 416 study. Although both studies monitored estrous cyclicity via vaginal smears, the Austin et al. paper specifically defines "irregular cyclers" as those rats who "missed" or skipped particular stages in the normal progression of proestrus, estrus, diestrus I, and diestrus II. No such definition was provided in the Luebker et al. study, so it is possible that subtle changes in estrous cyclicity were present in that study but not picked up. The difference between the two studies could also be due to the fact that PFOS was administered by injection in the Austin et al. study versus orally in the Luebker et al. study. However, this is unlikely since PFOS is well absorbed via the oral route of administration and undergoes no metabolic transformation.

Prenatal, peripubertal, or adult exposure to PFOA had adverse effects on mammary gland development and differentiation in mice. PFOA decreased mammary gland development in prenatally exposed CD-1 mice (White et al., 2007, 2009, 2011) and delayed differentiation into a lactating phenotype in their dams (White et al., 2007) at doses as low as 3–5 mg/kg/d administered during gestational days 1–17 (48–80 mg/kg TAD to the dam). At 18 months of age, abnormal histopathological findings were noted in the mammary glands of prenatally exposed CD-1 mice, including increased numbers of darkly staining foci, hyperplasia of the ductal epithelium, increased stromal epithelial densities, and/or inappropriate differentiation of ductal tissue (White et al., 2009). Some studies also report strain-dependent effects of PFOA on mammary gland development when administered around the time of puberty (postnatal day 21), with stimulatory effects noted with low-dose PFOA in C57Bl/6 mice and inhibitory effects noted with a higher dose of PFOA in the same strain or with all doses of PFOA in Balb/c mice (Yang et al., 2009). There are significant factors in the design of these studies that complicate interpretation of these results, that is, low power ($n < 6$/treatment group) in the Yang et al. study and, in the White et al. studies, use of a subjective, qualitative scoring method instead of quantitative morphometry or 5-bromo-2'-deoxyuridine (BrdU) staining of terminal end buds to evaluate mammary gland cell proliferation and development. As has been previously demonstrated, use of the qualitative method may give disparate results from quantitative morphometry (Hovey et al., 2011). However, the fact that the results of White et al. were also found by Yang et al. using BrdU staining of terminal end buds lends support to the validity of White et al.'s findings. As such, the available data indicate that PFOA has adverse effects on mammary gland development and differentiation at least in mice. Although the available bioassay data in rats do not indicate treatment-related effects on the mammary gland (Biegel et al., 2001), the pharmacokinetics of PFOA in the female rat may make it less sensitive than the female mouse to the long-term toxicity of PFOA at this end point. Moreover, the rats in the bioassay were not dosed beginning *in utero*, possibly further diminishing the sensitivity of that study to detect toxicity at that end point. As such, further studies conducted in mice are needed to discern whether PFOA may have carcinogenic effects in the mammary gland.

The effects on mammary gland development appear to be mediated via estrogen receptor signaling, as these effects were abolished with ovariectomy (Zhao et al., 2010b). Indeed, PFOA and PFOS have been reported to be positive for estrogenic activity in the E-screen MCF-7 cell line at concentrations of ≥0.01 and 30 µg/mL, respectively (Henry and Fair, 2013), and PFOA reportedly enhanced the agonism of β-estradiol and the estrogen receptor and altered steroidogenic gene expression in H295R cells at ≥200 ng/mL (Du et al., 2013). However, Henry and Fair (2013) reported that PFOS and PFOA were more potent as antagonists in the E-screen assay, and Gao et al. (2013) reported that PFBS, PFBA, PFDoDA, and PFTriDA neither bound to nor activated the estrogen receptor.

Epidemiological evidence of effects of PFAS on female reproductive health is inconclusive. Impaired fecundity, measured as increased time to pregnancy, was associated with high serum PFOA levels in nonoccupationally exposed Danish women in one study (Fei et al., 2009), and increased incidences of endometriosis, which can impair fertility, were associated with PFOA and PFNA body burden in another (Louis et al., 2012). Serum PFHxS, PFOA, PFNA, PFDA, and PFOS were not associated with time to pregnancy in Danish couples (Vestergaard et al., 2012). Increased serum PFOS was associated with either the presence of infertility (La Rocca et al., 2012) or decreased fecundity (Louis et al., 2013). Dose-responsive increases in rates of menopause or hysterectomy with serum PFOA, PFNA, PFOS, and PFHxS in one study may be a case of reverse correlation, as menstruation is an important route of PFAS excretion in females (Taylor et al., 2014).

In summary, except for decreased mammary gland development in mice with PFOA and adverse effects of PFDoDA and possibly PFOS on estrous cyclicity in rats, very few adverse effects of PFAS on the female reproductive tract have been reported. Epidemiology studies reported decreased fertility or increased time to pregnancy, increased endometriosis, and increased risk of menopause with increased serum PFAS concentrations and may be confounded by reverse causality, as menstruation, lactation, and offloading of PFAS body burdens onto the products of conception during gestation are known to increase clearance of tissue PFAS in fertile, premenopausal females. As such, the effects of PFAS as a class on fertility and female reproductive organ health are unclear at the present time.

8.3.1.5 Effects on the Thyroid

Data on effects of PFAS on the thyroid are comparatively sparse, with the exception of studies conducted in rats with PFDA. Thyroid follicular cell hypertrophy, most likely a secondary response to increased clearance of serum thyroxine by hepatic enzymes, was noted in rats at TAD LOEL values of 2700 mg/kg PFBA in males (Van Otterdijk, 2007), 45,000 mg/kg NaPFHx in males and females (Slezak, 2007), and 132 mg/kg PFHxS in males (Butenhoff et al., 2009). Thyroid follicular cell adenoma was observed in male rats administered with 416.4 mg/kg PFOS TAD, but not in male rats administered with 718 mg/kg in the same study (Butenhoff et al., 2012b), calling into question the biological relevance of this finding. Decreased serum T4 and/or T3 was noted in cynomolgus monkeys after PFOA and PFOS administration at LOEL TAD values of 1638 mg/kg (Butenhoff et al., 2002) and 114.7 mg/kg (Seacat et al., 2002), respectively. These effects were not accompanied by changes in thyroid organ weights or adverse histopathological changes in the thyroid. No effects of PFOA or PFOS on serum thyroid hormone levels were noted in rats in guideline 90-day studies (Seacat et al., 2003; Perkins et al., 2004). Shorter-duration studies conducted in rats with PFOS noted dose-responsive decreases in serum total T4 and total T3 without effects on thyroid-stimulating hormone (TSH) levels at doses as low as 15 mg/kg TAD in female rats (Chang et al., 2008a; Yu et al., 2009). These effects were associated with upregulation of liver UDP1A1 RNA content and malic enzyme activity, along with increased fecal and urine I$^-$ content, implying that increased thyroid hormone turnover was responsible for the observed effects. PFDA, in contrast, induced marked hypothyroidism in rats after only a single injection at doses as low as 20 mg/kg (Van Rafelghem et al., 1987).

Rats are known to be sensitive to secondary effects on thyroid hormone production due to increased hepatic metabolism and clearance of thyroid hormone. This mechanism is thought to not be relevant to humans due to the presence of transthyrethrin, which buffers serum thyroxine levels against changes in hepatic metabolic capacity (Wu and Farrelly, 2006). However, PFAS have been shown to bind competitively to transthyretin with IC50 values of 717–46,894 nM, the most potent of which is PFHxS (Weiss et al., 2009). Further, although PFHxS, PFOS, and PFCAs of C8–C12 in length did not possess agonist activity to the thyroid hormone receptor, all decreased proliferation in a T3-dependent cell line (M. Long et al., 2013), and another study reported that PFAS bind to human T3 receptor with K50 values of 193 μM for PFHxS, 42 μM for PFOA, 16 μM for PFOS, 36 μM for PFNA, and 6 μM for PFDA, the most potent of the tested compounds (Ren et al., 2015). PFBS and PFHxA did not bind to the T3 receptor in this study. Therefore, the thyroid effects seen in rodents cannot be dismissed as irrelevant to human exposures.

The effects of PFAS on the thyroid are mixed in epidemiological studies. Some found no association of serum PFOA levels with serum thyroxine levels in occupationally exposed healthy subjects (Olsen et al., 2003; Olsen and Zobel, 2007). Others found direct associations between serum PFOA levels and the incidence of thyroid disease in the general population (Melzer et al., 2010) and in highly exposed populations (C8 Science Panel, 2012c). Negative associations of serum PFHxS levels with free T4 were noted in two studies in the US general population (Wen et al., 2013; Webster et al., 2016); some studies report positive associations of various serum PFAS levels with serum thyroid hormone levels, the opposite of what has been noted in animals (Knox et al., 2011; Jain, 2013; Lin et al., 2013b; Lewis et al., 2015). While the finding of decreased serum thyroxine in monkeys may indicate an effect of PFOA on the thyroid in humans, the decreased serum thyroxine seen in the monkey study was only noted at dose levels that clearly exceeded the maximum tolerable dose (MTD) for that compound due to excess mortality noted at that dose. Therefore, the applicability of this finding to risk assessment of low-level PFAS exposure in human populations is unclear. Further studies will be needed to fully assess the toxic effects of PFAS on the thyroid.

8.3.1.6 Other Effects

Other effects noted with almost all of the PFCAs and PFSAs tested in rats, except for PFOS (Seacat et al., 2003), and in monkeys include renal tubular hypertrophy and/or increased kidney weights and decreased RBC numbers and hematocrit (IRDC, 1978; Butenhoff et al., 2002, 2004b, 2009; Seacat et al., 2002; Van Otterdijk, 2007; Chengelis et al., 2009; Cui et al., 2009; Lieder et al., 2009a, 2009b; Hirata-Koizumi et al., 2012; Kato et al., 2015). No effects of PFOA on the incidence of chronic kidney disease were noted in the Washington Works cohort (C8 Science Panel, 2012d), and no associations of serum PFOS or PFOA were noted with hematological parameters in an occupationally exposed cohort with measured serum PFOS and PFOA levels of 1.32 and 1.78 ppm, respectively (Olsen et al., 2003). It should be noted that the decreased RBC numbers in animal studies are usually reported at high dose levels associated with significant weight loss, decreased feed consumption, and often liver toxicity; therefore, it is unclear whether the anemia induced at this dose level is secondary toxicity or a direct suppressive effect of PFAS on RBC production or hemolytic effect. The renal tubular hypertrophy often co-occurs with hepatocellular hypertrophy and is consistent with xenobiotic metabolic enzyme induction associated with repeated exposure to fairly large doses of a xenobiotic.

PFOS reportedly caused tonic convulsions in rats and mice upon ultrasonic stimulation at ≥250 mg/kg TAD in rats and ≥125 mg/kg TAD in mice (Sato et al., 2009; Kawamoto et al., 2011) and decreased performance in the Morris water maze in mice at ≥193.5 mg/kg TAD (Fuentes et al., 2007; Y. Long et al., 2013). The effects on Morris water maze performance were associated with dose-responsive increases in the numbers of apoptotic neurons and decreased glutamate concentrations in the hippocampus and decreased dopamine and DOPAC concentrations in the caudate putamen (Y. Long et al., 2013). The neurotoxic effects seen with PFOS in rodents were also seen in cultured rat hippocampal neurons where prolonged treatment with PFOS inhibited neurite growth, decreased the length of the longest neurite, and increased postsynaptic current frequency (Liao et al., 2009). PFOS's effect on postsynaptic frequency was subsequently shown to be due to activation of L-type Ca channels (Liao et al., 2008). Other PFAS also produced these effects in a perfluorinated chain-length-dependent manner, with potency for perfluorotridecanoic acid (PFTrDA) > PFDoDA > PFOA; PFOS > PFHxS; and PFSAs > PFCAs (Liao et al., 2009). However, no effects on nervous system histopathology, brain weight, or parameters in a functional observational battery (FOB) have been reported in rats administered PFBA, PFBS, PFHxS, PFHxA, PFUnDA, PFDoDA, or PFOcDA. FOBs were not conducted in either the 90-day studies or the chronic studies conducted with PFOS or PFOA in rats, but no clinical signs of neurotoxicity, alterations in brain weights, or adverse brain histopathological lesions were reported in these studies

(Seacat et al., 2003; Perkins et al., 2004; Butenhoff et al., 2012a, 2012b). Therefore, the biological relevance of the findings of the specialized studies is unclear.

8.3.2 Carcinogenicity and Genotoxicity

Only PFOA and PFOS have been evaluated under conditions of a standard rodent bioassay. One study reported an increased incidence of hepatocellular adenomas in rats of both sexes fed with PFOS in the diet at a level of 20 ppm, corresponding to TAD values of 718.32 and 913.23 mg/kg in males and females, respectively (Butenhoff et al., 2012b). The same study reported an increased incidence of thyroid follicular cell adenomas in males administered with 20 ppm PFOS in the diet for 52 weeks, followed by 52 weeks without treatment, but not in males fed with 20 ppm PFOS in the diet for 104 weeks. Significantly increased combined incidences of hepatocellular carcinomas and hyperplastic nodules and incidences of Leydig cell adenomas were noted in male Sprague-Dawley rats fed PFOA at levels of 10,366 mg/kg TAD; no increases in neoplasia incidence were noted in females (Butenhoff et al., 2012b). An additional bioassay in which PFOA was also administered in the diet to male CD rats for 2 years at 0 or 10,366 mg/kg TAD noted significantly increased incidences of hepatocellular adenoma and carcinoma combined, pancreatic acinar cell adenomas and carcinomas combined, and Leydig cell adenomas with PFOA treatment (Biegel et al., 2001). In contrast, PFHxA did not increase the incidence of any neoplastic lesions in either male or female rats at doses of up to 73,000 mg/kg TAD in males and 146,000 mg/kg TAD in females (Klaunig et al., 2014).

Some studies have reported an association between PFOA exposure and kidney and testicular cancer in the Washington Works cohort (C8 Science Panel, 2012e) or prostate cancer in an occupationally exposed cohort (Gilliland and Mandel, 1993; Lundin et al., 2009). Another case-control study reported increased serum PFDA in prostate cancer cases versus controls and increased odds ratios of prostate cancer in subjects with high serum PFAS and a family history of prostate cancer (Hardell et al., 2014). Similarly, a case-control study conducted in Greenland Inuit women reported increased serum PFOS, PFOA, and PFAS in cancer cases versus controls (Bonefeld-Jorgensen et al., 2011). Other studies found no association between PFOA exposure with cancer incidence in either men or women (Leonard et al., 2008; Eriksen et al., 2009) from occupationally exposed cohorts or cohorts from the general population, or with blood prostate-specific antigen in men from the Washington Works population (Ducatman et al., 2015).

The neoplastic effects of PFOA and PFOS in rats are not mediated via direct damage to DNA, as PFAS are generally negative in genotoxicity assays (Buhrke et al., 2013). PFOA's neoplastic effects may be mediated via rodent-specific pathways. For instance, the hepatocellular adenomas induced by PFOA and PFOS in rats may be due to PPAR-α activation, and disruption of HPG axis signaling by PFOA via increased hepatic androgen clearance may be responsible for the induction of Leydig cell tumors in rats chronically administered PFOA (Cook et al., 1992; Liu et al., 1996). The human liver appears to be refractory to tumor induction by stimulation of PPAR-α, even by extremely potent, specific activators of PPAR-α, such as fibrates

(Rosen et al., 2009). The spontaneous incidence of Leydig cell tumors in male rats is 135,500- to 1,920,000-fold the incidence in humans, and several physiological factors make humans refractory to development of this tumor type, including human sex hormone binding globulin, which stabilizes blood androgen levels, and increased sensitivity and responsiveness of rodent Leydig cells to proliferative stimuli compared with human Leydig cells (Cook et al., 1999). The relationship between the increased testicular cancers seen in the Washington Works cohort and the Leydig cell tumors seen in rats is also unclear, and histological descriptions of tumor type were not provided in the epidemiology study.

While mechanisms of action of the induction of pancreatic tumors have not been as thoroughly characterized as those for the liver and testes tumors, sustained elevation of cholecystokinin in rodents is believed to play a role in the development of the observed acinar cell tumors (Klaunig et al., 2012). PFOA accumulates in the pancreas of mice, and dosing mice by gavage for 7 days with ≥2.5 mg of PFOA/kg/d (≥17.5 mg/kg TAD) results in dose-responsive increases in serum amylase and lipase, focal ductal hyperplasia, and increased pancreatic isoprostane levels (Kamendulis et al., 2014). If this same mechanism occurs in rats, it may also account for the observed pancreatic acinar tumors in chronically treated animals. In contrast, PFOA exposure has no effect on pancreatic cancer incidence in epidemiology studies and is not associated with plasma cholecystokinin levels in an occupationally exposed cohort (Olsen et al., 2000).

In summary, PFOA and PFOS both have been shown to be carcinogenic in the liver of rats; PFOA also produces tumors in the pancreas and testis of male rats, while PFOS may produce thyroid tumors in male rats. There does not appear to be strong evidence for carcinogenic effects of PFOS in humans. In contrast, a large epidemiological study in a highly exposed US population cohort concluded that PFOA exposure increased the incidence of kidney and testicular tumors in this population. The histological subtype of testicular tumor associated with increased PFOA exposure in this population is unknown; therefore, it is not possible to assess the congruence of this finding with the increased Leydig cell tumors in PFOA-exposed rats. Moreover, although PFOA exposure is associated with increased kidney weights, PFOA does not induce kidney tumors in chronically exposed rats. Therefore, it appears that even though PFOA appears to be carcinogenic in both humans and rats, the target organs for neoplastic action differ between the two species.

Given the fact that almost all of the LC-PFAS tested are PPAR-α agonists and/or induce xenobiotic-metabolizing enzymes in the liver, PFAS may be expected to induce some or all of the neoplastic effects seen with PFOA and PFOS in rodents. Indeed, repeated injections of 3 or 10 mg/kg of PFDA every 2 weeks for 54 weeks increased hepatic peroxisomal proliferation and cell proliferation in rats (Chen et al., 1994), indicating the potential of this compound to be carcinogenic. There is currently insufficient information to definitively classify other PFAS as human carcinogens. More complete characterization of their mechanisms of action in humans versus rodents and more data from chronic studies in other PFAS compounds in rodent models will be necessary in order to determine whether this class of compounds poses a carcinogenic hazard to humans.

8.4 COMMON MECHANISMS OF ACTION AND CLASS-BASED ANALYSIS

Several mechanisms have been proposed to explain the observed effects of PFAS in the liver, thyroid, testes, and immune system. The most frequently proposed mechanism is PFAS activation of PPAR enzymes, most frequently PPAR-α and PPAR-γ. As noted above, direct evidence of PPAR-α activation, in the form of the observation of peroxisomal proliferation in the liver and increased palmitoyl CoA oxidation in hepatic microsomes, has been noted in both rats and mice treated with almost every PFAS compound assessed in rodents, with the notable exception of PFBS. Both PFOA and PFOS activate murine PPAR-α, PPAR-β/δ, CAR, and pregnane X receptor (PXR) (Takacs and Abbot, 2007; Rosen et al., 2010; Elcombe et al., 2012). PFOA is a more potent agonist for murine PPAR-α than PFOS, with significant activation observed starting at 10 and 120 µM, respectively (Takacs and Abbott, 2007). PFOA and PFNA are approximately equipotent at murine PPAR-α, with potency rapidly decreasing in the order PFOA ~ PFNA > PFHxA > PFOS > PFHxS; low doses of PFAS in combination produce an additive activation of mouse PPAR-α, while high combined doses deviate from additivity (Wolf et al., 2014). PFOA is threefold more potent than PFBA at stimulating PPAR-α in rat and human liver cells (Bjork and Wallace, 2009), and PFOA both is more potent and produces greater activation of rat and human PPAR-α than endogenous fatty acids, such as oleate or octanoic acid (Intrasuksri et al., 1998). Human PPAR-α is comparatively insensitive to activation by PFAS, with significant activation occurring at 30 µM for PFOA; human PPAR-α is not activated by PFOS (Takacs and Abbott, 2007). PFOA and PFOS are more potent agonists at PPAR-α than PPAR-β/δ in mice and do not activate PPAR-γ at all; human PPAR-β/δ and PPAR-γ receptors are refractory to stimulation by either PFOA or PFOS (Takacs and Abbott, 2007).

Increased biomarkers of oxidative stress, such as isoprostanes, MDA, or TBA reactive substances, have also been noted in the liver and other tissues after PFAS administration (Permadi et al., 1992; Fang et al., 2010; Kamendulis et al., 2014; Yang et al., 2014), often associated with observation of apoptotic or necrotic changes. PFOA and PFDA were the most potent of the compounds tested by Permadi et al. for induction of TBA in mouse liver, while PFBA and perfluoroacetic acid were virtually inactive. One study reported that PFDA directly inhibited glutathione S-transferase (GST) activity and blocked GST mRNA induction by phenobarbital (Schramm et al., 1989), which would lead to increased oxidative stress. Increased ROS production was also noted in HepG2 cells treated with PFOS and PFOA, while PFBS and PFHxA did not induce ROS under the study conditions (Eriksen et al., 2010). Concomitantly, another study noted increased ROS production in HepG2 cells after treatment with PFHxS, PFOS, PFOA, PFNA, PFDA, and PFUnDA, but not PFDoDA, at doses as low as 2 µM (Wielsoe et al., 2015). This increase in ROS production may be due to disruption of mitochondrial membrane potential due to interference with mitochondrial respiration. One study reported that, with the exception of PFOS, all tested PFCAs at concentrations ≥20 µM stimulated state 4 respiration and inhibited state 3 respiration in isolated mitochondria, with potency increasing with perfluorinated chain length (Kleszczynski et al., 2009; Wallace et al., 2013). This effect induces formation of the mitochondrial permeability transition pore, which is the first step in induction of apoptotic cell death (Crompton, 1999). PFCAs have also been shown to disrupt Ca homeostasis in cells, inducing an accumulation of Ca in mitochondria in the following order of potency: PFDoDA > PFDA > PFOA > PFHxA (Kleszczynski and Skladanowski, 2011). Interference with mitochondrial bioenergetics stimulates mitochondrial biogenesis, and increased liver mitochondrial protein content and enzyme activity is a characteristic finding in the livers of rodents treated with PFCA but not PFOS (Permadi et al., 1992; Sohlenius et al., 1992; Berthiaume and Wallace, 2002). One study reported that PFOA treatment of rats increased PPAR-γ coactivator-1α gene expression in the liver, leading to a doubling of mtDNA copy number, providing further support to the hypothesis of involvement of disruption of mitochondrial homeostasis in the liver injury induced by PFCAs (Walters et al., 2009).

Additional proposed mechanisms of action are alterations in cell plasma membrane fluidity and inhibition of intercellular gap junction communication (Hu et al., 2002, 2003). PFOS and PFHxS, but not PFBS, dose responsively inhibited gap junction communication in a rat liver cell line and a dolphin kidney cell line at EC50 values for rat liver cells of 29.96 and 125.5 μM and for kidney cells, 25.51 and 85.63 μM, respectively. Three or 21 days of treatment with 5 mg/kg/d PFOS to attain a hepatic PFOS concentration of 125 ppb was sufficient to decrease gap junction communication in liver lobes *ex vivo* as measured by a dye-loading assay (Hu et al., 2002). PFOS, but not PFHxS or PFBS, increased membrane fluidity in fish leukocytes at concentrations ≥33 μM (Hu et al., 2003). Notably, the concentration at which PFOS inhibited gap junction communication in cell lines was extremely close to the lowest observed effect concentration (LOEC) for inhibition of mitochondrial respiration, implying that these two mechanisms could be acting simultaneously to induce the observed effects *in vivo*.

The European Chemicals Agency (ECHA) Read-Across Assessment Framework (RAAF) states that "substances whose physicochemical, toxicological, and ecotoxicological properties are likely to be similar or follow a regular pattern as a result of structural similarity may be considered as a group, or category of substances" (ECHA, 2015). The RAAF further states that the similarity may be due to the presence of a common functional group, common precursors, or breakdown products, or a constant pattern in the changing of properties across the group. The perfluorinated tail common to PFCAs and PFSAs represents the common functional moiety between PFCAs and PFSAs, and polyfluoroalkyl precursor substances can be oxidized in the environment to a mixture of PFSAs and PFCAs (Barzen-Hanson and Field, 2015). PFCAs and PFSAs, particularly the four compounds commonly found in human serum, also share several common biological properties when assessed in rodent or primate model systems: biopersistence in tissues, particularly liver; enzyme induction and peroxisomal proliferation in the liver; testicular toxicity; alteration of thyroid homeostasis; immunosuppression; and toxicity in the postnatal period. These effects have been noted in studies conducted with PFBS, PFHxS, and PFOS for PFSAs and PFBA, PFHxA, PFOA, PFNA, PFUnDA, PFDoDA, and PFOcDA in rats and/or mice in OECD guideline repeated-dose study designs. Common mechanisms of action for PFCAs and PFSAs include PPAR-α activation, interference with mitochondrial respiration and concomitant induction of mitochondrial depolarization and production of ROS, and induction of xenobiotic-metabolizing enzymes.

Potency analyses across the chemical class, which are critical for the justification of a read-across approach, are complicated by the multiple mechanisms of action through which PFAS exert their effects and the sparse coverage of the chemical space by the available toxicity data. Potency for effects in the liver generally increases with perfluorinated chain length in both PFCAs and PFSAs, peaking at a chain length of C8–C12 for PFCAs and C6–C8 for PFSAs. Potency for the other effects common to PFCAs and PFSAs cannot be said to follow a regular pattern, mainly due to the low chemical space coverage of studies that interrogate effects at the most sensitive apical end point for that organ system. Effects on immune system functional parameters, for example, have only been assessed for PFOA, PFNA, and PFOS, and effects on testes steroidogenesis, male reproductive hormones, and serum testicular biomarkers have only been assessed for PFOA, PFOS, PFNA, PFDA, and PFDoDA. Potency order for apical effects on one organ system may also be different than the order of potency for another system or for the overall order of potency for systemic toxicity. For example, the order of potency for induction of hepatocellular hypertrophy in male rats from 90-day studies is PFOS > PFOA > PFBA > NaPFHxA > PFBS > PFHxA, while the order of potency for systemic toxicity overall from those same studies is PFOA > PFOS > PFBA > PFHxA, with neither PFBS nor NaPFHx reaching a no observed adverse effect level (NOAEL) in the respective studies. The TAD approach used to compare potency herein is extremely crude and does not take into account differences in achieved tissue concentrations due to toxicokinetic factors. Use of a TAD approach may therefore lead to errors in estimation of the difference in potency between two compounds. Qualitative differences in apical effects seen with different PFAS compounds (i.e., pancreatic and testicular tumors with chronic PFOA administration but not PFOS administration in male rats) also complicate use of a read-across approach. A recent published analysis concluded that use of a toxic equivalency factor (TEF) approach similar to that used for the assessment of polychlorinated biphenyls is not appropriate for use with PFAS due to the factors discussed in this section (Peters and Gonzalez, 2011). The author concurs with this analysis and further concludes that these same factors preclude the use of a read-across approach in the quantitative risk assessment for PFCAs or PFSAs, for which no data from animal models are available. The current toxicity database for PFSAs and PFCAs might be used, however, to identify critical end points of concern and design appropriate *in vivo* studies for use in the safety assessment of the PFCA or PFSA in question. This would constitute a type of qualitative read-across approach to identify data gaps and appropriate ways to address those gaps and would be predicated upon the conduct of further toxicity studies to fill in the gaps in the toxicity database. Once sufficient data have been generated in this fashion, there may be a role for traditional read-across approaches in PFAS risk assessment.

8.5 REGULATORY AGENCY RISK ASSESSMENTS FOR PFAS

Several national and international regulatory agencies have published provisional tolerable daily intake (pTDI) and/or drinking water guideline (DWG) values for

PFOS and PFOA or the sum of these two compounds. The state of New Jersey has published a DWG value for PFNA, and the Minnesota Department of Health has published DWG values for PFBS and PFBA, in addition to ones for PFOA and PFOS. Health effect guideline values and DWG values are summarized in Table 8.2. The derivations of these values are summarized as follows:

- ·Health effect guideline values: pTDIs and reference doses (RfDs)
  - Health Protection Agency, UK (DWI, 2009; Health Protection Agency, 2012)
    – PFOA TDI: 1.5 µg/kg bw/d (EFSA, 2008).
    – PFOS TDI: 0.3 µg/kg bw/d, based on application of 100-fold uncertainty factor (factors of 10 each for intraspecies sensitivity and interspecies extrapolation) to the NOAEL value of 30 µg/kg bw/d from a 26-week study conducted with PFOS in cynomolgus monkeys (Seacat et al., 2002).

TABLE 8.2
Health Effect Guideline Values

Compound	Daily Intake Value	Drinking Water Guideline Value	Source
PFBA	RfD: 2.9 µg/kg/d	13.5 µg/L	MDH (2011a)
PFBS	RfD: 1.44 µg/kg/d	7 µg/L	MDH (2011b)
PFOA	TDI: 1.5 µg/kg bw/d	N/C	EFSA (2008)
	TDI: 0.1 µg/kg bw/d	See Σ PFOA and PFOS	DWC (2006)
	RfD: 0.02 µg/kg bw/d	See Σ PFOA and PFOS	USEPA (2016)
	RfD: 0.077 µg/kg bw/d	0.3 µg/L	MDH (2009a)
	TDI: 1.5 µg/kg bw/d		BfR (2008)
PFOS	TDI: 0.3 µg/kg bw/d	See Σ PFOA and PFOS	Health Protection Agency (2012)
	TDI: 0.15 µg/kg bw/d	N/C	EFSA (2008)
	TDI: 0.1 µg/kg bw/d	N/C	BfR (2008)
	TDI: 0.1 µg/kg bw/d	See Σ PFOA and PFOS	DWC (2006)
	RfD: 0.02 µg/kg bw/d	See Σ PFOA and PFOS	USEPA (2016)
	RfD: 0.08 µg/kg bw/d	0.3 µg/L	MDH (2009b)
PFNA	BMDL10 target human serum level: 4.93 µg PFNA/mL	0.013 µg/L	New Jersey Drinking Water Quality Institute (2015)
Σ PFOS and PFOA	See individual TDIs above	0.3 µg/L	DWI (2009)
		0.070 µg/L	USEPA (2016)
		0.02 µg/L	Vermont Department of Health (2016)
		0.3 µg/L	DWC (2006)

Note: N/C = not calculated.

- BfR, Germany (BfR, 2008)
  - PFOS TDI: 0.1 µg/kg bw/d, based on application of a 1000-fold uncertainty factor (10 for intraspecies sensitivity and 100 for interspecies extrapolation) to a NOAEL of 100 µg/kg bw/d from a two-generation reproductive toxicity study with PFOS in rats (Luebker et al., 2005).
- Federal Environment Agency, Germany (DWC, 2006)
  - PFOS TDI: 0.1 µg/kg bw/d, based on application of a 300-fold uncertainty factor (10 each for intraspecies sensitivity and interspecies extrapolation and 3 for kinetic differences between rodents and humans) to a NOAEL of 0.025 mg/kg/d from a chronic study conducted with PFOS in rats (Butenhoff et al., 2012b).
  - PFOA TDI: 0.1 µg/kg bw/d, based on application of a 1000-fold uncertainty factor (10 each for intraspecies sensitivity and interspecies extrapolation and 10 for kinetic differences between rodents and humans) to the lower bound of the range of NOAELs for PFOA effects in animals of 0.1 mg/kg/d.
- EFSA (2008)
  - PFOA TDI: 1.5 µg/kg bw/d, based on application of a 200-fold uncertainty factor (10 each for intraspecies sensitivity and interspecies extrapolation and 2 for database incompleteness) to a benchmark dose modeling (BMDL) 10 value of 0.3 mg/kg bw/d for liver effects in rodents.
  - PFOS TDI: 150 ng/kg bw/d, based on application of a 200-fold uncertainty factor (10 each for intraspecies sensitivity and interspecies extrapolation and 2 for database incompleteness) to a NOAEL of 30 µg/kg bw/d from a 26-week study conducted in cynomolgus monkeys with PFOS (Seacat et al., 2002).
- US Environmental Protection Agency (2016)
  - PFOA: RfD of 0.00002 mg/kg bw/d, based on application of a 300-fold uncertainty factor (10 each for intraspecies sensitivity and extrapolation from a LOAEL to a NOAEL value and 3 for interspecies extrapolation) to a LOAEL value of 0.0053 mg/kg/d for decreased ossification in phalanges and accelerated puberty in male pups observed in a pre- and postnatal developmental toxicity study conducted in mice with PFOA (Lau et al., 2006). The LOAEL value is derived from serum values measured in the study and assumes that these values represent 60% of steady-state concentrations for PFOA.
  - PFOS: RfD of 0.00002 mg/kg bw/d, based on application of a 30-fold uncertainty factor (10 for intraspecies sensitivity and 3 for interspecies extrapolation) to a NOAEL value of 0.00051 mg/kg/d for decreased pup weights noted in a two-generation reproductive toxicity study conducted in rats with PFOS (Luebker et al., 2005). The NOAEL value is derived from pharmacokinetic modeling.

- ·DWG values
  - Minnesota Department of Health (chronic noncancer health limit guideline values)
    - PFBS (MDH, 2011b): 7 µg/L, based on an RfD of 0.0014 mg/kg/d, a relative source contribution (RSC) of 0.2, and a drinking water intake of 0.043 L/kg/d. The RfD is derived from a NOAEL of 60 mg/kg/d from a 90-day study in rats (Lieder et al., 2009a). The RfD is converted to a human equivalent dose (HED) via division by a factor of 142 to account for toxicokinetic differences between rats and humans and a total uncertainty factor of 300 (3 for interspecies extrapolation, 10 for intraspecies variability, 3 for database insufficiency, and 3 for extrapolation from subchronic to chronic exposure).
    - PFBA (MDH, 2011a): 13.5 µg/L, based on an RfD of 0.0029 mg/kg/d, an RSC of 0.2, and a drinking water intake of 0.043 L/kg/d. The RfD is derived from a NOAEL of 6.9 mg/kg/d from a 90-day study in rats (Van Otterdijk, 2007). The RfD is converted to a HED via division by a factor of 8 to account for toxicokinetic differences between rats and humans and a total uncertainty factor of 300 (3 for interspecies extrapolation, 10 for intraspecies variability, and 10 for database insufficiency). Of note, the online listing of the guideline values incorrectly lists the PFBA value as 7 µg/L.
    - PFOA (MDH, 2009a): 0.3 µg/L, based on an RfD of 0.00007 mg/kg/d, an RSC of 0.2, and a drinking water intake of 0.053 L/kg/d for the first 19 years of life. The RfD is derived from a BMDL10 serum concentration value of 23 mg/L from a 26-week study conducted in cynomolgus monkeys with PFOA (Butenhoff et al., 2002). The RfD is converted to a HED of 0.0023 mg/kg/d, and a total uncertainty factor of 30 (3 for interspecies extrapolation, 10 for intraspecies variability) is applied.
    - PFOS (MDH, 2009b): 0.3 µg/L, based on an RfD of 0.00008 mg/kg/d, an RSC of 0.2, and a drinking water intake of 0.049 L/kg/d for the first 27 years of life. The RfD is derived from a BMDL serum concentration value of 35 mg/L from a 26-week study conducted in cynomolgus monkeys with PFOS (Seacat et al., 2002). The RfD is converted to a HED of 0.0025 mg/kg/d, and a total uncertainty factor of 30 (3 for interspecies extrapolation, 10 for intraspecies variability) is applied.
  - Drinking Water Inspectorate (DWI, 2009), UK
    - PFOS and PFOA: 0.3 µg/L in Tier 1, based on the Health Protection Agency's TDI for PFOS of 0.3 µg/kg bw/d, assuming that 10% of the TDI for PFOS is attributable to drinking water consumption, and assuming daily consumption of 1 L of drinking water for a 10 kg child.
  - USEPA (2016): Lifetime health advisory levels for drinking water consumption
    - Sum of PFOA and PFOS: 70 ng/L (ppt), based on RfDs for PFOA and PFOS of 0.00002 mg/kg bw/d, an RSC value of 0.2, and a 90th

percentile estimate of daily drinking water intake for lactating women of 0.054 L/kg/d. This value assumes that 100% of PFOA and PFOS exposure comes from drinking water.
- New Jersey Drinking Water Quality Institute (2015)
  – PFNA: 13 µg/L, based on a derived target human serum concentration of 4.9 ng/mL, an RSC of 0.5, and a 200-fold conversion ratio for levels in human serum to drinking water levels. The target human serum level is based on a BMDL10 value of 4.93 µg PFNA/mL serum for increased liver weight in mouse dams on GD17 from a developmental toxicity study (Das et al., 2015) and an uncertainty factor of 1000 (10 each for intraspecies sensitivity, interspecies extrapolation, and database uncertainty).
- Vermont Department of Health (2016)
  – Sum of PFOA and PFOS: 0.02 µg/L, based on the USEPA's oral RfD for PFOA and PFOS of 0.00002 mg/kg/d, a body weight–adjusted water intake rate of 0.175 L/kg bw/d for the first year of life, a hazard quotient value of 1, and an RSC value of 0.2.
- Drinking Water Commission, German Federal Environment Agency (DWC, 2006)
  – Sum of PFOS and PFOA: 0.3 µg/L, based on pTDI values of 0.1 µg/kg bw/d each for PFOS and PFOA, a 10% RSC value from drinking water, and the assumption that a 70 kg adult consumed 2 L of water daily.

Instead of a health-based guidance value, such as an RfD or TDI, one may calculate the margin of exposure (MOE) between exposure to a compound and a NOEL or LOEL for critical effects seen in animal studies. This MOE is then compared with an "ideal" MOE, which is the product of all the uncertainty factors a risk assessor would normally use in conjunction with the study in question, such as safety factors for intraspecies sensitivity and interspecies extrapolation. Health Canada has calculated MOEs between serum and liver PFOS levels at critical effect levels in animal studies conducted with PFOS and serum levels reported in adults and children in Canada and liver PFOS levels from occupationally exposed humans (Health Canada, 2006). Health Canada calculated MOEs of 220–496 and 143–371, respectively, between the levels of PFOS seen in adults and children and between serum concentrations of PFOS associated with microscopic changes in the liver seen in a chronic study conducted in rats with PFOS (Butenhoff et al., 2012b). Health Canada calculated MOEs of 230–518 and 149–387 between serum levels of PFOS in adults and children and serum concentrations of PFOS in monkeys that were associated with thymic atrophy, decreased serum lipids, T3, and bilirubin in a 26-week study (Seacat et al., 2002). Of note, the risk assessment guideline values for PFOS are most often based on the results of the cynomolgus monkey study or the 2-year bioassay in rats, except the values calculated by the German BfR and the USEPA, which are based on the two-generation reproductive toxicity study in rats. Guideline values for PFOA are based on a range of different effects, including liver effects in rodents, effects in cynomolgus monkeys, developmental toxicity in mice, or an overall assessment of systemic toxicity potency. This author could not locate any regulatory agency guideline values for PFHxS, despite the high prevalence of exposure to this compound, nor have guideline values been derived for the extremely potent PFCAs, such as PFDoDA. The lack of standard toxicity studies conducted with these compounds and the uncertainty as to the extent of human exposure are possible reasons for these omissions. Finally, although guideline values for the sum of PFOA and PFOS concentrations in drinking water have been calculated, guideline values for drinking water exposures to PFAS mixtures other than PFOS + PFOA have not been calculated, likely due to the factors implicated in the difficulty of using a read-across approach in PFAS risk assessment.

8.6 PFAS "DARK MATTER," CONCLUSIONS, AND FUTURE PROSPECTS

This chapter has focused on the human health risk assessment of drinking water exposure to impurities in groundwater deriving from AFFF-contaminated sites. The majority of the identified impurities are the environmentally persistent PFCAs and PFSAs, which this chapter has subsequently discussed at length. However, Barzen-Hanson and Field (2015) have identified several additional PFAS compounds present in AFFF-contaminated groundwater that are not PFCAs or PFSAs and for which no toxicity data of any kind are available. In addition, studies have found that the PFAS impurities identified by traditional analytical methods, such as liquid chromatography–tandem mass spectrometry (LC-MS/MS), only represent a fraction of the total organic fluorine present in AFFF-contaminated groundwater. Assessment of AFFF-contaminated water in the total oxidizable precursor (TOP) assay revealed the presence of significant concentrations of PFCA and/or PFSA precursors, subsequently termed PFAS "dark matter" (Deeb et al., 2016). This dark matter has not even been characterized as to its chemical identity, much less as to its toxicological effects. The additional uncertainty regarding the identity and toxicological properties of this dark matter, the data gaps in the toxicological data space for PFAS, and the difficulties inherent in the assessment of exposure to complex mixtures of toxicologically active substances combine to make human health risk assessment of drinking water exposure to AFFF degradation products extremely challenging.

This chapter summarized the data from toxicity studies conducted in mammalian animal models with PFSAs and PFCAs in order to provide a snapshot view of the state of PFAS toxicology and to provide a broad outline of the structure of the PFAS class as a whole. Future work refining the mechanisms of action relevant to the toxicity of PFAS in humans and animal models and providing data on the apical effects of PFAS compounds for which, at present, no data exist are critical for the science of PFAS toxicology to progress. Hopefully, this work will then inform future efforts by regulatory agencies to address the risk to human health from AFFF-contaminated groundwater.

REFERENCES

Abe, T., Takahashi, M., Kano, M. et al. 2016. Activation of nuclear receptor CAR by an environmental pollutant perfluorooctanoic acid. *Arch. Toxicol.* DOI: 10.1007/s00204-016-1888-3.

Austin, M.E., Kasturi, B.S., Barber, M. et al. 2003. Neuroendocrine effects of perfluorooctane sulfonate in rats. *Environ. Health Perspect.* 111: 1485–89.

Barzen-Hanson, K.A., and Field, J.A. 2015. Discovery and implications of C2 and C3 perfluoroalkyl sulfonates in aqueous film-forming foams and groundwater. *Environ. Sci. Technol. Lett.* 2: 95–99.

Benskin, J.P., DeSilva, A.O., Martin, L.J. et al. 2009. Disposition of perfluorinated acid isomers in Sprague-Dawley rats, Part I: Single dose. *Environ. Toxicol. Chem.* 28: 542–54.

Berthiaume, J., and Wallace, K.B. 2002. Perfluoroctanoate, perfluorooctanesulfonate, and N-ethyl perfluoroctanesulfonamido ethanol: Peroxisome proliferation and mitochondrial biogenesis. *Toxicol. Lett.* 129: 23–32.

BfR. 2008. Gesundheitliche Risiken durch PFOS und PFOA in Lebensmitteln sind nach dem derzeitigen wissenschaftlichen Kenntnisstand unwahrscheinlich. Stellungnahme 004/2009 des BfR vom 11. September. http://www.bfr.bund.de/cm/343/gesundheitliche_risiken_durch_pfos_und_pfoa_in_lebensmitteln.pdf.

Biegel, L.B., Hurtt, M.E., Frame, S.R. et al. 2001. Mechanism of extrahepatic tumor induction by peroxisome proliferators in male CD rats. *Toxicol. Sci.* 60: 44–55.

Biegel, L.B., Liu, R.C.M., Hurtt, M.E. et al. 1995. Effects of ammonium perfluoroctanoate on Leydig cell function: *In vitro, in vivo*, and *ex vivo* studies. *Toxicol. Appl. Pharmacol.* 134: 18–25.

Bijland, S., Rensen, P.C.N., Pieterman, E.J. et al. 2011. Perfluoroalkyl sulfonates cause alkyl chain length-dependent hepatic steatosis and hypolipidemia mainly by impairing lipoprotein production in APOE*3-Leiden.CETP mice. *Toxicol. Sci.* 123: 290–303.

Bjork, J.A., and Wallace, K.B. 2009. Structure-activity relationships and human relevance for perfluoroalkyl acid-induced transcriptional activation of peroxisome proliferation in liver cell cultures. *Toxicol. Sci.* 11: 89–99.

Bogdanska, J., Sundstrom, M., Bergstrom, U. et al. 2014. Tissue distribution of 35S-labelled perfluorobutanesulfonic acid in adult mice following dietary exposure for 1–5 days. *Chemosphere* 98: 28–36.

Bonefeld-Jorgensen, E.C., Long, M., Bossi, R. 2011. Perfluorinated compounds are related to breast cancer risk in Greenland Inuit: A case control study. *Environ. Health* 10: 88.

Bookstaff, R.C., Moore, R.W., Ingall, G.B. et al. 1990. Androgenic deficiency in male rats treated with perfluorodecanoic acid. *Toxicol. Appl. Pharmacol.* 104: 322–33.

Borges, T., Robertson, L.W., Peterson, R.E. et al. 1992. Dose-related effects of perfluorodecanoic acid on growth, feed intake, and hepatic peroxisomal β-oxidation. *Arch. Toxicol.* 66: 18–22.

Botelho, S.C., Saghafian, M., Pavlova, S. et al. 2015. Complement activation is involved in the hepatic injury caused by high-dose exposure of mice to perfluoroctanoic acid. *Chemosphere* 129: 225–31.

Boujrad, N., Vidic, B., Gazouli, M. et al. 2000. The peroxisome proliferator perfluorodecanoic acid inhibits the peripheral-type benzodiazepine receptor (PBR) expression and hormone-stimulated mitochondrial transport and steroid formation in Leydig cells. *Endocrinology* 141: 3137–48.

Buck, R.C., Franklin, J., Berger, U. et al. 2011. Perfluoroalkyl and polyfluoroalkyl substances in the environment: Terminology, classification and origins. *Integr. Environ. Assess. Manag.* 7: 513–41.

Buhrke, T., Kibellus, A., and Lampen, A. 2013. In vitro toxicological characterization of perfluorinated carboxylic acids with different chain lengths. *Toxicol. Lett.* 218: 97–104.

Butenhoff, J.L., Kennedy, G.L., Chang, S.-C. et al. 2012a. Chronic dietary toxicity and carcinogenicity study with ammonium perfluorooctanoate in Sprague-Dawley rats. *Toxicology* 298: 1–13.

Butenhoff, J.L., Chang, S.-C., Olsen, G.W. et al. 2012b. Chronic dietary toxicity and carcinogenicity study with potassium perfluorooctanesulfonate in Sprague-Dawley rats. *Toxicology* 293: 1–15.

Butenhoff, J.L., Chang, S.-C., Ehresman, D.J. et al. 2009. Evaluation of potential reproductive and developmental toxicity of potassium perfluorohexanesulfonate in Sprague-Dawley rats. *Reprod. Toxicol.* 27: 331–41.

Butenhoff, J.L., Kennedy, G.L., Hinderliter, P.M. et al. 2004a. Pharmacokinetics of perfluorooctanoate (PFOA) in cynomolgus monkeys. *Toxicol. Sci.* 82: 394–406.

Butenhoff, J.L., Kennedy, G.L., Frame, S.R. et al. 2004b. The reproductive toxicity of ammonium perfluorooctanoate (APFO) in the rat. *Toxicology* 196: 95–116.

Butenhoff, J., Costa, G., Elcombe, C. et al. 2002. Toxicity of ammonium perfluorooctanoate in male cynomolgus monkeys after oral dosing for 6 months. *Toxicol. Sci.* 69: 244–57.

C8 Science Panel. 2012a. Probable link evaluation for liver diseases. October 29. http://www.c8sciencepanel.org/pdfs/Probable_Link_C8_Liver_29Oct2012.pdf (accessed June 17, 2013).

C8 Science Panel. 2012b. Probable link evaluation of infectious disease. July 30. http://www.c8sciencepanel.org/pdfs/Probable_Link_C8_Infections_30Jul2012.pdf (accessed June 17, 2013).

C8 Science Panel. 2012c. Probable link evaluation of thyroid disease. July 30. http://www.c8sciencepanel.org/pdfs/Probable_Link_C8_Thyroid_30Jul2012.pdf (accessed June 17, 2013).

C8 Science Panel. 2012d. Probable link evaluation for chronic kidney disease. October 29. http://www.c8sciencepanel.org/pdfs/Probable_Link_C8_Kidney_29Oct2012.pdf (accessed June 17, 2013).

C8 Science Panel. 2012e. Probable link evaluation of cancer. April 15. http://www.c8sciencepanel.org/pdfs/Probable_Link_C8_Cancer_16April2012_v2.pdf (accessed June 17, 2013).

Chang, S.-C., Thibodeaux, J.R., Eastvold, M.L. et al. 2008a. Thyroid hormone status and pituitary function in adult rats given oral doses of perfluorooctanesulfonate (PFOS). *Toxicology* 243: 330–39.

Chang, S.-C., Das, K., Ehresman, D.J. et al. 2008b. Comparative pharmacokinetics of perfluorobutyrate in rats, mice, monkeys, and humans and relevance to human exposure via drinking water. *Toxicol. Sci.* 104: 40–53.

Chen, H., Huang, C., Wilson, M.W. et al. 1994. Effect of the peroxisome proliferators ciprofibrate and perfluorodecanoic acid on hepatic cell proliferation and toxicity in Sprague-Dawley rats. *Carcinogenesis* 15: 2847–50.

Chengelis, C.P., Kirkpatrick, J.B., Radovsky, A. et al. 2009. A 90-day repeated dose oral (gavage) toxicity study of perfluorohexanoic acid (PFHxA) in rats (with functional observational battery and motor activity determinations). *Reprod. Toxicol.* 27: 342–51.

Cook, J.C., Klinefelter, G.R., Hardisty, J.F. et al. 1999. Rodent Leydig cell tumorigenesis: A review of the physiology, pathology, mechanisms, and relevance to humans. *Crit. Rev. Toxicol.* 29: 169–261.

Cook, J.C., Murray, S.M., Frame, S.R. et al. 1992. Induction of Leydig cell adenomas by ammonium perfluorooctanoate: A possible endocrine-related mechanism. *Toxicol. Pharmacol.* 113. 209–17.

Corsini, E., Sangiovanni, E., Avogadro, A. et al. 2012. In vitro characterization of the immunotoxic potential of several perfluorinated compounds (PFCs). *Toxicol. Appl. Pharmacol.* 258: 248–55.

Crompton, M. 1999. The mitochondrial permeability transition pore and its role in cell death. *Biochem. J.* 341: 233–49.

Cui, L., Zhou, Q., Liao, C. et al. 2009. Studies on the toxicological effects of PFOA and PFOS on rats using histological observation and chemical analysis. *Arch. Environ. Contam. Toxicol.* 56: 338–49.

Curran, I., Hierlihy, S.L., Liston, V. et al. 2008. Altered fatty acid homeostasis and related toxicologic sequelae in rats exposed to dietary potassium perfluorooctanesulfonate (PFOS). *J. Toxicol. Environ. Health A* 71: 1526–41.

Das, K.P., Grey, B.E., Rosen, M.B. et al. 2015. Developmental toxicity of perfluorononanoic acid in mice. *Reprod. Toxicol.* 51: 133–44.

D'eon, J.C., and Mabury, S.A. 2007. Production of perfluorinated carboxylic acids (PFCAs) from the biotransformation of perfluoroalkyl phosphate surfactants (PAPS): Exploring routes of human contamination. *Environ. Sci. Technol.* 41: 4799–805.

Deeb, R., Leeson, A., TerMaath, S., Field, J. et al. 2016. SERDP and ESTCP webinar series: Per- and polyfluoroalkyl substances (PFAS): Analytical and characterization frontiers. https://www.serdp-estcp.org/Tools-and-Training/Webinar-Series/01-28-2016/PFAS-Webinar-Slides_January-2016.

De Silva, A., Benskin, J.P., Martin, L.J. et al. 2009. Disposition of perfluorinated acid isomers in Sprague-Dawley rats: Part 2: Subchronic dose. *Environ. Toxicol. Chem.* 28: 555–67.

DeWitt, J.C., Copeland, C.B., and Luebke, R.W. 2009. Suppression of humoral immunity by perfluorooctanoic acid is independent of elevated serum corticosterone concentration in mice. *Toxicol. Sci.* 109: 106–12.

Dong, G.-H., Zhang, Y.-H., Zheng, L. et al. 2009. Chronic effects of perfluorooctanesulfonate exposure on immunotoxicity in adult male C57Bl6 mice. *Arch. Toxicol.* 83: 805–15.

Dong, G.-H., Liu, M.-M., Wang, D. et al. 2011. Sub-chronic effects of perfluorooctanesulfonate (PFOS) on the balance of type 1 and type 2 cytokine in adult C57BL6 mice. *Mol. Toxicol.* 85: 1235–1244.

Du, G., Huang, H., Hu, J. et al. 2013. Endocrine-related effects of perfluorooctanoic acid (PFOA) in zebrafish, H295R steroidogenesis, and receptor reporter gene assay. *Chemosphere* 91: 1099–1106.

Ducatman, A., Zhang, J., and Fan, H. 2015. Prostate-specific antigen and perfluoroalkyl acids in the C8 Health Study population. *J. Occup. Environ. Med.* 57: 111–14.

DWC (Drinking Water Commission). 2006. Provisional evaluation of PFT in drinking water with the guide substances perfluorooctanoic acid (PFOA) and perfluoroctane sulfonate (PFOS) as examples. https://www.umweltbundesamt.de/sites/default/files/medien/pdfs/pft-in-drinking-water.pdf.

DWI (Drinking Water Inspectorate). 2009. Guidance on the water supply (water quality) regulations 2000 specific to PFOS (perfluorooctane sulphonate) and PFOA (perfluoroctanoic acid) concentrations in drinking water. http://www.dwi.gov.uk/stakeholders/information-letters/2009/10_2009annex.pdf.

ECHA (European Chemicals Agency). 2015. Read-Across Assessment Framework (RAAF). ECHA-15-R-EN. https://echa.europa.eu/documents/10162/13628/raaf_en.pdf.

EFSA (European Food Safety Authority). 2011. Scientific report of EFSA: Results of the monitoring of perfluoroalkylated substances in food in the period 2000–2009. *EFSA J.* 9: 2016–50.

EFSA (European Food Safety Authority). 2008. Scientific opinion of the panel on contaminants in the food chain on perfluorooctane sulfonate (PFOS), perfluoroctanoic acid (PFOA) and their salts. http://www.efsa.europa.eu/sites/default/files/scientific_output/files/main_documents/653.pdf.

Elcombe, C.R., Elcombe, B.M., Foster, J.R. et al. 2012. Hepatocellular hypertrophy and cell proliferation in Sprague-Dawley rats from dietary exposure to potassium perfluoroctanesulfonate results from increased expression of xenosensor nuclear receptors PPARα and CAR/PXR. *Toxicology* 293: 16–29.

Eriksen, K.T., Raaschou-Nielsen, O., McLaughlin, J.K. et al. 2013. Association between plasma PFOA and PFOS levels and total cholesterol in a middle-aged Danish population. *PLoS One* 8: e56969.

Eriksen, K.T., Raaschou-Nielsen, O., Sorensen, M. et al. 2010. Genotoxic potential of the perfluorinated chemicals PFOA, PFOS, PFBS, PFNA, and PFHxA in human HepG2 cells. *Mutat. Res. Genet. Toxicol. Environ. Mutagen.* 700: 39–43.

Eriksen, K.T., Sorensen, M., McLaughlin, J.K. et al. 2009. Perfluorooctanoate and perfluorooctane sulfonate plasma levels and risk of cancer in the general Danish population. *J. Natl. Cancer Inst.* 101: 605–9.

Fairley, K.J., Purdy, R., Kearns, S. et al. 2007. Exposure to the immunosuppressant, perfluorooctanoic acid, enhances the murine IgE and airway hyperreactivity response to ovalbumin. *Toxicol. Sci.* 97: 375–83.

Fang, X., Gao, G., Xue, H. et al. 2012a. In vitro and in vivo studies of the toxic effects of perfluorononanoic acid on rat hepatocytes and Kupffer cells. *Environ. Toxicol. Pharmacol.* 34: 484–94.

Fang, X., Zou, S., Zhao, Y. et al. 2012b. Kupffer cells suppress perfluorononanoic acid-induced hepatic peroxisome proliferator-activated receptor α expression by releasing cytokines. *Arch. Toxicol.* 86: 1515–25.

Fang, X., Feng, Y., Wang, J. et al. 2010. Perfluorononanoic acid-induced apoptosis in rat spleen involves oxidative stress and the activation of caspase-independent death pathway. *Toxicology* 267: 54–59.

Fang, X., Feng, Y., Shi, Z. et al. 2009. Alterations of cytokines and MAPK signaling pathways are related to the immunotoxic effects of perfluorononanoic acid. *Toxicol. Sci.* 108: 367–76.

Fang, X., Zhang, L., Feng, Y. et al. 2008. Immunotoxic effects of perfluorononanoic acid in mice. *Toxicol. Sci.* 105: 312–21.

Fei, C., McLaughlin, J.K., Lipworth, L. et al. 2009. Maternal levels of perfluorinated chemicals and subfecundity. *Hum. Reprod.* 24: 1200–5.

Feng, Y., Fang, X., Shim Z. et al. 2010. Effects of PFNA exposure on expression of junction-associated molecules and secretory function in rat Sertoli cells. *Reprod. Toxicol.* 30: 429–37.

Feng, Y., Shi, Z., Fang, X. et al. 2009. Perfluorononanoic acid induces apoptosis involving the Fas death receptor signaling pathway in rat testis. *Toxicol. Lett.* 190: 224–30.

Fisher, M., Arbuckle, T.E., Wade, M. et al. 2013. Do perfluoroalkyl substances affect metabolic function and plasma lipids? Analysis of the 2007–2009, Canadian Health Measures Survey (CHMS) Cycle 1. *Environ. Res.* 121: 95–103.

Fitz-Simon, N., Fletcher, T., Luster, M.I. et al. 2013. Reductions in serum lipids with a 4-year decline in serum perfluorooctanoic acid and perfluorooctanesulfonic acid. *Epidemiology* 24: 569–76.

Frisbee, S.J., Shankar, A., Knox, S.S. et al. 2010. Perfluorooctanoic acid, perfluorooctanesulfonate, and serum lipids in children and adolescents. *Arch. Pediatr. Adolesc. Med.* 164: 860–9.

Fromme, H., Tittlemier, S.A., Volkel, W. et al. 2009. Perfluorinated compounds—Exposure assessment for the general population in western countries. *Int. J. Hyg. Environ. Health* 212: 239–70.

Fu, Y., Wang, T., Fu, Q. et al. 2014. Associations between serum concentrations of perfluoroalkyl acids and serum lipid levels in a Chinese population. *Ecotoxicol. Environ. Saf.* 106: 246–52.

Fuentes, S., Vicens, P., Colomina, M.T. et al. 2007. Behavioral effects in adult mice exposed to perfluorooctane sulfonate (PFOS). *Toxicology* 242: 12329.

Gallo, V., Leonardi, G., Genser, B. et al. 2012. Serum perfluorooctanoate (PFOA) and perfluorooctane sulfonate (PFOS) concentrations and liver function biomarkers in a population with elevated PFOA exposure. *Environ. Health Perspect.* 120: 655–60.

Gannon, S.A., Johnson, T., Nabb, D.L. et al. 2011. Absorption, distribution, metabolism, and excretion of [1-^{14}C]-perfluorohexanoate ([^{14}C]-PFHx) in rats and mice. *Toxicology* 283: 55–62.

Gao, Y., Li, X., and Guo, L.-H. 2013. Assessment of estrogenic activity of perfluoroalkyl acids based on ligand-induced conformation state of human estrogen receptor. *Environ. Sci. Technol.* 47: 634–41.

Gilliland, F.D., and Mandel, J.S. 1996. Serum perfluorooctanoic acid and hepatic enzymes, lipoproteins, and cholesterol: A study of occupationally exposed men. *Am. J. Ind. Med.* 29: 560–68.

Gilliland, F.D., and Mandel, J.S. 1993. Mortality among employees of a perfluorooctanoic acid production plant. *J. Occup. Med.* 35: 950–54.

Gleason, J.A., Post, G.B., and Fagliano, J.A. 2015. Associations of perfluorinated chemical serum concentrations and biomarkers of liver function and uric acid in the US population (NHANES) 2007–2010. *Environ. Res.* 136: 8–14.

Goecke, C.M., Jarnot, B.M., and Reo, N.V. 1992. A comparative toxicological investigation of perfluorocarboxylic acids in rats by fluorine-19 NMR spectroscopy. *Chem. Res. Toxicol.* 5: 512–19.

Grandjean, P., Anderson, E.W., Budtz-Jorgensen, E. et al. 2012. Serum vaccine antibody concentrations in children exposed to perfluorinated compounds. *JAMA* 307: 391–97.

Hansen, S., Vestergren, R., Herzke, D. et al. 2016. Exposure to per- and polyfluoroalkyl substances through the consumption of fish from lakes affected by aqueous film-forming foam emissions: A combined epidemiological and exposure modeling approach. The SAMINOR 2 Clinical Study. *Environ. Int.* 94: 272–82.

Hardell, E., Karrman, A., van Bavel, B. et al. 2014. Case-control study on perfluorinated alkyl acids (PFAAs) and the risk of prostate cancer. *Environ. Int.* 63: 35–39.

Harrison, E.H., Lane, J.S., Luking, S. et al. 1988. Perfluoro-n-decanoic acid: Induction of peroxisomal β-oxidation by a fatty acid with dioxin-like toxicity. *Lipids* 23: 115–19.

Haug, L.S., Huber, S., Becher, G. et al. 2011. Characterization of human exposure pathways to perfluorinated compounds—Comparing exposure estimates with biomarkers of exposure. *Environ. Int.* 37: 687–93.

Health Canada. 2006. State of the science report for a screening health assessment: Perfluorooctane sulfonate (PFOS)—Its salts and its precursors that contain the C8F17SO2 or C8F17SO3 moiety. HC publication no. 4447.

Health Protection Agency. 2012. The public health significance of perfluorooctane sulphonate (PFOS). https://www.gov.gg/chttphandler.ashx?id=83403&p=0.

Henry, N.D., and Fair, P.A. 2013. Comparison of in vitro cytotoxicity, estrogenicity and anti-estrogenicity of triclosan, perfluorooctane sulfonate and perfluorooctanoic acid. *J. Appl. Toxicol.* 33: 265–72.

Hirata-Koizumi, M., Fujii, S., Furukawa, M. et al. 2012. Repeated dose and reproductive/developmental toxicity of perfluorooctadecanoic acid in rats. *J. Toxicol. Sci.* 37: 63–79.

Hoffman, K., Webster, T.F., Bartell, S.M. et al. 2011. Private drinking water wells as a source of exposure to perfluorooctanoic acid (PFOA) in communities surrounding a fluoropolymer production facility. *Environ. Health Perspect.* 119: 92–97.

Hovey, R.C., Coder, P.S., Wold, J.C. et al. 2011. Quantitative assessment of mammary gland development in female Long-Evans rats following in utero exposure to atrazine. *Toxicol. Sci.* 119: 380–90.

Hu, W., Jones, P.D., DeCoen, W. et al. 2003. Alterations in cell membrane properties caused by perfluorinated compounds. *Comp. Biochem. Physiol. C* 135: 77–88.

Hu, W., Jones, P.D., Upham, B.L. et al. 2002. Inhibition of gap junctional intercellular communication by perfluorinated compounds in rat liver and dolphin kidney epithelial cell lines in vitro and Sprague-Dawley rats in vivo. *Toxicol. Sci.* 68: 429–36.

Hu, X.C., Andrews, D.Q., Lindstrom, A.B. et al. 2016. Detection of poly- and perfluoroalkyl substances (PFAS) in U.S. drinking water linked to industrial sites, military fire training areas, and wastewater treatment plants. *Environ. Sci. Technol. Lett.* 3: 344–50.

Innes, K.E., Ducatman, A.M., Luster, M.I. et al. 2011. Association of osteoarthritis with serum levels of the environmental contaminants perfluorooctanoate and perfluorooctane sulfonate in a large Appalachian population. *Am. J. Epidemiol.* 174: 440–50.

Intrasuksri, U., Rangwala, S.M., O'Brien, M. et al. 1998. Mechanisms of peroxisome proliferation by perfluoroctanoic acid and endogenous fatty acids. *Gen. Pharmacol.* 31: 187–97.

IRDC. (International Research and Development Corporation). 1978. Ninety day subacute rhesus monkey toxicity study. Compound: Fluorad® Flurochemical FC-143.

Jain, R.B. 2013. Association between thyroid profile and perfluoroalkyl acids: Data from NHANES 2007–2008. *Environ. Res.* 126: 51–59.

Joensen, U.N., Bossi, R., Leffers, H. et al. 2009. Do perfluoroalkyl compounds impair human semen quality? *Environ. Health Perspect.* 117: 923–27.

Kamendulis, L.M., Wu, Q., Sandusky, G.E. et al. 2014. Perfluorooctanoic acid exposure triggers oxidative stress in the mouse pancreas. *Toxicol. Rep.* 1: 513–21.

Kato, H., Fujii, S., Takahashi, M. et al. 2015. Repeated dose and reproductive/developmental toxicity of perfluorododecanoic acid in rats. *Environ. Toxicol.* 30: 1244–63.

Kato, K., Wong, L., Jia, L.T. et al. 2011. Trends in exposure to polyfluoroalkyl chemicals in the US population: 1999–2008. *Environ. Sci. Technol.* 45: 8037–45.

Kawamoto, K., Sato, I., Tsuda, S. et al. 2011. Ultrasonic-induced tonic convulsion in rats after subchronic exposure to perfluorooctane sulfonate (PFOS). *J. Toxicol. Sci.* 36: 55–62.

Kawashima, Y., Kobayashi, H., Miura, H. et al. 1995. Characterization of hepatic responses of rat to administration of perfluorooctanoic acid and perfluorodecanoic acid at low levels. *Toxicology* 99: 169–78.

Kawashima, Y., Uy-Yu, N., and Kozuka, H. 1989. Sex-related difference in the enhancing effects of perfluoro-octanoic acid on stearoyl-CoA desaturase and its influence on the acyl composition of phospholipid in the rat liver. *Biochem. J.* 263: 897–904.

Kielsen, K., Shamim, Z., Ryder, L.P. et al. 2016. Antibody response to booster vaccination with tetanus and diphtheria in adults exposed to perfluorinated alkylates. *J. Immunotoxicol.* 13: 270–73.

Kjeldsen, L.S., and Bonefeld-Jorgensen, E.C. 2013. Perfluorinated compounds affect the function of sex hormone receptors. *Environ. Sci. Pollut. Res.* 20: 8031–44.

Klaunig, J.E., Shinohara, M., Iwai, H. et al. 2014. Evaluation of the chronic toxicity and carcinogenicity of perfluorohexanoic acid (PFHA) in Sprague-Dawley rats. *Toxicol. Pathol.* 20: 1–12.

Klaunig, J.E., Hocevar, B.A., and Kamendulis, L.M. 2012. Mode of action and analysis of perfluorooctanoic acid (PFOA) tumorigenicity and human relevance. *Reprod. Toxicol.* 33: 410–18.

Kleszczynski, K., and Skladanowski, A.C. 2011. Mechanism of cytotoxic action of perfluorinated acids. II. Disruption of mitochondrial bioenergetics. *Toxicol. Appl. Pharmacol.* 235: 182–90.

Kleszczynski, K., Stepnowski, P., and Skladanowski, A.C. 2009. Mechanism of cytotoxic action of perfluorinated acids. III. Disturbance of Ca^{2+} homeostasis. *Toxicol. Appl. Pharmacol.* 251: 163–68.

Knox, S.S., Jackson, T., Frisbee, S.J. et al. 2011. Perfluorocarbon exposure, gender, and thyroid function in the C8 Health Project. *J. Toxicol. Sci.* 36: 403–10.

Kudo, N., and Kawashima, Y. 2003. Induction of triglyceride accumulation in the liver of rats by perfluorinated fatty acids with different carbon chain lengths: Comparison with induction of peroxisomal β-oxidation. *Biol. Pharm. Bull.* 26: 47–51.

Kudo, N., Yamazaki, T., Sakamoto, T. et al. 2011. Effects of perfluorinated fatty acids with different carbon chain length on fatty acid profiles of hepatic lipids in mice. *Biol. Pharm. Bull.* 34: 856–64.

Kudo, N., Sakai, A., Mitsumoto, A. et al. 2007. Tissue distribution and hepatic subcellular distribution of perfluoroctanoic acid at low dose are different from those at high dose in rats. *Biol. Pharm. Bull.* 30: 1535–40.

Kudo, N., Suzuki-Kakajima, E., Mitsumoto, A. et al. 2006. Responses of the liver to perfluorinated fatty acids with different carbon chain length in male and female mice: In relation to induction of hepatomegaly, peroxisomal β-oxidation, and microsomal 1-acyglycerophosphocholine acyltransferase. *Biol. Pharm. Bull.* 29: 1952–57.

La Rocca, C., Alessi, E., Bergamasco, B. et al. 2012. Exposure and effective dose biomarkers for perfluorooctane sulfonic acid (PFOS) and perfluorooctanoic acid (PFOA) in infertile subjects: Preliminary results of the PREVIENI project. *Int. J. Hyg. Environ. Health* 215: 206–11.

Lau, C., Thibodeaux, J.R., Hanson, R.G. et al. 2006. Effects of perfluorooctanoic acid exposure during pregnancy in the mouse. *Toxicol. Sci.* 90: 510–18.

Lefebvre, D.E., Curran, I., Armstrong, C. et al. 2008. Immunomodulatory effects of dietary potassium perfluorooctane sulfonate (PFOS) exposure in adult Sprague-Dawley rats. *J. Toxicol. Environ. Health A* 71: 1516–25.

Leonard, R.C., Kreckmann, K.H., Sakr, C.J. et al. 2008. Retrospective cohort mortality study of workers in a polymer production plant including a reference population of regional workers. *Ann. Epidemiol.* 18: 15–22.

Lewis, R.C., Johns, L.E., and Meeker, J.D. 2015. Serum biomarkers of exposure to perfluorinated compounds in relation to serum testosterone and measures of thyroid function among adults and adolescents from NHANES 2011–2012. *Int. J. Environ. Res. Public Health* 12: 6098–114.

Li, Y., Ramdhan, D.H., Naito, H. et al. 2011. Ammonium perfluorooctanoate may cause testosterone reduction by adversely affecting testis in relation to PPARα. *Toxicol. Lett.* 205: 265–72.

Liao, C., Wang, T., Cui, L. et al. 2009. Changes in synaptic transmission, calcium current, and neurite growth by perfluorinated compounds are dependent on the chain length and functional group. *Environ. Sci. Technol.* 43: 2099–104.

Liao, C.-Y., Li, X.-Y., Wu, B. et al. 2008. Acute enhancement of synaptic transmission and chronic inhibition of synaptogenesis induced by perfluoroctane sulfonate through mediation of voltage-dependent calcium channel. *Environ. Sci. Technol.* 42: 5335–41.

Lieder, P.H., Chang, S.-C., York, R.G. et al. 2009a. Toxicological evaluation of potassium perfluorobutanesulfonate in a 90-day oral gavage study with Sprague-Dawley rats. *Toxicology* 255: 45–52.

Lieder, P.H., York, R.G., Hakes, D.C. et al. 2009b. A two-generation reproduction study with potassium perfluorobutanesulfonate (K+PFBS) in Sprague-Dawley rats. *Toxicology* 259: 33–45.

Lin, C.-Y., Lin, L.-Y., Wen, T.-W. et al. 2013a. Association between levels of serum perfluorooctane sulfonate and carotid artery intima-media thickness in adolescents and young adults. *Int. J. Cardiol.* 168: 3309–16.

Lin, C.-Y., Wen L.-L., Wen, T.-W. et al. 2013b. The associations between serum perfluorinated chemicals and thyroid function in adolescents and young adults. *J. Hazard. Mater.* 244–45: 637–44.

Lin, C.-Y., Lin, L.-Y., Chiang, C.-K. et al. 2010. Investigation of the associations between low-dose serum perfluorinated chemicals and liver enzymes in US adults. *Am. J. Gastroenterol.* 105: 1354–63.

Liu, R.C.M., Hurtt, M.E., Cook, J.C. et al. 1996. Effect of peroxisome proliferator, ammonium perfluorooctanoate (C8), on hepatic aromatase activity in adult male Crl:CD BR (CD) rats. *Fund. Appl. Toxicol.* 30: 220–28.

Long, M., Ghisari, M., and Bonefield-Jorgensen, E.C. 2013. Effects of perfluoroalkyl acids on the function of the thyroid hormone and aryl hydrocarbon receptor. *Environ. Sci. Pollut. Res.* 20: 8045–56.

Long, Y., Wang, Y., Ji, G. et al. 2013. Neurotoxicity of perfluorooctane sulfonate to hippocampal cells in adult mice. *PLoS One* 8: 1–9.

Lopez-Doval, S., Salgado, R., Pereiro, N. et al. 2014. Perfluorooctane sulfonate effects on the reproductive axis in adult male rats. *Environ. Res.* 134: 158–68.

Louis, G.M.B., Peterson, M., Chen, Z. et al. 2012. Perfluorochemicals and endometriosis: The ENDO study. *Epidemiology* 23: 799–805.

Louis, G.M.B., Sundaram, R., Schisterman, E.F.M. et al. 2013. Persistent environmental pollutants and couple fecundity: The LIFE study. *Environ. Health Perspect.* 121: 231–36.

Luebker, D.J., Case, M.T., York, R.G. et al. 2005. Two-generation reproductive and cross-foster studies of perfluorooctanesulfonate (PFOS) in rats. *Toxicology* 215: 126–48.

Luebker, D.J., Hansen, K.J., Bass, N.M. et al. 2002. Interactions of fluorochemicals with rat liver fatty acid-binding protein. *Toxicology* 176: 175–85.

Lundin, J.I., Alexander, B.H., Olsen, G.W. et al. 2009. Ammonium perfluorooctanoate production and occupational mortality. *Epidemiology* 20: 921–28.

MDH (Minnesota Department of Health), Health Risk Assessment Unit, Environmental Health Division. 2011a. 2011 health risk limits for groundwater: Perfluorobutyrate. http://www.health.state.mn.us/divs/eh/risk/guidance/gw/pfba.pdf.

MDH (Minnesota Department of Health), Health Risk Assessment Unit, Environmental Health Division. 2011b. 2011 health risk limits for groundwater: Perfluorobutane sulfonate. http://www.health.state.mn.us/divs/eh/risk/guidance/gw/pfbs.pdf.

MDH (Minnesota Department of Health), Health Risk Assessment Unit, Environmental Health Division. 2009a. Health risk limits for groundwater 2008 rule revision: Perfluorooctanoic acid (PFOA) and its salts. http://www.health.state.mn.us/divs/eh/risk/guidance/gw/pfoa.pdf.

MDH (Minnesota Department of Health), Health Risk Assessment Unit, Environmental Health Division. 2009b. Health risk limits for groundwater 2008 rule revision: Perfluorooctane sulfonate (PFOS) and its salts. http://www.health.state.mn.us/divs/eh/risk/guidance/gw/pfos.pdf.

Melzer, D., Rice, N., Depledge, M.H. et al. 2010. Association between serum perfluorooctanoic acid (PFOA) and thyroid disease in the NHANES study. *Environ Health Perspect.* 118: 686–92.

Minata, M., Harada, K.H., Karrman, A. et al. 2010. Role of peroxisome proliferator-activated receptor-α in hepatobiliary injury induced by ammonium perfluorooctanoate in mouse liver. *Ind. Health* 48: 96–107.

Mollenhauer, M.A.M., Bradshaw, S.G., Fair, P.A. et al. 2011. Effects of perfluorooctane sulfonate (PFOS) exposure on markers of inflammation in female B6C3F1 mice. *J. Environ. Sci. Health A* 46: 97–108.

Nakagawa, T., Ramdhan, D.H., Tanaka, N. et al. 2012. Modulation of ammonium perfluorooctanote-induced hepatic damage by genetically-different PPARα in mice. *Arch. Toxicol.* 86: 63–74.

Nelson, D.L., Frazier, D.E., Ericson, J.E. et al. 1992. The effects of perfluorodecanoic acid (PFDA) on humoral, cellular, and innate immunity in Fischer 344 rats. *Immunopharmacol. Immunotoxicol.* 14: 925–38.

Nelson, J.W., Hatch, E.E., and Webster, T.F. 2010. Exposure to polyfluoroalkyl chemicals and cholesterol, body weight, and insulin resistance in the general US population. *Environ. Health Perspect.* 118: 197–202.

New Jersey Drinking Water Quality Institute, Health Effects Subcommittee. 2015. Health-based maximum contaminant level support document: Perfluorononanoic acid (PFNA).

OECD (Organisation for Economic Cooperation and Development). 2015. Guidelines for testing of chemicals. Section 4: Health effects. Test no. 422: Combined repeated dose and toxicity study with the reproductive/developmental toxicity screening test. http://www.oecd.org/environment/test-no-422-combined-repeated-dose-toxicity-study-with-the-reproduction-developmental-toxicity-screening-test-9789264242715-en.htm.

OECD (Organisation for Economic Cooperation and Development). 2001. OECD guidelines for testing of chemicals. Section 4: Health effects. Test no. 416: Two-generation reproduction toxicity. http://www.oecd.org/chemicalsafety/risk-assessment/1948466.pdf.

Ohmori, K., Kudo, N., Katayama, K. et al. 2003. Comparison of the toxicokinetics between perfluorocarboxylic acids with different carbon chain length. *Toxicology* 184: 135–40.

Okada, E., Sasaki, S., Kashino, I. et al. 2014. Prenatal exposure to perfluoroalkyl acids and allergic disease in early childhood. *Environ. Intl.* 65: 127–34.

Olsen, G.W., and Zobel, L.R. 2007. Assessment of lipid, hepatic, and thyroid parameters with serum perfluorooctanoate (PFOA) concentrations in fluorochemical production workers. *Int. Arch. Occup. Environ. Health* 81: 231–46.

Olsen, G.W., Chang, S.-C., Noker, P.E. et al. 2009. A comparison of the pharmacokinetics of perfluorobutanesulfonate (PFBS) in rats, monkeys, and humans. *Toxicology* 256: 65–74.

Olsen, G.W., Burris, J.M., Ehresman, D.J. et al. 2007. Half-life of serum elimination of perfluorooctanesulfonate, perfluorohexanesulfonate, and perfluoroctanoate in retired fluorochemical production workers. *Environ. Health Perspect.* 115: 1298–305.

Olsen, G.W., Burris, J.M., Burlew, M.M. et al. 2003. Epidemiologic assessment of worker serum perfluorooctane sulfonate (PFOS) and perfluoroctanoate (PFOA) concentrations and medical surveillance examinations. *J. Occup. Environ. Med.* 45: 260–70.

Olsen, G.W., Burris, J.M., Burlew, M.M. et al. 2000. Plasma cholecystokinin and hepatic enzymes, cholesterol, and lipoproteins in ammonium perfluorooctanoate production workers. *Drug Chem. Toxicol.* 23: 603–20.

Olsen, G.W., Gilliland, F.D., Burlew, M.M. et al. 1998. An epidemiologic investigation of reproductive hormones in men with occupational exposure to perfluorooctanoic acid. *J. Occup. Environ. Med.* 40: 614–22.

Peden-Adams, M.M., Keller, J.M., EuDaly, J.G. et al. 2008. Suppression of humoral immunity in mice following exposure to perfluorooctane sulfonate. *Toxicol. Sci.* 104: 144–54.

Perkins, R.G., Butenhoff, J.L., Kennedy, G.L. et al. 2004. 13-Week dietary toxicity study of ammonium perfluorooctanoate (APFO) in male rats. *Drug Chem. Toxicol.* 27: 361–78.

Permadi, H., Lundgren, B., Andersson, K. et al. 1993. Effects of perfluoro fatty acids on peroxisome proliferation and mitochondrial size in mouse liver: Dose and time factors and effect of chain length. *Xenobiotica* 23: 761–70.

Permadi, H., Lundgren, B., Andersson, K. et al. 1992. Effects of perfluoro fatty acids on xenobiotic-metabolizing enzymes, enzymes which detoxify reactive forms of oxygen, and lipid peroxidation in mouse liver. *Biochem. Pharmacol.* 44: 1183–91.

Peters, J.M., and Gonzalez, F.J. 2011. Why toxic equivalency factors are not suitable for perfluoroalkyl chemicals. *Chem. Res. Toxicol.* 24: 1601–9.

Qazi, M.R., Hassan, M., Nelson, B.D. et al. 2013. Sub-acute, moderate-dose, but not short-term, low-dose dietary pre-exposure of mice to perfluorooctanoate aggravates concanavalin A-induced hepatitis. *Toxicol. Lett.* 219, 1–7.

Qazi, M.R., Bogdanska, J., Butenhoff, J.L. et al. 2009. High-dose, short-term exposure of mice to perfluorooctanesulfonate (PFOS) or perfluorooctanoate (PFOA) affects the number of circulating neutrophils differently, but enhances the inflammatory responses of macrophages to lipopolysaccharide (LPS) in a similar fashion. *Toxicology* 262: 207–14.

Qui, L., Zhang, X., Zhang, X. et al. 2013. Sertoli cell is a potential target for perfluorooctane sulfonate-induced reproductive dysfunction in male mice. *Toxicol. Sci.* 135: 229–40.

Raymer, J.H., Michael, L.C., Studabaker, W.B. et al. 2012. Concentrations of perfluorooctane sulfonate (PFOS) and perfluoroctanoate (PFOA) and their association with human semen quality measurements. *Reprod. Toxicol.* 33: 419–27.

Ren, X.-M., Zhang, Y.-F., Guo, L.-H. et al. 2015. Structure-activity relations in binding of perfluoroalkyl compounds to human thyroid hormone T3 receptor. *Arch. Toxicol.* 89: 233–42.

Rockwell, C.E., Turkey, A.E., Cheng, X. et al. 2013. Acute immunotoxic effects of perfluorononanoic acid (PFNA) in C57Bl/6 mice. *Clin. Exp. Pharmacol.* S4: 1–9.

Rosen, M.B., Schmid, J.R., Corton, J.C. et al. 2010. Gene expression profiling in wild-type and PPARα-null mice exposed to perfluoroctane sulfonate reveals PPARα-independent effects. *PPAR Res.* DOI: 10.1155/2010/794739.

Rosen, M.B., Lau, C., and Corton, J.C. 2009. Does exposure to perfluoroalkyl acids present a risk to human health? *Toxicol. Sci.* 111: 1–3.

Rosen, M.B., Lee, J.S., Ren, H. et al. 2008. Toxicogenomic dissection of the perfluorooctanoic acid transcript profile in mouse liver: Evidence for the involvement of nuclear receptors PPARα and CAR. *Toxicol. Sci.* 103: 46–56.

Rotander, A., Karrman, A., Toms, L.-M.M. et al. 2015a. Novel fluorinated surfactants tentatively identified in firefighters using liquid chromatography quadropole time-of-flight tandem mass spectrometry and a case-control approach. *Environ. Sci. Technol.* 49: 2434–42.

Rotander, A., Toms, L.-M.M., Aylward, L. et al. 2015b. Elevated levels of PFOS and PFHxS in firefighters exposed to aqueous film forming foam (AFFF). *Environ. Int.* 82: 28–34.

Sakr, C.J., Kreckmann, K.H., Green, J.W. et al. 2007a. Cross-sectional study of lipids and liver enzymes related to a serum biomarker of exposure (ammonium perfluorooctanoate or APFO) as part of a general health survey in a cohort of occupationally exposed workers. *J. Occup. Environ. Med.* 49: 1086–96.

Sakr, C.J., Leonard, R.C., Kreackmann, K.H. et al. 2007b. Longitudinal study of serum lipids and liver enzymes in workers with occupational exposure to ammonium perfluorooctanoate. *J. Occup. Environ. Med.* 49: 872–79.

Sato, I., Kawamoto, K., Nishikawa, Y. et al. 2009. Neurotoxicity of perfluorooctane sulfonate (PFOS) in rats and mice after single oral exposure. *J. Toxicol. Sci.* 34: 569–74.

Schramm, H., Friedberg, T., Robertson, L.W. et al. 1989. Perfluorodecanoic acid decreases the enzyme activity and the amount of glutathione S-transferases proteins and mRNAs in vivo. *Chem. Biol. Interact.* 70: 127–43.

Seacat, A.M., Thomford, P.J., Hansen, K.J. et al. 2003. Sub-chronic dietary toxicity of potassium perfluorooctanesulfonate in rats. *Toxicology* 183: 117–31.

Seacat, A.M., Thomford, P.J., Hansen, K.J. et al. 2002. Subchronic toxicity studies on perfluorooctanesulfonate potassium salt in cynomolgus monkeys. *Toxicol. Sci.* 68: 249–64.

Shankar, A., Xiao, J., and Ducatman, A. 2012. Perfluorooctanoic acid and cardiovascular disease in US adults. *Arch. Intern. Med.* 172: 1397–403.

Sheng, N., Li, J., Liu, H. et al. 2014. Interaction of perfluoroalkyl acids with human fatty acid-binding protein. *Arch. Toxicol.* 90: 217–27.

Shi, Z., Feng, Y., Wang, J. et al. 2010. Perfluorododecanoic acid-induced steroidogenic inhibition is associated with steroidogenic acute regulatory protein and reactive oxygen species in cAMP-stimulated Leydig cells. *Toxicol. Sci.* 114: 285–94.

Shi, Z., Zhang, H., Ding, L. et al. 2009. The effect of perfluorododecanoic acid on endocrine status, sex hormones, and expression of steroidogenic genes in pubertal female rats. *Reprod. Toxicol.* 27: 352–59.

Shi, Z., Zhang, H., Liu, Y. et al. 2007. Alterations in gene expression and testosterone synthesis in the testes of male rats exposed to perfluorododecanoic acid. *Toxicol. Sci.* 98: 206–15.

Singer, S.S., Andersen, M.E., and George, M.E. 1990. Perfluoro-n-decanoic acid effects on enzymes of fatty acid metabolism. *Toxicol. Lett.* 54: 39–46.

Singh, T.S.K., Lee, S., Kim, H.-H. et al. 2012. Perfluorooctanoic acid induces mast cell-mediated allergic inflammation by the release of histamine and inflammatory mediators. *Toxicol. Lett.* 210: 64–70.

Slezak, B.P. 2007. Sodium perfluorohexanoate (H-27268): Subchronic toxicity 90-day gavage study in rats with one-generation reproduction evaluation. Haskell Laboratory for Health and Environmental Sciences, Newark, DE.

Sohlenius, A.-K., Lundgren, B., and DePierre, J.W. 1992. Perfluorooctanoic acid has persistent effects on peroxisome proliferation and related parameters in mouse liver. *J. Biochem. Toxicol.* 7: 205–12.

Son, H.-Y., Lee, S., Tak, E.-N. et al. 2009. Perfluorooctanoic acid alters T lymphocyte phenotypes and cytokine expression in mice. *Environ. Toxicol.* 24: 580–88.

Specht, I.O., Hougaard, K.S., Spano, M. et al. 2012. Sperm DNA integrity in relation to exposure to environmental perfluoroalkyl substances—A study of spouses of pregnant women in three geographical regions. *Reprod. Toxicol.* 33: 577–83.

Steenland, K., Tinker, S., Frisbee, S. et al. 2009. Association of perfluorooctanoic acid and perfluorooctane sulfonate with serum lipids among adults living near a chemical plant. *Am. J. Epidemiol.* 170: 1268–78.

Straus, D.S., and Glass, C.K. 2007. Anti-inflammatory actions of PPAR ligands: New insights on cellular and molecular mechanisms. *Trends Immunol.* 28: 551–58.

Sundstrom, M., Chang, S.-C., Noker, P.E. et al. 2012. Comparative pharmacokinetics of perfluorohexanesulfonate (PFHxS) in rats, mice, and monkeys. *Reprod. Toxicol.* 33: 441–51.

Takacs, M.L., and Abbott, B.D. 2007. Activation of mouse and human peroxisome proliferator-activated receptors (α, β/δ, γ) by perfluorooctanoic acid and perfluorooctane sulfonate. *Toxicol. Sci.* 95: 108–17.

Takahashi, M., Ishida, S., Hirata-Koizumi, M. et al. 2014. Repeated dose and reproductive/developmental toxicity of perfluoroundecanoic acid in rats. *J. Toxicol. Sci.* 39: 97–108.

Tan, X., Xie, G., Sun, X. et al. 2013. High fat diet feeding exaggerates perfluorooctanoic acid-induced liver injury in mice via modulating multiple metabolic pathways. *PLoS One* 8: 1–15.

Tatum-Gibbs, K., Wambaugh, J.F., Das, K.P. et al. 2011. Comparative pharmacokinetics of perfluorononanoic acid in rat and mouse. *Toxicology* 281: 48–55.

Taylor, K.W., Hoffman, K., Thayer, K.A. et al. 2014. Perfluoroalkyl chemicals and menopause among women 20–65 years of age (NHANES). *Environ. Health Perspect.* 122: 145–50.

Toft, G., Jonsson, B.A.G., Lindh, C.H. et al. 2012. Exposure to perfluorinated compounds and human semen quality in Arctic and European populations. *Hum Reprod.* 27: 2532–40.

USEPA (U.S. Environmental Protection Agency). 2009. Long-chain perfluorinated chemicals action plan, December 30. https://www.epa.gov/sites/production/files/2016-01/documents/pfcs_action_plan1230_09.pdf.

USEPA. 2016. Lifetime health advisories and health effects support documents for perfluoropctanoic acid and perfluorooctanesulfonate. *Federal Register* 81(101): 33250–33251.

Uy-Yu, N., Kawashima, Y., Horii, S. et al. 1990. Effects of chronic administration of perfluorooctanoic acid in fatty acid metabolism in rat liver: Relationship among stearoyl-coenzyme A desaturase, 1-acylglycerophosphocholine acyltransferase, and acyl composition of microsomal phosphatidylcholine. *J. Pharmacobiodyn.* 13: 581–90.

Vanden Heuvel, J.P., Thompson, J.T., Frame, S.R. et al. 2006. Differential activation of nuclear receptors by perfluorinated fatty acid analogs and natural fatty acids: A comparison of human, mouse, and rat peroxisome-proliferator activated receptor-α, -β, and -γ, liver X-receptor-β, and retinoid X receptor-α. *Toxicol. Sci.* 92: 476–89.

Van Otterdijk, F.M. 2007. Repeated dose 90-day oral toxicity study with MTDID 8391 by daily gavage in the rat followed by a 3-week recovery period. NOTOX B.V., Hertogenbosch, The Netherlands.

Van Rafelghem, M.J., Inhorn, S.L., and Peterson, R.E. 1987. Effects of perfluorodecanoic acid on thyroid status in rats. *Toxicol. Appl. Pharmacol.* 87: 430–39.

Vermont Department of Health. 2016. Environmental health: Perfluorooctanoic acid (PFOA) and perfluorooctanesulfonic acid (PFOS). Vermont drinking water health advisory. Memorandum Vose/Schwer. June 22.

Vestergaard, S., Nielsen, F., Anserssson, A. et al. 2012. Association between perfluorinated compounds and time to pregnancy in a prospective cohort of Danish couples attempting to conceive. *Hum. Reprod.* 27: 873–80.

Wallace, K.B., Kissling, G.E., Melnick, R.L. et al. 2013. Structure-activity relationships for perfluoroalkane-induced in vitro interference with rat liver mitochondrial respiration. *Toxicol. Lett.* 222: 257–64.

Walters, M.W., Bjork, J.A., and Wallace, K.B. 2009. Perfluorooctanoic acid stimulated mitochondrial biogenesis and gene transcription in rats. *Toxicology* 264: 10–15.

Wan, H.T., Zhao, Y.G., Wong, M.H. et al. 2011. Testicular signaling is the potential target of perfluoroctanesulfonate-mediated subfertility in male mice. *Biol. Reprod.* 84: 1016–23.

Wang, J., Zhang, Y., Zhang, W. et al. 2012. Association perfluorooctanoic acid with HDL cholesterol and circulating miR-26b and miR-199-3p in workers of a fluorochemical plant and nearby residents. *Environ. Sci. Technol.* 46: 9274–81.

Wang, Y., Wang, L., Li, J. et al. 2014. The mechanism of immunosuppression by perfluorooctanoic acid in BALB/c mice. *Toxicol. Res.* 3: 205–15.

Webster, G.M., Rauch, S.A., Ste Marie, N. et al. 2016. Cross-sectional associations of serum perfluoroalkyl acids and thyroid hormones in U.S. adults: Variation according to TPOAb and iodine status (NHANES 2007–2008). *Environ. Health Perspect.* 124: 935–42.

Weaver, Y.M., Ehresman, D.J., Butenhoff, J.L. et al. 2010. Roles of rat renal organic anion transporters in transporting perfluorinated carboxylates with different chain lengths. *Toxicol. Sci.* 113: 305–14.

Weiss, J.M., Andersson, P.L., Lamoree, M.H. et al. 2009. Competitive binding of poly- and perfluorinated compounds to the thyroid hormone transport protein transthyretin. *Toxicol. Sci.* 109: 206–16.

Wen, L.-L., Lin, L.-Y., Su, T.-C. et al. 2013. Association between serum perfluorinated chemicals and thyroid function in U.S. adults: The National Health and Nutrition Examination Survey 2007–2010. *J. Clin. Endocrinol. Metab.* 98: E1456–64.

White, S.S., Stanko, J.P., Kato, K. et al. 2011. Gestational and chronic low-dose PFOA exposures and mammary gland growth and differentiation in three generations of CD-1 mice. *Environ. Health Perspect.* 119: 1070–76.

White, S.S., Kato, K., Jia, L.T. et al. 2009. Effects of perfluorooctanoic acid on mouse mammary gland development and differentiation resulting from cross-foster and restricted gestational exposures. *Reprod. Toxicol.* 27: 289–98.

White, S.S., Calafat, A.M., Kuklenyik, Z. et al. 2007. Gestational PFOA exposure of mice is associated with altered mammary gland development in dams and female offspring. *Toxicol. Sci.* 96: 133–44.

Wielsoe, M., Long, M., Ghisari, M. et al. 2015. Perfluoroalkylated substances (PFAS) affect oxidative stress biomarkers in vitro. *Chemosphere* 129: 239–45.

Wolf, C.J., Rider, C.V., Lau, C. et al. 2014. Evaluating the additivity of perfluoroalkyl acids in binary combination on peroxisome proliferator-activated receptor-α activation. *Toxicology* 316: 43–54.

Wolf, C.J., Schmid, J.E., Lau, C. et al. 2012. Activation of mouse and human peroxisome proliferator-activated receptor-alpha (PPARα) by perfluoroalkyl acids (PFAAs): Further investigation of C4-C12 compounds. *Reprod. Toxicol.* 33: 546–51.

Wolf, D.C., Moore, T., Abbott, B.D. et al. 2008. Comparative hepatic effects of perfluorooctanoic acid and WY 14,643 in PPAR-α knockout and wild-type mice. *Toxicol. Pathol.* 36: 632–39.

Wu, K.M., and Farrelly, J.G. 2006. Preclinical development of new drugs that enhance thyroid hormone metabolism and clearance: Inadequacy of using rats as an animal model for predicting human risks in an IND and NDA. *Am. J. Ther.* 13: 141–44.

Yamazaki, Y., Kakizaki, S., Horiguichi, N. et al. 2007. The role of the nuclear receptor constitutive androstane receptor in the pathogenesis of non-alcoholic steatohepatitis. *Gut* 56: 565–74.

Yang, B., Zou, W., Hu, Z. et al. 2014. Involvement of oxidative stress and inflammation in liver injury caused by perfluoroctanoic acid exposure in mice. *Biomed. Res. Int.* 2: 1–7.

Yang, C., Tan, Y.S., Harkema, J.R. et al. 2009. Differential effects of peripubertal exposure to perfluorooctanoic acid on mammary gland development in C57Bl6, Balb/c mouse strains. *Reprod. Toxicol.* 27: 299–306.

Yang, Q., Abedi-Valugerdi, M., Xie, Y. et al. 2002a. Potent suppression of the adaptive immune response in mice upon dietary exposure to the potent peroxisome proliferator, perfluorooctanoic acid. *Int. Immunopharmacol.* 2: 389–97.

Yang, Q., Xie, Y., Alexson, S.E.H. et al. 2002b. Involvement of the peroxisome proliferator-activated receptor alpha in the immunomodulation caused by peroxisome proliferators in mice. *Biochem. Pharmacol.* 63: 1893–900.

Yang, Q., Xie, Y., Eriksson, A.M. et al. 2001. Further evidence for the involvement of inhibition of cell proliferation and development in thymic and splenic atrophy induced by the peroxisome proliferator perfluorooctanoic acid in mice. *Biochem. Pharmacol.* 62: 1133–40.

York, R.G., Kennedy, G.L., Olsen, G.W. et al. 2010. Male reproductive system parameters in a two-generation reproduction study of ammonium perfluoroctanoate in rats and human relevance. *Toxicology* 271: 64–72.

Yu, W.-G., Liu, W., and Lin, Y.-H. 2009. Effects of perfluorooctane sulfonate on rat thyroid hormone biosynthesis and metabolism. *Environ. Toxicol. Chem.* 28: 990–96.

Zhang, Y., Beesoon, S., Zhu, L. et al. 2013. Biomonitoring of perfluoroalkyl acids in human urine and estimates of biological half-life. *Environ. Sci. Technol.* 47: 10619–27.

Zhang, H., Lu, Y., Luo, B. et al. 2014. Proteomic analysis of mouse testis reveals perfluorooctanoic acid-induced reproductive dysfunction via direct disturbance of testicular steroidogenic machinery. *J. Proteome Res.* 13: 3370–85.

Zhang, H., Shi, Z., Liu, Y. et al. 2008. Lipid homeostasis and oxidative stress in the liver of male rats exposed to perfluorododecanoic acid. *Toxicol. Appl. Pharmacol.* 227: 16–25.

Zhao, B., Hu, G.-X., Chu, Y. et al. 2010b. Inhibition of human and rat 3β-hydroxysteroid dehydrogenase and 17β-hydroxysteroid dehydrogenase 3 activities by perfluoroalkylated substance. *Chem. Biol. Interact.* 188: 38–43.

Zhao, Y., Tan, Y.S., Haslam S.Z. et al. 2010a. Perfluorooctanoic acid effects on steroid hormone and growth factor levels mediate stimulation of peripubertal mammary gland development in C57Bl/6 mice. *Toxicol. Sci.* 115: 214–24.

Zheng, L., Dong, G.-H., Jin, Y.-H. et al. 2009. Immunotoxic changes associated with a 7-day oral exposure to perfluorooctanesulfonate (PFOS) in adult male C57Bl/6 mice. *Arch. Toxicol.* 83: 679–89.

9 Perfluoroalkyl Substance Toxicity from Early-Life Exposure

David Klein and Joseph M. Braun

CONTENTS

9.1 Introduction 171
9.2 Animal Studies to Determine PFAS Toxicity 173
    9.2.1 Obesity and Hepatic Hypertrophy 174
    9.2.2 Developmental Delay 176
9.3 Epidemiological Data to Determine Human Toxicity 179
    9.3.1 Impact of PFAS on Birth Weight 179
    9.3.2 Association between PFAS Exposure and Breastfeeding 181
    9.3.3 Impact of PFAS on Obesity 183
    9.3.4 PFAS and Neurodevelopment 185
    9.3.5 PFAS as an Endocrine Disruptor 186
    9.3.6 Challenges in Making Stronger Interferences 186
9.4 Conclusion 188
References 189

9.1 INTRODUCTION

Perfluoroalkyl substances (PFAS) are man-made compounds that have been used in a variety of industrial applications, including the formation of nonstick surfaces, firefighting foams, stain- and water-resistant surfaces, and textiles. PFAS are ubiquitous in the environment due to their high levels of international production, widespread use in home and food products, and resistance to thermal, chemical, and biological degradation. Their long environmental half-life ensures persistence even after use is discontinued. PFAS can be detected in the blood of populations across the globe due to widespread use for 70 years. There have also been numerous studies linking these compounds to adverse health outcomes. Therefore, there is tremendous interest in understanding the human health impact these compounds have since they are persistent and ubiquitous, and there is ample evidence suggesting that they interfere with physiological processes, especially during developmentally sensitive periods, including fetal development.

    PFAS are readily absorbed into the blood, not metabolized, and primarily eliminated, albeit slowly, via urine in both laboratory animals and humans (Ophaug and Singer 1980; Vanden Heuvel et al. 1991; Kudo et al. 2002; Zhang et al. 2015).

PFAS preferentially accumulate in the liver, and body burdens directly correlate to exposure (Lau et al. 2003; Thibodeaux et al. 2003). Serum levels at birth are similar to those of the dam, indicative that PFAS can equilibrate through the placenta (Thibodeaux et al. 2003; Apelberg et al. 2007; Midasch et al. 2007; Kummu et al. 2015), serving as a route of fetal exposure (Manzano-Salgado et al. 2015; Verner et al. 2015). In early childhood, the serum level surpasses that of adult populations partially due to exposure from breast milk (Olsen et al. 2004). The biological half-lives of PFAS in humans are long, typically in the range of years for the longer-chain PFAS. Perfluorooctanoic acid (PFOA) and perfluorooctanesulfonate (PFOS), two of the most well studied PFAS, have half-lives of about 3.5 and 8.5 years, respectively (Olsen et al. 2007; Brede et al. 2010; Steenland et al. 2010). The half-life of PFAS varies considerable between species, with rats exhibiting a sex difference, potentially complicating interpretations of animal studies. For example, PFOA has a half-life of 20–40 days in cynomolgus monkeys (Butenhoff et al. 2004; Loccisano et al. 2012); male rats, 4–6 days (Loccisano et al. 2011; Kim et al. 2016); female rats, 4–6 hours; and mice, 7–21 days, depending on strains (Lou et al. 2009; Tucker et al. 2015). The cause of the variance in half-life is unknown, but human transporters involved in enterohepatic recirculation may be a contributing factor since many of these proteins are known to transport PFAS (Thibodeaux et al. 2003; Zhao et al. 2012). This recirculation is especially relevant to adult exposure since it does not appear to occur in newborns in either humans or rodents (Belknap et al. 1981). The extent to which rodents undergo enterhepatic recirculation is currently unknown. PFOA has been shown to be a substrate of several organic anion transporters and play a key role in renal elimination (Kudo et al. 2002; Zhao et al. 2012). The rapid clearance of PFOA and PFOS in female rats makes them an inappropriate model for *in utero* exposure since the dam would need to be dosed multiple times a day to maintain constant exposure to the offspring. For this reason, mice are typically used as a model of *in utero* exposure in recent studies. Considering that the long half-life plays a significant role in the human toxicity of PFAS, selecting an animal model with an appropriate half-life is imperative to compare animal studies to humans. The long half-life is an important factor of PFAS toxicity since it poses a risk for bioaccumulation. This highlights the potential for chronic toxicity and the necessity to understand the impact of long-term exposure.

This chapter discusses the evidence of PFAS toxicity from animal studies and epidemiology data and notes how they complement one another. Animal studies are advantageous in that they allow for controlled studies with known levels of exposure, which allows for accurate assessment of causality, insight into mechanism(s) of action, and better assessment of the impact of high dose exposures. However, toxicokinetic parameters can also vary tremendously between species—hence the different half-lives of PFOA. Animals also have varying sensitivities to toxicants compared with humans that can manifest as health outcomes in lab animals that are irrelevant to humans. One example of this is Leydig cell adenomas, which are the most common testicular neoplasm in rats but are extraordinary rare in humans (Wakui et al. 2008; Meyts et al. 2013). Several toxicants, including 1,3-butadiene, trichloroethylene, and cadmium, have been reported to increase the incidence of Leydig cell tumors in rat models, but not in humans (Waalkes et al. 1988; Owen and Glaister 1990;

Divine and Hartman 2001; Verougstraete et al. 2003). Another major weakness of animal studies is that they require extrapolation to humans. For example, lab rodents mature much faster than humans. Mice have a gestational age of 18–22 days, first show signs of puberty between 4 and 6 weeks, and are ready to breed by 8 weeks. Rodents' shorter gestational and prepuberty development make subtle changes to delaying development easy to detect, but the quantitative impact on humans challenging to predict. Despite these development differences, animal studies can still be useful in detecting potential developmental toxicants warranting further investigation in human studies.

One of the benefits of employing epidemiological studies is that the data directly investigate humans, which eliminates the extrapolation required for animal studies. It is also possible to obtain data from larger populations than what can be practically achieved with animal studies, granting epidemiological studies more statistical power. However, such studies are not as well suited to determine causation as animal studies due to the uncontrolled nature of human data. Additionally, human exposure is generally lower and the health effects of PFAS may be conflated by correlates of both PFAS exposure and health; these other correlates could include other xenobiotics, which makes assessing the impact of a particular toxicant more difficult. Both animal and epidemiological studies are required for a comprehensive view of the impact of PFAS on public health. Animals directly assess the effects of the toxicants on a living organism, and epidemiology directly studies the impact on human populations.

9.2 ANIMAL STUDIES TO DETERMINE PFAS TOXICITY

In May 2016, the US Environmental Protection Agency (USEPA) published advisory levels for PFOA and PFOS at 70 parts per trillion (ppt) in drinking water to account for the chronic effects of these toxicants (USEPA 2016). The basis of this advisory level relied heavily on animal studies that reported a variety of toxicity, including skeletal variations (delay in ossification of phalanges), testicular cancer, and persistent liver effects (Figure 9.1). Skeletal variations had a particularly large impact on the advisory level since it was reported to occur at the lowest dose (Lau et al. 2006). Delayed mammary gland development was also discussed but was ultimately

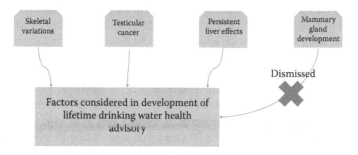

FIGURE 9.1 Illustration demonstrating the animal studies informing the USEPA drinking water advisory for PFOA and PFOS exposure.

dismissed due to an unknown mode of action, strain differences in mice, and unclear functional significance. This section reviews the literature surrounding adiposity and developmental (including both phalanges ossification and mammary gland development) toxicity since these are primarily observed effects resulting from early-life or prenatal exposure, as well as other toxicities reported from early-life exposure.

The mechanism for PFAS toxicity has not yet been determined, but PFOA is known to be an agonist for peroxisome proliferator–activated receptor alpha (PPARα) and is suspected to partially explain toxicity (Vanden Heuvel et al. 2006; Abbott et al. 2009; Mattsson et al. 2015). PPARα is a ligand-activated nuclear receptor that regulates genes important for lipid catabolism, inhibition of arachidonic acid metabolism, and lipid homeostasis (Abbott et al. 2009). Activation of PPARα is linked to three cancers in rodents: hepatocellular adenoma, pancreatic acinar cell adenoma, and Leydig cell adenoma, all of which are associated with PFOA exposure in animals (Andersen et al. 2007). However, the clinical relevance of these cancers to humans is questionable (Klaunig et al. 2012). *In vitro* assays investigating PPARα sensitivity to PFOA between different species have had mixed results, but generally indicate that human and rodent PPARα have similar affinities to PFOA (Maloney and Waxman 1999; Vanden Heuvel et al. 2006; Wolf et al. 2006, 2012). PPARα is highly expressed in the liver (more so in rodents than humans) and is detected at various levels during gestational development in both rodents (gestational day [GD] 5 in mice) and humans (Holden and Tugwood 1999; Keller et al. 2000; Abbott 2009; Abbott et al. 2010). While the endogenous function of PPARα in the liver is maintaining lipid homeostasis, its role in gestational development is less clear. Mice that are genetic knockouts for PPARα do not suffer from obvious birth defects and are fertile, which indicates that it is not essential for reproductive development (Lee et al. 1995). Additionally, PPARα knockout mice are insensitive to the gestational effects of PFOA but are still sensitive to PFOS developmental toxicity, indicating different mechanisms of action that vary in dependence of PPARα activation (Abbott 2009; Abbott et al. 2010). Overall, there is a relationship with PFAS and PPARα that may partially explain the toxicity, especially with adiposity, but how much it impacts the human liver, which has lower PPARα levels, and developmental toxicity is unclear.

9.2.1 Obesity and Hepatic Hypertrophy

There has been significant interest in the impact of early-life exposure on adulthood obesity. Adult rodents exposed to PFAS will exhibit weight loss presumably through increased fat oxidation (Andersen et al. 2007). Several studies have shown that PFAS can impact lipid homeostasis later in life, although the results do not always follow a typical dose–response paradigm. For example, mice exposed to PFOA *in utero* were shown to have increased serum insulin, leptin, and body weight at lower levels (0.01–0.1 mg/kg/d) compared with controls, but these increases were not observed at higher exposures (1 mg/kg/d) (Hines et al. 2009; Quist et al. 2015). Other studies have reported weight loss through a PPARα mechanism and a PPARα-independent increase in relative liver weight following PFAS exposure (Yang et al. 2002; Xie et al. 2002). The weight loss following PFOA exposure can be recovered in adult mice by eliminating exposure, but this is not the case for mice exposed *in utero* (Xie et al.

2002; Hines et al. 2009). This suggests that prenatal exposure to PFOA may induce irreversible changes in hepatic physiology that may last a lifetime despite cessation of PFOA exposure. Microarray data revealed that *in utero* PFOA exposure altered the expression of genes related to fat catabolism (Rosen et al. 2007), which may be indicative of a physiological switch in lipid homeostasis that can have latent consequences if exposed during gestation.

Conflicting with this notion, other studies have reported that the lower body weight of exposed mice recovers to control levels after 6.5 weeks (Lau et al. 2003; Wolf et al. 2006). It was suspected that some of these differences can be influenced by alterations in suckling behavior as PFOA can be detected in breast milk. However, exposure to PFOA during GDs 7–17 is sufficient to reduce body weight and delay development even without exposure to contaminated milk (Wolf et al. 2006). Additionally, pups born from an unexposed dam but nursed by an exposed female were shown to have an increased relative liver weight indicative of liver hypertrophy but did not experience neonatal mortality or developmental delay (Seacat et al. 2002; Lau et al. 2003; Wolf et al. 2006).

A recent study further addressed the impact of prenatal PFOA exposure altering lipid homeostasis with coexposure to a high-fat diet (Quist et al. 2015). Pregnant mice were exposed to PFOA and then the offspring were challenged with a high-fat diet. In agreement with another study, mice exposed to PFOA had increased relative liver weight in a dose-dependent manner at low doses (0.3–1.0 mg/kg/d) for the first few weeks of life (Quist et al. 2015; Tucker et al. 2015). This hypertrophy was determined to be primarily due to gestational exposure, followed by lactational exposure extending the toxicity (Fenton et al. 2009; Tucker et al. 2015). Typically, hepatocellular hypertrophy is reversible upon cessation from toxicant exposure. However, centrilobular hypertrophy from prenatal PFOA exposure could be observed even at 91 days postbirth even when serum concentrations were expected to be undetectable (Macon et al. 2011; Quist et al. 2015). This is indicative of an early-life exposure causing liver changes that result in hypertrophy that may not be reversible. While it is common for hepatic hypertrophy to originate from peroxisome proliferation, which implicates PPARα activation (Wolf et al. 2008), it is suspected that centrilobular hypertrophy from PFOA exposure is due to a mitochondrial defect (Quist et al. 2015). Interestingly, while this study did find alterations in cholesterol biosynthesis and fatty acid metabolism, there was no change in weight gain, which set it apart from other studies (Quist et al. 2015).

A mechanism proposed to explain alterations in offspring body weight implicates thyroxine since PFOS has been shown to cause perturbation of the thyroid axis (Biegel et al. 1995). Pregnant mice and rats exposed to PFOS have reduced serum thyroxine without a feedback induction of thyroid-stimulating hormone (TSH) (Thibodeaux et al. 2003). Thyroid hormone deficiency in pregnant mammals is alarming since the fetus relies on maternal hormones until late pregnancy (Lucas et al. 1988; Glinoer et al. 2001). Small reductions in gestational thyroid hormones may have long-lasting implications on neurodevelopment and basal metabolic rate of the offspring (Lucas et al. 1988; Haddow et al. 1999). While thyroxine reduction seemed like a promising mechanism of action for the developmental toxicity of PFAS, many epidemiological studies could not find a positive association between PFAS and thyroid disorder, which questions the relevance of thyroid disorder to humans (Bloom et al. 2010; Ji et al. 2012; Lin et al. 2013; Wen et al. 2013; Shrestha et al. 2015).

While most animal studies have focused on PFOA and PFOS, there has been a growing interest in perfluorononanoic acid (PFNA) since human exposures have been shown to be increasing in many regions (Kato et al. 2011). The half-life of PFNA is much longer than that of PFOA in mice (~30 days), and a half-life for humans has not be conclusive, although it is suspected to be on the order of years due to its low clearance (Tatum-Gibbs et al. 2011; Zhang et al. 2013). The toxicity of PFNA is similar to that of PFOA or PFOS, which is expected since it is also a long-chain perfluorinated compound. PFNA is shown to cause a dose-dependent increase in absolute and relative liver weights in mice following gestational exposure that is recovered by 70 days. Neonatal survival rates were also markedly decreased at 5 mg/kg/d exposure, but not at lower doses (Das et al. 2015). Pups exposed to PFNA *in utero* had reduced body weight compared with controls long after PFNA was no longer detectable in serum or liver (275 days). PFNA exhibited a great and more consistent delay in puberty for mice at 3 mg/kg/d and higher doses (Das et al. 2015). There are relatively few studies investigating PFNA toxicity, but the current literature suggests that it exhibits similar toxicity to PFOA, which is concerning giving its rising prevalence in human serum.

9.2.2 Developmental Delay

The reproductive and developmental effects of PFAS are critical to establish since pregnant mothers and young children are considered sensitive populations. The USEPA conducted several animal studies on the long-term impact of gestational exposure to PFAS that have been influential in the development of advisory levels (Lau et al. 2006; Abbott 2009; Quist et al. 2015; Tucker et al. 2015). The primary developmental effects noted for PFAS toxicity are reduced birth weight, neonatal fatality, and developmental delay. Importantly, many of these effects occur independent of breastfeeding and result from exposure as late as GDs 15–17 (Wolf et al. 2006). One of the earliest studies determining the developmental toxicity of these compounds was by Lau et al. (2003). This study reported that exposing pregnant rats to 2 mg/kg/d PFOS resulted in a significant increase in neonatal fatality, with higher exposures (5–10 mg/kg/d) resulting in death hours after birth. Pups that were exposed to 2 mg/kg/d PFOS *in utero* and survived a few days after birth still had reduced weight gain and had delayed eye opening, an indicator of hindered development. Another study reported that 3.2 mg/kg/d gestational exposure resulted in total mortality for the offspring within 24 hours of birth and a 34% fatality rate for offspring exposed to 1.6 mg/kg/d (Luebker et al. 2005). Mice exposed to PFOA and PFNA *in utero* had reduced birth weight in a dose-dependent manner that was PPARα dependent (Wolf et al. 2006, 2010; Hines et al. 2009). Exposing pregnant mice to 5 mg/kg/d PFOA reduced postnatal survival and delayed development, while 10 mg/kg/d increased the incidence of prenatal loss (Lau et al. 2006). Developmental toxicity, such as reduced birth weight, delayed ossification, and decreased neonatal survival, is also observed with PFOS gestational exposure (Lau et al. 2003; Wolf et al. 2006).

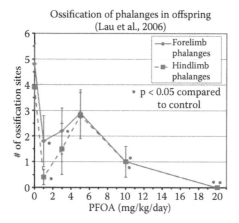

FIGURE 9.2 Graph representing data found in Lau et al. (2006) showing the reduction in ossification sites in hindlimb and forelimb phalanges correlating to the PFOA concentration of the dam (Gleason et al. 2016). (Courtesy of the New Jersey Drinking Water Quality Institute.)

One of the hallmark studies investigating developmental delay in mice determined that there was a marked delay of ossification of phalanges of the forelimb and hindlimb at low doses (1 mg/kg/d) (Lau et al. 2006), which heavily influenced the USEPA health advisory of PFOS plus PFOA at 70 ppt in drinking water. An interesting aspect of the reduction in ossification sites is the atypical response: as the dose increases, the delay in ossification returns toward control levels (Figure 9.2). The mechanism for this dose–response curve is unknown, although endocrine disruptors have been known to exhibit similar nonmonotonic dose responses (Vandenberg et al. 2012). An explanation is that a toxicant is able to stimulate multiple pathways at different doses that have counteracting effects. For example, a low dose elicits a measurable toxic response, but at higher concentrations, a compensatory mechanism becomes activated that ameliorates the toxicity. From a regulatory standpoint, this has some interesting implications since the model used to calculate health advisories relies on a no observed adverse effect level (NOAEL). When one cannot be experimentally determined, the lowest observed adverse effect dose is typically divided by 10 to estimate a NOAEL. However, this model assumes a diminishing response as the dose decreases, which is not the case with a nonmonotonic profile like what was observed in the delay in phalanges ossification. It is important to note that by adulthood, the phalanges had developed normally, thus indicating that the toxicity does not have long-term implications. In summary, low doses of *in utero* exposure caused a delay in phalange ossification, but at this time, the molecular mechanism for the unusual dose–response curve or the health relevance is unknown.

Another milestone in development is the onset of puberty, the delay or acceleration of which is indicative of toxicity relating to homeostasis of hormones. For rodents, the onset is defined as the preputial separation (separation of the foreskin) in males and vaginal opening or estrus in females (Lau et al. 2006). For male mice, prenatal PFOA exposure alters puberty onset in an atypical dose–response fashion. PFOA accelerated puberty onset as the dose decreased, while the highest dose (20 mg/kg/d) delayed male puberty (Figure 9.3). It is impossible to accurately ascertain a NOAEL

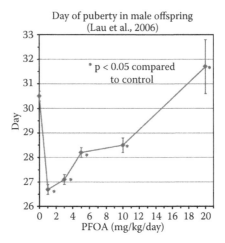

FIGURE 9.3 Graph depicting data from Lau et al. (2006) showing the start of puberty (prepubital separation) in male mice that were exposed to PFOA prenatally (Gleason et al. 2016). (Courtesy of the New Jersey Drinking Water Quality Institute.)

level since the effect becomes more pronounced as the dose approaches the low end of the experimental range. Female mice experience a similar effect: lowest doses (1 mg/kg/d) accelerated pubertal onset and higher doses delayed puberty, although the effect was not as dramatic as observed in males (Lau et al. 2006). Another study failed to find a change in vaginal opening at 1 mg/kg/d, so it is possible that level of exposure to PFOA produces only a minor effect on female puberty onset (Tucker et al. 2015).

One of the most controversial studies of PFAS toxicity is the impact on mammary gland development. Much of the mammary gland matures postnatally, with puberty initiating development. Hormones increase mammary stem cell populations, and terminal end buds migrate the epithelium into the fat pad, allowing for ductal branching within the gland (Tucker et al. 2015). Development of the mammary gland has been assessed based on the extensiveness of the branching and the number of terminal end buds via pathological scoring (Macon et al. 2011; Tucker et al. 2015). Prenatal exposure to low-dose PFOA stunts mammary epithelial growth in female mice (Macon et al. 2011). Some studies have noted that Balb/c mice are slightly more sensitive to peripubertal PFOA exposure than the C57Bl/6 strain, although this may not necessarily apply to prenatal exposure (Yang et al. 2009). CD-1 mice exposed prenatally to PFOA at low levels (0.01–1.0 mg/kg/d) exhibited a dose-dependent delay in mammary gland development that can be observed up to 84 days postbirth, although no differences in lactation were noted. C57Bl/6 mice were slightly more resistant, with toxic effects observed at 0.3 mg/kg/d (Macon et al. 2011; Tucker et al. 2015). Other studies report that PFOA reduces mRNA levels of placental lactogen-II and prolactin, like protein-E and protein-F, in a dose-dependent fashion (Suh et al. 2011), suggesting potential hormonal pathways by which PFOA could impact lactation. Interestingly, Tucker et al. (2015) did not find any alteration in estradiol or progesterone levels following PFOA exposure, implying that the delay in mammary gland development was not due to changes in hormone levels

(Zhao et al. 2012; Tucker et al. 2015). The functional impact of the stunted mammary gland development is unclear, but it is postulated that PFOA exposure may also influence expression of milk protein genes (White et al. 2006), thus adversely impacting the nutritional quality or taste of milk and making it less appealing to offspring.

As mentioned above, the USEPA did not use animal data on mammary gland development for the development of advisory levels due to unknown mode of action, strain differences, and unclear functional significance. Since hormone levels do not seem to correspond with alterations in mammary gland development, the mechanism of action for affecting mammary epithelia is unclear. Quantitative differences in PFOA sensitivities between CD-1 and C57Bl/6 have been noted; however, qualitatively they both showed low-level (0.01 and 0.3 mg/kg/d) PFOA exposure, causing a delay in mammary gland development, which implies that the strain difference is minor. The stunted development of the mammary gland also has not been related to reduced breast milk production or offspring malnourishment, calling into question the significance of this finding. However, these points are questioned by some state agencies since they could easily apply to phalanges ossification, which heavily informed advisory levels (Gleason et al. 2016). Overall, there are several animal studies that indicate that *in utero* exposure to PFAS poses a risk for developmental toxicity, although the mechanism of action and the functional impact of low-dose exposure are still unknown.

9.3 EPIDEMIOLOGICAL DATA TO DETERMINE HUMAN TOXICITY

9.3.1 Impact of PFAS on Birth Weight

Epidemiological data evaluating the impact of PFAS on birth weight have largely focused on PFOA and PFOS, although there are reports on PFNA, perfluorodecanoic acid (PFUA), and perfluorohexane sulfonic acid (PFHxS). Since pregnant women and the developing fetus are considered sensitive populations, there is tremendous interest in determining the effects of gestational PFAS exposure. Birth weight is a useful indicator of gestational health that is easy to measure and can corroborate human and animal studies. Several studies have evaluated the associations between PFAS concentrations in either maternal or neonatal cord serum and birth weight of offspring. While some published reports fail to find an association between any PFAS and changes in birth weight (Olsen et al. 2004; Grice et al. 2007; Savitz et al. 2012; Lee et al. 2016), many report at that least one PFAS concentration negatively affects birth weight (Apelberg et al. 2007; Fei et al. 2007; Stein et al. 2009; Washino et al. 2009; Chen et al. 2012; Maisonet et al. 2012; Whitworth et al. 2012; Bach et al. 2015). The variation in study results, which is typical in human data, can come from a number of sources, such as differences in study population, methodology, sample size, prevalence of other factors influencing fetal growth, misclassification of birth weight, and bias due to adjusting for confounding factors, such as gestational age (Apelberg et al. 2007; Grice et al. 2007). All the studies that assessed the impact of PFAS exposure on birth weight demonstrated either inverse associations or null effects.

In a systematic review of 18 publications and subsequent meta-analysis of 9 of these, each 1 ng/mL increase in serum PFOA concentration was associated with a

19 g decrease in birth weight (95% CI −30, −8) (Johnson et al. 2014). The results of these studies demonstrate an inverse association between maternal serum PFOA concentrations and birth weight, independent of gestational age (Fei et al. 2007; Maisonet et al. 2012; Whitworth et al. 2012). Maternal concentrations of PFOS have also been correlated to lower birth weights in many studies (Washino et al. 2009; Maisonet et al. 2012; Whitworth et al. 2012), and the serum concentrations are higher than those of PFOA (Konkel 2017). Other studies investigating cord blood concentrations have found inverse associations between PFOS and birth weight, as well as head circumference (Apelberg et al. 2007; Chen et al. 2012). No association was found for PFNA and PFUA in these studies. The associations of cord serum PFOA and birth weight have produced mixed results between studies (Apelberg et al. 2007; Chen et al. 2012). When comparing the effects of PFOA to PFOS, it is important to note that serum PFOA concentrations are almost an order of magnitude lower than serum PFOS concentrations; thus, reduced variation in serum PFOA concentrations could partially explain the variability in response to these PFAS.

One of the difficulties in assessing the impact of toxicants, such as PFAS, on birth weight is the abundance of confounding factors. While classic factors, like smoking during pregnancy, are typically accounted for, subtle lurking variables may unexpectedly influence the results. An example of such a variable is the effect of glomerular filtration rate (GFR) on both serum PFAS concentrations and birth weight (Verner et al. 2015). GFR is an indicator of kidney filtration, which is important for the elimination of toxicants and other endogenous compounds, such as creatinine. Interestingly, GFR increases by about 50% during the first half of pregnancy (Gibson 1973; Smith et al. 2008). This would be expected to cause a global increase in the excretion of compounds eliminated via urination (such as PFAS), resulting in lower serum PFAS concentrations among women with higher GFR than among women with lower GFR. Reductions in GFR are associated with reduced birth weight independent of toxicant presence (Gibson 1973; Morken et al. 2014). Taken together, if a woman's GFR did not increase during pregnancy, the birth weight would be anticipated to be lower but the concentrations of PFAS would also be increased compared with those of pregnant women with higher GFR (Figure 9.4). Thus, the inverse association between PFAS and birth weight may be partially a result of kidney function. This would mean that standard analysis would be anticipated to overestimate the impact of PFAS on birth weight (Verner et al. 2015). The effect of GFR on PFOA concentrations is an important issue to raise awareness of, but it should be noted that even after GFR has been corrected for, the concentration of PFAS still has a significant inverse association with birth weight, meaning GFR cannot completely explain a PFOA-induced decrease in birth weight (Verner et al. 2015).

Overall, there are copious reports indicating that PFAS, especially PFOA and PFOS, are associated with decreased birth weights and, on occasion, other pregnancy-related metrics, such as head circumference. As lurking variables such as GFR become identified and accounted for, a clearer picture of the impact of PFAS on fetal growth will emerge to better inform risk assessment.

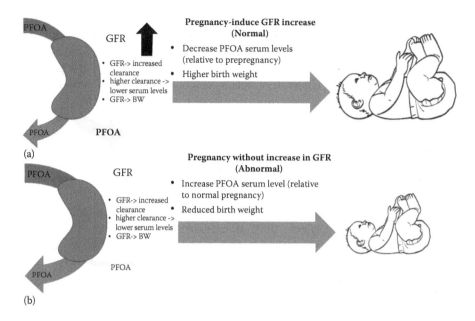

FIGURE 9.4 Illustration demonstrating the impact of the GFR increase during pregnancy on PFOA concentrations and birth weight. When GFR increases, serum PFAS levels are reduced due to an increase in renal elimination (a). When GFR is lower during pregnancy (b), then PFAS levels are higher than in women with higher GFR. In B, the birth weights of the infants are potentially lower than the birth weights of infants in A for reasons *independent* of PFAS concentrations. If GFR is not properly accounted for, then the impact of PFAS serum concentrations and birth weight might be overestimated. BW = birth weight.

9.3.2 Association between PFAS Exposure and Breastfeeding

The American Academy of Pediatrics (AAP) recommends 6 months of exclusive breastfeeding and continued partial breastfeeding for 12 months or longer (AAP 2012) to promote numerous short- and long-term health benefits for both mother and child (Ip et al. 2007). However, in the United States in 2012, only 29.2% of mothers continue any breastfeeding until 12 months (National Immunization Surveys 2015). Known barriers to breastfeeding include sociocultural, medical, and nutritional barriers; perceived inadequate milk supply; problems with latch; and insufficient support from health care providers, family, or workplace (Teich et al. 2014). However, a common reason for premature weaning is inadequate milk production (Konkel 2017). Exposure to environmental chemicals may also interfere with hormones regulating the initiation of lactation (e.g., prolactin) and milk production (e.g., oxytocin) (Rogan and Gladen 1985; Lew et al. 2009). One of the first toxicants investigated for its impact on lactations was dichlorodiphenyltrichloroethane (DDT), a pesticide that was widely used until the 1970s, when concerns of toxicity led to its use being banned in the United States (Konkel 2017). Dichlorodiphenyldichloroethylene (DDE), a breakdown product of DDT, was not shown to affect infant weight gain or frequency of illness, but DDE did appear to reduce the duration of breastfeeding in US and Mexican populations (Gladen and Rogan 1995; Rogan et al. 1987). Breastfeeding duration may be a sensitive end point to PFAS exposure given that animal studies suggest that PFOA exposure during pregnancy disrupts mammary gland differentiation and development (White et al. 2006; Yang et al. 2009; Tucker et al. 2015), delays epithelial involution, and may alter expression of placental prolactin family hormone and milk protein genes (Suh et al. 2011).

Few epidemiological studies have examined associations between prenatal PFAS exposure and duration of breastfeeding. Using data from the Danish National Birth Cohort, Fei et al. (2010) observed that greater concentrations of maternal plasma PFOA and PFOS during pregnancy were associated with a shorter duration of breastfeeding among multiparous women. However, that study did not control for prior breastfeeding duration, which is an important route of maternal PFAS excretion (Barbarossa et al. 2013; Mondal et al. 2014) and an important predictor of future breastfeeding success (Nagy et al. 2001; Whalen and Cramton 2010).

To ensure that the associations observed among multiparous women were not attributable to prior breastfeeding history, Romano et al. (2016) examined the associations of PFAS serum concentrations and duration of breastfeeding, controlling for prior breastfeeding history. This cohort had median serum PFOA concentrations about two times higher than those of pregnant women who participated in the National Health and Nutrition Examination Survey (NHANES) (Braun et al. 2016; Romano et al. 2016), a nationally representative sampling of the US general population (Jain 2013). The results of this study suggest that higher maternal serum PFOA concentrations during pregnancy are associated with shorter duration of breastfeeding in the HOME study cohort (Braun et al. 2016). This study observed an increased risk of quitting breastfeeding with increasing serum PFOA and PFOS concentrations during pregnancy, even after controlling for prior breastfeeding history. When prior breastfeeding was not included in multivariable models, the magnitude of the relative risk (RR) tended to be inflated, demonstrating the significance of including prior breastfeeding (Romano et al. 2016). There was not an association of either PFNA or PFHxS with duration of breastfeeding. Finally, a third study used two birth cohorts formed in the Faroe Islands, which is a self-governing community within the Danish kingdom (Timmermann et al. 2017). This study found that increased maternal PFAS concentrations were significantly associated with shorter breastfeeding duration. They also noted that primiparous (first-time mother) and multiparous women had similar trends, indicating that the previous breastfeeding history was not responsible for the observed associations (Timmermann et al. 2017).

In contrast to the study by Fei et al. (2010), Romano et al. (2016) did not observe a clear association between PFOS and duration of any breastfeeding, though risk of ending any breastfeeding by 3 months was marginally elevated among women in the highest PFOS quartile (>18.1 ng/mL). Although the median PFOA level (5.5 ng/mL) in the HOME study was comparable to the median in the Danish study (5.2 ng/mL), the median PFOS plasma concentration in the Danish study was substantially higher (33.3 ng/mL) (Fei et al. 2010) than that in the HOME study (13.9 ng/mL). The higher concentrations of PFOS in the Danish population may be necessary to adversely affect breastfeeding duration. Alternatively, the inverse association between maternal PFOS and duration of breastfeeding may have been confounded by lack of control for previous breastfeeding experience or PFOA coexposure. Analysis of the HOME study suggested that the association observed between PFOS and early termination of any breastfeeding by 3 months may have been driven by PFOA, as the association between PFOS and breastfeeding was attenuated when both PFOA and PFOS were included in the same model (Romano et al. 2016). The most prevalent PFAS compounds tend to be correlated with each other in human tissues. Other factors influencing differences in the results could include differences in the median duration of exclusive breastfeeding, differences in the definition of exclusive breastfeeding, or different maternity leave policies in the United States and Denmark. Such systemic factors, as well as environmental exposures, may play a role in the rates of exclusive breastfeeding. It is unknown whether PFOA exposure interferes with initiation of lactation, is related to poor quantity or quality of breast milk, or potentially decreases duration of breastfeeding through other mechanisms in humans.

Collectively, animal and human studies suggest that PFAS may have adverse effects on duration of breastfeeding. While confounding by prior history of breastfeeding is a concern, the available evidence suggests that this alone is not enough to explain the observed associations between PFAS and shorter breastfeeding duration. Although there are certainly social and cultural barriers to breastfeeding, observations highlight the relevance of emerging environmental risk factors, such as PFAS, that may adversely influence duration of breastfeeding.

9.3.3 IMPACT OF PFAS ON OBESITY

Childhood obesity is a major public health problem in the United States, where 17% of children are obese and another 15% are overweight (Ogden et al. 2014). Excess adiposity is difficult to reverse once established and increases the risk of cardiometabolic, pulmonary, and musculoskeletal disorders (Ebbeling et al. 2002). Although controversial, there is growing evidence that prenatal exposure to obesogens, chemicals that disrupt lipid metabolism, may play a role in childhood obesity risk by perturbing biological pathways involved in energy metabolism, appetite, or adipogenesis (Janesick and Blumberg 2012). PFAS are suspected obesogens (Alexander et al. 2008; Buck et al. 2011; White et al. 2011). Rodent and *in vitro* studies show that PFAS exposure may cause impaired glucose homeostasis, increased body weight, and altered adipocyte differentiation (Vanden Heuvel et al. 2006; Hines et al. 2009; Taxvig et al. 2012; Bastos Sales et al. 2013), although some studies suggest otherwise (Ngo et al. 2014).

Seven prospective epidemiological studies have examined the relationship between early-life PFAS exposure and adiposity in children or adults (Halldorsson et al. 2012; Maisonet et al. 2012; Andersen et al. 2013; Barry et al. 2014). Two studies reported no associations between prenatal or early-childhood PFAS exposure and child or adult adiposity (Andersen et al. 2013; Barry et al. 2014). One of these, a Danish study of 811 children, did not observe any associations between prenatal PFOA or PFOS concentrations and child body mass index (BMI) or waist circumference at 7 years of age (Andersen et al. 2013). In the other study of 8764 adults who lived near the DuPont Washington Works Plant and had exceptionally high PFOA exposure, Barry et al. (2014) did not find an association between estimated early-life PFOA exposure and BMI or risk of overweight or obesity. Five studies reported that prenatal PFAS exposure is associated with child adiposity or risk of being overweight or obese. In a study of 665 adults from Denmark, prenatal PFOA concentrations were associated with increased BMI and waist circumference at 20 years of age, with stronger associations among females (Halldorsson et al. 2012). Another study of 447 girls from the United Kingdom found that prenatal PFOS concentrations were associated with more rapid growth between birth and 20 months of age (Maisonet et al. 2012). Two other large studies reported that prenatal PFAS was associated with several different measures of child adiposity, including waist-to-hip ratio and dual x-ray absorptiometry–derived measures of adiposity (Hoyer et al. 2015; Mora et al. 2017). In a study of pregnant women and their children from Cincinnati, Ohio, higher PFOA concentrations observed during pregnancy were associated with greater BMI, waist circumference, and body fat in their children at 8 years of age (Braun et al. 2016). Finally, a study of Taiwanese mothers and their children reported that prenatal serum concentrations of some PFAS were associated with reductions in weight and height between birth and 11 years of age (Wang et al. 2016; Konkel 2017).

Misclassified child or adult adiposity could cause discrepant results across studies. The Andersen et al. (2013) and Barry et al. (2014) studies reported no association between PFAS exposure and adult or child BMI, which relied on self- or parent-reported weight and height. There are well-documented errors in self- and parent-reported anthropometry that could misclassify BMI and attenuate associations toward the null (Akinbami and Ogden 2009; Hattori and Sturm 2013). In contrast, studies reporting associations between prenatal PFAS exposure and increased adiposity or accelerated growth used standardized anthropometric measurements or clinical records for the majority of participants (Halldorsson et al. 2012; Maisonet et al. 2012; Braun et al. 2016; Mora et al. 2016). These more accurate measures of adiposity might have resulted in less misclassification and enhanced statistical precision. However, these studies are limited by reliance on less refined measures of adiposity. Future studies would benefit from using more detailed adiposity measures, including densitometry, bioelectric impedance, or dual energy x-ray absorptiometry.

Repeated assessments of weight and height may provide additional insight that might not be apparent in studies measuring adiposity at only one time point in childhood. A unique strength of the Braun et al. (2016) and Wang et al. (2016) studies is that they were able to examine trajectories of child adiposity. The examination of child growth trajectories remains relatively unexplored with regards to studies of PFAS and other chemical obesogens.

It is intriguing that PFAS are associated with lower size at delivery but greater adiposity later in life. This may be due to the growth-restrictive effect that PFAS have on the fetus and more rapid gains in adiposity that growth-restricted infants experience during childhood (Johnson et al. 2014). This is also similar to the phenomenon observed among tobacco smoke–exposed children who are smaller at birth, grow more rapidly than their unexposed peers, and are at increased risk of obesity later in life (Chen et al. 2005; Oken et al. 2008). Additional follow-up of these cohorts will allow for the determination of whether children with higher prenatal PFOA exposure continue to gain adiposity more rapidly and have excess adiposity at later ages.

Some biological mechanisms may explain discrepancies across study results. *In vitro* studies suggest that PFOA exposures above a certain level activate biological mechanisms, like the constitutive androstane receptor, that may dampen the obesogenic effects observed at lower exposures (Moreau et al. 2008; Peters and Gonzalez 2011). Thus, nonmonotonic associations may be one reason for discrepant findings across studies with different levels of PFOA (or other PFAS) exposure. The different effects of individual PFAS may be explained by the ability of these chemicals to act via distinct biological mechanisms. Rodent studies suggest that the effect of prenatal PFOA exposure on neonatal death, growth, and development is dependent on the PPARα, while prenatal PFOS exposure is not (Abbott et al. 2009). Thus, it is possible that individual PFAS act via different mechanisms despite their structural similarity.

Residual confounding effects from factors associated with both PFAS exposure and child adiposity may have biased results of epidemiological studies. This bias may arise if certain maternal dietary patterns or behaviors during pregnancy, such as packaged food consumption, were associated with both higher PFAS exposure and child dietary patterns that are associated with child adiposity (Fromme et al. 2009). Water consumption may be another confounder since water contamination is a likely source of PFOA exposure and may be associated with lifestyle factors that predict subsequent child adiposity. Finally, children have been assessed at different stages of pubertal development, which can influence their adiposity (Burt Solorzano and McCartney 2010). Future studies should carefully consider whether Tanner stage adjustment is appropriate since PFAS exposure may be associated with later pubertal development (Lopez-Espinosa et al. 2011).

In conclusion, prenatal PFAS exposure may increase the risk of obesity. These findings of PFAS-associated increases in adiposity may be consistent with studies showing PFAS-associated decreases in birth weight, as prior studies show that fetal growth deceleration and infancy growth acceleration are associated with increased adiposity and cardiometabolic risk markers in later childhood (Gishti et al. 2014; Perng et al. 2016).

9.3.4 PFAS AND NEURODEVELOPMENT

Eleven publications from prospective cohorts have examined the relations between prenatal PFAS exposure and cognitive abilities (Stein et al. 2013; Wang et al. 2015), attainment of developmental milestones (Forns et al. 2015), parent- or teacher-reported behaviors and executive function (Fei and Olsen 2010; Stein and Savitz 2011; Savitz et al. 2012; Braun et al. 2014; Forns et al. 2015; Vuong et al. 2016), psychomotor development (Fei and Olsen 2010), academic achievement (Lind et al. 2012), or risk of autism spectrum disorders, attention-deficit/hyperactivity disorder (ADHD), or cerebral palsy (Liew et al. 2014, 2015; Ode et al. 2014). With regard to the most commonly detected PFAS (PFOA, PFOS, PFNA, and PFHxS), these publications report inconsistent results. In a prospective birth cohort of 218 mother–child pairs, higher prenatal PFOS exposure was associated with worse parent-reported executive function (Vuong et al. 2016). In another prospective birth cohort, increasing prenatal PFOS and PFOA exposures was associated with 70% (95% CI 1.0, 2.8) and 110% (95% CI 1.2, 3.6) increased risk of cerebral palsy, respectively (Liew et al. 2014). Several studies report protective or null associations between prenatal PFAS exposure and child neurobehavior (Fei and Olsen 2010; Stein et al. 2013; Braun et al. 2014; Forns et al. 2015; Liew et al. 2015; Wang et al. 2015).

Four publications have examined the relations between childhood PFAS exposure and neurodevelopment (Hoffman et al. 2010; Stein and Savitz 2011; Stein et al. 2013, 2014). Two publications from cross-sectional studies report that children's serum PFAS concentrations were associated with increased prevalence of parent-reported ADHD or ADHD medication use (Hoffman et al. 2010; Stein and Savitz 2011). However, in a prospective cohort study with exceptionally high PFOA exposure, children's serum PFOA concentrations were not consistently associated with parent- or teacher-reported ADHD-related behaviors or other neuropsychological measurements (Stein et al. 2013; Liew et al. 2014).

The available body of evidence does not consistently suggest that early-life PFAS exposures are associated with neurodevelopment. While some studies suggest adverse neurobehavioral outcomes among children with elevated prenatal or childhood PFAS exposures, there are inconsistencies regarding which individual PFAS exposures may be associated with neurobehavior and whether there are heightened periods of vulnerability to PFAS exposures. The protective associations between early-life PFAS exposure and neurodevelopment are biologically plausible because *in vitro* studies report that PFOA and PFOS are agonists of PPARγ, and activation of this receptor may be neuroprotective (Kapadia et al. 2008). Additional studies with longitudinal measures of exposure and comprehensive assessment of neurodevelopment are warranted given the ubiquity and persistence of PFAS exposure.

9.3.5 PFAS AS AN ENDOCRINE DISRUPTOR

PFAS may act on a number of endocrine pathways to affect human health. PFOA and PFOS can bind to and activate the PPARα/γ to increase adipocyte differentiation and increase body fat (Vanden Heuvel et al. 2006; Taxvig et al. 2012; Bastos Sales et al. 2013). In addition, PFOA, PFOS, and PFHxS inhibit 11-β-hydroxysteroid dehydrogenase-2 to increase glucocorticoid concentrations in rodents, which might affect growth and brain development (Ye et al. 2014). Animal studies show that PFAS are capable of inducing changes in thyroid function, but results from human studies are not consistent (Chan et al. 2011; Boas et al. 2012). Future studies will need to consider if and how PFAS exposures affect epigenetic mechanisms, as some epidemiological studies suggest that PFOA and PFOS exposures are associated with lower global DNA cytosine methylation, higher long interspersed nuclear element-1 methylation, changes in the expression of genes involved in cholesterol metabolism, and changes in the methylation of specific genes (Guerrero-Preston et al. 2010; Fletcher et al. 2013). Changes in epigenetic biomarkers may be one of the "black boxes" between PFAS exposure and phenotypic changes in humans.

9.3.6 CHALLENGES IN MAKING STRONGER INTERFERENCES

Chemical mixtures, confounding factors, and periods of heightened vulnerability are potential challenges in making stronger inferences about the causal links between early-life PFAS exposures and the risk of childhood diseases. Exposure to a mixture of chemicals, including PFAS, occurs across the life span, yet researchers primarily examine exposures as if they occur in isolation from one another. This "one chemical at a time" approach has left us with insufficient knowledge about the individual, interactive, and cumulative health effects of PFAS mixtures. Epidemiological studies can address three broad questions related to endocrine-disrupting chemical (EDC) mixtures (Braun et al. 2016). First, studies may attempt to isolate the effect of one PFAS from another using standard multivariable regression techniques or more sophisticated methods (Braun et al. 2014). Second, multiple PFAS might have a synergistic association with health outcomes by disrupting the homeostasis of compensatory mechanisms. Finally, cumulative exposure to multiple PFAS could adversely affect child health when individual components of the mixture act via common biological pathways and cumulative exposure to these individual agents is sufficient to induce an adverse effect. Indeed, this seems quite plausible given that many PFAS act via common biological pathways in experimental systems. The use of more sophisticated statistical methods to address these three questions could help better resolve the potential health impacts of PFAS.

As is the case for all observational studies, there is the potential for confounding factors associated with both early-life EDC exposures and child health to bias study results. Socioeconomic factors are important determinants of childhood health and some chemical exposures. For instance, in the case of obesogens, many strong determinants of adiposity (e.g., diet) are correlated with lifestyle factors (e.g., maternal diet) that may also be associated with EDC exposures. Thus, to determine if PFAS have obesogenic effects independent of other predictors of obesity, it is necessary to measure and control for factors such as diet, physical activity, and breastfeeding. In addition, it is necessary to consider potential confounding effects from other PFAS (or other chemical) exposures since the effect of one PFAS may be misattributed to another correlated copollutant. This is especially pertinent since serum concentrations of individual PFAS tend to be moderately correlated with each other. Finally, it is imperative to note that it is not appropriate to adjust for variables that are caused by PFAS exposure and causes of poor childhood health (i.e., causal intermediates) since this "overadjustment bias" may mask causal associations (Schisterman et al. 2009). For instance, prenatal PFAS exposures are associated with lower birth weight and an increased risk of childhood obesity. Moreover, birth weight is a determinant of childhood obesity risk (Yu et al. 2011). Thus, adjusting for birth weight might bias associations between PFAS and risk of childhood obesity.

Finally, the potential effects of PFAS could be dependent on the timing of exposure given the possibility of unique periods of vulnerability to environmental stressors. For instance, the effect of PFAS exposures on neurodevelopment could depend on different biological mechanisms specific to prenatal (e.g., neurulation) and postnatal (e.g., synaptic pruning) neurodevelopmental processes (de Graaf-Peters and Hadders-Algra 2006), which could be a reason for some of the heterogeneity in the results of the studies discussed earlier. This limitation highlights the need to conduct prospective studies with serial measures of PFAS exposure across the life span, as well as for appropriate statistical methods to identify periods of heightened vulnerability (Sanchez et al. 2011).

9.4 CONCLUSION

PFAS represent a class of compounds of great interest in the field of environmental toxicology. Both epidemiological and animal data suggest that early-life or gestational exposure to PFAS can impact human health via changes in lipid metabolism that can alter body weight and delay development, particularly mammary gland development, which may indirectly impact breastfeeding reduction. The rodent studies are especially conclusive that late-gestation exposure to PFAS can drastically increase neonatal fatality and impact body and relative liver weight even after PFAS have been cleared from the body. PFAS exposure may also delay early development, including pubertal onset and phalanges ossification. Gestational exposure to PFAS has also been shown to stunt mammary gland development in mice and significantly alter the histology of breast tissue. This impact on the mammary gland is thought to be linked to epidemiology studies that associate PFOA exposure with a reduction in breastfeeding duration for infants. Rodent studies also implicate PPARα with hepatic hypertrophy and alterations in adiposity seen in human data. Perturbation of the thyroid axis may also explain cognitive issues, although the extent to which PFAS affect human neurodevelopment is unclear.

The mechanisms of action for PFAS are currently unclear and represent a critical gap that still needs to be addressed. Although PPARα activation may partially explain the mechanism for toxicity, further evidence to link this receptor to human toxicity is warranted. Knowledge of specific molecular mechanisms is important to not only enhance our knowledge of PFAS toxicity but also help inform strategies to ameliorate the toxicity or identify genetic populations at risk for greater toxicity. Such knowledge may also explain and grant credibility to unusual toxicity patterns, like nonmonotonic dose–response curves seen in developmental milestones (phalanges ossification and puberty onset) in mice.

Regulations surrounding PFAS exposure are currently in a dynamic state as state agencies determine appropriate actions to limit the toxicity of these persistent contaminants. These regulatory decisions depend heavily on the weight of evidence for the toxicity of early-life exposure to PFAS. Thus, emerging evidence, not discussed in this chapter, demonstrating that PFAS may have impacts on other organ systems (e.g., the immune system) is necessary to fully inform the risk assessment process (Dong et al. 2013; Stein et al. 2016; Timmermann et al. 2017). Finally, much of our understanding of PFAS toxicity is based on studies of PFOS and PFOA, which may not be representative of all the PFAS. There is especially a need to understand the toxicity of PFNA, which has seen an increase in human exposure and short-chained PFAS, like PFHxS, since they have been used to replace PFOA and PFOS. In summary, the evidence from both human and animal data demonstrates that high doses of many PFAS, particularly PFOA and PFOS, can have a lasting impact on health when exposed in early life, and additional studies on similar compounds investigating other potential health outcomes are required to fully understand the impact of these compounds on public health.

REFERENCES

AAP. 2012. Policy Statement Breastfeeding and the Use of Human Milk. *Pediatrics* 129 (3). doi:10.1542/peds.2011-3552.

Abbott, Barbara D. 2009. Review of the Expression of Peroxisome Proliferator-Activated Receptors Alpha (PPARα), Beta (PPARβ), and Gamma (PPARγ) in Rodent and Human Development. *Reproductive Toxicology* 27 (3): 246–57. doi: 10.1016/j.reprotox.2008.10.001.

Abbott, Barbara D., Cynthia J. Wolf, Kaberi P. Das, Robert D. Zehr, Judith E. Schmid, Andrew B. Lindstrom, Mark J. Strynar, and Christopher Lau. 2009. Developmental Toxicity of Perfluorooctane Sulfonate (PFOS) Is Not Dependent on Expression of Peroxisome Proliferator Activated Receptor-Alpha (PPARα) in the Mouse. *Reproductive Toxicology* 27 (3–4): 258–65. doi: 10.1016/j.reprotox.2008.05.061.

Abbott, Barbara D., Carmen R. Wood, Andrew M. Watkins, Kaberi P. Das, and Christopher S. Lau. 2010. Peroxisome Proliferator-Activated Receptors Alpha, Beta, and Gamma mRNA and Protein Expression in Human Fetal Tissues. *PPAR Research* 2010: 1–19. doi: 10.1155/2010/690907.

Akinbami, Lara J., and Cynthia L. Ogden. 2009. Childhood Overweight Prevalence in the United States: The Impact of Parent-Reported Height and Weight. *Obesity* 17 (8): 1574–80. doi: 10.1038/oby.2009.1.

Alexander, Jan, Guðjón Atli Auðunsson, Diane Benford, Andrew Cockburn, Jean-Pierre Cravedi, Eugenia Dogliotti, Alessandro Di Domenico et al. 2008. Perfluorooctane Sulfonate (PFOS), Perfluorooctanoic Acid (PFOA) and Their Salts: Scientific Opinion of the Panel on Contaminants in the Food Chain 1 (Question No. EFSA-Q-2004-163). *EFSA Journal* 653: 1–131.

Andersen, Camilla Schou, Chunyuan Fei, Michael Gamborg, Ellen A. Nohr, Thorkild I. A. Sørensen, and Jørn Olsen. 2013. Prenatal Exposures to Perfluorinated Chemicals and Anthropometry at 7 Years of Age. *American Journal of Epidemiology* 178 (6): 921–27. doi: 10.1093/aje/kwt057.

Andersen, M. E., J. L. Butenhoff, S.-C. Chang, D. G. Farrar, G. L. Kennedy, C. Lau, G. W. Olsen, J. Seed, and K. B. Wallace. 2007. Perfluoroalkyl Acids and Related Chemistries—Toxicokinetics and Modes of Action. *Toxicological Sciences* 102 (1): 3–14. doi: 10.1093/toxsci/kfm270.

Apelberg, Benjamin J., Frank R. Witter, Julie B. Herbstman, Antonia M. Calafat, Rolf U. Halden, Larry L. Needham, and Lynn R. Goldman. 2007. Cord Serum Concentrations of Perfluorooctane Sulfonate (PFOS) and Perfluorooctanoate (PFOA) in Relation to Weight and Size at Birth. *Environmental Health Perspectives* 115 (11): 1670–76. doi: 10.1289/ehp.10334.

Bach, Cathrine Carlsen, Bodil Hammer Bech, Nis Brix, Ellen Aagaard Nohr, Jens Peter Ellekilde Bonde, and Tine Brink Henriksen. 2015. Perfluoroalkyl and Polyfluoroalkyl Substances and Human Fetal Growth: A Systematic Review. *Critical Reviews in Toxicology* 45 (1): 53–67. doi: 10.3109/10408444.2014.952400.

Barbarossa, Andrea, Riccardo Masetti, Teresa Gazzotti, Daniele Zama, Annalisa Astolfi, Bruno Veyrand, Andrea Pession, and Giampiero Pagliuca. 2013. Perfluoroalkyl Substances in Human Milk: A First Survey in Italy. *Environment International* 51: 27–30. doi: 10.1016/j.envint.2012.10.001.

Barry, Vaughn, Lyndsey A. Darrow, Mitchel Klein, Andrea Winquist, and Kyle Steenland. 2014. Early Life Perfluorooctanoic Acid (PFOA) Exposure and Overweight and Obesity Risk in Adulthood in a Community with Elevated Exposure. *Environmental Research* 132: 62–69. doi: 10.1016/j.envres.2014.03.025.

Bastos Sales, L., J. H. Kamstra, P. H. Cenijn, L. S. van Rijt, T. Hamers, and J. Legler. 2013. Effects of Endocrine Disrupting Chemicals on In Vitro Global DNA Methylation and Adipocyte Differentiation. *Toxicology In Vitro* 27 (6): 1634–43. doi: 10.1016/j.tiv.2013.04.005.

Belknap, W. M., W. F. Balistreri, F. J. Suchy, and P. C. Miller. 1981. Physiologic Cholestasis II: Serum Bile Acid Levels Reflect the Development of the Enterohepatic Circulation in Rats. *Hepatology (Baltimore, Md.)* 1 (6): 613–16. http://www.ncbi.nlm.nih.gov/pubmed/7308994.

Biegel, L. B., J. C. Cook, J. C. O'Connor, M. Aschiero, A. J. Arduengo, and T. W. Slone. 1995. Subchronic Toxicity Study in Rats with 1-Methyl-3-Propylimidazole-2-Thione (PTI): Effects on the Thyroid. *Fundamental and Applied Toxicology* 27 (2): 185–94. http://www.ncbi.nlm.nih.gov/pubmed/8529813.

Bloom, Michael S., Kurunthachalam Kannan, Henry M. Spliethoff, Lin Tao, Kenneth M. Aldous, and John E. Vena. 2010. Exploratory Assessment of Perfluorinated Compounds and Human Thyroid Function. *Physiology & Behavior* 99 (2): 240–45. doi: 10.1016/j.physbeh.2009.02.005.

Boas, Malene, Ulla Feldt-Rasmussen, and Katharina M. Main. 2012. Thyroid Effects of Endocrine Disrupting Chemicals. *Molecular and Cellular Endocrinology* 355 (2): 240–48. doi: 10.1016/j.mce.2011.09.005.

Braun, Joseph M., Aimin Chen, Megan E. Romano, Antonia M. Calafat, Glenys M. Webster, Kimberly Yolton, and Bruce P. Lanphear. 2016. Prenatal Perfluoroalkyl Substance Exposure and Child Adiposity at 8 Years of Age: The HOME Study. *Obesity* 24 (1): 231–37. doi: 10.1002/oby.21258.

Braun, Joseph M., Amy E. Kalkbrenner, Allan C. Just, Kimberly Yolton, Antonia M. Calafat, Andreas Sjödin, Russ Hauser, Glenys M. Webster, Aimin Chen, and Bruce P. Lanphear. 2014. Gestational Exposure to Endocrine-Disrupting Chemicals and Reciprocal Social, Repetitive, and Stereotypic Behaviors in 4- and 5-Year-Old Children: The HOME Study. *Environmental Health Perspectives* 122 (5): 513–20. doi: 10.1289/ehp.1307261.

Brede, Edna, Michael Wilhelm, Thomas Göen, Johannes Müller, Knut Rauchfuss, Martin Kraft, and Jürgen Hölzer. 2010. Two-Year Follow-Up Biomonitoring Pilot Study of Residents' and Controls' PFC Plasma Levels after PFOA Reduction in Public Water System in Arnsberg, Germany. *International Journal of Hygiene and Environmental Health* 213 (3): 217–23. doi: 10.1016/j.ijheh.2010.03.007.

Buck, Robert C., James Franklin, Urs Berger, Jason M. Conder, Ian T. Cousins, Pim de Voogt, Allan Astrup Jensen, Kurunthachalam Kannan, Scott A. Mabury, and Stefan P. J. van Leeuwen. 2011. Perfluoroalkyl and Polyfluoroalkyl Substances in the Environment: Terminology, Classification, and Origins. *Integrated Environmental Assessment and Management* 7 (4): 513–41. doi: 10.1002/ieam.258.

Burt Solorzano, C. M., and C. R. McCartney. 2010. Obesity and the Pubertal Transition in Girls and Boys. *Reproduction* 140 (3): 399–410. doi: 10.1530/REP-10-0119.

Butenhoff, J. L., G. L. Kennedy, P. M. Hinderliter, P. H. Lieder, R. Jung, K. J. Hansen, G. S. Gorman, P. E. Noker, and P. J. Thomford. 2004. Pharmacokinetics of Perfluorooctanoate in Cynomolgus Monkeys. *Toxicological Sciences* 82 (2): 394–406. doi: 10.1093/toxsci/kfh302.

Chan, Emily, Igor Burstyn, Nicola Cherry, Fiona Bamforth, and Jonathan W. Martin. 2011. Perfluorinated Acids and Hypothyroxinemia in Pregnant Women. *Environmental Research* 111 (4): 559–64. doi: 10.1016/j.envres.2011.01.011.

Chen, A., Michael L. Pennell, Mark A. Klebanoff, Walter J. Rogan, and Matthew P. Longnecker. 2005. Maternal Smoking during Pregnancy in Relation to Child Overweight: Follow-up to Age 8 Years. *International Journal of Epidemiology* 35 (1): 121–30. doi: 10.1093/ije/dyi218.

Chen, Mei-Huei, Eun-Hee Ha, Ting-Wen Wen, Yi-Ning Su, Guang-Wen Lien, Chia-Yang Chen, Pau-Chung Chen, and Wu-Shiun Hsieh. 2012. Perfluorinated Compounds in Umbilical Cord Blood and Adverse Birth Outcomes. *PLoS One* 7 (8): e42474. doi: 10.1371/journal.pone.0042474.

Das, Kaberi P., Brian E. Grey, Mitchell B. Rosen, Carmen R. Wood, Katoria R. Tatum-Gibbs, R. Daniel Zehr, Mark J. Strynar, Andrew B. Lindstrom, and Christopher Lau. 2015. Developmental Toxicity of Perfluorononanoic Acid in Mice. *Reproductive Toxicology* 51: 133–44. doi: 10.1016/j.reprotox.2014.12.012.

De Graaf-Peters, Victorine B. and Mijna Hadders-Algra. 2006. Ontogeny of the human nervous system: What is happening when? *Early Human Development* 82 (4): 257–66. doi: 10.1016/j.earlhumdev.2005.10.013.

Divine, B. J., and C. M. Hartman. 2001. A Cohort Mortality Study among Workers at a 1,3 Butadiene Facility. *Chemico-Biological Interactions* 135–136: 535–53. http://www.ncbi.nlm.nih.gov/pubmed/11397411.

Dong, Guang-Hui, Kuan-Yen Tung, Ching-Hui Tsai, Miao-Miao Liu, Da Wang, Wei Liu, Yi-He Jin, Wu-Shiun Hsieh, Yungling Leo Lee, and Pau-Chung Chen. 2013. Serum Polyfluoroalkyl Concentrations, Asthma Outcomes, and Immunological Markers in a Case-Control Study of Taiwanese Children. *Environmental Health Perspectives* 121 (4): 507–13. doi: 10.1289/ehp.1205351.

Ebbeling, Cara B., Dorota B. Pawlak, and David S. Ludwig. 2002. Childhood Obesity: Public-Health Crisis, Common Sense Cure. *Lancet* 360 (9331): 473–82. doi: 10.1016/S0140-6736(02)09678-2.

Fei, Chunyuan, Joseph K. McLaughlin, Loren Lipworth, and Jørn Olsen. 2010. Maternal Concentrations of Perfluorooctanesulfonate (PFOS) and Perfluorooctanoate (PFOA) and Duration of Breastfeeding. *Scandinavian Journal of Work, Environment & Health* 36 (5): 413–21. http://www.ncbi.nlm.nih.gov/pubmed/20200757.

Fei, Chunyuan, Joseph K. McLaughlin, Robert E. Tarone, and Jørn Olsen. 2007. Perfluorinated Chemicals and Fetal Growth: A Study within the Danish National Birth Cohort. *Environmental Health Perspectives* 115 (11): 1677–82. doi: 10.1289/ehp.10506.

Fei, Chunyuan, and Jørn Olsen. 2010. Prenatal Exposure to Perfluorinated Chemicals and Behavioral or Coordination Problems at Age 7 Years. *Environmental Health Perspectives* 119 (4): 573–78. doi: 10.1289/ehp.1002026.

Fenton, Suzanne E., Jessica L. Reiner, Shoji F. Nakayama, Amy D. Delinsky, Jason P. Stanko, Erin P. Hines, Sally S. White, Andrew B. Lindstrom, Mark J. Strynar, and Syrago-Styliani E. Petropoulou. 2009. Analysis of PFOA in Dosed CD-1 Mice. Part 2: Disposition of PFOA in Tissues and Fluids from Pregnant and Lactating Mice and Their Pups. *Reproductive Toxicology* 27 (3–4): 365–72. doi: 10.1016/j.reprotox.2009.02.012.

Fletcher, Tony, Tamara S. Galloway, David Melzer, Paul Holcroft, Riccardo Cipelli, Luke C. Pilling, Debapriya Mondal, Michael Luster, and Lorna W. Harries. 2013. Associations between PFOA, PFOS and Changes in the Expression of Genes Involved in Cholesterol Metabolism in Humans. *Environment International* 57–58: 2–10. doi: 10.1016/j.envint.2013.03.008.

Forns, J., N. Iszatt, R. A. White, S. Mandal, A. Sabaredzovic, M. Lamoree, C. Thomsen, L. S. Haug, H. Stigum, and M. Eggesbø. 2015. Perfluoroalkyl Substances Measured in Breast Milk and Child Neuropsychological Development in a Norwegian Birth Cohort Study. *Environment International* 83: 176–82. doi: 10.1016/j.envint.2015.06.013.

Fromme, Hermann, Sheryl A. Tittlemier, Wolfgang Völkel, Michael Wilhelm, and Dorothee Twardella. 2009. Perfluorinated Compounds—Exposure Assessment for the General Population in Western Countries. *International Journal of Hygiene and Environmental Health* 212 (3): 239–70. doi: 10.1016/j.ijheh.2008.04.007.

Gibson, Helen M. 1973. Plasma Volume and Glomerular Filtration Rate in Pregnancy and Their Relation to Differences in Fetal Growth. *BJOG* 80 (12): 1067–74. doi: 10.1111 /j.1471-0528.1973.tb02981.x.

Gishti, Olta, Romy Gaillard, Rashindra Manniesing, Marieke Abrahamse-Berkeveld, Eline M. van der Beek, Denise H. M. Heppe, Eric A. P. Steegers et al. 2014. Fetal and Infant Growth Patterns Associated with Total and Abdominal Fat Distribution in School-Age Children. *Journal of Clinical Endocrinology & Metabolism* 99 (7): 2557–66. doi: 10.1210/jc.2013-4345.

Gladen, B. C., and W. J. Rogan. 1995. DDE and Shortened Duration of Lactation in a Northern Mexican Town. *American Journal of Public Health* 85 (4): 504–8. http://www.ncbi .nlm.nih.gov/pubmed/7702113.

Gleason, Jessie A., Keith R. Cooper, Judith B. Klotz, Gloria B. Post, and Geroge Van Orden. 2016. Health-Based Maximum Contaminant Level Support Document: Perfluorooctanoic Acid (PFOA). New Jersey Drinking Water Quality Institute Health Effects Subcommittee.

Glinoer, D., P. de Nayer, M. Bex, and Belgian Collaborative Study Group on Graves' Disease. 2001. Effects of L-Thyroxine Administration, TSH-Receptor Antibodies and Smoking on the Risk of Recurrence in Graves' Hyperthyroidism Treated with Antithyroid Drugs: A Double-Blind Prospective Randomized Study. *European Journal of Endocrinology* 144 (5): 475–83. http://www.ncbi.nlm.nih.gov/pubmed/11331213.

Grice, Mira M., Bruce H. Alexander, Richard Hoffbeck, and Diane M. Kampa. 2007. Self-Reported Medical Conditions in Perfluorooctanesulfonyl Fluoride Manufacturing Workers. *Journal of Occupational and Environmental Medicine* 49 (7): 722–29. doi: 10.1097/JOM.0b013e3180582043.

Guerrero-Preston, Rafael, Lynn R Goldman, Priscilla Brebi-Mieville, Carmen Ili-Gangas, Cynthia Lebron, Frank R. Witter, Ben J. Apelberg et al. 2010. Global DNA Hypomethylation Is Associated with In Utero Exposure to Cotinine and Perfluorinated Alkyl Compounds. *Epigenetics* 5 (6): 539–46. http://www.ncbi.nlm.nih.gov/pubmed /20523118.

Haddow, James E., Glenn E. Palomaki, Walter C. Allan, Josephine R. Williams, George J. Knight, June Gagnon, Cheryl E. O'Heir et al. 1999. Maternal Thyroid Deficiency during Pregnancy and Subsequent Neuropsychological Development of the Child. *New England Journal of Medicine* 341 (8): 549–55. doi: 10.1056/NEJM199908193410801.

Halldorsson, Thorhallur I., Dorte Rytter, Line Småstuen Haug, Bodil Hammer Bech, Inge Danielsen, Georg Becher, Tine Brink Henriksen, and Sjurdur F. Olsen. 2012. Prenatal Exposure to Perfluorooctanoate and Risk of Overweight at 20 Years of Age: A Prospective Cohort Study. *Environmental Health Perspectives* 120 (5): 668–73. doi: 10.1289/ehp.1104034.

Hattori, Aiko, and Roland Sturm. 2013. The Obesity Epidemic and Changes in Self-Report Biases in BMI. *Obesity* 21 (4): 856–60. doi: 10.1002/oby.20313.

Hines, Erin P., Sally S. White, Jason P. Stanko, Eugene A. Gibbs-Flournoy, Christopher Lau, and Suzanne E. Fenton. 2009. Phenotypic Dichotomy Following Developmental Exposure to Perfluorooctanoic Acid (PFOA) in Female CD-1 Mice: Low Doses Induce Elevated Serum Leptin and Insulin, and Overweight in Mid-Life. *Molecular and Cellular Endocrinology* 304 (1–2): 97–105. doi: 10.1016/j.mce.2009.02.021.

Hoffman, Kate, Thomas F. Webster, Marc G. Weisskopf, Janice Weinberg, and Verónica M. Vieira. 2010. Exposure to Polyfluoroalkyl Chemicals and Attention Deficit/Hyperactivity Disorder in U.S. Children 12–15 Years of Age. *Environmental Health Perspectives* 118 (12): 1762–67. doi: 10.1289/ehp.1001898.

Holden, P. R., and J. D. Tugwood. 1999. Peroxisome Proliferator-Activated Receptor Alpha: Role in Rodent Liver Cancer and Species Differences. *Journal of Molecular Endocrinology* 22 (1): 1–8. http://www.ncbi.nlm.nih.gov/pubmed/9924174.

Hoyer, Birgit B., Cecilia H. Ramlau-Hansen, Martine Vrijheid, Damaskini Valvi, Henning S. Pedersen, Valentyna Zviezdai, Bo A.G. Jonsson, Christian H. Lindh, Jens P. Bonde, Gunnar Toft. 2015. Anthropometry in 5- to 9-Year-Old Greenlandic and Ukrainian Children in Relation to Prenatal Exposure to Perfluorinated Alkyl Substances. *Environmental Health Perspectives* 123 (8): 841–6.

Ip, Stanley, Mei Chung, Gowri Raman, Priscilla Chew, Nombulelo Magula, Deirdre DeVine, Thomas Trikalinos, and Joseph Lau. 2007. Breastfeeding and Maternal and Infant Health Outcomes in Developed Countries. *Evidence Report/Technology Assessment* (153): 1–186. http://www.ncbi.nlm.nih.gov/pubmed/17764214.

Jain, Ram B. 2013. Effect of Pregnancy on the Levels of Selected Perfluoroalkyl Compounds for Females Aged 17–39 Years: Data from National Health and Nutrition Examination Survey 2003–2008. *Journal of Toxicology and Environmental Health, Part A* 76 (7): 409–21. doi: 10.1080/15287394.2013.771547.

Janesick, A., and B. Blumberg. 2012. Obesogens, Stem Cells and the Developmental Programming of Obesity. *International Journal of Andrology* 35 (3): 437–48. doi: 10.1111/j.1365-2605.2012.01247.x.

Ji, Kyunghee, Sunmi Kim, Younglim Kho, Domyung Paek, Joon Sakong, Jongsik Ha, Sungkyoon Kim, and Kyungho Choi. 2012. Serum Concentrations of Major Perfluorinated Compounds among the General Population in Korea: Dietary Sources and Potential Impact on Thyroid Hormones. *Environment International* 45: 78–85. doi: 10.1016/j.envint.2012.03.007.

Johnson, Paula I., Patrice Sutton, Dylan S. Atchley, Erica Koustas, Juleen Lam, Saunak Sen, Karen A. Robinson, Daniel A. Axelrad, and Tracey J. Woodruff. 2014. The Navigation Guide – Evidence-based medicine meets environmental health: Systematic review of human evidence for PFOA effects on fetal growth. *Environmental Health Perspectives* 122 (10): 1040–51. doi: 10.1289/ehp.1307893.

Kapadia, Ramya, Jae-Hyuk Yi, and Raghu Vemuganti. 2008. Mechanisms of Anti-Inflammatory and Neuroprotective Actions of PPAR-Gamma Agonists. *Frontiers in Bioscience* 13: 1813–26. http://www.ncbi.nlm.nih.gov/pubmed/17981670.

Kato, Kayoko, Lee-Yang Wong, Lily T. Jia, Zsuzsanna Kuklenyik, and Antonia M. Calafat. 2011. Trends in Exposure to Polyfluoroalkyl Chemicals in the U.S. Population: 1999–2008. *Environmental Science & Technology* 45 (19): 8037–45. doi: 10.1021/es1043613.

Keller, J. M., P. Collet, A. Bianchi, C. Huin, P. Bouillaud-Kremarik, P. Becuwe, H. Schohn, L. Domenjoud, and M. Dauça. 2000. Implications of Peroxisome Proliferator-Activated Receptors (PPARS) in Development, Cell Life Status and Disease. *International Journal of Developmental Biology* 44 (5): 429–42. http://www.ncbi.nlm.nih.gov/pubmed/11032176.

Kim, Sook-Jin, Seo-Hee Heo, Dong-Seok Lee, In Gyun Hwang, Yong-Bok Lee, and Hea-Young Cho. 2016. Gender Differences in Pharmacokinetics and Tissue Distribution of 3 Perfluoroalkyl and Polyfluoroalkyl Substances in Rats. *Food and Chemical Toxicology* 97: 243–55. doi: 10.1016/j.fct.2016.09.017.

Klaunig, James E., Barbara A. Hocevar, and Lisa M. Kamendulis. 2012. Mode of Action Analysis of Perfluorooctanoic Acid (PFOA) Tumorigenicity and Human Relevance. *Reproductive Toxicology* 33 (4): 410–18. doi: 10.1016/j.reprotox.2011.10.014.

Konkel, Lindsey. 2017. Mother's Milk and the Environment: Might Chemical Exposures Impair Lactation? *Environmental Health Perspectives* 125 (1). doi: 10.1289/ehp.125-A17.

Kudo, Naomi, Masanori Katakura, Yasunori Sato, and Yoichi Kawashima. 2002. Sex Hormone-Regulated Renal Transport of Perfluorooctanoic Acid. *Chemico-Biological Interactions* 139 (3): 301–16. http://www.ncbi.nlm.nih.gov/pubmed/11879818.

Kummu, M., E. Sieppi, J. Koponen, L. Laatio, K. Vähäkangas, H. Kiviranta, A. Rautio, and P. Myllynen. 2015. Organic Anion Transporter 4 (OAT 4) Modifies Placental Transfer of Perfluorinated Alkyl Acids PFOS and PFOA in Human Placental Ex Vivo Perfusion System. *Placenta* 36 (10): 1185–91. doi: 10.1016/j.placenta.2015.07.119.

Lau, C., Julie R. Thibodeaux, Roger G. Hanson, Michael G. Narotsky, John M. Rogers, Andrew B. Lindstrom, and Mark J. Strynar. 2006. Effects of Perfluorooctanoic Acid Exposure during Pregnancy in the Mouse. *Toxicological Sciences* 90 (2): 510–18. doi: 10.1093/toxsci/kfj105.

Lau, C., Julie R. Thibodeaux, Roger G. Hanson, John M. Rogers, B. E. Grey, M. E. Stanton, J. L. Butenhoff, and L. A. Stevenson. 2003. Exposure to Perfluorooctane Sulfonate during Pregnancy in Rat and Mouse. II: Postnatal Evaluation. *Toxicological Sciences* 74 (2): 382–92. doi: 10.1093/toxsci/kfg122.

Lee, Eung-Sun, Sehee Han, and Jeong-Eun Oh. 2016. Association between Perfluorinated Compound Concentrations in Cord Serum and Birth Weight Using Multiple Regression Models. *Reproductive Toxicology* 59: 53–59. doi: 10.1016/j.reprotox.2015.10.020.

Lee, S. S., T. Pineau, J. Drago, E. J. Lee, J. W. Owens, D. L. Kroetz, P. M. Fernandez-Salguero, H. Westphal, and F. J. Gonzalez. 1995. Targeted Disruption of the Alpha Isoform of the Peroxisome Proliferator-Activated Receptor Gene in Mice Results in Abolishment of the Pleiotropic Effects of Peroxisome Proliferators. *Molecular and Cellular Biology* 15 (6): 3012–22. http://www.ncbi.nlm.nih.gov/pubmed/7539101.

Lew, Betina J., Loretta L. Collins, Michael A. O'Reilly, and B. Paige Lawrence. 2009. Activation of the Aryl Hydrocarbon Receptor during Different Critical Windows in Pregnancy Alters Mammary Epithelial Cell Proliferation and Differentiation. *Toxicological Sciences* 111 (1): 151–62. doi: 10.1093/toxsci/kfp125.

Liew, Zeyan, Beate Ritz, Eva Cecilie Bonefeld-Jørgensen, Tine Brink Henriksen, Ellen Aagaard Nohr, Bodil Hammer Bech, Chunyuan Fei et al. 2014. Prenatal Exposure to Perfluoroalkyl Substances and the Risk of Congenital Cerebral Palsy in Children. *American Journal of Epidemiology* 180 (6): 574–81. doi: 10.1093/aje/kwu179.

Liew, Zeyan, Beate Ritz, Ondine S. von Ehrenstein, Bodil Hammer Bech, Ellen Aagaard Nohr, Chunyuan Fei, Rossana Bossi, Tine Brink Henriksen, Eva Cecilie Bonefeld-Jørgensen, and Jørn Olsen. 2015. Attention Deficit/Hyperactivity Disorder and Childhood Autism in Association with Prenatal Exposure to Perfluoroalkyl Substances: A Nested Case-Control Study in the Danish National Birth Cohort. *Environmental Health Perspectives* 123 (4): 367–73. doi: 10.1289/ehp.1408412.

Lin, Chien-Yu, Li-Li Wen, Lian-Yu Lin, Ting-Wen Wen, Guang-Wen Lien, Sandy H. J. Hsu, Kuo-Liong Chien et al. 2013. The Associations between Serum Perfluorinated Chemicals and Thyroid Function in Adolescents and Young Adults. *Journal of Hazardous Materials* 244–45: 637–44. doi: 10.1016/j.jhazmat.2012.10.049.

Lind, P. Monica, Vendela Roos, Monika Rönn, Lars Johansson, Håkan Ahlström, Joel Kullberg, and Lars Lind. 2012. Serum Concentrations of Phthalate Metabolites Are Related to Abdominal Fat Distribution Two Years Later in Elderly Women. *Environmental Health* 11 (1): 21. doi: 10.1186/1476-069X-11-21.

Loccisano, Anne E., Jerry L. Campbell, Melvin E. Andersen, and Harvey J. Clewell. 2011. Evaluation and Prediction of Pharmacokinetics of PFOA and PFOS in the Monkey and Human Using a PBPK Model. *Regulatory Toxicology and Pharmacology* 59 (1): 157–75. doi: 10.1016/j.yrtph.2010.12.004.

Loccisano, Anne E., Jerry L. Campbell, John L. Butenhoff, Melvin E. Andersen, and Harvey J. Clewell. 2012. Evaluation of Placental and Lactational Pharmacokinetics of PFOA and PFOS in the Pregnant, Lactating, Fetal and Neonatal Rat Using a Physiologically Based Pharmacokinetic Model. *Reproductive Toxicology* 33 (4): 468–90. doi: 10.1016/j.reprotox.2011.07.003.

Lopez-Espinosa, Maria-Jose, Tony Fletcher, Ben Armstrong, Bernd Genser, Ketan Dhatariya, Debapriya Mondal, Alan Ducatman, and Giovanni Leonardi. 2011. Association of Perfluorooctanoic Acid (PFOA) and Perfluorooctane Sulfonate (PFOS) with Age of

Puberty among Children Living Near a Chemical Plant. *Environmental Science & Technology* 45 (19): 8160–66. doi: 10.1021/es1038694.

Lou, Inchio, John F. Wambaugh, Christopher Lau, Roger G. Hanson, Andrew B. Lindstrom, Mark J. Strynar, R. Dan Zehr, R. Woodrow Setzer, and Hugh A. Barton. 2009. Modeling Single and Repeated Dose Pharmacokinetics of PFOA in Mice. *Toxicological Sciences* 107 (2): 331–41. doi: 10.1093/toxsci/kfn234.

Lucas, A., R. Morley, T. J. Cole, M. F. Bamford, A. Boon, P. Crowle, J. F. Dossetor, and R. Pearse. 1988. Maternal Fatness and Viability of Preterm Infants. *British Medical Journal (Clinical Research Ed.)* 296 (6635): 1495–97. http://www.ncbi.nlm.nih.gov/pubmed/3134083.

Luebker, Deanna J., Raymond G. York, Kristen J. Hansen, John A. Moore, and John L. Butenhoff. 2005. Neonatal Mortality from in Utero Exposure to Perfluorooctanesulfonate (PFOS) in Sprague–Dawley Rats: Dose–Response, and Biochemical and Pharamacokinetic Parameters. *Toxicology* 215 (1–2): 149–69. doi: 10.1016/j.tox.2005.07.019.

Macon, Madisa B., LaTonya R. Villanueva, Katoria Tatum-Gibbs, Robert D. Zehr, Mark J. Strynar, Jason P. Stanko, Sally S. White, Laurence Helfant, and Suzanne E. Fenton. 2011. Prenatal Perfluorooctanoic Acid Exposure in CD-1 Mice: Low-Dose Developmental Effects and Internal Dosimetry. *Toxicological Sciences* 122 (1): 134–45. doi: 10.1093/toxsci/kfr076.

Maisonet, Mildred, Metrecia L. Terrell, Michael A. McGeehin, Krista Yorita Christensen, Adrianne Holmes, Antonia M. Calafat, and Michele Marcus. 2012. Maternal Concentrations of Polyfluoroalkyl Compounds during Pregnancy and Fetal and Postnatal Growth in British Girls. *Environmental Health Perspectives* 120 (10): 1432–37. doi: 10.1289/ehp.1003096.

Maloney, E. K., and D. J. Waxman. 1999. Trans-Activation of PPARalpha and PPARgamma by Structurally Diverse Environmental Chemicals. *Toxicology and Applied Pharmacology* 161 (2): 209–18. doi: 10.1006/taap.1999.8809.

Manzano-Salgado, Cyntia B., Maribel Casas, Maria-Jose Lopez-Espinosa, Ferran Ballester, Mikel Basterrechea, Joan O Grimalt, Ana-María Jiménez et al. 2015. Transfer of Perfluoroalkyl Substances from Mother to Fetus in a Spanish Birth Cohort. *Environmental Research* 142: 471–78. doi: 10.1016/j.envres.2015.07.020.

Mattsson, Anna, Anna Kärrman, Rui Pinto, and Björn Brunström. 2015. Metabolic Profiling of Chicken Embryos Exposed to Perfluorooctanoic Acid (PFOA) and Agonists to Peroxisome Proliferator-Activated Receptors. *PloS One* 10 (12): e0143780. doi: 10.1371/journal.pone.0143780.

Meyts, Ewa Rajpert-De, Niels E. Skakkebaek, and Jorma Toppari. 2013. Testicular Cancer Pathogenesis, Diagnosis and Endocrine Aspects. MDText.com, Inc. http://www.ncbi.nlm.nih.gov/books/NBK278992/.

Midasch, O., H. Drexler, N. Hart, M. W. Beckmann, and J. Angerer. 2007. Transplacental Exposure of Neonates to Perfluorooctanesulfonate and Perfluorooctanoate: A Pilot Study. *International Archives of Occupational and Environmental Health* 80 (7): 643–48. doi: 10.1007/s00420-006-0165-9.

Mondal, Debapriya, Rosana Hernandez Weldon, Ben G. Armstrong, Lorna J. Gibson, Maria-Jose Lopez-Espinosa, Hyeong-Moo Shin, and Tony Fletcher. 2014. Breastfeeding: A Potential Excretion Route for Mothers and Implications for Infant Exposure to Perfluoroalkyl Acids. *Environmental Health Perspectives* 122 (2): 187–92. doi: 10.1289/ehp.1306613.

Mora, Ana María, Emily Oken, Sheryl L. Rifas-Shiman, Thomas F. Webster, Matthew W. Gillman, Antonia M. Calafat, Xiaoyun Ye, and Sharon K. Sagiv. 2017. Prenatal Exposure to Perfluoroalkyl Substances and Adiposity in Early and Mid-Childhood. *Environmental Health Perspectives*. 125 (3): 467–73. doi: 10.1289/EHP246.

Moreau, Amélie, Marie José Vilarem, Patrick Maurel, and Jean Marc Pascussi. 2008. Xenoreceptors CAR and PXR Activation and Consequences on Lipid Metabolism,

Glucose Homeostasis, and Inflammatory Response. *Molecular Pharmaceutics* 5 (1): 35–41. doi: 10.1021/mp700103m.

Morken, Nils-Halvdan, Gregory S. Travlos, Ralph E. Wilson, Merete Eggesbø, and Matthew P. Longnecker. 2014. Maternal Glomerular Filtration Rate in Pregnancy and Fetal Size. *PLoS One* 9 (7): e101897. doi: 10.1371/journal.pone.0101897.

Nagy, E., H. Orvos, A. Pál, L. Kovács, and K. Loveland. 2001. Breastfeeding Duration and Previous Breastfeeding Experience. *Acta Paediatrica (Oslo, Norway: 1992)* 90 (1): 51–56. http://www.ncbi.nlm.nih.gov/pubmed/11227334.

National Immunization Surveys. 2015. National Immunization Survey Breastfeeding among U.S. Children Born 2002–2012. Division of Nutrition, Physical Activity, and Obesity, National Center for Chronic Disease Prevention and Health Promotion, Centers for Disease Control and Prevention.

Ngo, Ha Thi, Ragna Bogen Hetland, Azemira Sabaredzovic, Line Småstuen Haug, and Inger-Lise Steffensen. 2014. In Utero Exposure to Perfluorooctanoate (PFOA) or Perfluorooctane Sulfonate (PFOS) Did Not Increase Body Weight or Intestinal Tumorigenesis in Multiple Intestinal Neoplasia (Min/+) Mice. *Environmental Research* 132: 251–63. doi: 10.1016/j.envres.2014.03.033.

Ode, Amanda, Karin Källén, Peik Gustafsson, Lars Rylander, Bo A. G. Jönsson, Per Olofsson, Sten A. Ivarsson, Christian H. Lindh, and Anna Rignell-Hydbom. 2014. Fetal Exposure to Perfluorinated Compounds and Attention Deficit Hyperactivity Disorder in Childhood. *PLoS One* 9 (4): e95891. doi: 10.1371/journal.pone.0095891.

Ogden, Cynthia L., Margaret D. Carroll, and Katherine M. Flegal. 2014. Prevalence of Obesity in the United States. *JAMA* 312 (2): 189. doi: 10.1001/jama.2014.6228.

Oken, E., E. B. Levitan, and M. W. Gillman. 2008. Maternal Smoking during Pregnancy and Child Overweight: Systematic Review and Meta-Analysis. *International Journal of Obesity* 32 (2): 201–10. doi: 10.1038/sj.ijo.0803760.

Olsen, Geary W., Jean M. Burris, David J. Ehresman, John W. Froehlich, Andrew M. Seacat, John L. Butenhoff, and Larry R. Zobel. 2007. Half-Life of Serum Elimination of Perfluorooctanesulfonate, Perfluorohexanesulfonate, and Perfluorooctanoate in Retired Fluorochemical Production Workers. *Environmental Health Perspectives* 115 (9): 1298–305. doi: 10.1289/ehp.10009.

Olsen, Geary W., Timothy R. Church, Eric B. Larson, Gerald van Belle, James K. Lundberg, Kristen J. Hansen, Jean M. Burris, Jeffrey H. Mandel, and Larry R. Zobel. 2004. Serum Concentrations of Perfluorooctanesulfonate and Other Fluorochemicals in an Elderly Population from Seattle, Washington. *Chemosphere* 54 (11): 1599–611. doi: 10.1016/j.chemosphere.2003.09.025.

Ophaug, R. H., and L. Singer. 1980. Metabolic Handling of Perfluorooctanoic Acid in Rats. *Proceedings of the Society for Experimental Biology and Medicine* 163 (1): 19–23. http://www.ncbi.nlm.nih.gov/pubmed/7352143.

Owen, P. E., and J. R. Glaister. 1990. Inhalation Toxicity and Carcinogenicity of 1,3-Butadiene in Sprague-Dawley Rats. *Environmental Health Perspectives* 86: 19–25. http://www.ncbi.nlm.nih.gov/pubmed/2401255.

Perng, Wei, Hanine Hajj, Mandy B. Belfort, Sheryl L. Rifas-Shiman, Michael S. Kramer, Matthew W. Gillman, and Emily Oken. 2016. Birth Size, Early Life Weight Gain, and Midchildhood Cardiometabolic Health. *Journal of Pediatrics* 173: 122–30.e1. doi: 10.1016/j.jpeds.2016.02.053.

Peters, Jeffrey M., and Frank J. Gonzalez. 2011. Why Toxic Equivalency Factors Are Not Suitable for Perfluoroalkyl Chemicals. *Chemical Research in Toxicology* 24 (10): 1601–9. doi: 10.1021/tx200316x.

Quist, Erin M., Adam J. Filgo, Connie A. Cummings, Grace E. Kissling, Mark J. Hoenerhoff, and Suzanne E. Fenton. 2015. Hepatic Mitochondrial Alteration in CD-1 Mice

Associated with Prenatal Exposures to Low Doses of Perfluorooctanoic Acid (PFOA). *Toxicologic Pathology* 43 (4): 546–57. doi: 10.1177/0192623314551841.

Rogan, W. J., and B. C. Gladen. 1985. Study of Human Lactation for Effects of Environmental Contaminants: The North Carolina Breast Milk and Formula Project and Some Other Ideas. *Environmental Health Perspectives* 60: 215–21. http://www.ncbi.nlm.nih.gov/pubmed/3928347.

Rogan, W. J., B. C. Gladen, J. D. McKinney, N. Carreras, P. Hardy, J. Thullen, J. Tingelstad, and M. Tully. 1987. Polychlorinated Biphenyls (PCBs) and Dichlorodiphenyl Dichloroethene (DDE) in Human Milk: Effects on Growth, Morbidity, and Duration of Lactation. *American Journal of Public Health* 77 (10): 1294–97. http://www.ncbi.nlm.nih.gov/pubmed/3115123.

Romano, Megan E., Yingying Xu, Antonia M. Calafat, Kimberly Yolton, Aimin Chen, Glenys M. Webster, Melissa N. Eliot, Cynthia R. Howard, Bruce P. Lanphear, and Joseph M. Braun. 2016. Maternal Serum Perfluoroalkyl Substances during Pregnancy and Duration of Breastfeeding. *Environmental Research* 149: 239–46. doi: 10.1016/j.envres.2016.04.034.

Rosen, Mitchell B., Julie R. Thibodeaux, Carmen R. Wood, Robert D. Zehr, Judith E. Schmid, and Christopher Lau. 2007. Gene Expression Profiling in the Lung and Liver of PFOA-Exposed Mouse Fetuses. *Toxicology* 239 (1–2): 15–33. doi: 10.1016/j.tox.2007.06.095.

Sanchez, Brisa N., Howard Hu, Heather J. Litman, and Martha M. Tellez-Rojo. 2010. Statistical Methods to Study Timing of Vulnerability with Sparsely Sampled Data on Environmental Toxicants. *Environmental Health Perspectives* 119 (3): 409–15. doi: 10.1289/ehp.1002453.

Sanchez, Brisa N., Howard Hu, Heather J. Litman, Martha M. Tellez-Rojo. 2011. Statistical Methods to Study Timing of Vulnerability with Sparsely Sampled Data on Environmental toxicants. *Environmental Health Perspectives* 119 (3): 409–25. doi: 10.1289/ehp.1002453.

Savitz, David A., Cheryl R. Stein, Scott M. Bartell, Beth Elston, Jian Gong, Hyeong-Moo Shin, and Gregory A. Wellenius. 2012. Perfluorooctanoic Acid Exposure and Pregnancy Outcome in a Highly Exposed Community. *Epidemiology* 23 (3): 386–92. doi: 10.1097/EDE.0b013e31824cb93b.

Schisterman, Enrique F., Stephen R. Cole, and Robert W. Platt. 2009. Overadjustment Bias and Unnecessary Adjustment in Epidemiologic Studies. *Epidemiology* 20 (4): 488–95. doi: 10.1097/EDE.0b013e3181a819a1.

Seacat, Andrew M., Peter J. Thomford, Kris J. Hansen, Geary W. Olsen, Marvin T. Case, and John L. Butenhoff. 2002. Subchronic Toxicity Studies on Perfluorooctanesulfonate Potassium Salt in Cynomolgus Monkeys. *Toxicological Sciences* 68 (1): 249–64. http://www.ncbi.nlm.nih.gov/pubmed/12075127.

Shrestha, Srishti, Michael S. Bloom, Recai Yucel, Richard F. Seegal, Qian Wu, Kurunthachalam Kannan, Robert Rej, and Edward F. Fitzgerald. 2015. Perfluoroalkyl Substances and Thyroid Function in Older Adults. *Environment International* 75: 206–14. doi: 10.1016/j.envint.2014.11.018.

Smith, M., Moran, P., Ward, M., and Davison, J. 2008. Assessment of glomerular filtration rate during pregnancy using the MDRD formula. *BJOG* 115: 109–12.

Steenland, Kyle, Tony Fletcher, and David A. Savitz. 2010. Epidemiologic Evidence on the Health Effects of Perfluorooctanoic Acid (PFOA). *Environmental Health Perspectives* 118 (8): 1100–1108. doi: 10.1289/ehp.0901827.

Stein, Cheryl R., Kathleen J. McGovern, Ashley M. Pajak, Paul J. Maglione, and Mary S. Wolff. 2016. Perfluoroalkyl and Polyfluoroalkyl Substances and Indicators of Immune Function in Children Aged 12–19 Y: National Health and Nutrition Examination Survey. *Pediatric Research* 79 (2): 348–57. doi: 10.1038/pr.2015.213.

Stein, Cheryl R., and David A. Savitz. 2011. Serum Perfluorinated Compound Concentration and Attention Deficit/Hyperactivity Disorder in Children 5–18 Years of Age. *Environmental Health Perspectives* 119 (10): 1466–71. doi: 10.1289/ehp.1003538.

Stein, Cheryl R., David A. Savitz, and David C. Bellinger. 2013. Perfluorooctanoate and Neuropsychological Outcomes in Children. *Epidemiology* 24 (4): 590–99. doi: 10.1097/EDE.0b013e3182944432.

Stein, Cheryl R., David A. Savitz, and David C. Bellinger. 2014. Perfluorooctanoate Exposure in a Highly Exposed Community and Parent and Teacher Reports of Behaviour in 6-12-Year-Old Children. *Paediatric and Perinatal Epidemiology* 28 (2): 146–56. doi: 10.1111/ppe.12097.

Stein, Cheryl R., David A. Savitz, and Marcelle Dougan. 2009. Serum Levels of Perfluorooctanoic Acid and Perfluorooctane Sulfonate and Pregnancy Outcome. *American Journal of Epidemiology* 170 (7): 837–46. doi: 10.1093/aje/kwp212.

Suh, Chun Hui, Nam Kyoo Cho, Chae Kwan Lee, Chang Hee Lee, Dae Hwan Kim, Jeong Ho Kim, Byung Chul Son, and Jong Tae Lee. 2011. Perfluorooctanoic Acid-Induced Inhibition of Placental Prolactin-Family Hormone and Fetal Growth Retardation in Mice. *Molecular and Cellular Endocrinology* 337 (1–2): 7–15. doi: 10.1016/j.mce.2011.01.009.

Tatum-Gibbs, Katoria, John F. Wambaugh, Kaberi P. Das, Robert D. Zehr, Mark J. Strynar, Andrew B. Lindstrom, Amy Delinsky, and Christopher Lau. 2011. Comparative Pharmacokinetics of Perfluorononanoic Acid in Rat and Mouse. *Toxicology* 281 (1–3): 48–55. doi: 10.1016/j.tox.2011.01.003.

Taxvig, Camilla, Karin Dreisig, Julie Boberg, Christine Nellemann, Ane Blicher Schelde, Dorthe Pedersen, Michael Boergesen, Susanne Mandrup, and Anne Marie Vinggaard. 2012. Differential Effects of Environmental Chemicals and Food Contaminants on Adipogenesis, Biomarker Release and PPARγ Activation. *Molecular and Cellular Endocrinology* 361 (1–2): 106–15. doi: 10.1016/j.mce.2012.03.021.

Teich, Alice S, Josephine Barnett, and Karen Bonuck. 2014. Women's Perceptions of Breastfeeding Barriers in Early Postpartum Period: A Qualitative Analysis Nested in Two Randomized Controlled Trials. *Breastfeeding Medicine* 9 (1): 9–15. doi: 10.1089/bfm.2013.0063.

Thibodeaux, J. R., R. G. Hanson, J. M. Rogers, B. E. Grey, B. D. Barbee, J. H. Richards, J. L. Butenhoff, L. A. Stevenson, and C. Lau. 2003. Exposure to Perfluorooctane Sulfonate during Pregnancy in Rat and Mouse. I: Maternal and Prenatal Evaluations. *Toxicological Sciences* 74 (2): 369–81. doi: 10.1093/toxsci/kfg121.

Timmermann, Clara Amalie Gade, Esben Budtz-Jørgensen, Tina Kold Jensen, Christa Elyse Osuna, Maria Skaalum Petersen, Ulrike Steuerwald, Flemming Nielsen, Lars K. Poulsen, Pál Weihe, and Philippe Grandjean. 2017. Association between Perfluoroalkyl Substance Exposure and Asthma and Allergic Disease in Children as Modified by MMR Vaccination. *Journal of Immunotoxicology* 14 (1): 39–49. doi: 10.1080/1547691X.2016.1254306.

Tucker, Deirdre K., Madisa B. Macon, Mark J. Strynar, Sonia Dagnino, Erik Andersen, and Suzanne E. Fenton. 2015. The Mammary Gland Is a Sensitive Pubertal Target in CD-1 and C57Bl/6 Mice Following Perinatal Perfluorooctanoic Acid (PFOA) Exposure. *Reproductive Toxicology (Elmsford, N.Y.)* 54: 26–36. doi: 10.1016/j.reprotox.2014.12.002.

U.S. Environmental Protection Agency. 2016. Drinking Water Health Advisory for Perfluorooctane Sulfonate (PFOS). EPA Document Number 822-R-16-004. Office of Water, Health and Ecological Criteria Division.

Vandenberg, Laura N., Theo Colborn, Tyrone B. Hayes, Jerrold J. Heindel, David R. Jacobs, Duk-Hee Lee, Toshi Shioda et al. 2012. Hormones and Endocrine-Disrupting Chemicals: Low-Dose Effects and Nonmonotonic Dose Responses. *Endocrine Reviews* 33 (3): 378–455. doi: 10.1210/er.2011-1050.

Vanden Heuvel, J. P., B. I. Kuslikis, M. J. Van Rafelghem, and R. E. Peterson. 1991. Tissue Distribution, Metabolism, and Elimination of Perfluorooctanoic Acid in Male and Female Rats. *Journal of Biochemical Toxicology* 6 (2): 83–92. http://www.ncbi.nlm.nih.gov/pubmed/1941903.

Vanden Heuvel, J. P., Jerry T. Thompson, Steven R. Frame, and Peter J. Gillies. 2006. Differential Activation of Nuclear Receptors by Perfluorinated Fatty Acid Analogs and Natural Fatty Acids: A Comparison of Human, Mouse, and Rat Peroxisome Proliferator-Activated Receptor-Alpha, -Beta, and -Gamma, Liver X Receptor-Beta, and Retinoid X Receptor-Alpha. *Toxicological Sciences* 92 (2): 476–89. doi: 10.1093/toxsci/kfl014.

Verner, Marc-André, Anne E. Loccisano, Nils-Halvdan Morken, Miyoung Yoon, Huali Wu, Robin McDougall, Mildred Maisonet et al. 2015. Associations of Perfluoroalkyl Substances (PFAS) with Lower Birth Weight: An Evaluation of Potential Confounding by Glomerular Filtration Rate Using a Physiologically Based Pharmacokinetic Model (PBPK). *Environmental Health Perspectives* 123 (12). doi: 10.1289/ehp.1408837.

Verougstraete, Violaine, Dominique Lison, and Philippe Hotz. 2003. Cadmium, Lung and Prostate Cancer: A Systematic Review of Recent Epidemiological Data. *Journal of Toxicology and Environmental Health, Part B* 6 (3): 227–56. doi: 10.1080/10937400306465.

Vuong, Ann M., Kimberly Yolton, Glenys M. Webster, Andreas Sjödin, Antonia M. Calafat, Joseph M. Braun, Kim N. Dietrich, Bruce P. Lanphear, and Aimin Chen. 2016. Prenatal Polybrominated Diphenyl Ether and Perfluoroalkyl Substance Exposures and Executive Function in School-Age Children. *Environmental Research* 147: 556–64. doi: 10.1016/j.envres.2016.01.008.

Waalkes, M. P., A. Perantoni, M. R. Bhave, and S. Rehm. 1988. Strain Dependence in Mice of Resistance and Susceptibility to the Testicular Effects of Cadmium: Assessment of the Role of Testicular Cadmium-Binding Proteins. *Toxicology and Applied Pharmacology* 93 (1): 47–61. http://www.ncbi.nlm.nih.gov/pubmed/3354001.

Wakui, Shin, Tomoko Muto, Yasuko Kobayashi, Kenta Ishida, Masataka Nakano, Hiroyuki Takahashi, Yoshihiko Suzuki, Masakuni Furusato, and Hiroshi Hano. 2008. Sertoli-Leydig Cell Tumor of the Testis in a Sprague-Dawley Rat. *Journal of the American Association for Laboratory Animal Science: JAALAS* 47 (6): 67–70. http://www.ncbi.nlm.nih.gov/pubmed/19049257.

Wang, Jun-Long, Shi Yang, Ru-Hui Tian, Zi-Jue Zhu, Ying Guo, Qing-Qing Yuan, Zu-Ping He, and Zheng Li. 2015. [Isolation, Culture, and Identification of Human Spermatogonial Stem Cells]. *Zhonghua Nan Ke Xue* (*National Journal of Andrology*) 21 (3): 208–13. http://www.ncbi.nlm.nih.gov/pubmed/25898550.

Wang, Yan, Margaret Adgent, Pen-Hua Su, Hsiao-Yen Chen, Pau-Chung Chen, Chao A. Hsiung, and Shu-Li Wang. 2016. Prenatal Exposure to Perfluorocarboxylic Acids (PFCAs) and Fetal and Postnatal Growth in the Taiwan Maternal and Infant Cohort Study. *Environmental Health Perspectives* 124 (11): 1794–800. doi: 10.1289/ehp.1509998.

Washino, Noriaki, Yasuaki Saijo, Seiko Sasaki, Shizue Kato, Susumu Ban, Kanae Konishi, Rie Ito et al. 2009. Correlations between Prenatal Exposure to Perfluorinated Chemicals and Reduced Fetal Growth. *Environmental Health Perspectives* 117 (4): 660–67. doi: 10.1289/ehp.11681.

Wen, Li-Li, Lian-Yu Lin, Ta-Chen Su, Pau-Chung Chen, and Chien-Yu Lin. 2013. Association between Serum Perfluorinated Chemicals and Thyroid Function in U.S. Adults: The National Health and Nutrition Examination Survey 2007–2010. *Journal of Clinical Endocrinology & Metabolism* 98 (9): E1456–64. doi: 10.1210/jc.2013-1282.

Whalen, Bonny, and Rachel Cramton. 2010. Overcoming Barriers to Breastfeeding Continuation and Exclusivity. *Current Opinion in Pediatrics* 22 (5): 1. doi: 10.1097 /MOP.0b013e32833c8996.

White, Sally S., Antonia M. Calafat, Z. Kuklenyik, L. Villanueva, R. D. Zehr, L. Helfant, M. J. Strynar et al. 2006. Gestational PFOA Exposure of Mice Is Associated with Altered Mammary Gland Development in Dams and Female Offspring. *Toxicological Sciences* 96 (1): 133–44. doi: 10.1093/toxsci/kfl177.

White, Sally S., Suzanne E. Fenton, and Erin P. Hines. 2011. Endocrine Disrupting Properties of Perfluorooctanoic Acid. *Journal of Steroid Biochemistry and Molecular Biology* 127 (1–2): 16–26. doi: 10.1016/j.jsbmb.2011.03.011.

Whitworth, Kristina W., Line S. Haug, Donna D. Baird, Georg Becher, Jane A. Hoppin, Rolv Skjaerven, Cathrine Thomsen et al. 2012. Perfluorinated Compounds and Subfecundity in Pregnant Women. *Epidemiology* 23 (2): 257–63. doi: 10.1097/EDE .0b013e31823b5031.

Wolf, Cynthia J., S. E. Fenton, Judith E. Schmid, Antonia M. Calafat, Z. Kuklenyik, X. A. Bryant, J. Thibodeaux et al. 2006. Developmental Toxicity of Perfluorooctanoic Acid in the CD-1 Mouse after Cross-Foster and Restricted Gestational Exposures. *Toxicological Sciences* 95 (2): 462–73. doi: 10.1093/toxsci/kfl159.

Wolf, Cynthia J., Judith E. Schmid, Christopher Lau, and Barbara D. Abbott. 2012. Activation of Mouse and Human Peroxisome Proliferator-Activated Receptor-Alpha (PPARα) by Perfluoroalkyl Acids (PFAAs): Further Investigation of C4–C12 Compounds. *Reproductive Toxicology* 33 (4): 546–51. doi: 10.1016/j.reprotox.2011.09.009.

Wolf, Cynthia J., M. L. Takacs, Judith E. Schmid, Christopher Lau, and Barbara D. Abbott. 2008. Activation of Mouse and Human Peroxisome Proliferator-Activated Receptor Alpha by Perfluoroalkyl Acids of Different Functional Groups and Chain Lengths. *Toxicological Sciences* 106 (1): 162–71. doi: 10.1093/toxsci/kfn166.

Wolf, Cynthia J., Robert D. Zehr, Judy E. Schmid, Christopher Lau, and Barbara D. Abbott. 2010. Developmental Effects of Perfluorononanoic Acid in the Mouse Are Dependent on Peroxisome Proliferator-Activated Receptor-Alpha. *PPAR Research* 2010: 1–11. doi: 10.1155/2010/282896.

Xie, Yi, Qian Yang, and Joseph W DePierre. 2002. The Effects of Peroxisome Proliferators on Global Lipid Homeostasis and the Possible Significance of These Effects to Other Responses to These Xenobiotics: An Hypothesis. *Annals of the New York Academy of Sciences* 973: 17–25. http://www.ncbi.nlm.nih.gov/pubmed/12485828.

Yang, Chengfeng, Ying S. Tan, Jack R. Harkema, and Sandra Z. Haslam. 2009. Differential Effects of Peripubertal Exposure to Perfluorooctanoic Acid on Mammary Gland Development in C57Bl/6 and Balb/c Mouse Strains. *Reproductive Toxicology* 27 (3–4): 299–306. doi: 10.1016/j.reprotox.2008.10.003.

Yang, Wei-Shiung, Chi-Yuan Jeng, Ta-Jen Wu, Sachiyo Tanaka, Tohru Funahashi, Yuji Matsuzawa, Jao-Ping Wang, Chi-Ling Chen, Tong-Yuan Tai, and Lee-Ming Chuang. 2002. Synthetic Peroxisome Proliferator-Activated Receptor-Gamma Agonist, Rosiglitazone, Increases Plasma Levels of Adiponectin in Type 2 Diabetic Patients. *Diabetes Care* 25 (2): 376–80. http://www.ncbi.nlm.nih.gov/pubmed/11815513.

Ye, Feng, Masahiro Tokumura, Md Saiful Islam, Yasuyuki Zushi, Jungkeun Oh, and Shigeki Masunaga. 2014. Spatial Distribution and Importance of Potential Perfluoroalkyl Acid Precursors in Urban Rivers and Sewage Treatment Plant Effluent—Case Study of Tama River, Japan. *Water Research* 67: 77–85. doi: 10.1016/j.watres.2014.09.014.

Yu, Mu-Xue, Xiao-Shan Qiu, Su-E Feng, Qing-Ping Mo, Xiao-Ying Xie, Zhen-Yu Shen, and Yong-Zhou Liu. 2011. [The Effect of Birth Weight and Early Growth on Body Fat Composition and Insulin Sensitivity]. *Zhonghua Yu Fang Yi Xue Za Zhi* [*Chinese Journal of Preventive Medicine*] 45 (7): 633–38. http://www.ncbi.nlm.nih.gov /pubmed/22041569.

Zhang, Tao, Hongwen Sun, Xiaolei Qin, Zhiwei Gan, and Kurunthachalam Kannan. 2015. PFOS and PFOA in Paired Urine and Blood from General Adults and Pregnant Women: Assessment of Urinary Elimination. *Environmental Science and Pollution Research* 22 (7): 5572–79. doi: 10.1007/s11356-014-3725-7.

Zhang, Yifeng, Sanjay Beesoon, Lingyan Zhu, and Jonathan W Martin. 2013. Biomonitoring of Perfluoroalkyl Acids in Human Urine and Estimates of Biological Half-Life. *Environmental Science & Technology* 47 (18): 10619–27. doi: 10.1021/es401905e.

Zhao, Yong, Ying S. Tan, Mark J. Strynar, Gloria Perez, Sandra Z. Haslam, Chengfeng Yang, and S. Z. Haslam. 2012. Perfluorooctanoic Acid Effects on Ovaries Mediate Its Inhibition of Peripubertal Mammary Gland Development in Balb/c and C57Bl/6 Mice. *Reproductive Toxicology* 33 (4): 563–76. doi: 10.1016/j.reprotox.2012.02.004.

10 PFAS Isomers
Characterization, Profiling, and Toxicity

Yun Xing, Gabriel Cantu, and David M. Kempisty

CONTENTS

10.1 Introduction 203
10.2 Isomer Characterization and Profiling... 205
10.3 Isomer-Specific Analysis for PFAS Source Tracking in the Environment... 214
    10.3.1 Isomer Analysis as a Source-Tracking Tool
        in Environmental Samples... 214
    10.3.2 Enantiomer Analysis as a Tracking Tool for Direct
        versus Indirect Exposure..215
10.4 Isomer-Specific Biotransformation and Bioaccumulation........................... 218
    10.4.1 Bioaccumulation in Fish and Wildlife.. 218
    10.4.2 Isomer-Specific Biotransformation of PFAS Precursors 219
10.5 PFAS Isomer Pattern in Humans..222
    10.5.1 Isomer Pattern in Human Serum, Blood, Plasma,
        and Other Matrices 222
    10.5.2 Isomer-Specific Transplacental Transfer and Implications
        on Developmental Toxicity..226
10.6 Isomer-specific Toxicity and Interactions with Biomolecules 229
    10.6.1 Isomer-Specific Toxicity..229
    10.6.2 Isomer-Specific Interactions with Biological Molecules 231
10.7 Conclusions and Future Perspective 233
References..233

10.1 INTRODUCTION

Historically, poly- and perfluoroalkyl substances (PFAS) were produced by two main methods: electrochemical fluorination (ECF) (3M, 1995) and telomerization. The ECF process accounted for 80%–90% of the total PFAS production between 1947 and 2002 (Prevedouros et al., 2006); however, telomerization generally is the preferred mode of synthesis in recent times. The two major manufacturing techniques for perfluorochemicals can be distinguished based on the isomeric profile of their products. ECF results in a product containing both linear and branched isomers, while telomerization typically yields an isomerically pure, linear product (Benskin et al., 2010a). For example, historic ECF perfluorooctanoic acid (PFOA) from 3M Company had a consistent isomer composition of 78 ± 1.2% linear and 22 ± 1.2% various branched isomers, based on 18 production lots over 20 years (Benskin et al., 2010b). Perfluorooctanesulfonic acid (PFOS)–related chemicals (i.e., PFOS and its derivatives) are produced solely by the ECF method (Martin et al., 2010). The percentage of linear PFOS (L-PFOS) present in commercial products is mostly in the range of 67%–82% (Benskin et al., 2010a; Zhang et al., 2013; Gao et al., 2015; Jiang et al., 2015). The isomer composition of the commercial PFOS products can be up to 30% of total PFOS. Moreover, some of these isomers (specifically those that are branched chain) are chiral, resulting in an environmental fate and behavior that may vary as a function of its isomeric and enantiomeric composition. Theoretically, hundreds of PFOS isomers with the elemental composition $C_8F_{17}SO_3^-$ are possible. However, fluorine-19 nuclear magnetic resonance (^{19}F-NMR) analysis has indicated that perfluoromonomethyl and perfluoroisopropyl isomers are the most abundant, with the majority of the rest being tert-perfluorobutyl and germinal-substituted perfluoreodimethyl compounds. Figure 10.1 shows the structure of the 11 most prevalent isomers found in technical PFOS products.

Perfluoroalkyl chain branching can affect the physical and chemical properties of these chemicals, which may influence their environmental transport and degradation, partitioning, bioaccumulation, pharmacokinetics, and toxicity. Traditional

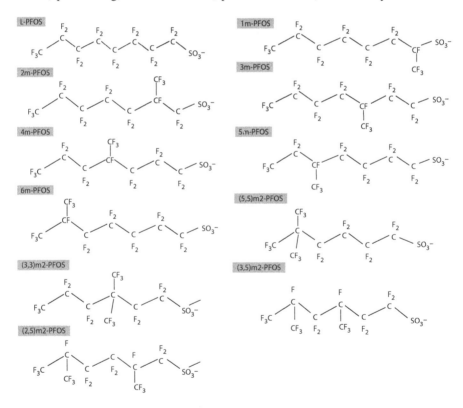

FIGURE 10.1 Structure of the prevalent PFOS isomers in technical mixtures and in samples. (Adapted from Riddell N, Arsenault G, Benskin JP, Chittim B, Martin JW, McAlees A, McCrindle R, *Environ Sci Technol*, 2009;43(20):7902–8.)

PFAS analysis, however, has avoided the distinction of isomers and considered the total amount only, potentially leading to inaccuracies in human and environmental risk assessments. On the regulatory side, it is important to know to what extent human and environmental exposure is from historical products (i.e., ECF production) versus currently manufactured fluorochemicals (i.e., telomerization production). These questions can only be answered when PFAS are considered individual isomers. Research developments within the last decade have led to the emergence of isomer-specific technologies and the adoption of these technologies in PFAS analysis of environmental and biological samples. Several studies have highlighted the usefulness of isomer-specific analysis for manufacturing (Wang et al., 2009) and exposure source elucidation (Benskin et al. 2010b), and for potential improvements in analytical accuracy (Riddell et al., 2009). The importance of isomer-specific analysis is further exemplified by evidence of isomer-specific toxicity (Loveless et al., 2006) and differences in pharmacokinetics (Ross et al., 2012), along with highly variable PFOS isomer profiles in humans (Kärrman et al., 2007; Liu et al., 2015) and wildlife (Houde et al., 2008; Gebbink et al., 2010).

When considering PFAS isomers, interesting and incomplete results emerge, some of which are conflicting and sometimes with sparse numbers of studies. In this chapter, we provide an overview of important topics pertaining to PFAS isomer research to date. Five areas in particular are focused on: (1) isomer characterization and profiling, (2) isomer-specific analysis in environmental PFAS source tracking, (3) isomer-specific biotransformation and bioaccumulation, (4) variety of isomer profiles in humans, and (5) isomer-specific toxicity and interaction with biomolecules. The chapter ends with a summary of the five focus areas and provides a future perspective on PFAS isomer research.

10.2 ISOMER CHARACTERIZATION AND PROFILING

The most commonly used analytical instrumentation for the analysis of PFAS is high-performance liquid chromatography with tandem mass spectrometry (HPLC-MS/MS). Although multiple isomers are thought to be present in environmental samples, traditional HPLC methods, called the total PFAS ("ΣPFAS") method, often avoid isomer separation and use the entire broadened peak for PFAS quantification. This method can lead to systematic quantification bias of unknown proportions since various branched isomers are expected to have different ionization and fragmentation during electrospray ionization and distinct collision-induced dissociation patterns in MS/MS. Since the C8 PFAS (PFOS and PFOA) were the most prominent PFAS found in the environment, most of the isomer-specific methodologies were focused on these particular compounds.

The first systematic HPLC-MS/MS study on PFOS isomer separation and identification in a technical PFOS product was carried out by Langlois and Oehme (2006). In this study, they found that a perfluorinated phenyl (PFP) column (Thermo Electron, 150 mm column length, 2.1 mm inner diameter [i.d.], 5 mm particles, 100 Å pore size) and an X-Terra C18 column (Waters, 100 mm length, 3.0 mm i.d., 3.5 mm particles, 125 Å pore size) were the most suitable for the separation of PFOS isomers. Both ion trap (IT) and a triple quadrupole (TQ) MS/MS were able to identify

10 PFOS isomers, with IT MS/MS giving the best results. Geminal diperfluoromethyl isomers eluted first from the best-suited HPLC phase (PFP), followed by perfluoromonomethyl isomers and finally the linear isomer. Subsequent ^{19}F-NMR analysis of the purified isomer fractions allowed the elucidation of structures and assignment of retention times of the branched isomers. The relative abundance of product ions of PFOS isomers in a technical product is summarized in Table 10.1, and it is clear that significant differences exist among the isomers with regard to fragmentation of the molecular ion m/z 499. The m/z 499 → 99 transition used for PFOS quantification was only the base peak for linear isomers, and its abundance in the spectra of branched isomers, such as 3m-, 4m-, and 5m-PFOS, was relatively low. This meant both a higher detection limit and different response factors for these isomers compared with their linear counterparts. Moreover, among the diperfluoromethyl isomers, only F^4 gives a product ion with an m/z 99; all the other diperfluoromethyl isomers can only be detected if their abundant ions at m/z 130 and 180 are included in the quantification. According to ^{19}F-NMR data, technical PFOS contains about 20%–30% of perfluoromethyl isomers and less than 2% of diperfluoromethyl compounds. These results highlighted the need for quality reference standards of the corresponding isomers for accurate quantification of isomers in a technical product.

Benskin et al. (2007) developed a comprehensive method to simultaneously separate and detect PFAS and PFAS precursor isomers using LC-MS/MS. A linear perfluorooctyl stationary phase and acidified mobile phase increased separation efficiency, relative to alkyl stationary phases, for the many perfluoroalkyl carboxylate (PFCA), perfluoroalkyl sulfonate (PFSA), and perfluorooctane sulfonamide (FOSA) isomers, and in combination with their distinct MS/MS transitions, allowed full resolution of most isomers in standards. Utilizing the absence of the 9-series and 0-series productions, several PFOS ($C_8F_{17}SO_3^-$) isomers were structurally elucidated. In human serum, only perfluorooctanesulfonamide (PFOSA, $C_8F_{17}SO_2NH_2$) and PFOS consisted of significant quantities of branched isomers, whereas PFCAs were predominantly linear. Interferences that coeluted with the m/z 499 → 80 transition of PFOS on alkyl stationary phases were simultaneously separated and identified as taurodeoxycholate isomers, the removal of which permitted the use of the more sensitive m/z 80 product ion and a resulting 20-fold decrease in PFOS detection limits compared with the m/z 499 → 99 transition (0.8 pg vs. 20 pg using m/z 80 and 99, respectively). Interferences in human serum that caused a 10- to 20-fold overreporting of perfluorohexane sulfonate ($C_6F_{13}SO_3^-$, PFHxS) concentrations on alkyl stationary phases were also simultaneously separated from linear PFHxS and identified as endogenous steroid sulfates. PFOSA isomers, generated by incubating N-ethylperfluorooctane sulfonamide (N-Et-FOSA) isomers with human liver microsomes, had different rates of metabolism, suggesting that the perfluoroalkyl branching pattern may affect the biological properties of individual isomers.

Riddell et al. (2009) performed a systematic study comparing the differences between the total PFOS (ΣPFOS) method and the isomer-specific analysis method using real human serum samples acquired through participation in the 2006 PFAS intercalibration study (Figure 10.2). HPLC-MS/MS chromatograms of serum samples were first generated using isomer-specific methods. Three distinct standard curve quantification techniques were then applied to these chromatograms: (1) ΣPFOS

TABLE 10.1
Relative Abundance (%) of Product Ions of PFOS Isomers in a Technical Product Obtained by Fragmentation of the Molecular Anions m/z 499 by IT MS/MS (40% CE) and TQ MS/MS (40% CE)

Isomer	Type	Mass (m/z)																	
		499	430	419	380	369	330	319	280	269	261	230	219	180	169	130	119	99	80
F^1	IT	–	–	8	–	–	11	59	100	19	–	–	–	–	–	*	*	*	*
	TQ	11	–	–	–	–	12	–	–	20	–	–	24	11	–	100	–	–	12
F^2	IT	–	–	–	–	3	–	–	44	7	–	100	–	100	–	*	*	*	*
	TQ	–	–	–	–	–	–	22	–	18	–	32	–	–	–	40	14	–	36
F^3	IT	–	–	100	14	–	29	–	7	–	8	–	7	18	6	*	*	–	*
	TQ	–	–	–	–	–	–	–	7	–	6	–	–	–	–	100	–	–	100
F^4	IT	32	–	–	45	–	5	–	100	–	–	26	18	10	–	*	–	*	*
	TQ	9	–	–	–	–	–	23	–	–	10	22	75	42	25	100	58	45	–
M^5 (3-CF_3-PFOS)	IT	53	3	–	–	–	7	5	100	–	3	–	34	7	–	*	*	*	*
M^6 (4-CF_3-PFOS)	IT	55	3	–	–	–	100	–	16	–	–	39	5	19	–	*	*	*	*
M^{5-6}	TQ	5	–	–	–	–	5	–	28	–	–	42	6	68	15	100	7	15	90
M^7 (1-CF_3-PFOS)	IT	30	–	100	–	–	–	–	–	–	–	–	–	–	–	*	*	*	*
	TQ	–	–	–	–	–	–	–	–	10	–	–	21	–	26	–	14	100	–
M^{7-8} (1-CF_3 and 5-CF_3-PFOS)	IT	53	3	100	36	–	–	–	57	–	–	19	3	–	–	*	*	*	*
	TQ	–	–	–	–	–	–	–	–	–	–	33	5	12	–	100	–	13	16

(Continued)

TABLE 10.1 (CONTINUED)
Relative Abundance (3%) of Product Ions of PFOS Isomers in a Technical Product Obtained by Fragmentation of the Molecular Anions m/z 499 by IT MS/MS (40% CE) and TQ MS/MS (40% CE)

Isomer	Type	Mass (m/z)																
		499	430	419	380	369	330	319	280	269	261	230	219	180	169	130	119	99	80
M^9 (6-CF$_3$-PFOS)	IT	100	3	—	—	—	48	—	10	—	—	3	—	3	9	*	*	*	*
	TQ	22	—	—	—	—	28	—	29	—	—	89	6	34	100	60	—	15	48
L^{10} (L-PFOS)	IT	100	—	—	—	—	—	—	—	—	—	—	—	—	5	*	*	*	*
	TQ	26	—	—	—	—	—	—	5	—	—	25	—	9	14	5	21	100	77

Assignment	Proposed Structure	Product Ions
0-series	[C$_m$F$_{2m}$SO$_3$]$^-$, 1 ≤ m ≤ 7	130*, 180, 230, 280, 330, 380, 430
9-series	[C$_n$F$_{2n+1}$]$^-$, 1 ≤ n ≤ 7	69*, 119*, 169, 219, 269, 319, 369, 419
1-series	[C$_p$F$_{2p-1}$SO$_3$]$^-$	261
Other	[FSO$_3$]$^-$, [SO$_3$]$^-$	99*, 80*

Source: Adapted from Langlois I, Oehme M, *Rapid Commun Mass Spectrom*, 2006;20(5):844–50.

Note: Also included is the assignment of ions to different series. The relative uncertainty of the abundances is ca. 1%–2%. — = not detected; * = not detected due to the operational low mass cutoff of the IT instrument; F = first eluting isomers; M = middle eluting isomers; L = linear isomer.

PFAS Isomers

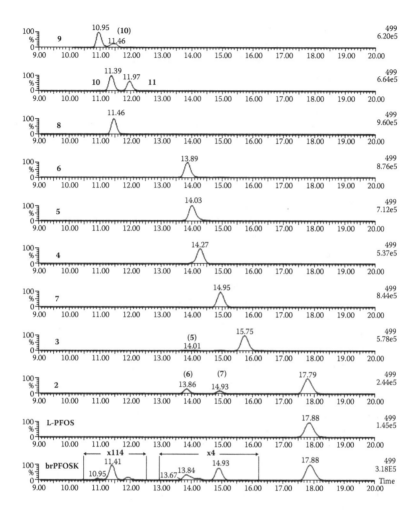

FIGURE 10.2 Chromatograms of individual PFOS isomers and branched K-PFOS obtained in SIM mode (m/z 499) on a Waters Acquity UPLC using a shield RP18 column at a concentration of 500 ng/mL. (Reprinted with permission from Riddell N, Arsenault G, Benskin JP, Chittim B, Martin JW, McAlees A, McCrindle R, *Environ Sci Technol*, 2009;43(20):7902–8.)

quantification using m/z 499 → 99 and m/z 499 → 80 transitions and summing the response of all isomers relative to the technical branched potassium perfluorooctane sulfonate (K-PFOS) standard; (2) quantification of linear and total branched isomers separately using m/z 499 → 99 and m/z 499 → 80 transitions relative to a L-PFOS and branched K-PFOS standard, respectively; and (3) quantification of the linear and each of the major perfluoromonomethyl PFOS isomers individually using separate calibration curves generated from the known isomer composition of branched K-PFOS (determined by ^{19}F-NMR) and isomer-specific transitions. The first major finding was that the chromatograms showed striking differences in the branched isomer profiles (linear/branched ratio, relative profile of the branched isomers) of the two human serum samples. This finding suggested that the two serum samples (A and B) had different exposure sources and highlighted the importance of isomer-specific analysis as a source-tracking tool. Depending on the sample and the particular MS/MS transition chosen for ΣPFOS analysis (i.e., 499 → 80 or 499 → 99), ΣPFOS concentrations may be over- or underestimated compared with the isomer-specific analysis. Specifically, the m/z 99 product ion produced a higher ΣPFOS concentration than m/z 80 for serum A, whereas in serum B the m/z 99 product ion produced a lower concentration relative to m/z 80. The quantification bias of the ΣPFOS method was attributed to the differences in the in-source fragmentation and MS/MS dissociation of isomers. For example, isomers 2, 9, 10, and 11 showed a substantially lower response factor than the linear isomer in the selected ion monitoring mode (m/z 499). Ultraperformance liquid chromatography (UPLC) MS/MS results showed that under MS/MS conditions optimized for L-PFOS, isomers 9 and 10 did not fragment to m/z 99, while isomer 2 did not fragment to m/z 80. Careful examination of the product ion spectra of branched isomers revealed that the preferred collision-induced dissociation pathways for some of the branched isomers can be very different from that of the linear isomer. This has significant implications since some of these isomers may possess unique biological properties. For instance, isomer 2 had been found to be more bioaccumulative than L-PFOS in rats, but would not be quantified when the m/z 80 product ion is used. Lastly, the decision to use pure linear or technical PFOS calibration standards was suggested to contribute to the interlaboratory discrepancies.

LC-MS/MS is the most commonly used method for PFAS analysis, but it tends to be a lengthy process (95–115 min) or is limited to single compounds (Benskin et al., 2012). Benskin et al. developed a rapid (<23 min) HPLC-MS/MS method for simultaneous characterization of 24 per- and polyfluoroalkyl compounds, including PFAS for which branched isomers had not been previously reported (Benskin et al., 2012). The method utilized a fused-core pentafluorophenylpropyl stationary phase based on prior success in resolving PFAS isomers using fluorinated stationary phases. Technical standards (containing branched and linear isomers) of PFHxS, PFOSA, PFOS, and PFOA were used in method development. Landfill leachate samples were used to determine the PFAS isomer profiles in real samples. PFAS isomer separation and quantification were accomplished by LC-MS/MS using a Dionex HPLC coupled to an API 5000Q TQ mass spectrometer operated under negative-ion, multiple-reaction monitoring mode. The most effective mobile phase for isomer separation and speed was found to be 100% methanol and 20 mM ammonium formate/20 mM formic acid in water. This new method achieved comparable chromatographic separation to, and was approximately four times faster (<23 min vs. 95 min) than, previous comprehensive isomer-specific HPLC-MS/MS methods. For PFAS in which isomeric mixtures were unavailable, a combination of multiple product ions and retention times within 4 min of the linear isomer peak was used for confirmation of branched isomers in leachate. For perfluorobutane sulfonate (PFBS), perfluorobutyric acid (PFBA), perfluoropropionic anhydride, and fluorotelomer acids, only linear isomers were observed. For perfluorohexanoic acid (PFHxA), perfluoroheptanoic acid (PFHpA), perfluorodecanoic acid (PFDA), and perfluoroundecanoic acid (PFUnDA), one branched isomer peak was observed per compound (in addition to linear), while for perfluorononanoate (PFNA), two branched isomer peaks were observed. This method provided a means not only of separating the major PFSA, PFCA, and FOSA isomers without compromising run times or target lists, but also of separating isomers of PFAS for which no technical standards of isomer mixture were available.

Isomer-specific gas chromatography–mass spectrometry (GC-MS) methods have also been developed to analyze PFCAs, PFSAs, or certain FOSAs. Chu et al. (2009) developed a method that incorporated cleanup by solid-phase extraction (SPE) weak anion exchange cartridges and in-port derivatization–GC/MS to identify and quantitatively determine L-PFOS and branched (monotrifluoromethyl and bistrifluoromethyl) isomers in PFOS technical product and in environmentally relevant biological samples. Tetrabutylammonium hydroxide was used for derivatization via an *in situ* pyrolytic alkylation reaction that occurred in the GC injector and generated butyl PFOS isomer derivatives. In addition to L-PFOS, 10 branched PFOS isomers were identified in the technical product. The environmental relevance of branched PFOS isomers, in addition to L-PFOS, was evidenced by the presence of six branched PFOS and L-PFOS in representative herring gull and double-crested cormorant egg, and polar bear liver and plasma samples from the Great Lakes and Arctic, respectively (Houde et al., 2011). For all PFOS isomers in the technical product and biota samples, the method demonstrated high sensitivity, with the limit of detection (LOD) ranging from 0.05 to 0.25 ng/mL, with the exception of L-PFOS, where the LOD was 1.46 ng/mL. For the biotic samples, the method detection limits (MDLs) were slightly higher than the LODs and ranged from 0.09 to 0.46 ng/g wet weight (w.w.) with the exception of L-PFOS (MDL = 6.87 ng/g w.w.).

High-throughput analysis procedures are highly desirable for large-scale epidemiological studies. This has been achieved by the introduction of automated PFAS analysis methods. These automated methods are predominantly based on on-line and column-switching extraction or automated on-line extraction. A high-throughput procedure that employs protein precipitation and extraction using 96-well plates together with LC-MS/MS was presented by Flaherty et al. (2005). This method was developed and validated for analysis of only PFOA and later used for analysis of PFOA and nine additional PFAS in an epidemiological study involving 69,030 human serum samples from the United States (Frisbee et al., 2009). Using similar technology (96-well plate coupled with column-switching HPLC-MS/MS), Salihovic et al. (2013) developed an automated column-switching UPLC-MS/MS method for the determination of PFCAs (C5, C6, C7, C8, C9, C10, C11, C12, and C13), PFSAs (C4, C6, C8, and C10), PFOSA, and five groups of structural PFOS isomers in human serum or plasma. The analytical procedure involves rapid protein precipitation using 96-well plates, followed by an automated sample cleanup using an on-line trap column to remove many potentially interfering sample components. The mobile phase gradient was also adjusted to allow for further separation onto the analytical column and subsequent mass detection of target analytes. The method was linear ($R^2 < 0.995$) at concentrations ranging from 0.01 to 60 ng/mL, with MDLs ranging between 0.01 and 0.17 ng/mL depending on the analyte. The developed method was precise, with repeatability ($n = 7$) and reproducibility ($n = 103$) coefficients of variation between 2% and 20% for most compounds, including PFOS (2% and 8%) and its structural isomers (2%–6% and 4%–8%). The method was in conformity with a standard reference material. The column-switching HPLC-MS/MS method has been successfully applied for the determination of perfluoroalkyl substances, including structural PFOS isomers in human plasma from an epidemiological study. More recent developments on rapid high-throughput analysis of PFAS in biological samples focus on incorporating rapid protein precipitation and on-line SPE with HPLC-MS/MS (Poothong et al., 2017; Rogatsky et al., 2017).

^{19}F-NMR spectroscopy is often used in combination with HPLC-MS/MS for structure analysis and confirmation of the isolated isomer elutes. It has also been used independently to determine the isomer composition in PFAS. The advantage of NMR spectroscopy over LC-MS is that it allows the investigation of differences in the composition of a large number of samples without *a priori* knowledge of relevant NMR signals. Using this technique, Vyas et al. (2007) identified the characteristic signals in PFOS, namely, the signal of terminal CF_3 groups of a linear $CF_3CF_2CF_2$ chain, the CF_3 and CF groups of isopropyl groups (i.e., terminally branched perfluorocarbon chains with the general structure $(CF_3)_2CFCF_2$-), the signals of CF_3 and CF groups of internally branched isomers with the general structure $-CF_2CF(CF_3)CF_2-$, and the CF_3 groups of a terminal perfluorinated t-butyl group $(CF_3)_3CCF_2$-. Their results showed that PFOS contains on average 73% of the linear, 10% of the isopropyl-branched, and 18% of internally branched PFOS isomers and 0.3% of the tertiary butyl isomer, which is in good agreement with the isomer composition of ECF PFOS. Other findings included

1. Compared with PFOS, shorter-chain PFAS (e.g., PFBS derivatives) are relatively pure and contain only the linear isomer.
2. PFOS fluorides typically contain a significantly higher amount of branched isomers than PFOS fluoride–derived products (i.e., PFOS potassium salts). The latter was attributed to the observation that L-PFOS fluoride is not easily separated from its isomers, whereas L-PFOS derivatives, especially PFOS amides, apparently can be more easily separated from the branched isomer.
3. PFOS potassium salts can vary in their isomer composition depending on the source. This has important implications for toxicity studies since many of them used PFOS potassium salt.

Many PFOS isomers, including the environmentally relevant monomethyl-branched isomers 1m-, 3m-, 4m-, and 5m-PFOS, have chiral centers, each of which corresponds to two stereoisomers (enantiomers). Enantiomeric analysis measures the enantiomer fraction (EF) (i.e., the relative abundance of the two enantiomers). Unless production of a specific enantiomer is sought, the relative abundance of each enantiomer is equal in commercially produced chemicals, and the chiral signature is said to be racemic (EF = 0.5). Consequently, observations of chiral signatures that deviate significantly from racemic in environmental or biological matrices are strong evidence of biodegradation or metabolism of a precursor material, and hence indicative of indirect exposure.

Enantiomer analysis of PFAS, however, used to present a significant challenge because PFOS and its precursors do not contain the usual functional groups that are known to interact with enantioselectors on chiral stationary phases (CSPs) or chiral ion pair additives. In addition, because all the metabolic reactions of any PFOS precursor occur on the $SO_2N(R)(R)$ moiety (Xu et al., 2004), a chiral center on the perfluorooctyl chain would not necessarily influence an enzymatic reaction rate enough to significantly influence the resulting EF of the substrate or its products. Wang et al. (2009) developed an HPLC-MS/MS technique for enantiomer separation and demonstrated enantioselective biotransformation of a PFOS precursor. Using $C_6F_{13}C^*F(CF_3)SO_2N(H)CH_2(C_6H_4)OCH_3$ (where * indicates the chiral carbon center) as the model 1m-PFOS precursor, they developed an HPLC-MS/MS technique that was able to achieve near-baseline enantioseparation of the two enantiomers of this precursor molecule. Key to the success is the careful selection of an effective CSP that recognizes the chirality of the chemical. In this case, it was found that only a CSP based on cellulose tris (3,5-dicholorophenylcarbamate; trade name Chiralpack IC) was suitable. Several chromatographic parameters were then optimized on Chiralpak IC: operation mode (normal phase or reversed phase), type and concentration of organic modifier and acid additives, and separation temperature. Enantiomer separation was not achieved in normal-phase mode but was achieved under reversed-phase conditions using methanol as the organic modifier (but not isopropanol or acetronitrile). An acid additive was also essential (without it, enantioseparation could not be accomplished), and 0.1% acetic acid in the mobile phase was found to be optimal, resulting in near-baseline separation of the enantiomers. Although isocratic elution conditions are most commonly used in chiral HPLC separations, the retention time was relatively long (~85 min) and the peaks were quite wide (~3 min peak width at half height). Addition of a weak gradient elution profile resulted in 25% less analysis time, a negligible loss of resolution, and significantly greater sensitivity.

In summary, tremendous progress had been made in isomer characterization and accurate profiling in various real samples (water, human serum, etc.). As discussed below, isomeric profiling by HPLC-MS/MS has been increasingly applied to analyzing environmentally relevant samples and serves as a tool to assess the contribution of different manufacturer processes. Isomer-specific analysis not only can reduce errors in the quantification of structural PFAS isomers (Benskin et al., 2010a), but also has been shown to be important for source identification, accumulation, and toxicity analysis (Benskin et al., 2010b; Miralles-Marco and Harrad, 2015). As it is getting more widely utilized and goes beyond C8 PFAS, it should be noted that different isomers may fragment differently from the traditional linear isomers; therefore, monitoring the traditional MS/MS transitions may miss or underestimate the quantity of some isomers. Accurate quantitative analysis of isomers depends on well-characterized technical standards of PFAS isomers and an optimized technique for less common PFAS and their isomers, especially the shorter-chain PFAS. In addition, rapid high-throughput techniques, such as LC-MS coupled with a multiwell plate technique, are highly desirable to large-scale epidemiological biomonitoring studies. Enantiomer analysis technologies have been developed and are expected to serve as a powerful tool for indirect exposure source tracking.

10.3 ISOMER-SPECIFIC ANALYSIS FOR PFAS SOURCE TRACKING IN THE ENVIRONMENT

As mentioned earlier, a major application of PFAS isomer-specific analysis is to quantitatively assess the contributions from ECF (mostly historical production) and telomerization (ongoing production) manufacturing processes. In addition, enantiomer analysis of isomers can also serve as a powerful tool to identify exposure from direct (PFAS and their derivatives) versus indirect (precursor materials) exposure sources. The following section consists first of a brief review of the isomer-specific analysis in various environmental samples, followed by the relatively more recent developments of enantiomer analysis.

10.3.1 Isomer Analysis as a Source-Tracking Tool in Environmental Samples

Benskin et al. (2010b) performed isomer profiling of PFOA in water samples from North America, Asia, and Europe. Their results showed, with the exception of three sites in Japan, that more than 80% of total PFOA was from ECF, with the rest attributable to telomerization processes. PFOA profiles in samples from China suggested exclusive contribution from local ECF production. In Tokyo Bay, ECF, linear telomer, and isopropyl telomer contributed 33%, 53%, and 14% of the total PFOA, respectively. PFOS isomer profiles were enriched in branched content (i.e., greater than 50% branched) in the Mississippi River, but in all other locations were similar or only slightly enriched in branched content relative to historical ECF PFOS. Overall, these data suggested that, with the exception of Tokyo Bay, ECF manufacturing has contributed to the bulk of contamination around these source regions, but other sources are also significant.

Chen X. et al. (2015) studied the isomer-specific partitioning behaviors of isomers of PFOA, PFOS, and PFOSA in the water-dissolved phase, suspended particulate matters (PMs), and sediments in Liao River Basin and Taihu Lake, China. Their results showed that the linear isomers displayed higher partition coefficients on particulate phases than the branched ones. Using isomer profiling, they were able to assess the relative contributions of various industrial origins for PFOA. In Liao River, when suspended PM was included in the water samples, there were contributions of PFOA from ECF production (~55%), linear telomer (~41%), and isopropyl telomer (~4%) sources. These results differ from the results based on the dissolved phase alone, which indicated a greater ECF contributing source (~70%) and a lower contribution from linear telomer sources (~26%). In Taihu Lake, the isomer profile of PFOA was influenced mainly by ECF (~88%) and partially by linear telomer sources (~12%). Jin et al. (2015) investigated the isomer profiles of PFAS in water and soil samples surrounding a Chinese fluorochemical manufacturing park. They have found that shorter-chain (C4–C6) PFAS were the predominant PFAS released at this site. Branched isomers of PFBS were also detected, accounting for 15%–27% of total PFBS found in water. More numerous branched isomers were detected for the longer-chain C9–C13 PFCAs (e.g., C_{12} PFCA had 16 branched isomers), including high proportions of one major branched isomer (likely isopropyl), possibly as impurities from isopropyl-PFOA manufacturing.

Yu et al. (2015) utilized GC-MS to study the distribution of seven PFOS isomers (linear and six branched monomethyl isomers, named 1m- and 6m-PFOS) in drinking water in China's Jiangsu Province, an area exposed to high emissions of PFOS. Compared with the 30% proportion of branched PFOS in technical PFOS, the levels of branched PFOS in drinking water increased to 31.8%–44.6% of total PFOS. Using a modified binding equilibrium model to assess the risk of thyroid hormone perturbation conferred by PFOS isomers, a health risk assessment of PFOS for in drinking water was performed. The risk quotients (RQs) of individual PFOS isomers indicated that L-PFOS contributed most to the risk among all the target PFOS isomers (83.0%–90.2% of the total PFOS RQ), and that risk from 6m-PFOS (5.2%–11.9% of the total PFOS RQ) was higher than that from other branched PFOS isomers. It was also found that the risks associated with PFOS in drinking water would be overestimated by 10.0%–91.7% if contributions from individual PFOS isomers were not considered. The results highlighted that the PFOS isomer profile and the toxicity of individual PFOS isomers were important factors in health risk assessment of PFOS and should be considered in the future risk assessments.

Shan et al. (2015) performed an HPLC-MS/MS isomer analysis in snow samples. In this study, various PFAS and the isomers of PFOA and PFOS were analyzed in fresh snow samples collected from 19 cities in northern China in 2013. The levels of total PFAS in the snow samples were 33.5–229 ng/L, suggesting heavy atmospheric pollution of PFAS in northern China. PFOA (9.08–107 ng/L), PFOS (3.52–54.3 ng/L), PFHpA (3.66–44.8 ng/L), and PFHxA (3.21–23.6 ng/L) were predominant, with a summed contribution of 82% to the total PFAS. The PM-associated PFAS contributed 21.5%–56.2% to the total PFAS in the snow, suggesting PM is vital for the transport and deposition of airborne PFAS. Partitioning of PFAS between PM and dissolved phases was dependent on the carbon chain length and end functional groups. Isomer profiles of PFOA and PFOS in the snow were in agreement with the signature of the historical 3M ECF products, suggesting that the ECF products were still produced and used in China. Further source analysis showed that the airborne PFAS in urban areas were mainly due to direct release rather than degradation of their precursors.

10.3.2 Enantiomer Analysis as a Tracking Tool for Direct versus Indirect Exposure

The presence of PFAS in the environment has been attributed to two major sources: direct and indirect (Prevedouros et al., 2006; Armitage et al., 2009; Paul et al., 2009). In the case of PFOS, direct sources are derived from the manufacture and application of PFOS and perfluorooctanesulfonyl fluoride (POSF) (Paul et al., 2009). By comparison, indirect sources are a consequence of chemical reaction impurities or the breakdown of so-called precursors, such as 2-(N-methylperfluorooctane sulfonamido) ethanol (N-Me-FOSE) and 2-(N-ethylperfluorooctane sulfonamido) ethanol (N-Et-FOSE). It has been estimated that 85% of indirect emissions occur via release from consumer products during use and disposal (3M, 2000). Since commercially produced PFAS are normally racemic (EF = 0.5), nonracemic chiral signatures are strong evidence of the biodegradation of a precursor material. Therefore, enantiomeric analyses of PFOS isomers have been developed to characterize the extent to which PFOS precursors (PreFOS) play a role in human or environmental exposure.

Using $C_6F_{13}C^*F(CF_3)SO_2N(H)CH_2(C_6H_4)OCH_3$ (where * indicates the chiral carbon center) as the model PFOS precursor, Wang et al. (2009) demonstrated for the first time that biotransformation of PFOS precursors could be enantioselective. They monitored the biodegradation rates of the enantiomers by human liver microsomes. Enantiomeric profiles of the precursor molecules at different time points after incubating with microsomes showed that both enantiomers could be significantly biodegraded within 50 min. A significant difference in degradation rates existed between the two enantiomers, with a rate constant of 6.5 ($\pm$ 0.3) $\times$ 10^{-2} min^{-1} and 5.2 ($\pm$ 0.3) $\times$ 10^{-2} min^{-1}, respectively. Accordingly, the half-lives were 10.6 and 13.3 min, respectively. This is the first confirmed indication that biotransformation of PFOS precursors may be enantioselective and the technique can be used as a source-tracking tool of PFOS in biological samples.

A similar technique was developed for α-perfluoromethyl branched PFOS (1m-PFOS, typically 2%–3% of total PFOS) and applied to EF analysis in biological samples (Wang et al., 2011). In blood and tissues of rodents exposed subchronically to electrochemical PFOS, 1m-PFOS was racemic (EF = 0.485–0.511) and no evidence for enantioselective excretion was found. 1m-PFOS in serum of pregnant women, from Edmonton, was significantly nonracemic, with a mean EF ($\pm$ standard deviation) of 0.432 $\pm$ 0.009, similar to a pooled North American serum. In a highly exposed Edmonton family (mother, father, and five children) living in a house where Scotchgard had been applied repeatedly to carpet and upholstery, EFs ranged from 0.35 to 0.43, significantly more nonracemic than in pregnant women. Semiquantitative estimates of percent serum 1m-PFOS coming from 1m-PreFOS biotransformation in both subpopulations were in reasonable agreement with model predictions of human exposure to PFOS from PreFOS. The data were overall suggestive that the measured nonracemic EFs were influenced by the relative extent of exposure to PreFOS. A similar enantioselective analysis was conducted on samples collected from the Lake Ontario ecosystem, including water, sediment, fish, and invertebrates (Asher et al., 2012). Concentrations of PFOS and PFOSA, PFOS isomer profiles, and EFs of 1m-PFOS were determined. Concentrations of PFOS and PFOSA were highest in slimy sculpin and *Diporeia*, and concentrations of the two compounds were often correlated. 1m-PFOS was racemic in sediment, water, sculpin, and rainbow smelt, but nonracemic in the top predator, lake trout, and all invertebrate species. Furthermore, EFs were correlated with the relative concentrations of PFOS and PFOSA in invertebrates. Overall, these empirical observations confirm previous suggestions that PFOS precursors contribute to PFOS in the food web, likely via sediment. Implications are that future PFOS exposures in this ecosystem will be influenced by an *in situ* source, and that the apparent environmental behavior of PFOS (e.g., bioaccumulation potential) can be confounded by precursors, adding complication to remedial and risk management decisions.

Naile et al. (2016) developed a method to analyze multiple chiral and achiral isomers of PFOA. By using tandem-column GC-MS with negative chemical ionization and selected-ion monitoring, they were able to observe the occurrence of nine chiral and achiral isomers of PFOA in both standards and various environmental samples (PFAS- and PFAS precursor–contaminated soils and river sediments). The enantiomers of two of the chiral isomers, P3 and P4, were chromatographically separated for the first time and observed in several environmental sediment and plant samples. Analysis of various environmental samples revealed nonracemic environmental occurrences of chiral PFOA isomers (such as P3 and P4), with the branched isomers being selectively enriched (relative to technical grade PFOA) in at least one river sediment sample. This enantioselective enrichment was attributed to several possible reasons: (1) unequal biotransformation of an ECF PFOA precursor, (2) differential degradation rates between PFOA enantiomers, and (3) differential sorption affinities among the isomers for natural chirally selective surfaces, such as minerals like quartz and natural organic matter. The same study also looked at the mass chromatogram of a standard mix of C7–C12 PFAS made from neat chemicals and identified the P3 isomer of PFOA, a C9 chiral isomer, and likely branched isomers of other homologs. Occurrence of the C9 isomers was also detected in the environmental samples. Based on these findings, the authors suggested that these isomers occur routinely in ECF-derived PFAS.

PFOS enantiomer profiles in serum have also been proposed as a biomarker of human exposure to PreFOS (Wang et al., 2009). This is because

1. Many branched PFOS and PreFOS isomers are chiral (Rayne et al., 2008) and were manufactured as racemic mixtures (i.e., 50:50) of two nonsuperimposable enantiomers.
2. *In vitro* biotransformation rates of 1m-PreFOS (i.e., an α-perfluoromethyl branched PreFOS isomer) to 1m-PFOS differ for the two enantiomers (Wang et al., 2009).
3. From rat studies, there is currently no evidence for differential absorption distribution or excretion of 1m-PFOS enantiomers (Wang et al., 2011).

Thus, the detection of significantly nonracemic 1m-PFOS EFs in humans (i.e., when the ratio of the two enantiomers is significantly different from 1:1) may point to the contribution of PreFOS, because direct PFOS exposure alone should always result in a racemic 1m-PFOS enantiomer signal (based on 1 and 3).

In summary, isomer-specific analyses have become a powerful tool for PFAS source tracking in the environment. The technique is now well established, and various types of real samples, including water, soil, sediment, and snow, can be analyzed. These studies have shown that ECF products contributed predominantly to PFOS contamination around the world, although regional differences are also very clear. Enantiomer analysis techniques have been developed and proved useful in differentiating direct versus indirect exposure through PFAS precursors.

10.4 ISOMER-SPECIFIC BIOTRANSFORMATION AND BIOACCUMULATION

10.4.1 BIOACCUMULATION IN FISH AND WILDLIFE

The PFOS isomer profile in fish and wildlife is typically enriched in the linear isomer (Houde et al., 2011). De Silva and Mabury (2004) determined the PFCA isomer profiles in liver samples of 15 polar bears from the Canadian Arctic and eastern Greenland by GC-MS. The PFOA isomer pattern in Greenland polar bear samples showed a variety of branched isomers, while only the linear PFOA isomer was determined in Canadian samples. Samples of both locations had primarily (>99%) linear isomers of PFNA and perfluorotridecanoate. Branched isomers of perfluorodecanoate, perfluoroundecanoate, and perfluorododecanoate were detected in the polar bear samples. Unlike the PFOA isomer signature, only a single branched isomer peak on the chromatograms was observed for these longer-chain PFCAs. The presence of branched isomers suggests some contribution from ECF sources. However, in comparison with the amount of branched isomers in the ECF PFOA standard, such minor percentages of branched PFCAs may suggest additional input from an exclusively linear isomer source.

Houde et al. (2008) studied the fractionation and bioaccumulation of PFOS isomers in water, sediment, and biota (invertebrates, forage fish, slimy sculpin, alewife, and lake trout) collected from Lake Ontario using HPLC-MS/MS. In addition to the L-PFOS, a total of six isomers were detected in all three types of samples. L-PFOS predominated in water with mean proportions ranging from 43% to 56% (compared with 77% for the standard) depending on the ion transition monitored. By contrast, L-PFOS represented a much higher proportion (88%) of total PFOS in all biota and the sediment samples (81%–89%) than the proportion reported in technical PFOS used (77%). This predominance of L-PFOS in the biota suggested a reduced uptake of branched isomers, more rapid elimination of the branched isomers, and/or a selective retention of the L-PFOS. The similarity in the PFOS isomer profile between sediment and biota also suggests that sediment is a source of PFOS contamination in this Lake Ontario food web and/or that the partitioning into organic material in general is favored for the linear isomer. The bioaccumulation factor between lake trout and water for L-PFOS and branched isomers differed remarkably. The tropic magnification of L-PFOS was also found to be greater than that for branched isomers (monomethyl- and dimethyl-PFOS). The differences in environmental behavior of the isomers imply that as fresh sources decline, the ultimate fate of PFOS will be for only the linear isomer to remain, at least in the biosphere. Similarly, in clam, shrimp, and fish collected in Georgia in the United States, PFOS isomer profiles in soft tissue, muscle, and liver ranged from 77% to 89% L-PFOS (Senthil Kumar et al., 2009). Powley et al. (2008) reported that PFOS isomer profiles in seals from the Canadian Arctic were composed of 96% linear isomer. Interestingly, the same study also observed 50% L-PFOS of the total PFOS in Arctic cod, lower than that of the technical mixture (around 70%).

Chu and Letcher (2009) examined PFOS isomer profiles (10 branched and 1 linear) in herring gull egg and polar bear liver and plasma samples. Again, L-PFOS was enriched in >90% of the ΣPFOS in bird egg and bear liver samples and >80% for the bear plasma samples, in comparison with the <80% in commercial PFOS product. Gennink and Letcher (2010) analyzed the PFOS isomer profiles in individual herring gull eggs (n = 13 per site) collected from 15 colonies across the Laurentian Great Lakes of North America. L-PFOS consistently dominated the isomer pattern in all eggs, comprising between 95.0% and 98.3% of the ΣPFOS concentration. L-PFOS was highly enriched in the gull eggs as the concentration ratios of Σ-branched PFOS to L-PFOS isomer were constant (overall average 0.038 [± 0.001]) and much lower than the technical PFOS (range 0.27–0.54). The highest proportions of L-PFOS were generally observed in the eggs from the lower lake (Erie and Ontario) colonies. All six mono(trifluoromethyl) branched isomers were detected in the eggs from all the colonies. Comparable to technical-grade PFOS (T-PFOS), the percentage of the mono(trifluoromethyl) isomer to ΣPFOS concentration decreased as the branch substitution was located closer to the sulfonate group. Di(trifluoromethyl) isomers were detected in >60% of the individual eggs per site. Relative to T-PFOS, and independent of colonial location, the high and consistent enrichment of L-PFOS in gull eggs is likely a function of several processes, including PFOS or precursor sources, and isomer-specific PFOS or precursor exposure, accumulation, biotransformation, retention, and/or elimination.

Wang et al. (2016a) examined the isomer-specific distribution and bioaccumulation potential of perfluoroethylenecyclohexanesulfonate (PFECHS), a cyclic PFAS anticorrosive agent used at airports and an additional source of PFAS. Surface water, sediments, and fish samples (carp) were collected in the surrounding area of Beijing International Airport and examined for the presence of PFECHS isomers. Five distinct peaks attributed to different isomers of PFECHS and perfluoropropylcyclopentanesulfonate (PFPCPeS) were identified in environmental samples. The sum of PFECHS and PFPCPeS isomers displayed logarithmically decreasing spatial trends in water (1.04–324 ng/L) and sediment samples (method limit of quantification (MLQ) 2.23 ng/g dw) with increasing distance from Beijing International Airport. PFECHS and PFPCPeS displayed the highest accumulation in liver, kidney, blood, and bladder, and average whole-body bioaccumulation factors (log $BAF_{whole\ body}$) were estimated to be 2.7 and 1.9, respectively. Isomer-specific differences in the tissue/blood distribution ratios and $BAF_{whole\ body}$ indicated that the ring structure and position of the sulfonic acid group affected the bioaccumulation potential of cyclic perfluoroalkyl acids. Based on the substantially lower sorption affinity for soil and sediment of cyclic compared with linear and branched PFAS, they will be transported more efficiently to groundwater and potentially lead to human exposure via drinking water. Lower hydrophobicity and steric effects were proposed to be the reasons for the relatively low bioaccumulation factors (BAFs) and the isomer-specific enrichments observed in fish tissues.

10.4.2 Isomer-Specific Biotransformation of PFAS Precursors

As discussed above, even though the overall trend is that linear isomers are preferentially enriched in fish and wildlife, there are a number of biomonitoring studies reporting that the percentage of branched PFOS isomers was much higher (up to 52%) than that in commercial ECF PFOS (Kärrman et al., 2007; Powley et al., 2008; Beesoon et al., 2011; Zhang et al., 2013; Liu et al., 2015). These observations cannot be explained based on what is known about the pharmacokinetics of branched PFOS isomers and are suspected to be the result of biotransformation of PreFOS. Understanding the biotransformation of precursors in the environment or inside living organisms can therefore assist in the interpretation of isomeric profiles (e.g., significantly enriched branched isomers) in biomonitoring studies.

Benskin et al. (2009) examined the relative isomer-specific biotransformation rates of a model PFOS precursor, N-Et-FOSA, with human microsomes and recombinant human cytochrome P450s (CYPs) 2C9 and 2C19. Using solid-phase microextraction–gas chromatography–electron capture detection to monitor N-Et-FOSA disappearance, and LC-MS to monitor product formation, they have shown that, in general, human microsomes and CYP isozymes transformed the branched isomers more rapidly than linear N-Et-FOSA. Among branched isomers, perfluoroalkyl branching geometry significantly influenced the rate of biotransformation. As a result, PFOS isomer patterns in biota exposed predominantly to precursors could be much different than expected from the isomer pattern of the precursor. While these data are suggestive that the relatively high abundance of branched PFOS isomers present in some humans or wildlife may be explained by substantial exposure to PFOS precursors, *in vivo* studies with other relevant PFOS precursors are warranted to validate this as a biomarker of exposure source.

Ross et al. (2012) evaluated this hypothesis by examining the isomer specific fate of PFOSA, a known PFOS precursor, in male Sprague-Dawley rats exposed to commercial PFOSA via food for 77 days (83.0 ± 20.4 ng/kg/day), followed by 27 days of depuration. Elimination half-lives of the two major branched PFOSA isomers (2.5 ± 1.0 days and 3.7 ± 1.2 days) were quicker than those for linear PFOSA (5.9 ± 4.6 days), resulting in a depletion of branched PFOSA isomers in blood and tissues relative to the dose. A corresponding increase in the total branched isomer content of PFOS, the ultimate metabolite, in rat serum was not observed. However, a significant enrichment of 5m-PFOS and a significant depletion of 1m-PFOS were observed, relative to authentic electrochemical PFOS. Although the data cannot be directly extrapolated to humans due to known differences in the toxicokinetics of PFOS in rodents and humans, it confirmed that *in vivo* exposure to commercially relevant PFOS precursors could result in a distinct PFOS isomer profile that may be useful as a biomarker of exposure source.

Peng et al. (2014) studied the biotransformation of perfluorooctane sulfonamide ethanol (FOSE)–based phosphate diester (diSPAP) (a potential PFOS precursor) in the common Japanese rice fish, medaka, after exposure in water for 10 days, followed by 10 days of depuration. PFOS, together with PFOSA, N-Et-FOSA, 2-(perfluorooctane sulfonamido) acetic acid (FOSAA), N-ethylperfluorooctane sulfonamido acetic acid (N-Et-FOSAA), FOSE, and N-Et-FOSE, was detected in medaka exposed to diSPAP, indicating the potential for biotransformation of diSPAP to PFOS via multiple intermediates. Branched isomers of diSPAP (B-diSPAP) were preferentially enriched in medaka exposed to diSPAP, with the proportion of branched isomers ranging from 0.56 to 0.80 and significantly greater than that in the water to which the medaka were exposed (0.36) ($p < 0.001$). This enrichment was due primarily to preferential uptake of branched diSPAP. Due to preferential metabolism of branched isomers, FOSAA and PFOSA exhibited greater fractions of branched isomers (BF) (>0.5) than those of N-Et-FOSA, N-Et-FOSAA, and N-Et-FOSE (<0.2). Such preferential metabolism of branched isomers along the primary pathway of metabolism and preferential accumulation of branched diSPAP led to enrichment of branched PFOS in medaka. Enrichment of branched PFOS was greater for 3-, 4-, and 5-perfluoromethyl PFOS (P3MPFOS, P4MPFOS, and P5MPFOS), for which values of BF were 0.58 ± 0.07, 0.62 ± 0.06, and 0.61 ± 0.05 (day 6), respectively; these values are 5.8-, 7.8-, and 6.4-fold greater than those of technical PFOS.

Chen M. et al. (2015) investigated the isomeric degradation of PFOSA by *in vivo* and *in vitro* tests using common carp as an animal model. In the *in vivo* tests, branched isomers of PFOSA and PFOS were eliminated faster than the corresponding linear (n-) isomers, leading to enrichment of linear PFOSA in the fish. In contrast, branched PFOS was enriched in the fish, suggesting that branched PFOSA isomers were preferentially metabolized to branched PFOS over linear PFOSA. This was confirmed by the *in vitro* test. The exception was 1m-PFOSA, which could be the most difficult to metabolize due to its α-branched structure, resulting in the deficiency of 1m-PFOS in the fish. The *in vitro* tests indicated that the metabolism mainly took place in the fish liver instead of its kidney, and it was mainly a Phase I reaction. These results were suggested to provide a plausible explanation for the enriched branched PFOS isomer profiles found in some wildlife studies, even though branched PFOS isomers had a higher clearance rate than the linear isomers.

An *in silico* study by Fu et al. (2015) has unveiled that in the enzymatic environment, the PFOS precursor, N-Et-PFOSA, is hydroxylated feasibly (reaction energy barriers $\beta E = 11.4-14.5$ kcal/mol) with an H atom transfer (HAT) from the ethyl Cα to Compound I. The HAT-derived Cα radical then barrierlessly combines with the OH radical to produce a ferric-ethanolamine intermediate. Subsequently, the ethanolamine intermediate decomposes via N-dealkylation to PFOSA and acetaldehyde products nonenzymatically with the assistance of water molecules. The rate-limiting O addition ($\beta E = 21.2-34.0$ kcal/mol) of Compound I to PFOSA initiated a novel deamination pathway that comprises O−S bond formation and S−N bond cleavage. The resulting hydroxylamine is then hydrolyzed to PFOS. In addition, the results reveal that both the N-dealkylation and deamination pathways are isomeric specific, consistent with the above-mentioned experimental observations.

In summary, bioaccumulation studies generally show enriched linear isomer in fish and wildlife, suggesting that branched isomers are either selectively degraded, metabolized, or less accumulated in animal tissues, while linear PFAS are selectively enriched via preferential bioconcentration, uptake, and accumulation in their respective food webs. Despite the overall trend of linear isomer enrichment in biota, there are also reports of enriched branched isomers in both wildlife and human samples. These observations are attributed to the isomer-specific biotransformation of PFAS precursors, the indirect sources of PFAS. *In vitro* studies have shown that branched isomers of PFOS precursors are biotransformed by human microsomes and CYP isozymes more rapidly than the linear isomer. Animal studies have confirmed that branched isomers of a PFOS precursor had a much shorter elimination half-life than the linear isomer. The study of biotransformation of PFAS precursors is expected to become increasingly more important as some temporal trend studies have shown that the percentage of branched isomers in the Swedish population has increased from 32% to 45% between 1996 and 2010 (Liu et al., 2015).

10.5 PFAS ISOMER PATTERN IN HUMANS

10.5.1 Isomer Pattern in Human Serum, Blood, Plasma, and Other Matrices

The first report of the presence of PFOS, PFOA, and other PFAS in samples of human blood purchased from biological supply companies emerged in 2001 (Hansen et al., 2001). They assumed that the "shoulders" of the HPLC signals of PFOS and PFOA in human sera were isomers, but no identification of the isomers was performed. Several branched PFOA isomers were detected in human blood and polar bear liver applying GC-MS (De Silva and Mabury, 2004, 2006). Four branched isomers out of the total of eight isomers found in a PFOA standard were detected in the human samples. Since then, as a considerable database concerning human exposure to PFAS has emerged, so has the number study of isomer-specific analysis in humans (Table 10.2).

Kärrman et al. (2007) analyzed PFOS isomer patterns in human serum and plasma samples from Sweden, the United Kingdom, and Australia by HPLC-MS/MS. They found similar isomer patterns typical for the ECF process in all samples. The L-PFOS was the major isomer found (58%–70%), followed by the monosubstituted PFOS isomers 1m-/6m-PFOS (18%–22%) and 3m-/4m-/5m-PFOS (13%–18%). Di-substituted isomers were also detected. The percentage of L-PFOS found in the serum and plasma samples was slightly lower than that in a standard PFOS product (76%–79%). The observed difference in the isomer pattern of human serum or plasma differed from the technical standard and may suggest a preferential elimination of L-PFOS and/or a preferential bioaccumulation of the branched PFOS isomers. Such discrimination can occur by biotic or abiotic degradation or by human uptake of ECF-based precursors of PFOS, the biotransformation of which has been shown to be preferential to branched isomers. It has been shown that the chain length of PFAS influences excretion (Kudo et al., 2001; Ohmori et al., 2003), bioaccumulation (Martin et al., 2003a, 2003b), and toxicity (Ohmori et al., 2003). Therefore, it is not unlikely that isomers also behave differently in the environment. Significantly higher content of L-PFOS (68%) in Swedish samples than in Australian and UK samples (59%) was also found, which may suggest differences in exposure sources for humans. Rylander et al. (2009) investigated the possibility of using dietary intake, age, gender, and body mass index as predictors of PFAS body burden in a study population from northern Norway (44 women and 16 men). The participants donated blood for the study and answered a detailed questionnaire about diet and lifestyle. PFOS (29 ng/mL), PFOA (3.9 ng/mL), PFHxS (0.5 ng/mL), PFNA (0.8 ng/mL), and perfluoroheptane sulfonate (PFHpS) (1.1 ng/mL) were detected in more than 95% of all samples. Fruits and vegetables were found to significantly reduce the concentrations of PFOS and PFHpS, whereas fatty fish to a smaller extent

TABLE 10.2
Linear Isomer Composition Profiles and Enantiomer Fractions of PFOS and Its Precursors in Various Matrices

Authors	Year	Country/Region	Study	Matrix	Sample Size (n)	Analytes
Benskin et al.	2007	Canada	Human	Serum	14	PFOS ($\approx$80% linear)
Kärrman et al.	2007	Sweden	Human	Serum/blood	17	PFOS (68% linear)
		United Kingdom			13	PFOS (59% linear)
		Australia			40	PFOS (59% linear)
Houde et al.	2008	Canada	Niagara/lake	Fish	22	PFOS (88%–93% linear)
				Water	NR	PFOS (43%–56% linear)
Haug et al.	2009	Norway	Human	Serum	57	PFOS (53%–78% linear)
Sharpe et al.	2010	Canada	–	Fish	NR	PFOS (>70% linear)
Beesoon et al.	2011	Canada	Human	Dust	18	PFOS ($\approx$70% linear)
				Serum	20	PFOS ($\approx$64% linear)
				Cord serum	20	PFOS ($\approx$54% linear)
Wang et al.	2011	Canada	Animals	Rats	3	1m-PFOS (EF PFOS $\approx$ 0.5)
			Human	Serum	8	1m-PFOS (EF PFOS $\approx$ 0.43)
			Human	Serum	7	1m-PFOS (EF PFOS = 0.35–0.43)
Asher et al.	2012	Canada	Lake	Aquatic species	67	PFOSA ($\approx$57% linear), PFOS (>90% linear)
				Water	2	PFOS (70% linear)
				Sediment	3	PFOS (>90% linear)
Ross et al.	2012	Canada	Animals	Blood	8	PFOSA ($\approx$78% linear)
				Blood	8	PFOS ($\approx$77% linear)
				Heart	8	PFOSA ($\approx$86% linear)
				Fat	8	PFOSA ($\approx$86% linear)
Zhang et al.	2013	China	Human	Serum	129	PFOS (48% linear)
Salihovic et al.	2015	Sweden	Human	Plasma	398	PFOS ($\approx$64% linear)
Gao et al.	2015	China	Human	Serum	36	PFOS ($\approx$63.3% linear), PFOA ($\approx$91.1% linear), PFOHxS ($\approx$92.7% linear)
Zhang et al.	2017	China	Human	Cord serum	321	PFOS ($\approx$75.16% linear), PFOA ($\approx$98.69% linear)

Source: Adapted and updated from Miralles-Marco A, Harrad S, *Environment International*, 2015; 77:148–59.

significantly increased the levels of the same compounds. Men had significantly higher concentrations of PFOS, PFOA, PFHxS, and PFHpS than women. There were significant differences in PFOS isomer pattern between genders, with women having the largest proportion of L-PFOS. PFOS, PFHxS, and PFHpS concentrations also increased with age.

Zhang et al. (2013) performed isomer analysis of PFOS, PFOA, and other PFAS in human serum from China. A total of 129 serum samples were collected in two typical cities in North China and analyzed by HPLC-MS/MS. Among all samples, total PFOS (ΣPFOS, mean 33.3 ng/mL) was the predominant PFAS, followed by PFHxS (2.95 ng/mL), total PFOA (ΣPFOA, 2.38 ng/mL), and PFNA (0.51 ng/mL). The level of ΣPFOS was higher than that in people from North America in recent years. The mean concentrations of ΣPFAS in the participants living in urban Shijiazhuang (59.0 ng/mL) and urban Handan (35.6 ng/mL) were significantly higher ($p < 0.001$ and $p = 0.041$, respectively) than those living in the rural district of Shijiazhuang (24.3 ng/mL). The young female subpopulation had the lowest ΣPFAS concentrations compared with older females and all males. On average, the proportion of L-PFOS was only 48.1% of ΣPFOS, which is much lower than what was present in technical PFOS from the major historical manufacturer, and which is also lower than data reported from any other countries. Moreover, the proportion of L-PFOS decreased significantly with increasing ΣPFOS concentration in the serum samples ($r = -0.694$, $p < 0.001$). Taken together, the data lend weight to previous suggestions that (1) high branched PFOS content in serum is a biomarker of exposure to PFOS precursors, and (2) people with the highest ΣPFOS concentrations are exposed disproportionately to high concentrations of PFOS precursors. On average, linear PFOA contributed 96.1% of ΣPFOA, significantly higher than that in technical PFOA (ca. 75%–80% linear), thus suggesting a mixed exposure to both ECF- and telomerization-produced PFAS. Additionally, the ΣPFOA is lower than in Americans, suggesting higher exposure to electrochemically fluorinated PFOA than in the United States.

Liu et al. (2015) studied the temporal trends of overall human serum PFOS level (ΣPFOS), as well as the isomer and enantiomer patterns in archived Swedish and American serum samples. Archived Swedish primiparous women's serum samples collected yearly during 1996–2010 (except 2003 and 2005) and archived American serum and plasma samples collected in 1974, 1989, 2000–2001, 2006, and 2010 were processed and analyzed by HPLC-MS/MS. Their results showed a temporal decline of total PFOS in both Swedish and American samples that corresponds to the phaseout of PFOS and PFOS precursors by the major manufacturer in the United States between 2000 and 2002. Isomer and enantiomer pattern analysis revealed a significant increase of branched PFOS in the Swedish samples between 2000 and 2010 and increased nonracemic 1m-PFOS EFs from 1996 to 2000. These observations could not be easily explained by existing data and were suggested to be affected by the increasing relative significance of indirect exposure pathways, environmental processes such as differential soil adsorption between linear and branched isomers, isomer-selective enrichment in diets such as fish, and exposure from nondietary sources such as consumer products. No statistically significant temporal trend for branched PFOS or 1m-PFOS EF was observed in American samples, but American males had significantly higher branched PFOS and significantly lower 1m-PFOS EF (i.e., more nonracemic) than females, and a similar significant difference was shown in the older age group relative to the younger age group.

Another temporal change study was performed by Gebbink et al. (2015a) on Swedish serum samples between 1997 and 2012. They found decreases in the relative abundance of linear PFHxS and PFOS isomers, while the PFOA isomer pattern was significantly enriched with the linear isomer. The relative abundance of linear PFOSA and methylperfluorooctane sulfonamide acetic acid (MeFOSAA) increased between 1997 and 2012; however, the increase was not significant for MeFOSAA. Branched ethylperfluorooctane sulfonamido acetic acid (EtFOSAA) isomers were only detected in serum pools from 1997 to 2002; therefore, no temporal changes were determined. The observed PFOA isomer pattern and its temporal trend in the serum samples could be explained by two factors:

1. The production of ECF PFOA was largely phased out by 2002, whereas the production of telomere-based PFOA and precursor chemicals is still ongoing.
2. Branched PFOA isomers are excreted faster than linear isomers.

This study showed that the PFOS isomer pattern of total PFOS intake largely reflects the isomer pattern in food since dietary intake is the dominant exposure pathway. In Swedish food, an increase in the relative abundance of the linear isomer in the PFOS isomer pattern has been reported, from 81% L-PFOS in 1999 to 91% in 2005 and 2010 (Gebbink et al., 2015b). In all serum samples of this study, the PFOS isomer pattern was enriched with branched isomers (33%–40%) relative to the isomer pattern reported in Swedish food. Furthermore, the serum samples showed a decreasing abundance of L-PFOS, which was opposite of the change in the isomer pattern in food. A faster biotransformation of branched precursors to branched PFOS relative to linear precursors could have contributed to an increasing abundance of branched PFOS over time. However, Gebbink et al. (2015a) recently estimated that precursors only contributed 16% to the total PFOS intake. Even when including the (known) precursor contribution, the isomer pattern of total PFOS intake cannot explain the PFOS isomer patterns and trends found in human serum (Gebbink et al., 2015a). This lack of agreement between the temporal changes in isomer pattern of intake and serum illustrates the incomplete understanding of isomer-specific pharmacokinetic processes of PFAS and their precursors.

Jiang et al. (2014) performed PFAS isomeric analysis of the blood samples from pregnant women in China. In addition, hematological parameters (white blood cell [WBC] count, red blood cell [RBC] count, hemoglobin, and platelet) and selected serum biochemistry biomarkers (total bilirubin, total protein, albumin, glucose, aspartate aminotransferase [AST], and alanine aminotransferase [ALT]) were analyzed and correlated to PFAS concentrations. WBC count was positively correlated with 1m-PFOS, 4m-PFOS, 3 + 5m-PFOS and PFHxS, but negatively correlated with the proportion of L-PFOS significantly ($p < 0.05$). The concentration of L-PFOS did not show any correlation to WBC count. RBC count and hemoglobin concentration were positively correlated with the concentrations of ΣPFOS, branched PFOS, 1m-PFOS, 3 + 5m-PFOS, and PFNA. A significantly positive correlation between total bilirubin and m$_2$-PFOS, PFHxA, PFHpA, PFDA, and PFUnDA was also observed, suggesting that PFCAs could be related to adverse liver or bile effects in pregnant women. Glucose was positively correlated with PFUdA ($r = 0.166$, $p < 0.05$), and total protein was positively correlated with PFHpA ($r = 0.226$, $p < 0.01$). Total protein concentration was correlated with branched PFOA and 5m-PFOA but not with linear PFOA. No significant correlation was observed between AST and ALT with any PFAA in this study. These findings indicate that the isomer composition of PFOS may affect human health.

Salihovic et al. (2015) studied the isomers of PFOS in the plasma samples of 398 individuals, a subset of the 1016 (507 women) 70-year-old participants in the Prospective Investigation of the Vasculature in Uppsala Seniors in Sweden. PFOS isomers were observed in all 398 samples. The median plasma concentration of branched PFOS isomers was 36% (6.75 ng/mL), and the median plasma concentration of L-PFOS was 64% (12.3 ng/mL). The median proportions of linear and branched PFOS isomers were similar to those in the technical product (70% L-PFOS) produced during the time of the sampling (2001–2004). The isomer pattern of PFOS also showed similarities to that previously observed in Sweden, Australia, Canada, and China, in which the mean concentrations of the branched isomers constituted from 30% to 42% of the total PFOS concentration (Kärrman et al., 2007; Beesoon et al., 2011; Jiang et al., 2014). Considerable interindividual differences existed in the proportion of branched versus L-PFOS isomers among the participants. The minimum (19%), median (36%), and maximum (54%) concentrations of branched PFOS isomers as detected in three different individuals had T-PFOS concentrations of 17.5, 29.3, 16.5 ng/mL, respectively. This might suggest different routes of exposure to PFOS isomers and PFOS precursors or an isomer-specific accumulation.

A study by Gao et al. (2015) examined the differential accumulation and elimination behavior of PFAS isomers in occupational workers. Serum and urine samples of 36 occupational workers from a fluorochemical manufacturing plant were analyzed to evaluate the body burden and elimination of linear and branched PFAS. Indoor dust, total suspended particles (TSPs), diet, and drinking water samples were also collected to trace the occupational exposure pathway to PFAS isomers. The geometric mean concentrations of PFOS, PFOA, and PFHxS isomers in the serum were 1386, 371, and 863 ng/mL, respectively. The linear isomer of PFOS, PFOA, and PFHxS was the most predominant PFAS in the serum, with mean proportions of 63.3, 91.1, and 92.7%, respectively, which were higher than the proportions in urine. The most important exposure routes to PFAS isomers in the occupational workers were considered to be the intake of indoor dust and TSPs. A renal clearance estimation indicated that branched PFAS isomers had a higher renal clearance rate than did the corresponding linear isomers. Molecular docking modeling implied that L-PFOS had a stronger interaction with human serum albumin (HSA) than branched isomers did, which could decrease the proportion of L-PFOS in the blood of humans via the transport of HSA.

10.5.2 Isomer-Specific Transplacental Transfer and Implications on Developmental Toxicity

Understanding the maternal–fetal transmission of PFAS is necessary to clearly understand the risks and mechanisms of human developmental toxicity. Multiple studies

(Inoue et al., 2004; Fei et al. 2007; Monroy et al., 2008; Hanssen et al., 2010; Beesoon et al., 2011; Needham et al., 2011) have attempted to understand the transplacental transfer of PFAS by analyzing PFAS concentration in maternal and cord blood samples from different populations. These studies have consistently found that cord blood has lower total PFOA and total PFOS than that does maternal blood. A few of these studies (Hanssen et al., 2010; Beesoon et al., 2011) made a distinction between branched and linear isomers in their analysis. The work by Hanssen et al. (2010) measured the maternal serum and cord blood samples of South American women. In maternal serum, PFOS was found to be the most abundant PFAS (1.6 ng/mL), followed by PFOA (1.3 ng/mL) and PFHxS (0.5 ng/mL). However, in cord blood PFOA was the most abundant compound (1.3 ng/mL), followed by PFOS (0.7 ng/mL) and PFHxS (0.3 ng/mL). This study also found a statistically greater relative abundance of L-PFOS in maternal serum than in cord serum relative to total branched PFOS isomers ($p < 0.05$ by Wilcoxon's signed-rank test).

Beesoon et al. (2011) performed isomer-specific analysis of maternal blood and cord samples from Canadian women. Similar to earlier studies, they found that total PFAS concentrations in maternal blood are significantly higher than those in cord blood. The majority of the PFAS were PFOS, PFOA, PFHxS, and PFNA. Transplacental transfer efficiencies (TTEs) (estimated by dividing the PFAS concentrations in cord serum at delivery by maternal serum concentration at 15 weeks of gestation) were found to be dependent on carbon chain length for both PFCAs and perfluorosulfonates; shorter-chain PFAS crossed the placenta more efficiently than did longer-chain PFAS. Isomeric analysis revealed that the percent branched content of total PFOS was consistently and significantly higher in cord serum than in corresponding maternal serum samples. Branched PFOS isomers contributed 27%–44% (median 36%) of total PFOS in maternal serum and 36%–54% (median 46%) in cord serum. A paired t-test indicated statistically greater proportions of branched PFOS in the cord serum ($p < 0.01$). Overall, all branched PFOS isomers were transferred more efficiently (median TTEs of the different branched isomers, 0.34–0.88) than was the linear isomer (median TTE 0.30). Specifically, among the perfluoromethyl PFOS-branched isomers, TTE increased as the branching point moved closer to the sulfonate moiety: 1m- > 3m- > 4m- ≈ 5m- > 6m-PFOS. In fact, for 1m-, 3m-, and particularly Σm_2-PFOS, the concentrations were sometimes higher in cord serum than in corresponding maternal serum (resulting in maximum TTE values >1.0). Branched PFOA isomers contributed 0.43%–4.3% (mean 1.9%) of total PFOA in maternal serum and 0.71%–5.7% (mean 2.2%) in cord serum. No structure–activity relationship was evident for PFOA isomers, but a paired t-test indicated significantly higher total branched PFOA isomers in cord serum than in maternal serum ($p = 0.02$). In some cases, the concentrations of 5m-, 4m-, and 3m-PFOA were higher in the cord serum than in corresponding maternal serum (resulting in maximum TTE > 1.0). The observed higher TTEs of branched isomers were inconsistent with their supposed higher hydrophilicity since hydrophilic molecules are less likely to cross the placental barrier (Van der Aa et al., 1998). An alternative possible explanation was the relative higher binding affinity of linear isomers to serum proteins, leaving a higher fraction of free branched isomers to cross the placenta.

Chen et al. (2017) assessed the occurrence and distribution of different PFHxS, PFOS, and PFOA isomers in maternal serum, umbilical cord serum, and placenta to gain a better understanding of TTE and prenatal exposure risks. Quantitative determination of isomer-specific concentrations of PFHxS, PFOS, and PFOA was carried out in samples of maternal serum ($n = 32$), cord serum ($n = 32$), and placenta ($n = 32$) from pregnant women in Wuhan, China. The results indicated that both linear and branched PFHxS, PFOS, and PFOA can be efficiently transported across the placenta, with exposure levels ordered maternal serum > cord serum > placenta. For PFOS isomers, the concentration ratios between cord serum and maternal serum (R_{CM}) were ordered linear < 6m- < 4m- < (3 + 5)m- < 1m-PFOS < Σm_2. The placenta/maternal ratio (R_{PM}) values exhibited a similar trend for branched PFOS isomers: 6m- < 4m- ≈ (3 + 5)m- < 1m-PFOS ≈ Σm_2. Conversely, PFOA isomers did not exhibit an obvious structure–activity relationship for R_{CM} and R_{PM}. L-PFHxS transported across the placenta to a greater extent than branched PFHxS.

Zhang et al. (2017) examined the profile of PFAS and isomers, including 17 linear PFAS and 10 branched PFOS and PFOA isomers, by using isotopic internal standards in umbilical cord serum samples from Guangzhou, China. The researchers collected 321 cord blood serum samples from July to October 2013 and analyzed the PFAS concentration with an isomer-specific PFAS analysis method. The results showed that 9 out of 17 PFAS (linear PFAS) were detected (>50% detection rate). PFHxS (median 3.87 ng/mL) was the most predominant, followed by total PFOS (median 2.99 ng/mL) and total PFOA (median 1.23 ng/mL) in cord serum. In addition, 1m-, 6m-, $\Sigma 3 + 4 + 5$m-PFOS and 6m-PFOA were the branched PFAS detected. The proportion of L-PFOS was 75.16% of ΣPFOS, similar to the proportion of ECF technical PFOS products. On the contrary, linear PFOA accounted for 98.69% ΣPFOA in cord serum samples.

Li et al. (2017) conducted a study including 321 pairs of mothers and their infants recruited from Guangzhou, China. Isomers of PFOS and PFOA, along with other PFAS levels in cord serum samples, were analyzed by HPLC-MS. The resulting data revealed that higher PFOS, PFOA, and isomers of PFOS were associated with lower birth weight. Per ln (natural log) unit (ng/mL), an increase in cord serum total branched PFOS isomers was associated with a 126.3 g (95% CI −195.9, −56.8) reduction in the weight of infants at birth, while an ln unit (ng/mL) increase of serum L-PFOS isomers was associated with a 57.2 g (95% CI −103.1, −11.3) reduction in the weight of infants at birth upon the subsequent adjustment for potential confounding variables. Notably, the association between cord PFAS level and birth weight was more pronounced in male infants. Furthermore, a positive association among branched PFOS isomers (1m-PFOS and $\Sigma 3 + 4 + 5$m-PFOS) and gestational age was found. No associations could be found among other PFAS in conjunction with gestational age or birth weight.

Human exposure to PFAS is complex, and several influential factors that affect the overall perfluoroalkyl exposure in an individual have been not been described in this review. Among these are social factors that include occupation, residency, diet, and lifestyle, as well as biological factors that include age, body size, metabolism, and gender (Vahter et al., 2007). For this reason, interpretation of biomonitoring data of PFAS isomers in human samples is difficult, and inconsistency between studies may occur. Moreover, the pharmacokinetics of individual isomers remain largely unknown and thus compound the interpretation of the isomer profiles in human matrices.

10.6 ISOMER-SPECIFIC TOXICITY AND INTERACTIONS WITH BIOMOLECULES

10.6.1 Isomer-Specific Toxicity

Based on results from animal and *in vitro* cell culture studies, the toxic effects of PFAS include interference with lipid and carbohydrate metabolism, increased liver weight, hepatomegaly, disruption of the neuroendocrine system, and decreased reproductive success (Beesoon and Martin, 2015). Most of these earlier laboratory toxicology studies, however, exposed test animals or cultured cells to T-PFOS, making no distinction among the isomers. These experiments might not have accurately reflected natural exposure conditions experienced by living organisms because they did not consider individual PFOS isomer kinetics and therefore may have led to poor estimations of exposure effects. After the pharmacokinetic differences between PFOS isomers were recognized, a limited number of studies started to address the issue of PFAS isomer content. These studies have shown that PFAS can elicit different molecular and physiological effects depending on the isomeric composition.

Hickey et al. (2009) studied the effects of 18 PFAS on mRNA expression in chicken embryo hepatocyte (CEH) cultures. The issue of isomer content was studied by comparing T-PFOS and L-PFOS. T-PFOS contained ~80% PFOS isomers, 62% of which were linear. The remaining 20% of T-PFOS consisted of various PFAS and inorganic salts typically found in commercial samples, while L-PFOS contained >98% of the pure chemical. CEH cultures were exposed to PFAS of concentrations ranging from 0.1 to 50 µM for 36 h and assessed for cell viability as well as mRNA expression of the following targeted genes: acyl coenzyme A (CoA) oxidase, a known peroxisome proliferator–activated receptor (PPAR) α target; liver fatty acid–binding protein (L-FABP), a transport protein known to bind PFAS; CYP1A4/5 and CYP4B1, regulators of xenobiotic metabolism; and 3-hydroxy-3-methylglutaryl-Coenzyme A (HMG-CoA) reductase and sterol regulatory element–binding protein 2 (SREBP2), important cholesterol metabolism genes. Results showed important differences in the effects of L-PFOS and T-PFOS formulations. Specifically, exposure to T-PFOS (50 µM) led to greater induction of target gene mRNA expression in CEH cultures than exposure to L-PFOS (50 µM): L-FABP (9.32-fold increase vs. 3.19), ACOX (3.00 vs. 1.11), CYP1A4 (12.85 vs. 2.96), CYP1A5 (5.98 vs. 2.17), and CYP4B1 (7.69 vs. 1.10). The authors attributed these differences in response to the isomeric difference or to other compounds present in the T-PFOS, 20% of which were various PFAS and inorganic salts.

Loveless et al. (2006) compared the responses of rats and mice exposed to linear/branched mixtures, as well as solely linear or solely branched ammonium perfluorooctanoate (APFO), which is used as a processing aid in the production of fluoropolymers and is a source of PFOA found in exposed workers and the general population. Three types of APFO were tested: linear/branched APFO by 3M

Company, consisting of 77.6% linear APFO; linear APFO; and 100% branched APFO (58:42 blend of the two most prevalent isomers) from DuPont. Rats and mice were given doses by oral gavage ranging from 0.3 to 30 mg/kg of either the linear/branched, linear or branched APFO for 14 days. Clinical signs, body weights, food consumption, selected hematology and serum lipid parameters, liver and kidney weights, hepatic peroxisomal β-oxidation, and serum PFOA concentrations were evaluated. Mean body weights were about 20% lower in rats and mice dosed with 30 mg/kg of linear/branched or linear APFO than in controls, and 3%–5% lower in animals dosed with 30 mg/kg of branched APFO. In rats, all three forms reduced lipids. In mice, all three forms reduced total and HDL cholesterol similarly, but triglycerides were increased at lower doses. Increased peroxisomal β-oxidation activity and serum PFOA concentrations were seen in both species, but these effects were least pronounced in rats dosed with the branched material. Analysis of serum PFOA isomer patterns showed a linear/branched isomer ratio greater than the linear/branched ratio of APFO administered. This increase became greater with dose and was suggested as a reflection of the altered absorption or preferential clearance of branched APFOs. Changes in isomer pattern were also found in the ratio between the two isomeric constituents in the 100% pure branched APFOs. In both rats and mice, the overall responses to the linear/branched and the linear forms of PFOA were similar, but the branched form appears to be less potent.

A similar phenomenon was found by O'Brien et al. (2011a) in chicken embryos exposed to T-PFOS. T-PFOS composed of 62.7% L-PFOS and 37.3% branched isomers, including six mono(trifluoromethyl)-branched isomers and four bis(trifluoromethyl)-branched isomers, were injected into eggs at concentrations of 0.1, 5, and 100 µg/g egg. Liver tissues from the embryos were collected at pipping, and hepatic isomer profiles were compared with that of the originally injected T-PFOS. All 11 PFOS isomers (linear and branched) were detected in most treated samples. The proportion of L-PFOS in the liver of all treated embryos was significantly higher than that in T-PFOS. The degree of difference was inversely proportional to dose, with the largest increase, approximately 20%, occurring in the 0.1 µg/g dose group. Concordantly, liver tissue from all treatment groups had lower proportions of branched isomers. Although all branched isomers were proportionately lower compared with T-PFOS at 0.1 µg/g, the degree to which all isomers were affected was not equal. When data were normalized to total branched isomer content, it became evident that certain isomers contributed less to the sum of all branched isomers compared with T-PFOS. This suggests a strong selection against the accumulation of these particular isomers in the liver. In contrast, a specific terminally branched isomer was proportionately higher than other branched isomers, relative to T-PFOS. It is clear from this evidence that the processing (e.g., absorption and clearance) of PFAS inside living animals is isomer specific, resulting in the differential effects on the biological end points studied.

In another study by O'Brien et al. (2011b), gene expression changes were studied in cultured CEH cells exposed to T-PFOS and L-PFOS. T-PFOS comprised 65% linear and 35% branched isomers (mostly mono(trifluoromethyl)-branched isomers), with the L-PFOS having >98% purity. CEH cells were treated with T-PFOS and L-PFOS at doses (1, 10, 20, 30, and 40 µM) that were enough to induce transcriptional effects

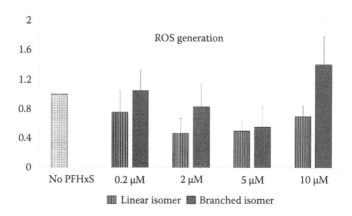

FIGURE 10.3 ROS expression in JEG3 cells exposed to linear and branched isomers of PFHxS.

but not enough to affect cell viability. After 24 h incubation with PFOS, the CEH cells were processed for microarray analysis. Results showed that both L-PFOS and T-PFOS elicited a transcriptional response in genes related to lipid metabolism and transport, oxidative stress response, and cellular growth and proliferation. At equal concentrations (10 µM), T-PFOS altered the expression of more transcripts (340, >1.5-fold change, $p < 0.05$) than L-PFOS (130 transcripts). Higher concentrations of L-PFOS (40 µM) were also less transcriptionally disruptive (217 transcripts) than T-PFOS at 10 µM. In almost all functional categories, pathways, and interaction networks examined in the present study, the impact of T-PFOS was much greater than that of L-PFOS. The authors proposed that the difference was likely due to the increased number of molecule shapes presented by the branched isomers of T-PFOS, allowing the chemical to interact with cellular machinery and thereby interfering with more signaling cascades, activating more receptors, and recruiting more transcription factors than the linear isomer of PFOS alone.

A preliminary study in our laboratory investigated the toxicological differences between branched and linear PFAS isomers *in vitro* using the JEG3 human placental cell line as a model (Cantu, 2017). Cells were exposed to linear and branched PFHxS for 24 h. Our results showed that at the dose range tested (0–10 µM), there was no significant difference between linear and branched isomers of PFHxS on cell viability as assessed by esterase activity. Reactive oxygen species (ROS) generation was statistically higher in JEG3 cells exposed to branched PFHxS isomers at corresponding concentrations. PPAR-α expression level was determined by immunofluorescence, and results showed no significant difference between linear and branched isomers (Figure 10.3).

10.6.2 Isomer-Specific Interactions with Biological Molecules

Studying the interactions between PFAS and biomolecules in the human body provides a mechanistic understanding of PFAS toxicity. For instance, Wang et al. (2016b) studied the conformational and functional behavior of hemoglobin in the presence of PFOS. Their results showed that PFOS acted as a structure destabilizer for hemoglobin and caused a decrease in its thermal stability. Apart from biomolecules in the blood, there are also reports of interactions between PFAS and human liver fatty acid–binding proteins (Sheng et al., 2016). However, as it is with other PFAS studies, most of these studies did not make a distinction between isomers. By far, the only study that has attempted to provide a mechanistic explanation of the different pharmacokinetic behavior of PFAS isomers through the study of their interactions with biomolecules was performed by Beesoon and Martin (2015). The authors studied the binding affinity of PFOS (linear and branched isomers) and PFOA (linear and branched isomers) to HSA. Using an ultrafiltration method, they estimated the dissociation constants (K_d) of linear and three individual branched isomers (3m-, 4m-, 5m-) of PFOS and PFOA to HSA. Their results showed a much lower dissociation constant of L-PFOS ($K_d = 8 [\pm 4] \times 10^{-8}$ M) than of linear PFOA ($K_d = 1 [\pm 0.9] \times 10^{-4}$ M), indicating a much stronger binding affinity of PFOS to HSA than PFOA. For both PFOS and PFOA, linear isomers are found to associate more strongly with albumin than the branched isomers. However, the relative effect was much more pronounced for PFOS isomers than for PFOA isomers. For PFOS, the K_d ($\pm$ SE) of the linear isomer, $8 (\pm 4) \times 10^{-8}$ M, was at least three orders of magnitude lower than that of any of the isomers investigated here: $4 (\pm 2) \times 10^{-4}$ M for 3m-PFOS, $8 (\pm 1) \times 10^{-5}$ M for 4m-PFOS, and $9 (\pm 5) \times 10^{-5}$ M for 5m-PFOS. For PFOA, the K_d of the linear isomer, $1 (\pm 0.9) \times 10^{-4}$ M, was only approximately three times smaller than that for 3m-PFOA, $4 (\pm 2) \times 10^{-4}$ M, and approximately half the estimate for 5m-PFOA, $3 (\pm 2) \times 10^{-4}$ M. The higher binding affinities of L-PFOS and PFOA to total serum protein were confirmed with both human serum and calf serum spiked with technical mixtures. Overall, these data provided a mechanistic explanation for the longer biological half-life of PFOS in humans than of PFOA, and for the higher TTEs and renal clearance of branched PFOS and PFOA isomers compared with the respective linear isomer.

In summary, even though only a limited number of toxicity studies have addressed the issue of PFAS isomers, the evidence is clear that PFAS toxicity is isomer specific. However, the results may vary depending on the experimental system used and the end points studied. While toxicological studies with living animals have shown that branched isomers are less potent based on selected end points, such as animal body or organ weight, serum lipid profiles, and bioaccumulation in the liver and kidneys, cell culture studies revealed that higher branched isomer composition resulted in higher inductions of gene transcription and greater generation of ROS. These apparent inconsistencies might be explained by the intrinsic differences between animal studies and cell culture studies. Cell culture studies are straightforward and involve only the direction interactions between the chemical and the cells. PFAS administered to a living animal, however, has to pass through tissues and the circulation system before reaching the target organ or cells. PFAS bioavailability at the target sites is subjected to the influences of PFAS absorption and clearance; therefore, the observed differences in selected end points are the combined effects of multiple isomer-specific processes. Future studies are needed to verify these early findings, and molecular interaction studies are needed to elucidate the fundamental mechanisms of PFAS toxicity.

10.7 CONCLUSIONS AND FUTURE PERSPECTIVE

Most commercially produced PFAS were manufactured as isomeric mixtures (ECF) or as a single isomer (telomerization). Studies have shown isomer-specific differences in partitioning, biotransformation, enrichments in the environment, and toxicity. The majority of research to date involving traditional PFAS quantification, however, did not make distinctions between isomers and only measured total PFAS. Recent advances in isomer characterization methods have allowed the accurate characterization of individual isomers, and isomer-specific analysis has become increasingly useful in source tracking. An increasing number of studies have highlighted the potential insights into PFAS' environmental fate that may be gained from better knowledge of the isomer- and enantiomer-specific behavior of both PFAS and its precursors. However, the lack of characterized isomeric mixtures (currently only available for PFOS and PFOA) has hampered the development of new isomer separation methods. Overall, the isomer profiles of PFOS in fish and wildlife are typically enriched in linear isomers (>70%). This is speculated to be a combined result of branched isomer-selective degradation, excretion, and linear isomer-selective bioconcentration, uptake, and bioaccumulation. Some field studies, however, have shown the enrichment of branched isomers in wildlife and humans, likely due to the preferential biotransformation of branched PFAS precursors. A limited number of toxicity studies have provided clear evidences that PFAS toxicity is isomer specific. However, the results may vary depending on the experimental system and the end points studied. While animal studies have shown that branched isomers are less potent, cell culture studies suggest otherwise. These inconsistencies may have to do with the differences between *in vitro* and *in vivo* studies. Some studies have shown that temporal changes of PFAS isomer profiles in human serum samples are correlated with changes in exposure (e.g., a decline in branched isomer content was observed after the phase-out of the ECF products). Meanwhile, there are still many discrepancies that cannot be explained by exposure alone, calling for more research on isomer-specific biotransformation and pharmacokinetics. Affinity analysis of PFAS isomers with human albumin protein revealed stronger adsorption of linear isomers than of branched isomers, providing a mechanistic explanation of the differential pharmacokinetics of branched versus linear isomers. More of this type of molecular interaction studies are needed in order to fully elucidate the fundamental mechanisms of PFAS actions in biological systems.

REFERENCES

3M Company Technical Bulletin. 3M The Leader in Electrofluorination; 3M Company: St. Paul, MN, 1995.

3M Company. Voluntary Use and Exposure Information Profile Perfluorooctanic Acid and Salts; U.S. EPA Administrative Record AR226-0595; U.S. Environmental Protection Agency: Washington, DC, 2000.

Armitage J, MacLeod M, Cousins IT. Modeling the global fate and transport of perfluorooctanoic acid (PFOA) and perfluorooctanoate (PFO) emitted from direct sources using a multispecies mass balance model. *Environ Sci Technol* 2009;43:1134–40.

Asher BJ, Wang Y, De Silva AO, Backus S, Muir DC, Wong CS, Martin JW. Enantiospecific perfluorooctane sulfonate (PFOS) analysis reveals evidence for the source contribution of PFOS-precursors to the Lake Ontario foodweb. *Environ Sci Technol* 2012;46(14):7653–60.

Beesoon S, Martin JW. Isomer-specific binding affinity of perfluorooctanesulfonate (PFOS) and perfluorooctanoate (PFOA) to serum proteins. *Environ Sci Technol* 2015;49(9):5722–31.

Beesoon S, Webster GM, Shoeib M, Harner T, Benskin JP, Martin JW. Isomer profiles of perfluorochemicals in matched maternal, cord, and house dust samples: Manufacturing sources and transplacental transfer. *Environ Health Perspect* 2011;119(11):1659–64.

Benskin JP, Bataineh M, Martin JW. Simultaneous characterization of perfluoroalkyl carboxylate, sulfonate, and sulfonamide isomers by liquid chromatography-tandem mass spectrometry. *Anal Chem* 2007;79(17):6455–64.

Benskin JP, De Silva AO, Martin JW. Isomer profiling of perfluorinated substances as a tool for source tracking: A review of early findings and future applications. *Rev Environ Contam Toxicol* 2010a;208:111–60.

Benskin JP, Holt A, Martin JW. Isomer-specific biotransformation rates of a perfluorooctane sulfonate (PFOS)-precursor by cytochrome P450 isozymes and human liver microsomes. *Environ Sci Technol* 2009;43(22):8566–72.

Benskin JP, Ikonomou MG, Woudneh MB, Cosgrove JR. Rapid characterization of perfluoralkyl carboxylate, sulfonate, and sulfonamide isomers by high-performance liquid chromatography-tandem mass spectrometry. *J Chromatogr A* 2012;1247:165–70.

Benskin JP, Yeung LW, Yamashita N, Taniyasu S, Lam PK, Martin JW. Perfluorinated acid isomer profiling in water and quantitative assessment of manufacturing source. *Environ Sci Technol* 2010b;44(23):9049–54.

Cantu GA. Toxicological differences between perfluoroalkyl substances (PFAs) isomers using developmental biomarkers. MS thesis, Air Force Institute of Technology, Dayton, OH, 2017.

Chen F, Yin S, Kelly BC, Liu W. Isomer-specific transplacental transfer of perfluoroalkyl acids: Results from a survey of paired maternal, cord sera, and placentas. *Environ Sci Technol* 2017;51(10):5756–63.

Chen M, Qiang L, Pan X, Fang S, Han Y, Zhu L. In vivo and in vitro isomer-specific biotransformation of perfluorooctane sulfonamide in common carp (*Cyprinus carpio*). *Environ Sci Technol* 2015;49(23):13817–24.

Chen X, Zhu L, Pan X, Fang S, Zhang Y, Yang L. Isomeric specific partitioning behaviors of perfluoroalkyl substances in water dissolved phase, suspended particulate matters and sediments in Liao River Basin and Taihu Lake, China. *Water Res* 2015;80:235–44.

Chu S, Letcher RJ. Linear and branched perfluorooctane sulfonate isomers in technical product and environmental samples by in-port derivatization-gas chromatography-mass spectrometry. *Anal Chem* 2009;81(11):4256–62.

De Silva AO, Mabury SA. Isolating isomers of perfluorocarboxylates in polar bears (Ursus maritimus) from two geographical locations. *Environ Sci Technol* 2004;38(24):6538–45.

De Silva AO, Mabury SA. Isomer distribution of perfluorocarboxylates in human blood: Potential correlation to source. *Environ Sci Technol* 2006;40(9):2903–9.

Fei C, McLaughlin JK, Tarone RE, Olsen J. Perfluorinated chemicals and fetal growth: A study within the Danish National Birth Cohort. *Environ Health Perspect* 2007;115(11):1677–82.

Flaherty JM, Connolly PD, Decker ER, Kennedy SM, Ellefson ME, Reagen WK, Szostek B. Quantitative determination of perfluorooctanoic acid in serum and plasma by liquid chromatography tandem mass spectrometry. *J Chromatogr B Analyt Technol Biomed Life Sci* 2005;819(2):329–38.

Frisbee SJ, Brooks AP Jr, Maher A, Flensborg P, Arnold S, Fletcher T, Steenland K, Shankar A, Knox SS, Pollard C, Halverson JA, Vieira VM, Jin C, Leyden KM, Ducatman AM. The C8 health project: Design, methods, and participants. *Environ Health Perspect* 2009;117(12):1873–82.

Fu Z, Wang Y, Wang Z, Xie H, Chen J. Transformation pathways of isomeric perfluorooctanesulfonate precursors catalyzed by the active species of P450 enzymes: In silico investigation. *Chem Res Toxicol* 2015;28(3):482–9.

Gao Y, Fu J, Cao H, Wang Y, Zhang A, Liang Y, Wang T, Zhao C, Jiang G. Differential accumulation and elimination behavior of perfluoroalkyl acid isomers in occupational workers in a manufactory in China. *Environ Sci Technol* 2015;49(11):6953–62.

Gebbink WA, Letcher RJ. Linear and branched perfluorooctane sulfonate isomer patterns in herring gull eggs from colonial sites across the Laurentian Great Lakes. *Environ Sci Technol* 2010;44(10):3739–45.

Gebbink WA, Berger U, Cousins IT. Estimating human exposure to PFOS isomers and PFCA homologues: The relative importance of direct and indirect (precursor) exposure. *Environ Int* 2015b;74:160–9.

Gebbink WA, Glynn A, Berger U, Temporal changes (1997-2012) of perfluoroalkyl acids and selected precursors (including isomers) in Swedish human serum. *Environ Pollut* 2015a;199:166–73.

Hansen KJ, Clemen LA, Ellefson ME, Johnson HO. Compound-specific, quantitative characterization of organic fluorochemicals in biological matrices. *Environ Sci Technol* 2001;35:766–70.

Hanssen L, Rollin H, Odland JO, Moe MK, Sandanger TM. Perfluorinated compounds in maternal serum and cord blood from selected areas of South Africa: Results of a pilot study. *J Environ Monitor* 2010;12(6):1355–61.

Haug LS, Thomsen C, Bechert G. Time trends and the influence of age and gender on serum concentrations of perfluorinated compounds in archived human samples. *Environ Sci Technol* 2009;43:2131–6.

Hickey NJ, Crump D, Jones SP, Kennedy SW. Effects of 18 perfluoroalkyl compounds on mRNA expression in chicken embryo hepatocyte cultures. *Toxicol Sci* 2009;111(2):311–20.

Houde M, Czub G, Small JM, Backus S, Wang X, Alaee M, Muir DC. Fractionation and bioaccumulation of perfluorooctane sulfonate (PFOS) isomers in a Lake Ontario food web. *Environ Sci Technol* 2008;42(24):9397–403.

Houde M, De Silva AO, Muir DC, Letcher RJ. Monitoring of perfluorinated compounds in aquatic biota: An updated review. *Environ Sci Technol* 2011;45(19):7962–73.

Inoue K, Okada F, Ito R, Kato S, Sasaki S, Nakajima S et al. Perfluorooctane sulfonate (PFOS) and related perfluorinated compounds in human maternal and cord blood samples: Assessment of PFOS exposure in a susceptible population during pregnancy. *Environ Health Perspect* 2004;112:1204–7.

Jiang W, Zhang Y, Yang L, Chu X, Zhu L. Perfluoroalkyl acids (PFAAs) with isomer analysis in the commercial PFOS and PFOA products in China. *Chemosphere* 2015;127:180–7.

Jiang W, Zhang Y, Zhu L, Deng J. Serum levels of perfluoroalkyl acids (PFAAs) with isomer analysis and their associations with medical parameters in Chinese pregnant women. *Environment International* 2014;64:40–47.

Jin H, Zhang Y, Zhu L, Martin JW. Isomer profiles of perfluoroalkyl substances in water and soil surrounding a Chinese fluorochemical manufacturing park. *Environ Sci Technol* 2015;49(8):4946.

Kärrman A, Langlois I, van Bavel B, Lindström G, Oehme M. Identification and pattern of perfluorooctane sulfonate (PFOS) isomers in human serum and plasma. *Environ Int* 2007;33(6):782–8.

Kudo N, Suzuki E, Katakura M, Ohmori K, Noshiro R, Kawashima Y. Comparison of the elimination between perfluorinated fatty acids with different carbon chain length in rats. *Chem Biol Interact* 2001;134:203–16.

Langlois I, Oehme M. Structural identification of isomers present in technical perfluorooctane sulfonate by tandem mass spectrometry. *Rapid Commun Mass Spectrom* 2006;20(5):844–50.

Li M, Zeng XW, Qian ZM, Vaughn MG, Sauvé S, Paul G et al. Isomers of perfluorooctanesulfonate (PFOS) in cord serum and birth outcomes in China: Guangzhou Birth Cohort Study. *Environ Int* 2017;102:1–8.

Liu Y, Pereira AS, Beesoon S, Vestergren R, Berger U, Olsen GW, Glynn A, Martin JW. Temporal trends of perfluorooctanesulfonate isomer and enantiomer patterns in archived Swedish and American serum samples. *Environ Int* 2015;75:215–22.

Loveless SE, Finlay C, Everds NE, Frame SR, Gillies PJ, O'Connor JC, Powley CR, Kennedy GL. Comparative responses of rats and mice exposed to linear/branched, linear, or branched ammonium perfluorooctanoate (APFO). *Toxicology* 2006;220(2–3):203–17.

Martin JW, Asher BJ, Beesoon S, Benskin JP, Ross MS. PFOS or PreFOS? Are perfluorooctane sulfonate precursors (PreFOS) important determinants of human and environmental perfluorooctane sulfonate (PFOS) exposure? *J Environ Monit* 2010;12(11):1979–2004.

Martin JW, Mabury SA, Solomon KR, Muir DC. Bioconcentration and tissue distribution of perfluorinated acids in rainbow trout (Onchorhynchus mykiss). *Environ Toxicol Chem* 2003a;22:196–204.

Martin JW, Mabury SA, Solomon KR, Muir DC. Dietary accumulation of perfluorinated acids in juvenile rainbow trout (Oncorhynchus mykiss). *Environ Toxicol Chem* 2003b;22:189–95.

Miralles-Marco A, Harrad S. Perfluorooctane sulfonate: A review of human exposure, biomonitoring and the environmental forensics utility of its chirality and isomer distribution. *Environ Int* 2015;77:148–59.

Monroy R, Morrison K, Teo K, Atkinson S, Kubwabo C, Stewart B et al. Serum levels of perfluoroalkyl compounds in human maternal and umbilical cord blood samples. *Environ Res* 2008;108(1):56–62.

Naile JE, Garrison AW, Avants JK, Washington JW. Isomers/enantiomers of perfluorocarboxylic acids: Method development and detection in environmental samples. *Chemosphere* 2016;144:1722–8.

Needham LL, Grandjean P, Heinzow B, Jørgensen PJ, Nielsen F, Patterson DG Jr. Partition of environmental chemicals between maternal and fetal blood and tissues. *Environ Sci Technol* 2011;45(3):1121–6.

O'Brien JM, Austin AJ, Williams A, Yauk CL, Crump D, Kennedy SW. Technical-grade perfluorooctane sulfonate alters the expression of more transcripts in cultured chicken embryonic hepatocytes than linear perfluorooctane sulfonate. *Environ Toxicol Chem* 2011a;30(12):2846–59.

O'Brien JM, Kennedy SW, Chu S, Letcher RJ. Isomer-specific accumulation of perfluorooctane sulfonate in the liver of chicken embryos exposed in ovo to a technical mixture. *Environ Toxicol Chem* 2011b;30(1):226–31.

Ohmori K, Kudo N, Katayama K, Kawashima Y. Comparison of the toxicokinetics between perfluorocarboxylic acids with different carbon chain length. *Toxicology* 2003;184:135–40.

Paul AG, Jones KC, Sweetman AJ. A first global production, emission, and environmental inventory for perfluorooctane sulfonate. *Environ Sci Technol* 2009;43:386–92.

Peng H, Zhang S, Sun J, Zhang Z, Giesy JP, Hu J. Isomer-specific accumulation of perfluorooctanesulfonate from (N-ethyl perfluorooctanesulfonamido) ethanol-based phosphate diester in Japanese medaka (*Oryzias latipes*). *Environ Sci Technol* 2014;48(2):1058–66.

Poothong S, Lundanes E, Thomsen C, Haug LS. High throughput online solid phase extraction-ultra high performance liquid chromatography-tandem mass spectrometry method for polyfluoroalkyl phosphate esters, perfluoroalkyl phosphonates, and other perfluoroalkyl substances in human serum, plasma, and whole blood. *Anal Chim Acta* 2017;957:10–9.

Powley CR, George SW, Russell MH, Hoke RA, Buck RC. Polyfluorinated chemicals in a spatially and temporally integrated food web in the Western Arctic. *Chemosphere* 2008;70:664–72.

Prevedouros K, Cousins IT, Buck RC, Korzeniowski SH. Sources, fate and transport of perfluorocarboxylates. *Environ Sci Technol* 2006;40(1):32–44.

Rayne S, Forest K, and Friesen, KJ. Congener-specific numbering systems for the environmentally relevant C4 through C8 perfluorinated homologue groups of alkyl sulfonates, carboxylates, telomer alcohols, olefins, and acids, and their derivatives. *J Environ Sci Health A Tox Hazard Subst Environ Eng* 2008;43:1391–1401.

Riddell N, Arsenault G, Benskin JP, Chittim B, Martin JW, McAlees A, McCrindle R. Branched perfluorooctane sulfonate isomer quantification and characterization in blood serum samples by HPLC/ESI-MS(/MS). *Environ Sci Technol* 2009;43(20):7902–8.

Rogatsky E, O'Hehir C, Daly J, Tedesco A, Jenny R, Aldous K. Development of high throughput LC/MS/MS method for analysis of perfluorooctanoic acid from serum, suitable for large-scale human biomonitoring. *J Chromatogr B Analyt Technol Biomed Life Sci* 2017;1049–50:24–9.

Ross MS, Wong CS, Martin JW. Isomer-specific biotransformation of perfluorooctane sulfonamide in Sprague-Dawley rats. *Environ Sci Technol* 2012;46(6):3196–203.

Rylander C, Brustad M, Falk H, Sandanger TM. Dietary predictors and plasma concentrations of perfluorinated compounds in a coastal population from northern Norway. *J Environ Public Health* 2009;2009:268219.

Salihovic S, Kärrman A, Lind L, Lind PM, Lindström G, van Bavel B. Perfluoroalkyl substances (PFAS) including structural PFOS isomers in plasma from elderly men and women from Sweden: Results from the Prospective Investigation of the Vasculature in Uppsala Seniors (PIVUS). *Environ Int* 2015;82:21–7.

Salihovic S, Kärrman A, Lindström G, Lind PM, Lind L, van Bavel B. A rapid method for the determination of perfluoroalkyl substances including structural isomers of perfluorooctane sulfonic acid in human serum using 96-well plates and column-switching ultra-high performance liquid chromatography tandem mass spectrometry. *J Chromatogr A* 2013;1305:164–70.

Senthil Kumar K, Zushi Y, Masunaga S, Gilligan M, Pride C, Sajwan KS. Perfluorinated organic contaminants in sediment and aquatic wildlife, including sharks, from Georgia, USA. *Mar Pollut Bull* 2009;58(4):621–9.

Shan G, Chen X, Zhu L. Occurrence, fluxes and sources of perfluoroalkyl substances with isomer analysis in the snow of northern China. *J Hazard Mater* 2015;299:639–46.

Sharpe RL, Benskin JP, Laarman AH, MacLeod SL, Martin JW, Wong CS, Goss GG. Perfluorooctane sulfonate toxicity, isomer-specific accumulation, and maternal transfer in zebrafish (Danio rerio) and rainbow trout (Oncorhynchus mykiss). *Environ Toxicol Chem* 2010;29(9):1957–66.

Sheng N, Li J, Liu H, Zhang A, Dai J. Interaction of perfluoroalkyl acids with human liver fatty acid-binding protein. *Arch Toxicol* 2016;90(1):217–27.

Vahter M, Gochfeld M, Casati B, Thiruchelvam M, Falk-Filippson A, Kavlock R, Marafante E, Cory-Slechta D. Implications of gender differences for human health risk assessment and toxicology. *Environ Res* 2007;104:70–84.

Van der Aa EM, Peereboom-Stegman JHJC, Noordhoek J, Gribnau FWJ, Russel FGM. Mechanisms of drug transfer across the human placenta. *Pharm World Sci* 1998;20(4):139–48.

Vyas SM, Kania-Korwel I, Lehmler H. Differences in the isomer composition of perfluoroctanesulfonyl (PFOS) derivatives. *J Environ Sci Health* 2007;Part A,42(3):249–55.

Wang Y, Arsenault G, Riddell N, McCrindle R, McAlees A, Martin JW. Perfluorooctane sulfonate (PFOS) precursors can be metabolized enantioselectively: Principle for a new PFOS source tracking tool. *Environ Sci Technol* 2009;43(21):8283–9.

Wang Y, Beesoon S, Benskin JP, De Silva AO, Genuis SJ, Martin JW. Enantiomer fractions of chiral perfluorooctanesulfonate (PFOS) in human sera. *Environ Sci Technol* 2011;45(20):8907–14.

Wang Y, Vestergren R, Shi Y, Cao D, Xu L, Cai Y, Zhao X, Wu F. Identification, tissue distribution, and bioaccumulation potential of cyclic perfluorinated sulfonic acids isomers in an airport impacted ecosystem. *Environ Sci Technol* 2016a;50(20):10923–32.

Wang Y, Zhang H, Kang Y, Fei Z, Cao J. The interaction of perfluorooctane sulfonate with hemoglobin: Influence on protein stability. *Chem Biol Interact* 2016b;254:1–10.

Xu L, Krenitsky DM, Seacat AM, Butenhoff JL, Anders MW. Biotransformation of N-ethyl-N-(2-hydroxyethyl)perfluorooctanesulfonamide by rat liver microsomes, cytosol, and slices and by expressed rat and human cytochromes P450. *Chem Res Toxicol* 2004;17:767–75.

Yu N, Wang X, Zhang B, Yang J, Li M, Li J, Shi W, Wei S, Yu H. Distribution of perfluorooctane sulfonate isomers and predicted risk of thyroid hormonal perturbation in drinking water. *Water Res* 2015;76:171–80.

Zhang Y, Beesoon S, Zhu L, Martin JW. Isomers of perfluorooctanesulfonate and perfluorooctanoate and total perfluoroalkyl acids in human serum from two cities in North China. *Environ Int* 2013;53:9–17.

Zhang YZ, Zeng XW, Qian ZM, Vaughn MG, Geiger SD, Hu LW, Lu L, Fu C, Dong GH. Perfluoroalkyl substances with isomer analysis in umbilical cord serum in China. *Environ Sci Pollut Res Int* 2017;24(15):13626–37.

Section IV

Remediation

11 Water Treatment Technologies for Targeting the Removal of Poly- and Perfluoroalkyl Substances

Eric R. V. Dickenson and Edgard M. Verdugo

CONTENTS

11.1 Introduction 241
11.2 Water and Wastewater Treatment 242
    11.2.1 Conventional Water Treatment 242
    11.2.2 Conventional Wastewater Treatment 243
    11.2.3 Adsorption 243
        11.2.3.1 Activated Carbon 243
        11.2.3.2 Anion Exchange 244
        11.2.3.3 Other Adsorbents 246
    11.2.4 Membrane Filtration 247
    11.2.5 Chemical Oxidation/Reduction 248
    11.2.6 Soil Aquifer Treatment 249
11.3 Summary 249
References 250

11.1 INTRODUCTION

By 2009, membrane filtration (i.e., nanofiltration [NF] and reverse osmosis [RO]) and adsorption processes (i.e., activated carbon and ion exchange) were accepted as the most effective established treatment processes for removing poly- and perfluoroalkyl substances (PFAS) (Rayne and Forest 2009; Lutze et al. 2012). Since PFAS have been identified as chemicals of concern in the environment, established and novel water treatment technologies have continued to be researched. Such treatment approaches include sorption, filtration, advanced oxidation, biodegradation, advanced reduction, and thermal destruction (Merino et al. 2016). Between 2009 and 2017, studies have discovered 455 new (emerging) PFAS with a wide distribution of electronegativity, hydrophobicity, and size (Xiao 2017). For many of these emerging

PFAS, we have limited to no understanding of their fate and removal by conventional and advanced water and wastewater treatment. This chapter summarizes the behavior of legacy and emerging PFAS and their transformation through established treatment technologies, includes comments on the mechanistic understanding, and highlights current advancements of promising novel treatment technologies for the physical separation or destruction of PFAS in water.

11.2 WATER AND WASTEWATER TREATMENT

11.2.1 CONVENTIONAL WATER TREATMENT

Quiñones and Snyder (2009) compared the concentration of eight perfluoroalkyl acids (PFAAs) (the combination of C ≥6 perfluoroalkyl carboxylic acids [PFCAs] and perfluoroalkyl sulfonic acids [PFSAs]) in the influents and effluents of several different drinking water treatment facilities. Three utilities used treatment trains consisting of various conventional treatment processes, including coagulation and flocculation, deep bed filtration, chloramination, medium-pressure ultraviolet (UV), chlorination, and ozonation. Regardless of the treatment train used, there was little to no attenuation of perfluorohexanoic acid (PFHxA), perfluorononanoic acid (PFNA), perfluorodecanoic acid (PFDA), perfluoroundecanoic acid (PFUnA), perfluorododecanoic acid (PFDoA), perfluorohexane sulfonic acid (PFHxS), perfluorooctanoic acid (PFOA), or perfluorooctane sulfonic acid (PFOS). Similarly, Appleman et al. (2014) compared the concentrations of 23 PFAS (prominently 9 PFCAs and 4 PFSAs) in the influents and effluents of 15 different drinking water treatment facilities. These utilities used treatment trains consisting of various conventional treatment processes, including coagulation and flocculation, dissolved air flotation, aeration packed towers, granular (deep bed) filtration, chloramination, chlorination, chlorine dioxide, potassium permanganate, UV photolysis, and ozonation. All conventional treatment operations except dissolved air flotation were ineffective at reducing PFAAs. Long-chain PFCAs and PFSAs were partially removed (49% and 29% removal for PFOS and PFNA, respectively) by dissolved air flotation, with removal performance diminishing with decreasing PFAA chain length. The results of these studies suggest that conventional treatment methods are not effective in removing PFAS.

Ozonation has been shown to produce PFCAs from precursors (Appleman et al. 2014; Rahman et al. 2014; Boiteux et al. 2017). One set of alleged precursors is the fluorotelomers (FTs), since they were removed during full-scale ozonation while simultaneously observing an increase in PFCAs (Boiteux et al. 2017). Anumol et al. (2016) determined that FTs 6:2 FTUCA and 8:2 FTUCA transformed primarily into PFHxA (6-C) and PFOA (8-C), respectively, during oxidation with ozone (O_3), ozone/hydrogen peroxide (O_3/H_2O_2), and UV/hydrogen peroxide (UV/H_2O_2), in order of decreasing efficacy, using bench-scale experiments containing spiked concentrations in ground and surface waters. Comparably, Pisarenko et al. (2015) demonstrated that molecular ozone was the primary oxidant responsible for the transformation of wastewater-occurring precursors into PFAAs, specifically PFHxA and perfluorobutane sulfonic acid (PFBS) in three different wastewaters.

11.2.2 CONVENTIONAL WASTEWATER TREATMENT

Dauchy et al. (2017) monitored the most comprehensive list of PFAS ($n = 51$) to date at a single wastewater treatment plant that utilized activated sludge, sludge flotation, and sand filtration downstream of a fluorochemical manufacturing facility. Their results showed that FTs were removed during wastewater treatment from a combination of sorption onto sludge, biodegradation, and volatilization during aeration and flotation.

There is convincing evidence that PFCAs and PFSAs are not biodegradable and are rather products of the biotransformation of unidentified and emerging PFAS (i.e., perfluorooctane sulfonamidoethanol [EtFOSE] and 8:2 fluorotelomer alcohol [FTOH]) (Liu and Mejia Avendaño 2013; Arvaniti and Stasinakis 2015; Xiao 2017).

Though adsorption of PFAAs to activated sludge has been reported (Pan et al. 2016; Dauchy et al. 2017), activated sludge (e.g., anaerobic, anoxic, and aerobic) does not effectively reduce the concentration of PFAA, and rather often leads to an increase in effluent concentration in the aqueous phase due to biological transformation (Dinglasan et al. 2004; Lee et al. 2010; Weiner et al. 2013; Harding-Marjanovic et al. 2015; Pan et al. 2016). Unlike activated sludge, however, flotation reduces PFAAs and FTs by 50% (Appleman et al. 2014; Dauchy et al. 2017), where better removal was observed for long-chain PFAAs and FTs. From the PFAS concentrations measured in sludge, Dauchy et al. (2017) determined that the removals of PFOA and PFNA during flotation occurred due to adsorption. In an unrelated study, Meng et al. (2014) demonstrated that air bubbles facilitated adsorption of PFOS on carbonaceous materials. Therefore, the possible reasons for seeing better removal from dissolved air flotation than from activated sludge are threefold. First, there are organic emulsions and colloids with adsorbed PFAAs that do not settle out, but are separated via flotation. Second, the air–water interface adsorbs or assists in the adsorption of long-chain PFAAs, allowing PFAAs to transport to the froth layer, where they become trapped. Third, the residence time (time allowed for biotransformation) is more limited in sludge flotation than it is in activated sludge treatment, and hence less bioformation of PFAAs from precursors is expected.

11.2.3 ADSORPTION

11.2.3.1 Activated Carbon

To date, multiple batch test studies on PFAA removal have been performed with granular activated carbon (GAC). Most studies have been limited to PFOS and PFOA (Lampert et al. 2007; Yu et al. 2009; Deng et al. 2010; Senevirathna et al. 2010; Schuricht et al. 2014), with the exceptions of Carter and Farrell (2010) and Inyang and Dickenson (2017), who also looked at PFBS and PFBA, respectively. These bench-scale studies revealed that GAC can be effective at removing PFAAs under certain conditions. The mechanism for adsorption of PFAAs with activated carbon has been reviewed (Du et al. 2014).

The breakthrough of PFAAs during GAC directly correlates with chain length (e.g., PFBA breakthrough is faster than PFOA) (Lampert et al. 2007; Carter and Farrell 2010; Senevirathna et al. 2010; Appleman et al. 2013, 2014; Inyang and Dickenson 2017) and is reached faster for PFCAs than for PFSAs of equal carbon chain length (e.g., PFOA breakthrough is faster than PFOS). The breakthrough behavior of PFAAs during GAC adsorption correlates to their hydrophobicity. The hydrophobicity of PFAAs increases with increasing C–F bond, and PFSAs contain one more C–F bond than their corresponding PFCAs (Appleman et al. 2014; Merino et al. 2016). GACs with higher surface areas and larger micropores that facilitate diffusion of PFAS, such as bamboo-derived activated carbon, are most effective at removing PFAAs (Merino et al. 2016). The concentration of short-chain PFAAs in the effluent of GAC-treated waters can be higher than that at the influent after an extended operating time and believed to be due to short-chain desorption caused by their exchange with other absorbable species with higher hydrophobicity, but biological formation from precursors could also be a contributing factor (Eschauzier et al. 2012; Gellrich and Knepper 2012; Appleman et al. 2014). The selectivity of GAC for long-chain PFAAs requires supplemental treatment strategies (i.e., ion exchange or NF) if short-chain PFAAs (e.g., PFBA) are targeted for removal.

Organic matter has a negative impact on the GAC adsorption of PFAS where quicker breakthrough occurs for organic matter in the form of dissolved organic carbon (DOC)–containing waters over lab-grade deionized water (Appleman et al. 2013). Organic matter is ubiquitous in source waters and is preferably adsorbed to GAC compared with most PFAAs; therefore, the competitive effect of organic matter on PFAA adsorption with GAC can be extensive in wastewater-impacted waters.

In 2007, GAC was employed in two public water systems located in Ohio and West Virginia for the treatment of PFOA from their potable water supply (Bartell et al. 2010). With a dual-contactor design, careful monitoring for breakthrough, and frequent adsorbent changes, these systems have proven highly effective at removing PFOA.

Takagi et al. (2008) looked at the effectiveness of rapid sand filtration, followed by GAC and then chlorination on PFOA and PFOS, and measured a drop from 92 to 4.1 ng/L and 4.5 to <0.1 ng/L, respectively. GAC was most likely responsible for most of this removal.

Powdered activated carbon (PAC) could be used on an as-needed basis to reduce the concentration of long-chain PFAAs (Yu et al. 2012). Compared with GAC, PAC has comparable sorption capacity and higher sorption kinetics (Yu et al. 2009). When PAC is added with a coagulant, the removal efficiency for PFOS and PFOA increases significantly (Pramanik et al. 2015). Like GAC, the efficacy of PAC for the removal of PFOS and PFOA diminishes in the presence of organic matter (Yu et al. 2012; Pramanik et al. 2015).

Similar in properties to activated carbon, biochars have gained recognition as alternative inexpensive hydrophobic adsorbents. Inyang and Dickenson (2017) showed that removals of PFOA with biochars (created from hardwood and pinewood) are as high as those with GAC in lab-grade water free of organic matter. Biochar performance for the adsorption of PFAAs suffers from competition with organic matter, and its performance in wastewater at pilot-scale was considerably inferior to that of GAC (Inyang and Dickenson 2017).

11.2.3.2 Anion Exchange

Lampert et al. (2007) investigated the removal of PFOA and PFOS using acrylic gel anion exchange (AIX) resins. They determined that Siemens A-714 AIX resin effectively removed both PFOA and PFOS (>99% removal) in 25 hours of contact in batch experiments, while the AIX resin Siemens A-244 achieved only 33% removal in the same contact time. Both resins were found to be more selective for PFOS than for PFOA. Yu et al. (2009) conducted batch experiments looking at the removal of PFOS and PFOA by the AIX resin Amberlite IRA-400. IRA-400 was found to have a higher capacity than GAC for both PFOS and PFOA and a higher capacity for PFOA than for PFOS. The removal mechanism was proposed to be a combination of anion exchange and hydrophobic interactions. The kinetics were slow, with the resin reaching equilibrium after 168 hours. Senevirathna et al. (2010) found that two AIX resins (IRA-400 and DowTM MarathonA) were more effective than GAC or nonionic resin adsorbents for the removal of PFOS. Deng et al. (2010) studied the removal of PFOS in batch tests with six different AIX resins. Polyacrylic resins featured faster kinetics and higher capacities than polystyrene resins. Carter and Farrell (2010) found that PFOS was more effectively removed than PFBS and suggested that this was due to hydrophobic interactions between the resin polymer backbone and the longer carbon chain of PFOS. Deng et al. (2010) and Carter and Farrell (2010) each found that conventional regeneration techniques were not sufficient to regenerate AIX resins that had adsorbed PFAS. Dudley (2012) studied the removal of eight PFCAs and PFSAs by four different resins and found that the acrylic, macroporous AIX resin exhibited faster PFAA removal, and that each resin had higher affinities for longer-chain PFAAs and PFSAs than for PFCAs. This study also observed that polystyrene strong base AIX resins achieved the highest PFAA removal compared with polyacrylic strong base and weak base AIX resins. Interestingly, PFAAs were more effectively removed from natural waters than from laboratory waters, possibly due to interactions between the PFAAs and natural organic matter (Dudley 2012). Further studies are warranted to look at other resins and to investigate removal kinetics and mechanisms for process optimization. Regeneration also needs to be investigated, as it will greatly affect the cost of implementing an ion exchange system intended for PFAA removal.

Appleman et al. (2014) observed removal of some PFAS at full-scale, particularly perfluoroheptanoic acid (PFHpA) (46% removal), PFOA (75% removal), and PFBS (81% removal), with the use of Purolite FerrlX A33e media (iron-infused AIX designed for anions). The authors also observed that PFSAs were preferably removed over PFCAs, and long-chain PFAAs exhibited more removal than short-chain ones for this ion exchange treatment.

Zaggia et al. (2016) compared three different AIX resins of increasing hydrophobicity, Purolite A520E, A600E, and A532E, using continuous-flow plug-flow reactors and drinking water with 1–430 ng/L of 12 PFAAs. Their results indicated that the sorption directly correlated to the level of hydrophobicity of the resin, and that the sorption capacity of their most hydrophobic AIX resin (A532E) resembled that of GAC media. Like GAC, PFBA and PFBS underwent breakthrough five times quicker than long-chain PFOA and PFOS during treatment with A532E. Unlike GAC, however, AIX resins, A520E and A600E but not A532E, were regenerable *in situ* with an ammonia hydroxide and ammonia chloride aqueous solution. AIX media A532E required an ammonia chloride water plus methanol solution, which the authors deemed impractical.

Gao et al. (2017) evaluated anion exchange resin IRA67 for potassium 2-[(6-chloro-1,1,2,2,3,3,4,4,5,5,6,6-dodecafluorohexyl)oxy]-1,1,2,2-tetrafluoroethanesulfonate (F-53B) and PFOS removal from constructed solutions and wastewater (chrome plating waste stream). The sorption capacity of PFOS (4.2 mmol/g at pH 3) and F-53B (5.5 mmol/g at pH 3) decreased with increasing pH, but only saw detrimental performance at pH >10. Because most PFAAs have a low pKa (<0), the pH_{pzc} must be higher than the pH of the solution to allow for greater contact with PFAAs. The mechanism involved in the removal of PFOS and F-53B was determined to be more than just ion exchange, and included hydrophobic interaction and the formation of micelles or hemimicelles. Like other AIX studies, the regeneration of IRA67 was most effective with water plus methanol ionic solutions. IRA67 demonstrated comparable performance for the removal of PFOS and F-53B in chrome plating waste stream and a clean solution of equal PFOS and F-53B concentrations.

11.2.3.3 Other Adsorbents

Wang et al. (2016) demonstrated that a covalent triazine-based framework (CTF) has higher sorption affinity and capacity toward PFAAs than PAC, single-walled carbon nanotubes, and Amberlite IRA-400 anion exchange resin. The superior sorption performance of CTF for PFAA adsorption was attributed to the favored electrostatic interaction between the protonated triazine and the negatively charged head groups of the PFAAs. CTF performance was higher for long-chain PFAAs and decreased with decreasing chain length and was considerably affected by pH, where lower pH exhibited higher PFAA adsorption (Wang et al. 2016). The pH dependency allowed CTF to be regenerable with alkaline water (pH >11). Unlike other carbonaceous materials, CTF performance for adsorption of PFOS and PFOA was not hindered by organic matter (Wang et al. 2016). Further research is needed to determine an effective method of immobilization for practical application (i.e., fixed-bed reactor).

Yan et al. (2014) synthesized a novel magnetic mesoporous carbon nitride (MMCN) adsorbent that is readily recovered in an external magnetic field. MMCN showed promising performance for the removal of PFOS and PFOA in bench-scale laboratory experiments with 454.55 and 370.37 mg/g adsorption capacities, respectively, and was attributed to electrostatic attraction and hydrophobic interaction (Yan et al. 2014). Further research is needed to determine its applicability in environmentally relevant water matrices.

Wang et al. (2014) showed that sorption of PFOS, PFHxS, and PFBS on permanently confined micelle arrays (PCMAs) reached equilibrium very quickly (within 5 min) but also exhibited low sorption capacities at 1.75, 1.25, and 1.45 mg/g, respectively. While pH and salt in solution showed limited effect on the sorption performance of PCMAs (Wang et al. 2014), the effect of natural organic matter may be detrimental given that PCMAs were tailored for organic contaminant removal (Wang et al. 2009; Huang and Keller 2013). However, PCMAs can be regenerated with methanol (Wang et al. 2014). and immobilized on a magnetic core (Mag-PCMAs) that is economical to synthesize (~$4/kg) and easily recoverable with an applied magnetic field, perhaps making PCMAs suitable for practical application (Wang et al. 2009; Huang and Keller 2013).

11.2.4 MEMBRANE FILTRATION

Research on the removal of PFAS using membrane filtration shows that RO and NF are effective at removing all analyzable PFAS, while microfiltration (MF) and ultrafiltration (UF) provide poor removal of PFAS. Tang et al. (2006) observed >99% removal of PFOS during RO using four different types of membranes. One NF study (Steinle-Darling and Reinhard 2008) measured the rejection capacities of NF membranes on a suite of PFAS. Flat-sheet membranes were tested originally as virgin membranes and then fouled with an alginate solution to observe the effect of a fouled layer. Average rejections were 99.3% by clean membranes and 95.3% by fouled membranes for the sulfonates. For perfluorooctane sulfonamide (FOSA), rejection decreased from about 93% to 43% because of fouling. This study only examined the effect of fouling on the PFSAs and FOSA, but not the PFCAs. While removal was generally good (93% or greater), the transmembrane pressure was not increased to maintain an identical permeate flux for the fouled and virgin membranes. Maintaining constant operational conditions (e.g., flux, cross-flow velocity, and recovery) is important for comparing contaminant rejection by fouled and unfouled membrane specimens due to their impact on the observed rejection values. While it has been shown (Bellona et al. 2010) that fouling, specifically cake-enhanced concentration polarization, can cause an increase in solute transport across the membrane surface, findings from the previous study may be explained by the fouled membranes being operated at a significantly lower flux than the virgin membranes. Another NF study (Appleman et al. 2013) measured the rejection of NF membranes on a suite of PFAAs that included PFCAs and PFSAs. All PFCAs were removed to below the method reporting limit (MRL) (>93% removal; MRL ranging from 26 to 75 ng/L) starting from an initial (spiked) concentration of a few hundred nanograms per liter in deionized or ground water. PFSAs were removed by 95% or greater and better performance was observed for groundwater after fouling had occurred. A similar study that looks only at the separation of PFHxA with NF (DowFilm NF270 membrane) achieved removals as high as 99.4% in industrial process waters using a recirculation configuration that demanded an operating pressure of 20 bar (Soriano et al. 2017).

Full-scale applications of membranes have shown that MF and UF are not effective at removing PFAAs, especially the smaller PFBA and PFBS. One full-scale water treatment plant demonstrated that MF alone (polypropylene membrane with 0.2-micron-rated pore size) did not reduce PFOA and PFOS (Appleman et al. 2014). A second plant utilizing MF (Microza MF Model UNA 620 A membrane with 0.1-micron nominal pore size) and UF (0.02-micron nominal pore size) in parallel partially removed PFAAs likely on the account of UF (Appleman et al. 2014). Removals were limited to long-chain PFAS, that is, PFOS (24% removal), PFDoA (44% removal), and FOSA (42% removal) (Appleman et al. 2014).

Full-scale application of RO and NF is effective at removing PFAAs. Boiteux et al. (2017) demonstrated that NF is effective at removing more than 41 PFAS, as PFCA and FT, in a full-scale water treatment plant to below detection levels. Appleman et al. (2014) and Thompson et al. (2011) examined reuse plants that utilize RO for the removal of PFAAs. All RO plants removed measurable PFAS to below the

MRLs, including short-chain PFBA and PFBS. There are a couple of limitations that come with the use of membrane filtration. First, RO is the most effective at removing PFAAs, but also the most expensive to implement and operate. Second, the use of RO and NF produces a concentrated waste stream that contains high levels (mg/L) of PFAS that could be difficult to treat or dispose of.

11.2.5 Chemical Oxidation/Reduction

The effects of indirect photolysis on 8:2 FTOH were studied by Gauthier and Mabury (2005). Evidence from their study suggests that 8:2 FTOH undergoes indirect photolysis with the hydroxyl radical as the main degradation agent. The photodegradation of 8:2 FTOH produced 8:2 fluorotelomer aldehyde (FTAL), 8:2 fluorotelomer acid (FTCA), 8:2 fluorotelomer unsaturated acid (FTUCA), PFOA, and PFNA. To understand the pathways of photolysis, the photodegradation of 8:2 FTUCA was examined, and it produced considerable quantities of PFOA. In a separate study performed by Plumlee et al. (2009), the effects of hydroxyl radical induced photolysis on N-EtFOSE, N-ethyl perfluorooctane sulfonamidoacetic acid (N-EtFOSAA), N-ethyl perfluorooctane sulfonamide (N-EtFOSA), and perfluorooctane sulfonamide acetic acid (FOSAA) were investigated. The results of this study indicate that the four compounds degrade and their final products are PFOA and FOSA, which did not undergo additional degradation. The photodegradation of these precursor compounds may be an explanation for the detection of PFOA in water sources in remote regions since the precursors might be more volatile than the end-product PFAAs.

PFAAs are generally resistant to oxidation. Advanced oxidation processes (AOPs), which utilize the hydroxyl radical, such as alkaline ozonation, peroxone, Fenton's reagent, and UV/hydrogen peroxide, have been shown ineffective toward PFOA and PFOS (Hori et al. 2004; Moriwaki et al. 2005; Schröder and Meesters 2005; Appleman et al. 2014). However, other oxidation/reduction technologies, such as photocatalytic oxidation, photochemical oxidation, photochemical reduction, photocatalysis, catalyzed H_2O_2 propagation, persulfate radical treatment, thermally induced reduction, and sonochemical pyrolysis, have been shown to be effective at degrading some PFAAs in water (LaZerte et al. 1953; Hori et al. 2004; Moriwaki et al. 2005; Yamamoto et al. 2007; Ochiai et al. 2011; Li et al. 2012; Mitchell et al. 2014). However, most of these technologies are not commonly employed in current drinking water treatment practices.

C–F bonds increase the reduction potential by making aliphatic compounds more electron poor than the C–H counterpart. This makes PFAS harder to oxidize but easier to reduce. Advanced reduction processes degrade contaminants utilizing aqueous electrons (hydrated electrons), hydrogen radical, and sulfite radicals. The reaction of PFAAs with hydrated electrons generated from UV photolysis (λ = 254 nm) of iodide was limited and resulted in negligible to little removal due to the interference (electron sink) of dissolved oxygen (Park et al. 2009). Its applicability is further limited due to by-product production of strong greenhouse gases, including hydrocarbons, CHF_3, and C_2F_6, that can be controlled utilizing alkaline conditions (Qu et al. 2010; Park et al. 2011). PFAS degradation with methods utilizing dithionite and sulfite combined with UV has shown more limited success (Merino et al. 2016).

Stratton et al. (2017) utilized aqueous electrons produced from plasma-based water treatment (PWT) to effectively treat groundwater containing PFOA and PFOS. According to this study, PWT had a higher defluorination efficiency (defluorination rate divided by the power density) than sonoloysis, UV/persulfate, and electrochemical treatment, but the comparison was made to data collected under different experimental conditions. The effectiveness of PWT comes from allowing adsorption of PFAAs to the gas–liquid interface where the hydrophobic fluorinated tail of PFAAs is directly exposed to plasma-generated electrons (Selma Mededovic et al. 2017).

Another emerging treatment technology capable of removing PFAAs but not currently applied in full-scale water or wastewater treatment plants is electrochemical oxidation/reduction. Soriano et al. (2017) electrochemically degraded PFHxA in NF concentrate at bench-scale with 90% removals with as little as 16 kWh/m^3 utilizing a p-type silicon boron-doped diamond (BDD) anode in an undivided cell. They determined that this application of electrochemical degradation fully mineralized PFHxA. BDD electrodes are more efficient at forming hydroxyl radicals when compared to other electrodes. As a result, BDD electrodes are commonly used for electrochemical degradation of organic contaminants (Merino et al. 2016). Because the use of BDD anodes has been shown to form perchlorate during anodic oxidation (Schaefer et al. 2015), a push has been made to use alternative electrolytic cell configurations. Schaefer et al. (2015) evaluated the use of Ti/RuO$_2$ anodes in a divided cell (using a Nafion 117 cation exchange membrane) for the decomposition of PFAAs, with primary focus on PFOS and PFOA in impacted groundwater. At a current density of 10 mA/cm^2, they reported a first-order rate constant of 45 × 10^{-5} and 70 × 10^{-5} L/min cm^2/mA. Defluorination was confirmed with 54% and 95% recovery as fluoride of PFOA and PFOS. Short-chain PFAAs, like PFBA and PFBS, were more recalcitrant than long-chain PFAAs.

11.2.6 SOIL AQUIFER TREATMENT

Only two studies on the effectiveness of soil aquifer treatment (SAT) systems were found. One involves riverbank filtration (RBF) with a 10-day subsurface travel time on drinking water that resulted in little attenuation (~10% removal) of PFOA and PFOS (Appleman et al. 2014). Snyder et al.'s (2010) study used SAT with a 2-year subsurface travel time on drinking water and observed a 28% decrease in PFOA and an increase in the concentration of PFOS.

11.3 SUMMARY

Conventional methods of wastewater treatment are ineffective at removing most PFAS. This means that wastewater treatment plants' effluents can contain varying quantities of these chemicals and their precursors when the effluents are released into the environment. Drinking water treatment plants downstream will then have contaminated influents. Because conventional methods of water treatment have also been shown to be ineffective, humans will be exposed from contaminated drinking water. The concentrations of these compounds in the drinking water will vary depending on the source of contamination—industrial or municipal wastewater effluents. The limited bench-, pilot-, and full-scale evaluations that have been done on advanced treatment methods have shown promising results for the removal of these chemicals, especially using GAC, AIX, and NF/RO membrane filtration. These methods, as well as other advanced treatment methods, need to be explored further to verify their effectiveness at removing emerging PFAS from contaminated water sources.

REFERENCES

Anumol, Tarun, Sonia Dagnino, Darcy R. Vandervort, and Shane A. Snyder. 2016. Transformation of polyfluorinated compounds in natural waters by advanced oxidation processes. *Chemosphere* 144:1780–1787. doi: http://dx.doi.org/10.1016/j.chemosphere.2015.10.070.

Appleman, Timothy D., Eric R. V. Dickenson, Christopher Bellona, and Christopher P. Higgins. 2013. Nanofiltration and granular activated carbon treatment of perfluoroalkyl acids. *Journal of Hazardous Materials* 260:740–746. doi: http://dx.doi.org/10.1016/j.jhazmat.2013.06.033.

Appleman, Timothy D., Christopher P. Higgins, Oscar Quiñones, Brett J. Vanderford, Chad Kolstad, Janie C. Zeigler-Holady, and Eric R. V. Dickenson. 2014. Treatment of poly-and perfluoroalkyl substances in US full-scale water treatment systems. *Water Research* 51:246–255.

Arvaniti, Olga S., and Athanasios S. Stasinakis. 2015. Review on the occurrence, fate and removal of perfluorinated compounds during wastewater treatment. *Science of the Total Environment* 524:81–92. doi: http://dx.doi.org/10.1016/j.scitotenv.2015.04.023.

Bartell, Scott M., Antonia M. Calafat, Christopher Lyu, Kayoko Kato, P. Barry Ryan, and Kyle Steenland. 2010. Rate of decline in serum PFOA concentrations after granular activated carbon filtration at two public water systems in Ohio and West Virginia. *Environmental Health Perspectives* 118 (2):222–228. doi: 10.1289/ehp.0901252.

Bellona, Christopher, Melissa Marts, and Jörg E. Drewes. 2010. The effect of organic membrane fouling on the properties and rejection characteristics of nanofiltration membranes. *Separation and Purification Technology* 74 (1):44–54. doi: http://dx.doi.org/10.1016/j.seppur.2010.05.006.

Boiteux, Virginie, Xavier Dauchy, Cristina Bach, Adeline Colin, Jessica Hemard, Véronique Sagres, Christophe Rosin, and Jean-François Munoz. 2017. Concentrations and patterns of perfluoroalkyl and polyfluoroalkyl substances in a river and three drinking water treatment plants near and far from a major production source. *Science of the Total Environment* 583:393–400. doi: https://doi.org/10.1016/j.scitotenv.2017.01.079.

Carter, Kimberly E., and James Farrell. 2010. Removal of perfluorooctane and perfluorobutane sulfonate from water via carbon adsorption and ion exchange. *Separation Science and Technology* 45 (6):762–767. doi: 10.1080/01496391003608421.

Dauchy, Xavier, Virginie Boiteux, Cristina Bach, Adeline Colin, Jessica Hemard, Christophe Rosin, and Jean-François Munoz. 2017. Mass flows and fate of per- and polyfluoroalkyl substances (PFASs) in the wastewater treatment plant of a fluorochemical manufacturing facility. *Science of the Total Environment* 576:549–558. doi: https://doi.org/10.1016/j.scitotenv.2016.10.130.

Deng, Shubo, Qiang Yu, Jun Huang, and Gang Yu. 2010. Removal of perfluorooctane sulfonate from wastewater by anion exchange resins: Effects of resin properties and solution chemistry. *Water Research* 44 (18):5188–5195. doi: http://dx.doi.org/10.1016/j.watres.2010.06.038.

Dinglasan, Mary Joyce A., Yun Ye, Elizabeth A. Edwards, and Scott A. Mabury. 2004. Fluorotelomer alcohol biodegradation yields poly- and perfluorinated acids. *Environmental Science & Technology* 38 (10):2857–2864. doi: 10.1021/es0350177.

Du, Ziwen, Shubo Deng, Yue Bei, Qian Huang, Bin Wang, Jun Huang, and Gang Yu. 2014. Adsorption behavior and mechanism of perfluorinated compounds on various adsorbents—A review. *Journal of Hazardous Materials* 274:443–454. doi: http://dx.doi.org/10.1016/j.jhazmat.2014.04.038.

Dudley, Leigh-Ann. 2012. Removal of perfluorinated compounds by powdered activated carbon, superfine powder activated carbon, and anion exchange resin. http://www.lib.ncsu.edu/resolver/1840.16/7654.

Eschauzier, Christian, Erwin Beerendonk, Petra Scholte-Veenendaal, and Pim De Voogt. 2012. Impact of treatment processes on the removal of perfluoroalkyl acids from the drinking water production chain. *Environmental Science & Technology* 46 (3):1708–1715.

Gao, Yanxin, Shubo Deng, Ziwen Du, Kai Liu, and Gang Yu. 2017. Adsorptive removal of emerging polyfluoroalky substances F-53B and PFOS by anion-exchange resin: A comparative study. *Journal of Hazardous Materials* 323, Part A:550–557. doi: https://doi.org/10.1016/j.jhazmat.2016.04.069.

Gauthier, Suzanne A., and Scott A. Mabury. 2005. Aqueous photolysis of 8:2 fluorotelomer alcohol. *Environmental Toxicology and Chemistry* 24 (8):1837–1846. doi: 10.1897/04-591R.1.

Gellrich, Vanessa, and Thomas P. Knepper. 2012. Sorption and leaching behavior of perfluorinated compounds in soil. In *Polyfluorinated Chemicals and Transformation Products*, edited by Thomas P. Knepper and Frank T. Lange, 63–72. Berlin: Springer.

Harding-Marjanovic, Katie C., Erika F. Houtz, Shan Yi, Jennifer A. Field, David L. Sedlak, and Lisa Alvarez-Cohen. 2015. Aerobic biotransformation of fluorotelomer thioether amido sulfonate (Lodyne) in AFFF-amended microcosms. *Environmental Science & Technology* 49 (13):7666–7674. doi: 10.1021/acs.est.5b01219.

Hori, Hisao, Etsuko Hayakawa, Hisahiro Einaga, Shuzo Kutsuna, Kazuhide Koike, Takashi Ibusuki, Hiroshi Kiatagawa, and Ryuichi Arakawa. 2004. Decomposition of environmentally persistent perfluorooctanoic acid in water by photochemical approaches. *Environmental Science & Technology* 38 (22):6118–6124. doi: 10.1021/es049719n.

Huang, Yuxiong, and Arturo A. Keller. 2013. Magnetic nanoparticle adsorbents for emerging organic contaminants. *ACS Sustainable Chemistry & Engineering* 1 (7):731–736. doi: 10.1021/sc400047q.

Inyang, Mandu, and Eric R. V. Dickenson. 2017. The use of carbon adsorbents for the removal of perfluoroalkyl acids from potable reuse systems. *Chemosphere* 184:168–175. doi: https://doi.org/10.1016/j.chemosphere.2017.05.161.

Lampert, David J., Michael A. Frisch, and Gerald E. Speitel. 2007. Removal of perfluorooctanoic acid and perfluorooctane sulfonate from wastewater by ion exchange. *Practice Periodical of Hazardous, Toxic, and Radioactive Waste Management* 11 (1):60–68. doi: 10.1061/(ASCE)1090-025X(2007)11:1(60).

LaZerte, J. D., L. J. Hals, T. S. Reid, and G. H. Smith. 1953. Pyrolyses of the salts of the perfluoro carboxylic acids. *Journal of the American Chemical Society* 75 (18):4525–4528. doi: 10.1021/ja01114a040.

Lee, Holly, Jessica D'eon, and Scott A. Mabury. 2010. Biodegradation of polyfluoroalkyl phosphates as a source of perfluorinated acids to the environment. *Environmental Science & Technology* 44 (9):3305–3310. doi: 10.1021/es9028183.

Li, Xiaoyun, Pengyi Zhang, Ling Jin, Tian Shao, Zhenmin Li, and Junjie Cao. 2012. Efficient photocatalytic decomposition of perfluorooctanoic acid by indium oxide and its mechanism. *Environmental Science & Technology* 46 (10):5528–5534. doi: 10.1021/es204279u.

Liu, Jinxia, and Sandra Mejia Avendaño. 2013. Microbial degradation of polyfluoroalkyl chemicals in the environment: A review. *Environment International* 61:98–114. doi: http://dx.doi.org/10.1016/j.envint.2013.08.022.

Lutze, Holger, Stefan Panglisch, Axel Bergmann, and Torsten C. Schmidt. 2012. Treatment options for the removal and degradation of polyfluorinated chemicals. In *Polyfluorinated Chemicals and Transformation Products*, edited by Thomas P. Knepper and Frank T. Lange, 103–125. Berlin: Springer.

Meng, Pingping, Shubo Deng, Xinyu Lu, Ziwen Du, Bin Wang, Jun Huang, Yujue Wang, Gang Yu, and Baoshan Xing. 2014. Role of air bubbles overlooked in the adsorption of perfluorooctanesulfonate on hydrophobic carbonaceous adsorbents. *Environmental Science & Technology* 48 (23):13785–13792. doi: 10.1021/es504108u.

Merino, Nancy, Yan Qu, Rula A. Deeb, Elisabeth L. Hawley, Michael R. Hoffmann, and Shaily Mahendra. 2016. Degradation and removal methods for perfluoroalkyl and polyfluoroalkyl substances in water. *Environmental Engineering Science* 33 (9):615–649. doi: 10.1089/ees.2016.0233.

Mitchell, Shannon M., Mushtaque Ahmad, Amy L. Teel, and Richard J. Watts. 2014. Degradation of perfluorooctanoic acid by reactive species generated through catalyzed H2O2 propagation reactions. *Environmental Science & Technology Letters* 1 (1):117–121. doi: 10.1021/ez4000862.

Moriwaki, Hiroshi, Youichi Takagi, Masanobu Tanaka, Kenshiro Tsuruho, Kenji Okitsu, and Yasuaki Maeda. 2005. Sonochemical decomposition of perfluorooctane sulfonate and perfluorooctanoic acid. *Environmental Science & Technology* 39 (9):3388–3392. doi: 10.1021/es040342v.

Ochiai, Tsuyoshi, Yuichi Iizuka, Kazuya Nakata, Taketoshi Murakami, Donald A. Tryk, Yoshihiro Koide, Yuko Morito, and Akira Fujishima. 2011. Efficient decomposition of perfluorocarboxylic acids in aqueous suspensions of a TiO2 photocatalyst with medium-pressure ultraviolet lamp irradiation under atmospheric pressure. *Industrial & Engineering Chemistry Research* 50 (19):10943–10947. doi: 10.1021/ie1017496.

Pan, Chang-Gui, You-Sheng Liu, and Guang-Guo Ying. 2016. Perfluoroalkyl substances (PFASs) in wastewater treatment plants and drinking water treatment plants: Removal efficiency and exposure risk. *Water Research* 106:562–570. doi: http://dx.doi.org/10.1016/j.watres.2016.10.045.

Park, Hyunwoong, Chad D. Vecitis, Jie Cheng, Wonyong Choi, Brian T. Mader, and Michael R. Hoffmann. 2009. Reductive defluorination of aqueous perfluorinated alkyl surfactants: Effects of ionic headgroup and chain length. *Journal of Physical Chemistry A* 113 (4):690–696. doi: 10.1021/jp807116q.

Park, Hyunwoong, Chad D. Vecitis, Jie Cheng, Nathan F. Dalleska, Brian T. Mader, and Michael R. Hoffmann. 2011. Reductive degradation of perfluoroalkyl compounds with aquated electrons generated from iodide photolysis at 254 nm. *Photochemical & Photobiological Sciences* 10 (12):1945–1953. doi: 10.1039/C1PP05270E.

Pisarenko, Aleksey N., Erica J. Marti, Daniel Gerrity, Julie R. Peller, and Eric R. V. Dickenson. 2015. Effects of molecular ozone and hydroxyl radical on formation of N-nitrosamines and perfluoroalkyl acids during ozonation of treated wastewaters. *Environmental Science: Water Research & Technology* 1 (5):668–678. doi: 10.1039/C5EW00046G.

Plumlee, Megan H., Kristopher McNeill, and Martin Reinhard. 2009. Indirect photolysis of perfluorochemicals: Hydroxyl radical-initiated oxidation of N-ethyl perfluorooctane sulfonamido acetate (N-EtFOSAA) and other perfluoroalkanesulfonamides. *Environmental Science & Technology* 43 (10):3662–3668. doi: 10.1021/es803411w.

Pramanik, Biplob Kumar, Sagor Kumar Pramanik, and Fatihah Suja. 2015. A comparative study of coagulation, granular- and powdered-activated carbon for the removal of perfluorooctane sulfonate and perfluorooctanoate in drinking water treatment. *Environmental Technology* 36 (20):2610–2617. doi: 10.1080/09593330.2015.1040079.

Qu, Yan, Chaojie Zhang, Fei Li, Jing Chen, and Qi Zhou. 2010. Photo-reductive defluorination of perfluorooctanoic acid in water. *Water Research* 44 (9):2939–2947. doi: http://dx.doi.org/10.1016/j.watres.2010.02.019.

Quiñones, Oscar, and Shane A. Snyder. 2009. Occurrence of perfluoroalkyl carboxylates and sulfonates in drinking water utilities and related waters from the United States. *Environmental Science & Technology* 43 (24):9089–9095. doi: 10.1021/es9024707.

Rahman, Mohammad Feisal, Sigrid Peldszus, and William B. Anderson. 2014. Behaviour and fate of perfluoroalkyl and polyfluoroalkyl substances (PFASs) in drinking water treatment: A review. *Water Research* 50:318–340. doi: http://dx.doi.org/10.1016/j.watres.2013.10.045.

Rayne, Sierra, and Kaya Forest. 2009. Perfluoroalkyl sulfonic and carboxylic acids: A critical review of physicochemical properties, levels and patterns in waters and wastewaters, and treatment methods. *Journal of Environmental Science and Health, Part A* 44 (12):1145–1199. doi: 10.1080/10934520903139811.

Schaefer, Charles E., Christina Andaya, Ana Urtiaga, Erica R. McKenzie, and Christopher P. Higgins. 2015. Electrochemical treatment of perfluorooctanoic acid (PFOA) and perfluorooctane sulfonic acid (PFOS) in groundwater impacted by aqueous film forming foams (AFFFs). *Journal of Hazardous Materials* 295:170–175. doi: http://dx.doi.org/10.1016/j.jhazmat.2015.04.024.

Schröder, Horst Fr., and Roland J. W. Meesters. 2005. Stability of fluorinated surfactants in advanced oxidation processes—A follow up of degradation products using flow injection–mass spectrometry, liquid chromatography–mass spectrometry and liquid chromatography–multiple stage mass spectrometry. *Journal of Chromatography A* 1082 (1):110–119. doi: http://dx.doi.org/10.1016/j.chroma.2005.02.070.

Schuricht, Falk, Wladimir Reschetilowski, Andy Reich, and Eckart Giebler. 2014. Elimination of perfluorinated surfactants—Adsorbent evaluation applying surface tension measurements. *Chemical Engineering & Technology* 37 (7):1121–1126. doi: 10.1002/ceat.201400025.

Selma Mededovic, Thagard, R. Stratton Gunnar, Dai Fei, L. Bellona Christopher, M. Holsen Thomas, G. Bohl Douglas, Paek Eunsu, and R. V. Dickenson Eric. 2017. Plasma-based water treatment: Development of a general mechanistic model to estimate the treatability of different types of contaminants. *Journal of Physics D: Applied Physics* 50 (1):014003.

Senevirathna, S. T. M. L. D., Shuhei Tanaka, Shigeo Fujii, Chinagarn Kunacheva, Hidenori Harada, Binaya Raj Shivakoti, and Risa Okamoto. 2010. A comparative study of adsorption of perfluorooctane sulfonate (PFOS) onto granular activated carbon, ion-exchange polymers and non-ion-exchange polymers. *Chemosphere* 80 (6):647–651. doi: http://dx.doi.org/10.1016/j.chemosphere.2010.04.053.

Snyder, Shane A., Mark J. Benotti, Fernando Rosario-Ortiz, Brett J. Vanderford, Jorg. E. Drewes, and Eric R. V. Dickenson. 2010. Comparison of chemical composition of reclaimed and conventional waters. Alexandria, VA: WateReuse Research Foundation.

Soriano, Álvaro, Daniel Gorri, and Ane Urtiaga. 2017. Efficient treatment of perfluorohexanoic acid by nanofiltration followed by electrochemical degradation of the NF concentrate. *Water Research* 112:147–156. doi: https://doi.org/10.1016/j.watres.2017.01.043.

Steinle-Darling, Eva, and Martin Reinhard. 2008. Nanofiltration for trace organic contaminant removal: Structure, solution, and membrane fouling effects on the rejection of perfluorochemicals. *Environmental Science & Technology* 42 (14):5292–5297. doi: 10.1021/es703207s.

Stratton, Gunnar R., Fei Dai, Christopher L. Bellona, Thomas M. Holsen, Eric R. V. Dickenson, and Selma Mededovic Thagard. 2017. Plasma-based water treatment: Efficient transformation of perfluoroalkyl substances in prepared solutions and contaminated groundwater. *Environmental Science & Technology* 51 (3):1643–1648. doi: 10.1021/acs.est.6b04215.

Takagi, Sokichi, Fumie Adachi, Keiichi Miyano, Yoshihiko Koizumi, Hidetsugu Tanaka, Mayumi Mimura, Isao Watanabe, Shinsuke Tanabe, and Kurunthachalam Kannan.

2008. Perfluorooctanesulfonate and perfluorooctanoate in raw and treated tap water from Osaka, Japan. *Chemosphere* 72 (10):1409–1412. doi: http://dx.doi.org/10.1016/j.chemosphere.2008.05.034.

Tang, Chuyang Y., Q. Shiang Fu, A. P. Robertson, Craig S. Criddle, and James O. Leckie. 2006. Use of reverse osmosis membranes to remove perfluorooctane sulfonate (PFOS) from semiconductor wastewater. *Environmental Science & Technology* 40 (23):7343–7349. doi: 10.1021/es060831q.

Thompson, Jack, Geoff Eaglesham, Julien Reungoat, Yvan Poussade, Michael Bartkow, Michael Lawrence, and Jochen F. Mueller. 2011. Removal of PFOS, PFOA and other perfluoroalkyl acids at water reclamation plants in South East Queensland Australia. *Chemosphere* 82 (1):9–17. doi: http://dx.doi.org/10.1016/j.chemosphere.2010.10.040.

Wang, Bingyu, Linda S. Lee, Chenhui Wei, Heyun Fu, Shourong Zheng, Zhaoyi Xu, and Dongqiang Zhu. 2016. Covalent triazine-based framework: A promising adsorbent for removal of perfluoroalkyl acids from aqueous solution. *Environmental Pollution* 216:884–892. doi: http://dx.doi.org/10.1016/j.envpol.2016.06.062.

Wang, Fei, Xingwen Lu, Kai Min Shih, Peng Wang, and Xiaoyan Li. 2014. Removal of perfluoroalkyl sulfonates (PFAS) from aqueous solution using permanently confined micelle arrays (PCMAs). *Separation and Purification Technology* 138:7–12. doi: http://dx.doi.org/10.1016/j.seppur.2014.09.037.

Wang, Peng, Qihui Shi, Yifeng Shi, Kristin K. Clark, Galen D. Stucky, and Arturo A. Keller. 2009. Magnetic permanently confined micelle arrays for treating hydrophobic organic compound contamination. *Journal of the American Chemical Society* 131 (1):182–188. doi: 10.1021/ja806556a.

Weiner, Barbara, Leo W. Y. Yeung, Erin B. Marchington, Lisa A. D'Agostino, and Scott A. Mabury. 2013. Organic fluorine content in aqueous film forming foams (AFFFs) and biodegradation of the foam component 6:2 fluorotelomermercaptoalkylamido sulfonate (6:2 FTSAS). *Environmental Chemistry* 10 (6):486–493. doi: https://doi.org/10.1071/EN13128.

Xiao, Feng. 2017. Emerging poly- and perfluoroalkyl substances in the aquatic environment: A review of current literature. *Water Research* 124:482–495. doi: http://dx.doi.org/10.1016/j.watres.2017.07.024.

Yamamoto, Takashi, Yukio Noma, Shin-ichi Sakai, and Yasuyuki Shibata. 2007. Photodegradation of perfluorooctane sulfonate by UV irradiation in water and alkaline 2-propanol. *Environmental Science & Technology* 41 (16):5660–5665. doi: 10.1021/es0706504.

Yan, Tingting, Huan Chen, Fang Jiang, and Xin Wang. 2014. Adsorption of perfluorooctane sulfonate and perfluorooctanoic acid on magnetic mesoporous carbon nitride. *Journal of Chemical & Engineering Data* 59 (2):508–515. doi: 10.1021/je400974z.

Yu, Jing, Lu Lv, Pei Lan, Shujuan Zhang, Bingcai Pan, and Weiming Zhang. 2012. Effect of effluent organic matter on the adsorption of perfluorinated compounds onto activated carbon. *Journal of Hazardous Materials* 225:99–106. doi: http://dx.doi.org/10.1016/j.jhazmat.2012.04.073.

Yu, Qiang, Ruiqi Zhang, Shubo Deng, Jun Huang, and Gang Yu. 2009. Sorption of perfluorooctane sulfonate and perfluorooctanoate on activated carbons and resin: Kinetic and isotherm study. *Water Research* 43 (4):1150–1158. doi: http://dx.doi.org/10.1016/j.watres.2008.12.001.

Zaggia, Alessandro, Lino Conte, Luigi Falletti, Massimo Fant, and Andrea Chiorboli. 2016. Use of strong anion exchange resins for the removal of perfluoroalkylated substances from contaminated drinking water in batch and continuous pilot plants. *Water Research* 91:137–146. doi: http://dx.doi.org/10.1016/j.watres.2015.12.039.

12 Oxidation and Reduction Approaches for Treatment of Perfluoroalkyl Substances

Blossom N. Nzeribe, Selma M. Thagard, Thomas M. Holsen, Gunnar Stratton, and Michelle Crimi

CONTENTS

12.1	Introduction	256
12.2	Chemical Oxidation Processes Using Activated Persulfate	257
	12.2.1 Activated Persulfate	257
	12.2.2 Persulfate Activation	259
	12.2.2.1 Iron Activation	259
	12.2.2.2 Alkaline pH-Activated Persulfate	260
	12.2.2.3 Heat-Activated Persulfate	260
	12.2.3 Mechanism of Decomposition	260
	12.2.4 Supporting Research	262
12.3	Electrochemical Oxidation	263
	12.3.1 Mechanism of Decomposition	266
	12.3.2 Supporting Research	267
12.4	Ultrasonication (Sonolysis)	269
	12.4.1 Mechanism of Decomposition	274
	12.4.2 Supporting Research	275
12.5	Chemical Reduction Processes	276
	12.5.1 Reductants	276
	12.5.2 Mechanism of Decomposition	279
	12.5.3 Supporting Research	281
12.6	Plasma Technology	282
	12.6.1 Mechanism of Decomposition	285
	12.6.2 Supporting Research	286
12.7	Implications	291
12.8	Conclusions	293
References		293

12.1 INTRODUCTION

A great concern among many nations that rely on groundwater as a source of drinking water is the contamination of their aquifers. Groundwater can be contaminated via point and non–point sources. Examples of point sources include landfill sites, underground storage tanks (chemical storage tanks), sewer lines, and hazardous industrial wastes. Non–point sources include agricultural activities (application and storage of herbicides, pesticides, manure, and fertilizers) and atmospheric deposition.

Per- and polyfluoroalkyl substances (PFAS) are chemicals that are used in a wide array of industrial applications due to their unique hydrophobic and hydrophilic properties, their ability to withstand extreme environmental conditions, and their high surface activity (Buck et al. 2012). They are mostly used in aqueous film-forming foams (AFFFs), surfactants, and lubricants (Moody and Field 2000) and were used by the US Department of Defense starting in the 1960s during firefighting operations and training at refineries and airports (Backe et al. 2013).

PFAS can also form in the environment from polyfluorinated precursors, such as fluorotelomer acrylates (FTAs), fluorotelomer alcohols (FTOHs), per-fluoroalkyl sulfonamids (FASA), and sulfonamido ethanols (Buck et al. 2011). They are highly recalcitrant to conventional water treatment processes and advanced oxidative processes (AOPs) due to the strong carbon-fluorine bonds (C–F) that exist within the compounds (Lee et al. 2012a). Therefore, PFAS are environmentally persistent, bioaccumulative (Herzke et al. 2007), and a potential risk to humans.

PFAS are usually classified into two major groups: perfluorinated sulfonic acids (PFSAs) and perfluorinated carboxylic acids (PFCAs). Perfluorooctanoic acid (PFOA) and perfluorooctanesulfonic acid (PFOS) are the two PFAS that have received the most attention due to their bioaccumulation potential, persistence, toxicity, and ubiquitous presence in the environment. PFOS was shortlisted as a persistent organic pollutant (POP) by the Stockholm Convention on Persistent Organic Pollutants in 2009, while PFOA was anticipated to be a likely carcinogen by the US Environmental Protection Agency (USEPA). In 2009, the USEPA established provisional health advisory (PHA) levels in drinking water for PFOA and PFOS at 0.4 and 0.2 µg/L, respectively (USEPA 2009). In May 2016, the USEPA released a lifetime health advisory level for combined concentrations of PFOA and PFOS of 0.07 µg/L (parts per trillion) (USEPA 2016a). Table 12.1 shows the most common PFAS, their molecular weight, and common acronyms.

This chapter focuses on the oxidative and reductive chemical processes (redox) that have been investigated for the treatment of PFAS in groundwater. Unlike most water treatment processes based on phase separation that produce a residual waste resulting in increased costs and long-term liability, chemical processes are aimed at destroying and mineralizing contaminants to water and carbon dioxide or transforming them to less toxic biodegradable products. The treatment processes discussed here include a chemical oxidation process using heat-activated persulfate, electrochemical oxidation, ultrasonication (US) (sonolysis), a reductive chemical process using zerovalent iron (ZVI), advanced reduction processes (ARPs), and plasma-based technology.

TABLE 12.1
Names and Molecular Weights of Most Common PFAS

PFAS	Acronyms	Molecular Weight (g/mol)	Carbon Chain Length (n)	Formula
Trifluoracetic acid	TFA	114.02	2	CF_3COOH
Perfluoropropanoic acid	PFPrA	164.03	3	C_2F_5COOH
Perfluorobutanoic acid	PFBA	214.03	4	C_3F_7COOH
Perfluorobutanesulfonate	PFBS	300.10	4	$C_4HF_9SO_3$
Perfluoropentanoic acid	PFPnA	264.05	5	C_4F_9COOH
Perfluorohexanoic acid	PFHxA	314.05	5	$C_5F_{11}COOH$
Perfluorohexanesulfonate	PFHxS	400.11	6	$C_6HF_{13}SO_3$
Perfluoroheptanoic acid	PFHpA	364.06	7	$C_6F_{13}COOH$
Perfluorooctanoic acid	PFOA	414.07	8	$C_7F_{15}COOH$
Perfluorooctanesulfonic acid	PFOS	500.13	8	$C_8HF_{17}SO_3$

12.2 CHEMICAL OXIDATION PROCESSES USING ACTIVATED PERSULFATE

Successful chemical oxidation processes degrade recalcitrant organic or inorganic compounds into nontoxic and/or biodegradable products. Common oxidants for groundwater remediation include potassium permanganate, chlorine, activated persulfate, catalyzed hydrogen peroxide, and ozone. Oxidants are mainly introduced to the subsurface via injection wells, where they react directly with target compounds or with water, forming free radicals (Siegrist et al. 2011). Chemical oxidation is the transfer of electrons from an electron donor (reductant) to an acceptor (oxidant), which results in a chemical transformation of a contaminant directly or the formation of unstable reactive species (free radicals), which then aggressively degrade contaminants. The free radical species are usually nonspecific and highly reactive. Free radical chemical oxidation processes typically involve a series of chain reactions: initiation, propagation, and termination (Petri et al. 2011a). Production of radicals from nonradical species occurs in the initiation stage. These radicals are consumed and form new radicals in the propagation stage until no more radicals are consumed or formed (termination stage), resulting in mineralized or oxidized contaminants. A drawback of *in situ* chemical oxidation (ISCO) is the scavenging of oxidants by nontarget compounds, such as co-contaminants, or subsurface constituents, such as natural organic matter or reactive metals.

12.2.1 ACTIVATED PERSULFATE

Perfluorinated compounds are too stable to be susceptible to direct electron transfer oxidation; however, researchers have investigated free radical oxidation processes since they are a more aggressive treatment approach. The majority of these studies have focused on activated persulfate. Persulfate ($S_2O_8^{2-}$) is a strong oxidant with a redox potential of 2.01 V (Petri et al. 2011). It is an anion that reacts through direct electron transfer or activation into reactive free radicals to degrade contaminants (Equation 12.1). In direct electron transfer, persulfate reacts directly with contaminants with no formation of free radicals. However, direct electron transfer is a slow and specific reaction.

$$S_2O_8^{2-} + 2e^- \rightarrow 2SO_4^{2-} \tag{12.1}$$

Unactivated persulfate has a long persistence in the subsurface (Tsitonaki et al. 2010) and can be transported a considerable distance prior to activation. This persistence provides enough contact time for interaction with contaminants, hence its widespread use for *in situ* remediation of groundwater. There is also less interaction between natural organic matter and persulfate in contaminated aquifers compared with other oxidants. Upon activation, persulfate generates sulfate radicals (2.6 V) (Equations 12.2 and 12.3).

$$S_2O_8^{2-} \xrightarrow{\text{heat}} 2SO_4^{\bullet-} \tag{12.2}$$

$$S_2O_8^{2-} \xrightarrow{\text{transition metal}} SO_4^{\bullet-} + SO_4^{2-} \tag{12.3}$$

The generated sulfate radical reacts with water to also form hydroxyl radicals (2.7 V) (Equation 12.4).

$$SO_4^{\bullet-} + H_2O \rightarrow OH^{\bullet} + SO_4^{2-} + H^+ \tag{12.4}$$

Sulfate radicals are very reactive with compounds from which they can draw an electron (electrophilic); thus, they react faster with electron donors than electron acceptors (Tsitonaki et al. 2010). Activated persulfate degrades most common groundwater contaminants, including trichloroethylene (TCE) (Liang and Lai 2008; Y. Lin et al. 2016), tetrachlorethylene (PCE) (Liang et al. 2007), herbicides (Liu et al. 2012b; Tan et al. 2012; Fan et al. 2015; Ji et al. 2015), volatile organic compounds (VOCs) (Huang et al. 2005; Liang et al. 2010; Huling et al. 2011), pharmaceuticals (Ji et al. 2014; Chen et al. 2016), pesticides (Khan et al. 2016), and some perfluorinated compounds (Park et al. 2016).

Various methods have been used to activate persulfate in ISCO treatment of groundwater to generate sulfate radicals, mostly via chemical or heat activation. Chemical activation involves the use of acids, bases, or transition metals, such as ferrous iron (Fe^{2+}) and ZVI, while heat activation includes processes such as sonolysis, ultraviolet (UV) irradiation, microwave heating, and hydrothermal activation.

The type of activation method plays a critical role in the rate of degradation of contaminants because each activation method involves different mechanisms and may be contaminant specific (Block et al. 2004; Crimi and Taylor 2007).

At high pH conditions, OH radicals are the dominant radicals formed, because the sulfate radicals initially produced are effectively scavenged by hydroxyl anion to form hydroxyl radical. While at low pH, there is a greater yield of sulfate radical, although hydroxyl radical is still formed. Treatment time, pH, persulfate concentration, and the presence of groundwater ions, such as chloride and carbonate, are important operating parameters that play a role in activated persulfate degradation of contaminants. Persulfate has limited interaction with natural organic matter, but reacts with groundwater ions, such as bicarbonate, carbonate, and chloride ions (Bennedsen et al. 2012).

12.2.2 Persulfate Activation

12.2.2.1 Iron Activation

The activation of persulfate with iron, especially ferrous iron and ZVI (ZVI/Fe^0), involves a one-electron transfer and formation of ferric iron (Fe^{3+}) and sulfate radical (SO$_4^{\bullet-}$) (Crimi and Taylor 2007) (Equation 12.5). Iron is mostly used in metal activation of persulfate because of its low toxicity, cost-effectiveness, and environmental friendliness (Rastogi et al. 2009).

$$S_2O_8^{2-} + Fe^{2+} \rightarrow SO_4^{\bullet-} + SO_4^{2-} + Fe^{3+} \tag{12.5}$$

Although iron has shown success in activating persulfate, low reaction efficiency due to scavenging of sulfate radical by excess ferrous iron (Equation 12.6) has limited the oxidizing potential of an iron-activated persulfate system (Liang et al. 2004a; Vicente et al. 2011). Thus, there is a need for a system that promotes the slow release of ferrous iron.

$$SO_4^{\bullet-} + Fe^{2+} \rightarrow SO_4^{2-} + Fe^{3+} \tag{12.6}$$

ZVI has been used to achieve the continuous slow release of Fe^{2+} (Oh et al. 2009; Rodriguez et al. 2014; Pardo et al. 2015). However, there is an intrinsic drawback to the use of ZVI in the subsurface due to the formation of a passive layer that lowers surface area (Guan et al. 2015). Its large size also makes ZVI hard to inject directly into the subsurface. The development of nanoscale ZVI (nZVI) has solved both problems (Li et al. 2006; Chang and Kang 2009). Studies have reported successful degradation of perfluorinated compounds by iron-activated persulfate. Lee et al. (2010) observed an enhanced decomposition and defluorination of PFOA (240.7 μM) at 90°C with a microwave-hydrothermal persulfate and ZVI system under acidic conditions. They suggested that the enhanced destruction might be due to the addition of ZVI, which led to increased formation of sulfate free radicals.

The use of chelating agents, such as citric acid, ethylenediaminetetraacetic acid (EDTA), and sodium thiosulfate (Na$_2$S$_2$O$_3$), to regulate ferrous iron release has also been investigated (Liang et al. 2004b; Furman et al. 2010; Zhou et al. 2013; Zhang et al. 2014), with citric acid being the most commonly used due to its biodegradability (Liang et al. 2004b). The activation energy associated with iron activation of persulfate has been reported to be less than that of thermal activation. Fordham and Williams (1951) reported an activation energy of 12 kcal/mol compared with that reported by Kolthoff and Miller (1951) of 33.5 kcal/mol for thermal activation.

12.2.2.2 Alkaline pH-Activated Persulfate

Alkaline pH-activated persulfate is one of the most commonly used persulfate activation processes for ISCO of groundwater contaminants. It involves the addition of a base, such as sodium or potassium hydroxide, to elevate the pH value above 10 (Furman et al. 2010; Petri et al. 2011b). Several mechanisms have been proposed for this process. A study by Kolthoff and Miller (1951) suggested that upon persulfate decomposition, two sulfate radicals are formed and are then transformed to OH•, while Singh and Venkatarao (1976) suggested that persulfate decomposes to peroxomonosulfate and then breaks down to sulfate and molecular oxygen. Another proposed mechanism is that sulfate and hydrogen peroxide anions are formed by base-catalyzed hydrolysis of persulfate (Equation 12.7), followed by reduction of another persulfate molecule by hydrogen peroxide to sulfate anion and sulfate radical (Equation 12.8), followed by oxidation of hydrogen peroxide to superoxide (Equation 12.9). The sulfate radical then oxidizes hydroxide-producing hydroxyl radicals as proposed by Furman et al. (2010).

$$S_2O_8^{2-} + 2H_2O \rightarrow 2SO_4^{2-} + 3H^+ + HO_2^- \tag{12.7}$$

$$HO_2^- + S_2O_8^{2-} \rightarrow SO_4^{2-} + SO_4^{\bullet-} + O_2^{\bullet-} \tag{12.8}$$

$$SO_4^{\bullet-} + OH^- \rightarrow SO_4^{2-} + OH^{\bullet} \tag{12.9}$$

The efficiency of alkaline activated persulfate in the subsurface is strongly dependent on pH, which can be difficult to manipulate and control.

12.2.2.3 Heat-Activated Persulfate

Under heat activation, persulfate decomposes into two sulfate radicals at a rate that is Arrhenius law dependent (Costanza et al. 2010; Qi et al. 2014). However, too high of a temperature results in an increased rate of formation of potential scavengers (anions). The sulfate radical anion is an electron acceptor that, upon accepting a single electron, is reduced into sulfate anion (Equation 12.10).

$$2SO_4^{\bullet-} + e^- \rightarrow SO_4^{2-} \tag{12.10}$$

12.2.3 MECHANISM OF DECOMPOSITION

Several researchers have proposed that during persulfate oxidation of PFCAs, PFCAs are attacked by sulfate radicals initiating a decarboxylation reaction where cleavage of the carbon-to-carbon (C–C) bonds occurs between the PFCAs perfluorinated backbone and the carboxyl functional group (COOH), forming unstable perfluoroalkyl radicals (C_nF_{2n+1}) (Hori et al. 2010; Lee et al. 2012a; Yin et al. 2016). The perfluoroalkyl radicals then react with water to form hydroxyl radicals, resulting in thermally unstable alcohol ($C_nF_{2n+1}OH$). The alcohol undergoes hydrogen fluoride (HF) elimination to form $C_{n-1}F_{2n-1}COF$, which further undergoes hydrolysis, resulting in a CF_2 unit–shortened PFCA ($C_{n-1}F_{2n-1}COOH$) (Yin et al. 2016). This series of reactions (decarboxylation and HF elimination) continues to form shorter-chain PFCAs (stepwise degradation) until all PFCAs are mineralized to fluoride and carbon dioxide.

Lee et al. (2012b) proposed another degradation pathway where PFOA is attacked by sulfate free radicals, initiating a decarboxylation reaction forming unstable perfluoroalkyl radicals. These radicals may then react with molecular oxygen (O_2) generated from water electrolysis to form a perfluoroperoxy radical, which combines with another perfluoroperoxy radical to produce a perfluoroalkoxy radical. The perfluoroalkoxy radical then reacts with other radicals, such as $H_2O^{\bullet}$, to form perfluorinated alcohols, which will then undergo HF elimination and hydrolysis as in Yin et al. (2016).

Yang et al. (2013) studied the defluorination of PFOS by persulfate oxidation and suggested that the oxidation of PFOS by sulfate radical occurs via desulfonation at the carbon-to-sulfur (C–S) bond, forming $C_8F_{17}^{\bullet}$. These radicals then reacts with water, forming unstable $C_8F_{17}OH$. The $C_8F_{17}OH$ then undergoes HF elimination to form $C_7F_{15}COF$, which undergoes hydrolysis, resulting in the formation of $C_7F_{15}COOH$ (PFOA), followed by subsequent stepwise loss of CF_2 units. However, others have reported that PFOS does not degrade by persulfate oxidation (Park et al. 2016). Figure 12.1 shows the degradation pathway of PFAS by heat-activated persulfate as proposed by Yin et al. (2016).

The decomposition of PFOA by activated persulfate typically follows a pseudo-first-order kinetics model (Lee et al. 2010; Liu et al. 2012a; Park et al. 2016; Yin et al. 2016), suggesting that increasing the initial persulfate concentration would increase

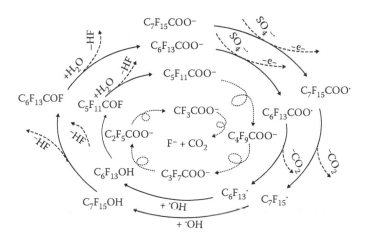

FIGURE 12.1 Proposed degradation pathway of PFOA by heat-activated persulfate. (Reprinted with permission from Yin, P. et al., *Int. J. Environ. Res. Public Health*, 13(6), 602, 2016.)

the decomposition rate. This has been found to occur up to a point where excess persulfate begins to scavenge sulfate radicals (Y. Lee et al. 2009; Zhao et al. 2014; B. Zhang et al. 2015).

12.2.4 SUPPORTING RESEARCH

A recent study by Park et al. (2016) investigated the oxidation of PFOA, 6:2 fluorotelomer sulfonate (6:2 FTSA), and PFOS by heat-activated persulfate under conditions suitable for *in situ* groundwater remediation. They found that PFOA was oxidized to below relevant environmental concentrations of 0.121–6.04 µm at a temperature of 50°C with an activation energy of 60.5 kJ/mol and followed the unzipping degradation pathway of PFCAs to shorter-chain-length compounds and fluoride, while 6:2 FTSA was oxidized to perfluoroheptanoic acid (PFHpA) and perfluorohexanoic acid (PFHxA). There was no transformation of PFOS even at a higher temperature (85°C and 90°C). Co-contaminants such as benzene, toluene, ethyl benzene, and xylene (BTEX) did not seem to have any significant effect on PFOA degradation rates. Oxidation in a soil slurry was not effective due to the presence of additional radical scavengers; therefore, frequent persulfate injections would be needed for effective ISCO of PFCAs. Costanza et al. (2010) and Liu et al. (2012a) reported a similar activation energy of 60 kJ/mol and the production of degradation intermediates (PFHpA, PFHxA, perfluoropentanoic acid [PFPnA], and perfluorobutanoic acid [PFBA]), indicating the sequential loss of CF_2 units. However, at a very high temperature, they reported that the sulfate radicals were intensively scavenged and an acidic pH led to inhibition of PFOA degradation.

This is in contrast to a study by Yin et al. (2016), who investigated the oxidation of PFOA with activated persulfate in groundwater under acidic conditions. They evaluated the impact of operating parameters such as pH, reaction temperature, time, and persulfate dosage under different experimental conditions. PFOA (20 µM) was effectively degraded in a pseudo-first-order manner at pH 2.0 with a reaction temperature and time of 50°C and 100 h, respectively. They reported higher degradation efficiency (89.9%) with increasing temperature, time, and persulfate dosage. Defluorination efficiency decreased at pH >5.0, which suggested that the sulfate free radical played the major role in PFOA degradation compared with the hydroxyl radical, since the sulfate radical is known to be the predominant radical at an acidic pH in activated persulfate systems. In addition, the authors noticed a decline in defluorination ratio when hydrochloric acid (HCl) was used to alter the solution pH when compared with sulfuric acid (H_2SO_4). This finding may result from the chloride ion (Cl^-) acting as a scavenger, thereby inhibiting the reaction between the sulfate radicals and PFOA. An activation energy of 56.2 kJ/mol was reported, which is similar to that reported by Lee et al. (2012a), who evaluated the decomposition of PFOA using a microwave-activated persulfate system at 90°C. They found that effective PFOA degradation was also observed at pH 2.5, while alkaline pH and the presence of chloride ions led to a slowdown in degradation rate due to scavenging of sulfate radicals. Lee Y. et al. (2009) previously reported an almost complete decomposition of PFOA and other PFCAs using a microwave-hydrothermal-induced persulfate system.

Lee et al. (2012b) also investigated the decomposition of PFOA at lower temperatures (20°C–40°C). They reported successful and almost complete decomposition of PFOA at 30°C (215 h) and 40°C (72 h), while at 20°C, 80.5% decomposition was reached after 648 h. They also reported that faster decomposition was achieved at acidic pH and increased persulfate concentration. Defluorination efficiencies of 97.8%, 75.8%, and 34.5% were obtained at 40°C, 30°C, and 20°C, respectively.

Yang et al. (2013) reported oxidative defluorination of PFOS (0.186 mM) with hydrothermal activated persulfate. Increasing the persulfate dose led to an increase in PFOS defluorination; however, after a certain point further increasing the persulfate dose had no effect. Defluorination was also seen to be favored by low pH (3.11). At the end of the reaction period (12 h), the defluorination efficiency was 22.5%. PFCAs with carbon-chain-length numbers 0–6 were detected as oxidation intermediates, with the concentrations of PFOA, PFHpA, and PFPA first increasing and later decreasing during the reaction period, while the concentrations of PFHxA, PFBA, perfluoropropanoic acid (PFPrA), and trifluoroacetic acid (TFA) increased with increasing reaction time. Perfluorohexanesulfonate (PFHxS) and perfluorobutanesulfonate (PFBS) were measured in low concentrations. Their concentration remained constant during the reaction period, which the authors suggest could be due to the recombination of $CF_3(CF_2)_5^{\bullet}$ and $CF_3(CF_2)_3^{\bullet}$ with the SO_3^- group. Table 12.2 gives a summary of activated persulfate technologies that have been investigated for the removal of PFAS from water.

12.3 ELECTROCHEMICAL OXIDATION

Electrochemical oxidation has been receiving increasing interest in recent years for water treatment due to its ability to degrade a wide range of recalcitrant organic compounds (Murugananthan et al. 2007; Carter and Farrell 2009; Radjenovic et al. 2011), including pharmaceuticals, dyes, petroleum products, and perfluorinated compounds (Faouzi et al. 2007; Zhuo et al. 2011; Sun et al. 2012; Antonin et al. 2015; Sopaj et al. 2015).

Electrochemical oxidation proceeds via direct and indirect anodic oxidation (Radjenovic and Seldak 2015). In direct electrolysis, contaminants are adsorbed onto and degraded directly at the electrode, while in indirect electrolysis contaminants are degraded in the bulk liquid in reactions with oxidizing agents formed at the electrode (Radjenovic and Sedlak 2015). Numerous materials have been used as electrodes, but most research on PFAS removal has been conducted using a boron-doped diamond (BDD) electrode due to its mechanical, chemical, and thermal stability (Trautmann et al. 2015). Some other electrodes, such as lead dioxide (PbO_2), titanium oxide (TiO_2), and tin oxide (SnO_2), have also shown the ability to treat PFAS-contaminated water (Ochiai et al. 2011; Zhuo et al. 2011; Zhao et al. 2013) when compared with other inactive electrode materials, such as iridium(VI) oxide (IrO_2) and platinum (Pt) (Niu et al. 2013). Operating conditions and parameters such as pH (Lin et al. 2012; Zhou et al. 2012), current density, electrolyte type (Song et al. 2010; Zhuo et al. 2012), electrode distance (Lin et al. 2012), initial PFAS concentration, and temperature are also important factors that influence electrochemical oxidation of PFAS (Niu et al. 2016). The superiority of the aforementioned BDD, PbO_2, TiO_2,

TABLE 12.2
Summary of Activated Persulfate Treatment of PFAS

Technology	Compound	Conditions	Efficiency % Degradation	Efficiency % Defluorination	Kinetics	Intermediates	Reference
Heat-activated persulfate	PFOA	0.24 µM, 42 mM Na$_2$S$_2$O$_8$, 50°C, 25 h, 40 mL	>90	21	Pseudo–first order	PFHpA, PFHxA, PFPnA, PFBA	Park et al. (2016)
	PFOS	0.92 µM, 60.5 mM Na$_2$S$_2$O$_8$, 85°C and 84 mM Na$_2$S$_2$O$_8$, 90°C, 40 mL	No transformation	N/A			
	6:2 fluorotelomer sulfonate	0.215 µM, 42 mM Na$_2$S$_2$O$_8$, 50°C, 50 h, 40 mL	98	N/A		PFHpA, PFHxA, PFPnA, PFBA	
	PFOA	5.0 µM, 20 mM Na$_2$S$_2$O$_8$, 85°C, pH 7.1, 30 h, 40 mL	93.5	43.6	Pseudo–first order	PFHpA, PFHeA, PFPnA, PFBA	Liu et al. (2012a)
	PFOA	20 µM, 2 mM S$_2$O$_8$, 50°C, pH 2.0, 100 h, 40 mL	89.9	23.9	Pseudo–first order	PFHpA, PFHeA, PFPnA, PFBA, PFPrA, TFA	Yin et al. (2016)
	PFOA	241.5 µM, 200 mM S$_2$O$_8$, 20°C–40°C, pH 2.5, 1000 mL, 72–648 h	80.5 (20°C, 648 h) 98.7 (30°C, 215 h) >99 (40°C, 72 h)	34.5 (20°C, 648 h) 75.5 (30°C, 240 h) 97.8 (40°C, 218 h)	Pseudo–first order	PFHpA, PFHeA, PFPnA, PFBA, PFPrA, TFA	Lee et al. (2012b)
Microwave-activated persulfate	PFOA	254 µM, 10 mM S$_2$O$_8$, 90°C, pH 2.5, 2450 MHz, 8 h, 50 mL	84.7	33.0	Pseudo–first order	PFHpA, PFHeA, PFPnA, PFBA, PFPrA, TFA	Lee et al. (2012a)

(*Continued*)

TABLE 12.2 (CONTINUED)
Summary of Activated Persulfate Treatment of PFAS

Technology	Compound	Conditions	Efficiency % Degradation	Efficiency % Defluorination	Kinetics	Intermediates	Reference
Microwave- and iron-activated persulfate	PFOA	240.7 µM, 5 mM S_2O_8, 3.6 mM ZVI, 90°C, pH 2.5, 2 h, 70 mL	67.6	22.5	Pseudo–first order	PFHpA, PFHeA, PFPnA, PFBA, PFPrA	Lee et al. (2010)
Microwave–hydrothermal-induced persulfate	PFOA	254 µM, 50 mM S_2O_8, 90°C, pH 2.0, 2450 MHz, 6 h, 50 mL	100	79.9	Pseudo–first order	PFHpA, PFHxA, PFPnA, PFBA, PFPrA, TFA	Y. Lee et al. (2009)
	PFHpA, PFHeA, PFPnA, PFBA, PFPrA, TFA	254 µM, 5 mM S_2O_8, 90°C, pH 2.0, 2450 MHz, 4 h, 50 mL	67.2, 67.5, 70.9, 74.5, 76.4, 77.9, respectively	23.2, 25.6, 37.8, 44.8, 57.5, 65.2, respectively			
Hydrothermal-activated persulfate	PFOS	186 µM, 18.5 mM S_2O_8, 20 °C, pH 3.11, 12 h, 100 mL	N/A	25.71	First order	PFOA, PFHxS, PFHpA, PFHxA, PFBS, PFPA, PFBA, PFPrA, TFA	Yang et al. (2013)

Note: Some of the reported degradation and defluorination efficiencies were estimated from the figures in the articles.
Abbreviation: N/A, not available.

and SnO_2 materials could be the result of their ability to generate more hydroxyl radicals, their high electron transfer ability (Zhou et al. 2012; Niu et al. 2013), and the ability of their atoms to remain in the same oxidation state during reactions while atoms in the active anode continuously cycle during oxidation (Chaplin 2014). Electrochemical oxidation offers several attractive advantages over other oxidation processes, such as the ability to operate at ambient temperature, no chemical requirement, and no waste generation (Radjenovic and Sedlak 2015).

There are associated environmental and cost risks, however, with electrochemical oxidation, as most electrodes contain toxic heavy metals, such as tin (Sn), cerium (Ce), lead (Pb), and antimony (Sb). Also, the formation of toxic by-products in electrochemical oxidation has been reported. Bagastyo et al. (2012) observed the presence of total trihalomethanes (tTHMs) and trihaloacetic acids (tHAAs) using a BDD electrode. tTHMs and tHAAs are disinfection by-products that are currently regulated in the United States (USEPA 2016b). The formation of perchlorate during electrochemical oxidation has also been reported in Bergmann and Rollin (2007), Bergmann et al. (2009), Azizi et al. (2011), and Donaghue and Chaplin (2013) using BDD anodes.

12.3.1 Mechanism of Decomposition

In general, degradation of PFAS by electrochemical oxidation is attributed to direct oxidation (Zhou et al. 2012). Studies have proposed three different pathways. The first pathway is assumed to be initiated by a direct, one-electron transfer from the carboxyl (COOH) or sulfonate (R-SO_3^-) group to the anode or electrode, forming PFAS radicals ($C_nF_{2n+1}COO^{\bullet}/C_nF_{2n+1}SO_3^{\bullet}$), which undergo decarboxylation or desulfonation to yield $C_nF_{2n+1}^{\bullet}$ (Schaefer et al. 2015; Niu et al. 2016). These combine with $OH^{\bullet}$ to form $C_nF_{2n+1}OH$, which is completely mineralized through further decomposition and hydrolysis. In the second pathway, the $C_{n-1}F_{2n-1}COF$ compound releases HF to form $C_{n-1}F_{2n-1}COF$, which then undergoes hydrolysis to form a perfluoroacid with one carbon atom removed ($C_{n-1}F_{2n-1}COO^-$), the cycle continuing to yield shorter-chained perfluorinated compounds (Chaplin 2014). In the third pathway, formed PFAS radicals react with O_2 (Niu et al. 2013) to yield peroxy radical species ($C_nF_{2n+1}OO^{\bullet}$), which then react with other peroxy radicals to form intermediate alcohol radicals ($C_nF_{2n+1}O^{\bullet}$) and decompose to n−1 perfluoroalkyl radicals ($C_{n-1}F_{2n+1}^{\bullet}$). These continue to combine with O_2, forming progressively shorter-chained compounds (losing CF_2 units in each cycle).

Zhuo et al. (2014) proposed that for electrochemical oxidation of 6:2 FTSA, its carbon-to-hydrogen (C–H) or C–C bond reacts with hydroxyl radicals, forming $C_6F_{13}COO^-$, which then undergo the CF_2 unzipping degradation pathway of PFCAs (hydrolysis and HF elimination). Carter and Farrell (2008) investigated the degradation of PFOS using BDD electrodes and proposed that degradation is initiated by a direct electron transfer at the anode surface. Liao and Farrell (2009) reported a similar mechanism with BDD electrodes for PFBS and reported the production of CO_2, SO_4^{2-}, F^-, and trace amounts of TFA as oxidation intermediates. Most studies have reported a pseudo-first-order kinetic model for electrochemical oxidation of

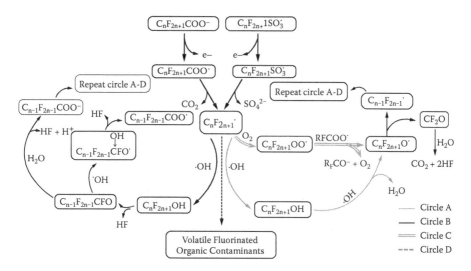

FIGURE 12.2 Proposed degradation pathways of PFAS in water by electrochemical oxidation. (Reprinted from Niu, J. et al., *Chemosphere*, 146, 526–538, 2016. Copyright 2016, with permission from Elsevier.)

PFAS (Trautmann et al. 2015). Figure 12.2 shows the possible PFAS degradation mechanisms by electrochemical oxidation.

12.3.2 Supporting Research

Schaefer et al. (2015) demonstrated the treatment potential of electrochemical oxidation of PFOA and PFOS in AFFF-impacted groundwater using a ruthenium(IV) oxide–coated titanium electrode (Ti/RuO$_2$). They found that increasing current density led to increased PFOA decomposition. For PFOS, decomposition slowed down after some time, suggesting the saturation of anode-reactive sites. However, no fluoride or shorter-chain-length PFAS were observed in solution. They further investigated the transformation products of PFAS using synthetic groundwater. Defluorination results showed that there was no fluoride loss in the membrane; rather, volatilization of HF or hypofluorous acid (HOF) occurred, with PFOS showing the highest fluoride recovery (98%), followed by PFOA (58%). Trace amounts of shorter-chain PFAS were measured in solution, which Schaefer et al. (2015) explained were formed by decomposition reactions of PFAS on the anode surface with limited release of shorter-chain PFAS from the electrode surface to the aqueous phase. Carter and Farrell (2008) reported a similar increase in PFOS decomposition using a Si/BDD anode with increasing current density. Complete decomposition of PFOS was observed at a current density of 20 mA/cm^2 using sodium perchlorate (NaClO$_4$) as an electrolyte in a parallel-plate flow-through reactor, while 50% removal was observed in a rotating disk electrode reactor. Liao and Farrell (2009) further investigated the degradation of PFBS using the same conditions as in Carter and Farrell (2008). They reported an almost complete removal and defluorination of PFBS, with TFA as the only identified intermediate.

Zhuo et al. (2012) investigated the effect of system conditions such as current density, initial pH, and initial PFAS concentration on PFAS removal using a BDD electrode. The PFAS investigated included PFBA, PFHxA, PFOA, PFDeA, PFBS, PFHxS, and PFOS. They reported a similar trend as Schaefer et al. (2015), where PFOA removal and defluorination increased with increasing current density, indicating nonsaturation of anode-reactive sites. An acidic pH of 3.0 promoted PFOA degradation when compared with alkaline solution. Kinetic studies on these PFAS showed increasing rate constants with increased chain lengths. The hydrophobicity of PFAS is known to increase with increasing chain length; hence, the increasing rate constants could be a result of their easier sorption onto the anode surface. Increasing the initial PFAS concentration also led to increased rate constants, which suggests that at higher concentrations, the ability of PFOA molecules to reach the electrode surface is greater than at low concentrations, resulting in greater removal rates. Zhuo et al. (2012). Ochiai et al. (2011) and Xiao et al. (2011) also reported similar removal efficiency and kinetics for PFOA with a BDD film electrode.

Trautmann et al. (2015) investigated the electrochemical degradation of PFAS-impacted groundwater in different media from a former firefighting training site containing high levels of dissolved organic carbon (DOC), reverse osmosis concentrates, and synthetic groundwater. Decomposition and defluorination of PFSAs were observed in the synthetic groundwater, with PFOS exhibiting greatest decomposition. For the high DOC levels, all PFAS investigated were found to be below the limit of quantification (LOQ) after treatment. PFBA was completely decomposed, while PFBS was not, which they attributed to the fact that PFSAs with the same carbon number as PFCAs are more difficult to degrade PFSAs possess higher hydrophobicity than the PFCAs. Also, the functional group of the PFSAs is more electrophilic, thus limiting the transfer of one electron to the electrode. This is also consistent with the finding of Zhou et al. (2012). The DOC concentration decreased by 96% and had no effect on PFAS decomposition. PFOA concentration was unchanged after 18 h of electrolysis in the reverse osmosis concentrates. The posttreatment PFCA concentrations were higher than their initial concentration, while the PFSAs decomposed in the order PFOS > PFHxS > PFBS > 6:2 FTSA, indicating the stepwise degradation mechanism of PFSAs (formation of shorter-chain length). Schaefer et al. (2015) and Trautmann et al. (2015) reported the formation of perchlorate and bromate in their study on the degradation of PFAS using BDD electrodes.

Zhuo et al. (2011) observed 93.3% removal efficiency while using a Ti/SnO$_2$-Sb anode with NaClO$_4$ as the electrolyte at a current density of 22.1 mA/cm^2 after 3 h. Zhuo et al. (2011) also investigated the use of Ti/SnO$_2$-Sb/Bi$_2$O$_3$ with the same current density and electrolyte and observed an almost complete removal of PFOA (>99%) and 63.8% defluorination efficiency after 2 h.

Niu et al. (2012) investigated the mineralization of PFCAs (PFBA, PFPnA, PFHxA, PFHpA, and PFOA) in aqueous media using a Ce-doped modified porous nanocrystalline PbO$_2$ film electrode and observed a 96.7% removal for PFOA with a defluorination efficiency of 81.7%. Lin et al. (2012) observed similar results for PFOA using a Ti/SnO$_2$-Sb/PbO$_2$ electrode (91.1% removal and 77.4% defluorination efficiency) and reported that increasing plate distance and initial PFOA concentration led to a decline in degradation efficiency while defluorination efficiency increased with increasing plate distance. Lin et al. (2012) suggested that it is possible that increasing plate distance resulted in a decline in the mass transfer of PFAS across the electrodes, thereby lowering the degradation efficiency, as the mass transfer rate of PFAS and the voltage across electrodes are dependent on plate distance. For the high defluorination efficiency, Lin et al. (2012) suggested that the C-F energy level of PFOA is very high, so in order for defluorination reactions to occur, a very high electrical potential is required in order to trigger electron transfer. It is also possible that increasing the plate distance led to an increase in voltage across the electrodes, hence increasing the defluorination rate.

Successful combinations of electrochemical with other treatment techniques have also been reported. Zhao et al. (2013) observed an increase in degradation ratio upon combining carbon aerogel/SnO_2-Sb with ultrasound. They reported that the removal ratio of PFOA increased from 47% (electrochemical oxidation) to 91% (electrochemical oxidation and ultrasound), which they suggest was due to the rapid mass transfer rate and reaction at the electrode surface upon addition of ultrasound. Table 12.3 shows a summary of electrochemical oxidation technologies that have been investigated for the removal of PFAS in water.

12.4 ULTRASONICATION (SONOLYSIS)

US has been successfully applied for rapid degradation of PFAS to F^-, SO_4^{2-}, and CO_2 (Vecitis et al. 2008). The sonochemical process relies on the propagation of acoustic waves in liquids at frequencies ranging between 20 and 1000 kHz (Furuta et al. 2004), which results in cavitation. Sonochemical degradation occurs via two mechanisms: (1) hydroxyl radical reaction due to the homolytic dissociation of water under extreme conditions and (2) pyrolysis (Rayaroth et al. 2016). During cavitation, cyclic formation, growth, and collapse of microbubbles result in an intense increase in temperature and pressure (5000 K and 2000 atm), and the generation of free radicals (Furuta et al. 2004; Chowdhury and Viraraghavan 2009). The cavitating bubbles have different active regions where chemical reactions occur: (1) bubble gas–liquid interface, (2) inside the bubble gas, and (3) in the bulk liquid region. Pyrolysis is said to occur more in the bubble gas, and hydrophobic compounds such as PFOA and PFOS tend to accumulate at the bubble–water interface due to high surface activity (Cheng et al. 2008; J. Lin et al. 2016), undergoing degradation via *in situ* pyrolysis (Cheng et al. 2008). Operating parameters, such as frequency (Campbell and Hoffman 2015), power density (Hao et al. 2014), solution temperature, type of sparge gas, and initial concentration of PFAS (Rodriguez-Freire et al. 2015), play a major role in the sonochemical degradation and defluorination rate of PFAS.

Increasing power leads to increased vibration amplitude, which can increase the number and rate of cavitation bubble generation and size of individual bubbles prior to the bubble collapse (Kanthale et al. 2008; Campbell and Hoffmann 2015). However, above a certain power level, the rate of reaction decreases (Adewuyi 2001). The sonochemical degradation rate is dependent on the availability of interfacial adsorption sites for the molecules, bubble collapse intensity, and the frequency of bubble oscillation. Therefore, when frequency increases, bubble oscillation accelerates, which creates more sites for adsorption, resulting in higher decomposition rates

TABLE 12.3
Summary of Electrochemical Oxidation Investigated for PFAS Treatment in Water

Technology			Conditions		Efficiency		Kinetics	Intermediates	Reference
Electrode	Compound	Current Density	Electrolyte		% Degradation	% F$^-$			
Ti/RuO$_2$/stainless steel	PFOA PFOS (Field GW)	10 mA/cm^2	3.52 mM Na$_2$SO$_4$	13 µg/L, 8 h, 200 W 18 µg/L, pH 20	90 90	N/A	First order	Perchlorate, F$^-$	Schaefer et al. (2015)
	PFOA PFOS (Synthetic GW)	1 mA/cm^2	3.52 mM Na$_2$SO$_4$	0.012 mM	>90	58% 98			
Si/BDD	PFBA PFBS PFHxA PFOA PFDeA PFHxS PFOS	23.24 mA/cm^2	11.4 mM NaClO$_4$	0.114 mM, pH 3.0, 32°C 2 h, 40 mL	PFOA 97.48	PFBA 80 PFBS 50 PFHxA 70 PFHxS 68 PFOA 60 PFOS 70	Pseudo– first order	PFHpA, PFHxA, PFPnA, PFBA, PFPrA, TFA, SO$_4^{2-}$	Zhuo et al. (2012)
Si/BDD	PFBS	10 mA/cm^2	10 mM NaClO$_4$	0.4 mM, pH 5.0, 22 °C 1 h, flow-through reactor, 350–400 mL 0.4 mM, pH 5.0, 22°C, 1 h, rotating disk electrode reactor, 350–400 mL	>90	91	N/A	TFA	Liao and Farrell (2009)

(Continued)

TABLE 12.3 (CONTINUED)
Summary of Electrochemical Oxidation Investigated for PFAS Treatment in Water

Technology			Conditions			Efficiency		Kinetics	Intermediates	Reference
Electrode	Compound	Current Density	Electrolyte			% Degradation	% F⁻			
BDD	PFOS	20 mA/cm²				≈38				
		20 mA/cm²	10 mM NaClO₄	0.4 mM, pH 4.0, 22°C, 30 h, flow-through reactor		100	N/A	Pseudo–first order	TFA, SO_4^{2-}, F⁻	Cater and Farrell (2008)
				0.4 mM pH 4.0, 22°C, 20 h, rotating disk electrode reactor						
			7.5 mM NaClO₄			50		Zeroth order		
BDD	PFOA	0.15 mA/cm²	10 mM NaClO₄	8 mM, 8 h, 300 mL		N/A	N/A	Pseudo–first order	PFHpA, PFHxA, PFPnA,	Ochiai et al. (2011)
BDD (hydrothermally enhanced)	PFOA	20 mA/cm²	0.05 M Na₂SO₄	0.48 mM 100°C, 6 h, 400 mL		93.6	90.2	First order	PFHxA, PFPnA, PFBA, PFPrA, TFA, oxalic acid	Xiao at al. (2011)

(Continued)

TABLE 12.3 (CONTINUED)
Summary of Electrochemical Oxidation Investigated for PFAS Treatment in Water

Technology		Conditions			Efficiency		Kinetics	Intermediates	Reference
Electrode	Compound	Current Density	Electrolyte		% Degradation	% F⁻			
BDD	PFBA PFPnA PFHxA PFHpA PFOA PFBS PFHxS	2.3 mA/cm²		0.7–79 µM, room temperature, 1000 mL GW from a former fire service training ground, 120 h	All were below LOQ except PFBS, ≈50 PFBS 83 PFHxS 92 PFOS 96 6:2 FTSA 60		Pseudo–first order	PFHpA, PFHxA, PFBA, PFPnA, perchlorate, bromate	Trautmann et al. (2015)
	PFOS 6:2 FTSA		0.1 M Na₂SO₄	GW from reverse osmosis concentrates, 18 h	PFOA unchanged PFBS 45 PFHxS 91 PFOS 98	66 for all 3			
Ti/SnO₂-Sb	PFOA	22.1 mA/cm²	NaClO₄	Synthetic GW 43 h 0.12 µM, 3 h, 25 mL	93.3		Pseudo–first order	PFHpA, PFHxA, PFPnA, PFBA, PFPrA, TFA	Zhuo et al. (2011)

(Continued)

TABLE 12.3 (CONTINUED)
Summary of Electrochemical Oxidation Investigated for PFAS Treatment in Water

Technology		Conditions			Efficiency				
Electrode	Compound	Current Density	Electrolyte		% Degradation	% F⁻	Kinetics	Intermediates	Reference
Ti/SnO$_2$-Sb	PFOA	5–40 mA/cm^2	10 mM NaClO$_4$	241 μM, 1.5 h, pH 5.0, 25°C, 100 mL	98.8	73.9	Pseudo–first order	PFHpA, PFHxA, PFPnA, PFBA, PFPrA, TFA	Lin et al. (2012)
Ti/SnO$_2$-Sb/PbO$_2$	PFOA	5–40 mA/cm^2	10 mM NaClO$_4$	241 μM, 1.5 h, pH 5.0, 25°C, 100 mL	91.1	77.4	Pseudo–first order	PFHpA, PFHxA, PFPnA, PFBA, PFPrA, TFA	Lin et al. (2012)
Ti/SnO$_2$-Sb/Bi	PFOA	22.1 mA/cm^2	NaClO$_4$	0.12 μM, 2 h, 25 mL	>99	63.8	First order	PFHpA, PFHxA, PFPnA, PFBA, PFPrA, TFA	Zhuo et al. (2011)
Ti/SnO$_2$-Sb/PbO$_2$-Ce	PFBA PFPnA PFHxA PFHpA PFOA	20 mA/cm^2	10 mM NaClO$_4$	100 mg/L PFCAs, 1.5 h, 25°C, 100 mL	PFBA 31.8 PFPnA 41.4 PFHxA 78.2 PFHpA 97.9 PFOA 96.7	PFBA 14.9 PFPnA 27.5 PFHxA 62.5 PFHpA 86.2 PFOA 81.7	Pseudo–first order	PFHpA, PFHxA, PFPnA, PFBA, PFPrA, TFA, Pb⁻	Niu et al. (2012)
Carbon aerogel/SnO$_2$-Sb with ultrasound	PFOA	20 mA/cm^2	0.1 M Na$_2$SO$_4$	0.24 μM, 5 h, 25°C, 60 mL	91	18	First order	PFHpA, PFHxA, PFPnA, PFBA, PFPrA, TFA	Zhao et al. (2013)

Note: Some of the reported degradation and defluorination efficiencies were deducted from the figures in the articles.
Abbreviations: GW, groundwater; N/A, not available.

(Adewuyi 2001). The degradation rate may be enhanced via addition of catalysts or additives such as persulfate (Hao et al. 2014), sulfate ions (Lin et al. 2015), and periodate (Lee et al. 2016).

12.4.1 Mechanism of Decomposition

Degradation of PFCAs by US occurs most likely via thermal decomposition, which involves (1) the adsorption of PFAS onto the collapsing bubble–water interface (Vecitis et al. 2009) and (2) cleavage of a C–C bond with the COOH group, which leads to the formation of perfluorocarbon intermediates. These intermediates undergo pyrolysis in the bubble vapor (Hao et al. 2014). Rodriguez-Freire et al. (2015) proposed that the sonolytic degradation of PFOS occurs via formation of shorter-chain PFSAs and not by the cleavage of the sulfonic head. A similar pathway was observed by Moriwaki et al. (2005), who detected PFHpS and PFHxS as intermediates of sonochemical degradation of PFOS. This proposed mechanism is, however, in contrast to a study performed by Vecitis et al. (2008), who suggested that the initial step for PFAS degradation is a rapid cleavage of the C–C or C–S bond, followed by pyrolytic degradation (complete mineralization yielding F^- with no formation of shorter-chain compounds). A similar degradation mechanism was proposed by Lee et al. (2016), who suggested that the PFOA decomposition occurs via total mineralization with no stepwise degradation, or if there is, the reaction is very rapid.

When combined with a catalyst, the reaction rate accelerates through the formation of free radicals, which then react with the ionized PFAS existing as anionic compounds via electronic transfer, which then undergo sonochemical degradation. For example, Lee et al. (2016) investigated the sonochemical degradation of PFOA using the oxidant periodate (IO_4^-). They proposed a pathway where IO_4^- undergoes sonolysis to form $IO_3^\bullet$, which then reacts with anionic PFOA via electron transfer to form perfluoroalkyl radicals. These radicals are sonodegraded to CF_2 and CF_3, which would then react with H_2O, HO, O atom, and H in the bubble gas to yield HF, CO, and CO_2. Figure 12.3 shows a schematic diagram of PFOS degradation by sonolysis.

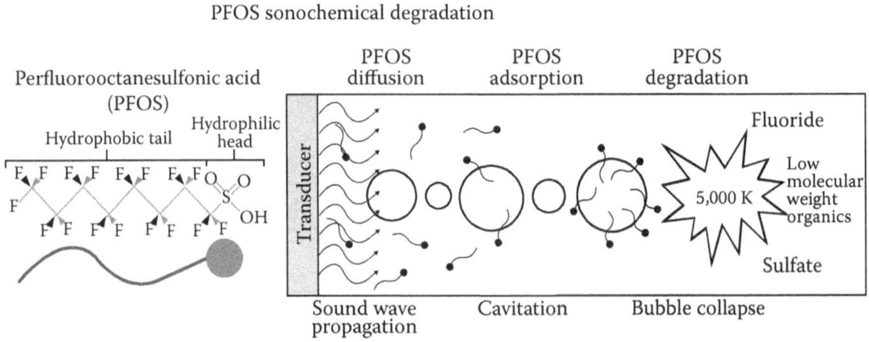

FIGURE 12.3 Sonochemical degradation of PFOS in water. (Reprinted from Rodriguez-Freire, L. et al., *J. Hazard. Mater.*, 300, 662–669, 2015. Copyright 2015, with permission from Elsevier.)

12.4.2 SUPPORTING RESEARCH

Moriwaki et al. (2005) investigated the sonochemical degradation of PFAS (20 µM) at a frequency of 200 kHz in an argon atmosphere. They reported 85% and 60% removal of PFOA and PFOS, respectively. The degradation rate constants determined for PFOS and PFOA during a 1 h sonication time were 0.016 and 0.032 min^{-1} with half-lives of 43 and 22 min, respectively. A similar study by Vecitis et al. (2008) reported half-lives of 16.9 and 25.7 min for PFOA and PFOS, respectively, with a defluorination efficiency of 95% for both compounds and 100% recovery of SO_4^{2-} for PFOS at a frequency of 354 kHz. Cheng et al. (2009) investigated the kinetic effect of matrix inorganics on the sonochemical degradation of PFAS in groundwater at a frequency of 612 kHz at 10°C under an argon atmosphere. Rate constants of 0.014 and 0.029 min^{-1} were obtained for PFOS and PFOA, respectively. At pH 3, there was enhanced degradation of PFOA and PFOS, while at pH >4 there was no observed effect on degradation rate. Bicarbonate was observed to cause a decrease in the removal rate of the PFAS, which may be due to partitioning and interaction of the ions with the water–bubble interface.

J. Lin et al. (2016) examined the effects of surfactants on the sonochemical degradation of PFOA. After a 2 h treatment, they observed an increase in PFOA removal efficiency (79%) under acidic conditions when hexadecyl trimethyl ammonium bromide (CTAB) (cationic surfactant) was present. On the contrary, sodium dodecyl sulfate (SDS) (anionic surfactant) inhibited PFOA degradation. The observed effects were explained by the competition between SDS and PFOA for adsorption sites at the bubble–water interface compared with CTAB, which increased the surface concentration of PFOA by reducing the surface energy of the bubble–water interface.

In another study by Lin et al. (2015), the addition of sulfate ions during ultrasonic treatment led to 100% PFOA removal, and 99.8% defluorination efficiency after 2 h, while 46% removal and 9% defluorination efficiency were obtained with ultrasound-alone. They reported that the enhanced PFOA removal by the US–sulfate system occurred via two reactions: (1) a direct reaction where PFOA was destroyed at the bubble–water interface and (2) an indirect reaction via the formation of SO_4^-, which then decomposed PFOA to F$^-$, CO_2, and SO_4^{2-}. The two resulting reactions therefore led to acceleration of PFOA degradation. Solution pH did not affect the removal efficiency, while ambient temperature (25°C) promoted decomposition when compared with 35°C and 45°C. The latter phenomenon was explained by the reduction of the surface tension at the bubble–water interface at higher temperatures, which resulted in poorer PFOA adsorption. Lin and coauthors also demonstrated that the rates of PFOA decomposition and defluorination were heavily dependent on the sulfate dose. Enhanced sonochemical decomposition with persulfate as an additive was also found in Hao et al. (2014), who observed that adding persulfate as an oxidant intensified the sonochemical degradation of ammonium perfluorooctanoate (APFO) at pH 6.0. They proposed that the reaction between $SO_4^{-\bullet}$ (persulfate by-product) and APFO at the bubble gas–water interface was the main pathway for the sonochemical degradation of APFO.

A study by Lee et al. (2016) reported 96.5% and 95.7% decomposition and defluorination efficiency, respectively, after 2 h of ultrasonic treatment when periodate

(IO_4^-) was added to solution under an argon atmosphere, as opposed to when only ultrasound was used (26.2% decomposition and 6.5% defluorination). IO_4^- accelerated the reaction rate via formation of effective $IO_3^{\bullet}$ radicals, which react with PFOA via electron transfer. Low pH promoted decomposition and defluorination due to acid catalysis. Under acidic conditions, IO_4^- reacts with hydrogen ions to from $IO_3^{\bullet}$ radicals, while under basic conditions, $OH^{\bullet}$ and $IO_4^{\bullet}$ are the resultant radicals, which are weaker radicals than $IO_3^{\bullet}$ (Lee et al. 2016). In addition, under an acidic pH, the bubble gas–water interface becomes more positively charged, thereby attracting the negatively charged PFOA to the surface and increasing the PFOA degradation rate. An O_2 atmosphere was found to inhibit the degradation rate by generating hydroxyl radicals, which reacted with IO_4^- to form less effective $IO_4^{\bullet}$, which acted as a scavenger of the $IO_3^{\bullet}$ radicals.

The rate of PFOS degradation has been shown to increase with the applied frequency (Rodriguez-Freire et al. 2015). Hao et al. (2014) and Campbell and Hoffman (2015) have reported improved degradation rates of PFAS while increasing power density during sonication. Table 12.4 shows the sonochemical technologies that have been used in treating PFAS-contaminated water.

12.5 CHEMICAL REDUCTION PROCESSES

Like oxidation processes, reduction processes involve either direct electron transfer to treat contaminants or the generation of reactive free radicals, which then degrade contaminants. Reductants for treating groundwater typically include ZVI, ferrous iron, and sodium dithionite. These can transfer electrons directly or react to form reducing radicals, such as a hydrogen radical ($H^{\bullet}$) and hydrated electron (e_{aq}^-) (Liu et al. 2016). Chemical reduction processes are dependent on solution pH (Qu et al. 2014), concentration of contaminants (Hawley et al. 2004), temperature (Hori et al. 2006), reductant concentration (Zhao et al. 2012; Vellanki et al. 2013), and presence of groundwater ions.

12.5.1 REDUCTANTS

ZVI or nanoscale zerovalent iron (nZVI) is an inexpensive groundwater remediation technology and is the most commonly used reductant for *in situ* groundwater remediation, as it is a strong reducing agent capable of successfully reducing major groundwater contaminants such as TCE (Phillips et al. 2010; Truex et al. 2011), arsenic (Sun et al. 2006; K. Lee et al. 2009), and hexavalent chromium (Franco et al. 2009; Němeček et al. 2014) in laboratory- and field-scale tests. ZVI can serve as a sorbent and/or a reductant. Recently, nZVI has had increased attention due to its higher reactivity and surface area compared with the micro-sized ZVI. In addition, the small size of nZVI makes its injection to the source of contamination faster and more effective. nZVI has successfully been applied for the remediation of chlorinated groundwater sites (Elliott et al. 2008; Lacina et al. 2015) and heavy metal–contaminated groundwater (Li et al. 2014; Fu et al. 2015). However, a limitation in the use of nZVI is its tendency to aggregate, which lowers its reactive surface area (Phenrat et al. 2007; Stefaniuk et al. 2016).

TABLE 12.4
Summary of Sonochemical Technologies Investigated for Treatment of PFAS in Water

			Conditions			Efficiency					
Technology	Compound	Frequency (kHz)	Power Density		Degradation (%)	Defluorination (%)	Rate Constants (min^{-1})	Kinetics	Intermediates	Reference
Ultrasound	PFOA	200	3 W/cm^2	24 μM (PFOA) 20 μM (PFOS) pH 3.5–4.89, 20°C, 1 h, argon atmosphere	85	N/A	0.032	Pseudo–first order	$CF_3(CF_2)_nCOO^-$ (n = 0–6) $C_7F_{15}SO_3^-$ and $C_6F_{13}SO_3^-$ Fluoride	Moriwaki et al. (2005)
	PFOS				60		0.016			
Ultrasound	PFOA (Milli-Q)	612	250 W/L	0.24 μM (PFOA) 0.20 μM (PFOS) pH 7.9, 10°C, 2 h, argon atmosphere	N/A	N/A	0.0366	Pseudo–first order	N/A	Cheng et al. (2009)
	PFOS (Milli-Q)						0.0192			
	PFOA (GW)						0.0291			
	PFOS (GW)						0.0135			
Ultrasound	PFOA (Milli-Q)	354	250 W/L	0.24 μM (PFOA) 0.20 μM (PFOS) pH 6.9–7.9, 10°C, 2 h, argon atmosphere	N/A	N/A	0.047	Pseudo–first order	N/A	Cheng et al. (2008)
	PFOS (Milli-Q)						0.024			
	PFOA (LGW)						0.021			
	PFOS (LGW)						0.0094			
Ultrasound and CTAB	PFOA	40	150	0.12 mM (PFOA), 0.12 mM (CTAB) pH 4.0, 25°C, 2 hrs	79	13	N/A	N/A	N/A	J. Lin et al. (2016)
Ultrasound and sulfate		40 kHz	150 W	117 mM PFOA, 25 mM SO_4^{2-}, pH 4.3 25°C, 1.5 hrs, air atmosphere	100	99.8	N/A	Pseudo–first order	N/A	Lin et al. (2015)

(Continued)

TABLE 12.4 (CONTINUED)
Summary of Sonochemical Technologies Investigated for Treatment of PFAS in Water

Technology	Compound	Conditions				Efficiency					Reference
		Frequency (kHz)	Power Density			Degradation (%)	Defluorination (%)	Rate Constants (min^{-1})	Kinetics	Intermediates	
Ultrasound	PFOS	25–1 MHz	8 W/cm^2	100 μM PFOS, pH 2.0, 30°C 5.5 h		N/A	100	N/A	Pseudo–first order (500 kHz) Zero-order (1 MHz)	SO_4^{2-} (68.6%), F$^-$	Rodriguez-Freire et al. (2015)
Ultrasound and periodate	PFOA	40 kHz	120 W	170.1 μM (PFOA), 45 mM PI, pH 3.9, 25°C, 2 h, nitrogen atmosphere		96.5%	95.7%	0.0222	Pseudo–first order	PFPrA, TFA	Lee et al. (2016)
Ultrasound and persulfate	APFO	20 kHz	300 W	46.4 μM APFO, 10 mM $S_2O_8^{2-}$ pH 6.5, 25°C, 2 h, air atmosphere		51.2%	11.15%	N/A	N/A	N/A	Hao et al. (2014)
Ultrasound	PFOA PFOS	354 kHz	250 W/L	0.24 μM (PFOA) 0.20 μM (PFOS) pH 7.0–8.0, 10°C, 2 h, argon atmosphere		N/A	>90	N/A	Pseudo–first order	CO, CO_2, F$^-$, and SO_4^{2-}	Vecitis et al. (2008)

Note: Some of the reported degradation and defluorination efficiencies were estimated from the figures in the articles.
Abbreviations: LGW, landfill groundwater; PI, periodate; TFA.

ARPs have also been investigated for the reductive degradation of groundwater contaminants. ARPs involve the combination of activation methods such as ultrasound, UV, microwaves, and electron beam with reducing agents (reductants), such as ferrous iron, sulfide, sulfite, iodide, and dithionite, to generate very reactive reducing radicals that mineralize contaminants to less toxic products (Vellanki et al. 2013). The oxidizing hydroxyl radical (OH•), the reducing hydrogen radical (H•), and the hydrated electron (e_{aq}^-) are the most reactive free radicals produced during ARP. ARP-induced degradation rates are dependent on the initial solution pH and reductant concentration (Vellanki et al. 2013).

12.5.2 Mechanism of Decomposition

Generally, the removal of contaminants by ZVI in reductive processes involves the mass transfer of contaminants to the ZVI surface, their adsorption and reaction (transformation of contaminants into less toxic or nontoxic species) on the ZVI surface, and the desorption and mass transfer of by-products into solution (Arvaniti et al. 2015). Since the reduction of contaminants by ZVI is a surface-mediated electron transfer process, the surface properties of ZVI impact contaminant reactivity. The degradation pathway of PFAS using ARP differs from that of oxidizing agents in that the C–F bond adjacent to COOH of PFAS is cleaved by the hydrated electron (Song et al. 2013) rather than the C–C or C–S bond.

Arvaniti et al. (2014) found that PFOS removal using Mg-aminoclay (MgAC)-coated nZVI occurred via adsorption of PFOS to the ZVI surface, followed by reduction. The MgAC-coated nZVI was said to be more stable in water than the uncoated nZVI due to the inhibition of nZVI aggregation by electrostatic repulsion, which made the nZVI particles stay dispersed as individual particles (Arvaniti et al. 2014). At the same time, its cationic surface attracted the negatively charged PFOS anions to its surface, thus leading to a higher reactivity toward reductive degradation (Arvaniti et al. 2014). A similar decomposition mechanism for PFOS using ZVI was also found in Hori et al. (2006), who suggested that adsorption of PFOS onto ZVI played a major role in PFOS decomposition as fluoride, and not PFOS, was detected in the treatment solution after 6 h. X-ray photoelectron spectroscopy indicated that a C–F bond was still associated with the ZVI surface. Hori et al. (2006) further increased the reaction period to 15 h, and the resulting spectrum showed only a peak of fluoride and no peak for a C–F bond, suggesting complete mineralization of PFOS on the ZVI surface.

Qu et al. (2010) investigated the reductive defluorination of PFOA in a UV-KI system and indicated that hydrated electrons were the critical radicals in the reductive decomposition of PFOA. They proposed that hydrated electrons act as nucleophiles, which leads to the reductive cleavage of the α C–F bonds, resulting in fluorine elimination from PFOA forming $C_nF_{2n-1}H_2COOH$. Under UV irradiation, $C_nF_{2n-1}H_2COOH$ forms free radicals such as the COOH radical, CH_2 carbene, and $C_{n-1}F_{2n-1}$. The formed COOH radical and $C_{n-1}F_{2n-1}$ then recombine to form $C_nF_{2n-1}COOH$, while the reactive CH_2 carbene form CH_3 radical, which recombines with COOH radical to form CH_3COOH. They proposed that under UV irradiation, there was a cleavage of the C–C bond between the COOH group and the perfluoroalkyl group as shorter-chain intermediates were detected in solution. Qu et al. (2014) therefore concluded that two reactions are responsible for the reductive defluorination of PFOA: (1) direct photolysis by UV irradiation and (2) photoreduction by hydrated electrons.

Zhao et al. (2012) postulated that the photo-induced electron is the sole reductant in PFOA decomposition. They proposed that PFOA is first ionized to H^+ and $C_7F_{15}COO^-$ in water, and then $C_7F_{15}COO^-$ absorbs onto the surface of beta gallium oxide (β-Ga_2O_3) where it reacts with the photo-induced electron forming $C_7F_{15}^{\bullet}$. The $C_7F_{15}^{\bullet}$ rapidly reacts with water to form unstable $C_7F_{15}OH$, which undergoes HF elimination and hydrolysis to form $C_7F_{15}COOH$, followed by the formation of shorter-chain PFCAs.

Park et al. (2009) reported that the reduction of a C–F bond occurs at a reduction potential of <–2.7 V, making defluorination of PFAS by hydrated electrons (E = –2.9 V) relatively easy. The kinetic study conducted by Park et al. (2009) found that for the PFCAs, generation of fluoride was not chain length dependent, whereas it was for PFSAs. They proposed a reaction pathway where the PFAS react with the hydrated electron to yield perfluoroalkyl radical anions ($C_nF_{2n+1}X^{2\bullet-}$), where X stands for COO^- for PFCAs and SO_3^- for PFSAs, which rapidly decompose via fluoride elimination to form $C_nF_{2n}X^{\bullet-}$. The $C_nF_{2n}X^{\bullet-}$ reacts further with the hydrated electron to form carbanions ($C_nF_{2n}X^{2-}$) or with iodide radicals ($I^{\bullet}$, $I_2^{\bullet-}$, $I_3^{\bullet-}$) to form perfluoroalkyl iodide carboxylates or sulfonates ($C_nF_{2n}IX^-$), which are reconverted to $C_nF_{2n}X^{\bullet-}$ and I^- via reaction with hydrated electrons or photolytic homolysis of the carbon-iodide (C–I) bond (Park et al. 2009). The initially formed $C_nF_{2n}X^{2-}$ undergoes protonation, resulting in an H/F exchange. Park et al. (2009) suggested that the formed product from the H/F exchange may retain the anionic COO^- or SO_3^- terminal group, which remains in solution and proceeds via repeated H/F exchanges, resulting in the formation of shorter-chain intermediates.

Figure 12.4 shows the reductive defluorination pathway of PFOA in a UV–sulfite system as proposed by Song et al. (2013).

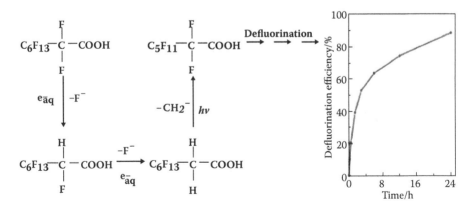

FIGURE 12.4 Reductive defluorination pathway of PFOA in a UV–sulfite system. (From Song, Z. et al., *J. Hazard. Mater.*, 262, 332–338, 2013. Copyright 2013, with permission from Elsevier.)

12.5.3 SUPPORTING RESEARCH

Arvaniti et al. (2014) compared three types of nZVI for the reductive degradation of PFAS. They observed a significant removal of PFOS (≥70%) with MgAC-coated nZVI, and no significant removal for the commercial iron powder and uncoated nZVI. PFOS removal was enhanced at low temperature, acidic pH, and increased nZVI concentration. F^- was the only by-product identified, suggesting removal via sorption and reductive degradation. ZVI in subcritical water (350°C) was used to successfully degrade PFOS with no formation of shorter-chain PFCAs (Hori et al. 2006). A >99% degradation efficiency and 46.7% defluorination efficiency were reached after 6 h.

Ochoa-Herrera et al. (2008) reported the reductive degradation of PFOS with vitamin B_{12} as a catalyst and Ti(III)–citrate as a bulk reductant. They observed 18% and 71% defluorination of technical PFOS and PFOS-branched isomers at pH 9.0 and 70°C after 120 and 168 h, respectively. Increasing solution pH and temperature led to an increasing degradation rate.

Park et al. (2011) examined the decomposition of PFOA and PFOS with iodide photolysis at 254 nm. They reported that the degradation rate was dependent on the iodide concentration, initial PFAS concentration, head group type, and chain length. pH had no significant effect on the degradation rate. A similar study by Qu et al. (2014) investigated the reductive decomposition of PFOA with a UV-KI system. They found that the PFOA removal was higher at pH 10.0 than at pH 5.0, which they attributed to the generation of more hydrated electrons by recycling of I^- in an alkaline condition. It should be noted that Park et al. (2011) conducted their experiments in an open system, which can allow entry of oxygen, which may affect reduction efficiency, while that of Qu et al. (2014) was in a closed system. Qu et al. (2010) previously reported 93.9% PFOA removal at 6 h and 98.8% defluorination after 14 h with a UV-KI system at pH 9.0 under anaerobic conditions. They reported that increasing KI concentration initially increased PFOA removal but then decreased PFOA removal with further KI increase. This was due to generation of hydrated electrons and triiodide (I_3^-), which is also an oxidant. Upon photolysis of iodide anions, the generated iodide radical ($I^{\bullet}$) reacts with iodide ion (I^-) to form diiodine anions ($I_2^{\bullet-}$), which will then react with I^- to form I_3^- (Qu et al. 2010). So, increasing the KI concentration will generate increased amounts of hydrated electron and triiodide. But once a certain KI concentration is met, the additional triiodide will scavenge the available hydrated electron and no longer be available for PFOA decomposition (Qu et al. 2010).

C. Zhang et al. (2015) investigated the effect of temperature and ionic strength on the reductive decomposition of PFOA with a UV-KI system under a nitrogen atmosphere. It was found that increasing temperature increased PFOA decomposition. Temperature effects were also observed in the formation of intermediates, as there was a decline in the concentration of shorter-chain PFAS and intermediates with increasing temperature. A positive correlation between PFOA degradation rate and ionic strength was observed. A study by Song et al. (2013) found complete removal of PFOA after 1 h and 88.5% defluorination after 24 h by hydrated electrons generated from a UV–sulfite system under a nitrogen atmosphere. The reaction was favored by an alkaline pH of 10.0 and increasing SO_3^{2-} concentration. They also observed the formation and decomposition of shorter-chain PFCAs (C_2–C_7) as intermediates. Vellanki et al. (2013) also reported successful degradation of PFOA with a UV–sulfite system with increasing pH.

Zhao et al. (2012) demonstrated the photoreduction of PFOA with β-Ga_2O_3 removal; defluorination efficiencies were 98.8% and 31.6%, respectively, after 3 h in the presence of thiosulfate ($S_2O_3^{2-}$) and under a nitrogen atmosphere. The decomposition efficiency was observed to decrease with increasing pH and improved with the addition of sulfite. Zhao et al. (2012) attributed the pH effect on decomposition efficiency related to the surface charge state of β-Ga_2O_3. At pH <9, the β-Ga_2O_3 surface is positively charged, thereby increasing adsorption of anionic PFOA onto the β-Ga_2O_3 surface, while at pH >9, the β-Ga_2O_3 surface is negatively charged, repelling PFOA. Table 12.5 gives a summary of reductive processes that have been investigated for PFAS degradation in water.

12.6 PLASMA TECHNOLOGY

Plasma processing plays an important role in various industrial applications, such as semiconductor fabrication, polymer functionalization, chemical synthesis, and toxic waste management (van Veldhuizen and Rutgers 2002; Yang et al. 2012). Plasma is a gaseous state of matter consisting of ions, atoms, atomic fragments, and free electrons. Depending on the energy or temperature of the electrons compared with the temperature of the background gas, plasmas can be classified as thermal or nonthermal. In thermal plasma, an example of which is an electrical arc, sufficient energy is available to allow the plasma constituents (electrons, ions, and neutrals) to be in thermal equilibrium, with species at temperatures typically exceeding several thousand kelvin. Nonthermal plasmas are typically formed using less power and are characterized by strong nonequilibrium between the electrons whose energy varies from ~1 to ~10 eV (1 eV = 11,600 K), with the background gas temperatures ranging from ambient to approximately 1000 K. An example of a nonthermal plasma is a dielectric barrier discharge used for commercial ozone generation. Plasmas of interest for water treatment are typically nonthermal and can be generated by direct current, alternating current, pulsed discharges, and radiofrequency and microwave power supply sources (Fridman and Kennedy 2004; Locke and Thagard 2017). Compared with traditional AOPs, plasma electrical discharges can simultaneously oxidize and reduce organic molecules. When plasma is formed in close proximity to or in contact with aqueous media, electron collisions in the plasma with vaporized water molecules result in the generation of oxidants (e.g., OH• and H_2O_2), reductive species such as e_{aq}^-, UV light, and heat, among other chemical processes (Locke et al. 2006; Joshi and Thagard 2013; Locke and Thagard 2017). In the presence of oxygen or air, significant concentrations of ozone can also be formed. When generated above the plasma–liquid interface, ozone diffuses into the liquid and directly reacts with solutes (Lukes et al. 2004; Lukes and Locke 2005). While the oxidation process is based predominantly on OH• attack, the reduction most likely involves simultaneous reactions of aqueous electrons, hot plasma electrons, and background gas ions. The relative amounts of reactive species can be controlled by adjusting the input

TABLE 12.5
Summary of Reductive Processes for the Degradation of PFAS in Water

Technology	Compound	Conditions	Efficiency Degradation %	Efficiency Defluorination %	Kinetics	Intermediates	Reference
Mg-aminoclay nZVI	PFOS	40 mg/L; 10–1000 mg/L nZVI; pH 3, 5, and 7; 20°C and 55 °C, 1 h	95 (pH 3.0, 20°C)	N/A	N/A	F−	Arvaniti et al. (2014)
ZVI in subcritical water	PFOS	372 μM PFOS, 9.60 mmol ZVI, 350°C, 6 h, argon atmosphere	>99	51.4	Pseudo-first order	CHF_3	Hori et al. (2006)
Ti(III)–citrate/ vitamin B_{12}	PFOS	0.5 mM PFOS, 70°C, pH 9, 36 mM Ti(III)–citrate, 260 μM vitamin B_{12}, 120–168 h	N/A	18 (technical PFOS, 168 h) 71 (branched isomers PFOS, 120 h)	N/A	N/A	Ochoa-Herrera et al. (2008)

Advanced Reduction Processes

Technology	Compound	Wavelength (nm)	Atmosphere	Operating Parameters	Degradation %	Defluorination %	Kinetics	Intermediates	Reference
UV-KI	PFOA PFOS	254	Argon	24 μM, 20 μM, 10 mM KI, 2.5 h, pH 1.5–12	≈13 ≈35	N/A 57 μmol	Pseudo-first order	$C_xF2_x + 1I$	Park et al. (2011)
UV-KI	PFOA PFOS	54	Argon	0.24 μM 0.20 μM, 10 mM KI, 2.5 h	≈15 ≈35	1.4 17	Pseudo-first order	C_1–C_8 iodinated and noniodinated gaseous intermediates, with the heaviest being perfluorohexyl iodide ($C_6F_{13}I$) for PFOA and perfluorooctyl iodide ($C_8F_{17}I$) for PFOS	Park et al. (2009)

(*Continued*)

TABLE 12.5 (CONTINUED)
Summary of Reductive Processes for the Degradation of PFAS in Water

Technology	Compound		Conditions	Efficiency % Degradation	Efficiency % Defluorination	Kinetics	Intermediates	Reference
UV-KI	PFOA	254	Nitrogen	0.025 mM, pH 10, closed system, 0.25 mM KI, 6 h	97.9	>70	Pseudo–first order	PFHpA, PFHxA, PFPnA, PFBA, PFPrA, TFA, fluorinated and iodinated hydrocarbons	Qu et al. (2014)
UV-KI	PFOA	254	Nitrogen	0.025 mM, pH 9, closed system, 0.25 mM KI, 6 h	93.9 (6 h)	98.8 (14 h)	First order	PFHpA, PFHxA, PFPnA, PFBA, PFPrA, TFA, fluorinated and iodinated hydrocarbons	Qu et al. (2010)
UV-KI	PFOA	254	Nitrogen	0.02 mM, pH 9, 0.60 mM KI, 6 h, 40°C, 6 h	>99	>80%	First order	PFHpA, PFHxA, PFPnA, PFBA, PFPrA, TFA,	B. Zhang et al. (2015)
UV-SO_3^{2-}	PFOA	254	Nitrogen	0.02 mM, pH 10.3, 10 mM -SO_3^{2-}, 25°C	100 (1 h)	88.5 (24 h)	NA	C_2–C_7, fluorinated alkyl sulfonates	Song et al. (2013)
UV-SO_3^{2-}	PFOA	253.7 and 311 nm	Nitrogen	pH 2.2, 7.2, and 11.8; 20 h	≈0–10	≈31.1–35.9	N/A	N/A	Vellanki et al. (2013)
UV-β-Ga_2O_3	PFOA	254	Nitrogen	3 h	98.8	31.6	Pseudo–first order	N/A	Zhao et al. (2012)

Note: Some of the reported degradation and defluorination efficiencies were estimated from the figures in the articles.
Abbreviation: N/A, not available.

power, thereby allowing the plasma system to be optimized for removing a broad variety of organic pollutants. Furthermore, the ability of the plasma process to produce significant amounts of long-lived oxidative species can reduce the requirement for externally supplied chemicals, such as H_2O_2 and O_3, if those are to be utilized as additional oxidative sources (Sunka et al. 1999).

Plasma can be generated directly in a liquid, entirely in a gas phase, or in a gas phase such that the plasma propagates along the interface between the gas and the liquid phases (Locke and Thagard 2017). Various forms of electrical discharge plasma-based AOPs have been used for water treatment applications. These include different types of reactors, electrode configurations, electrode materials, discharge phases, and additives. Reviews of these configurations and their bench-scale performances have been reported in the literature (Locke et al. 2006, 2012; Lukes et al. 2012; Bruggeman and Locke 2013; Joshi and Thagard 2013; Jiang et al. 2014; Locke and Thagard 2017). Examples of successfully developed pilot-scale electrical discharge water treatment systems include the multiple-point high-voltage electrode reactor (Even-Ezra et al. 2009), commercial Symbios plasma reactor (Symbios), and multistage gas–liquid electrical discharge column reactor (Holzer and Locke 2008).

Regardless of the reactor type, the effectiveness of the plasma-based water treatment has been shown to depend on the input power, pulse repetition frequency, electrode material, solution pH, conductivity, temperature, pollutant structure, and concentration (Ruma et al. 2013; Jiang et al. 2014). Numerous studies have shown that electrical discharge is viable for use in the degradation of USEPA-regulated contaminants and unregulated compounds, such as bisphenol A (Wang et al. 2008), polychlorinated biphenyls (Sahni et al. 2005), TCE (Even-Ezra et al. 2009), organic dyes (Sugiarto et al. 2003; Mededovic et al. 2008; Yano et al. 2009), and phenols (Sharma et al. 1993; Grymonpre et al. 1999; Sunka et al. 1999; Lukes et al. 2005; Yang et al. 2009). Electrical discharges can be also applied to inactivate bacteria and viruses in water (Moreau et al. 2008).

12.6.1 MECHANISM OF DECOMPOSITION

Takeuchi et al. (2013) proposed a two-step degradation pathway for PFAS (PFOA) wherein PFAS molecules first adsorb onto a bubble–liquid interface where they encounter plasma at a high temperature (2000 K). Upon contact with plasma, a thermal cleavage of their C–C bonds occurs generating fluorocarbon radicals in the bubbles. Due to the oxidation and reduction of the fluorocarbon radicals by the plasma-reactive species (hydrogen and hydroxyl radicals), CO, CO_2, and HF are formed. They, therefore, concluded that the main reaction pathway for PFAS degradation by plasma treatment is direct thermal decomposition to gaseous products.

For DC plasma generated within gas bubbles, Hayashi et al. (2015) reported complete decomposition of PFOA and PFOS. They proposed a decomposition mechanism different from that of Takeuchi et al. (2013), where the most energized ions in the plasma collide with the negatively charged portion of the PFOA molecule residing on the water surface stripping an electron from the carboxyl group which results in unstable radicals (1st stage). A decarboxylation reaction occurs forming fluorocarbon radicals and carbon dioxide (2nd stage). The fluorocarbon radical instantly reacts with water, which results to a shortened carbon chain PFAS, and F⁻ release into the treatment solution (3rd stage). The shortened carbon chain PFAS undergoes the 1st stage again (unzipping pathway). Hayashi et al. (2015) suggested that in addition to the earlier reactions proposed (1st stage to 3rd stage), it is likely that the C–C bonds of the most energized ions are cleaved to form TFA ($CF_3\bullet$) and PFHxA ($C_6F_{12}\bullet$) radicals. The authors suggested this because higher levels of TFA and PFHxA were measured in the treatment solution compared to PFHpA that is one carbon shorter than PFOA. For PFOS, they stipulated a degradation pathway that involved first the formation of PFOA and a decomposition process similar to that in PFOA.

Yasuoka et al. (2011) investigated the decomposition of PFAS (PFOA and PFOS) using DC plasma generated within gas (oxygen) bubbles. They reported successful defluorination and removal of PFOS and proposed that degradation of PFOS occur by PFOS reacting with a positive species (an unidentified species they termed M⁺) at the bubble gas–liquid interface. They suggested that applying a positive voltage to the reactor led to a cathode drop near the surface of the solution causing the positive species to collide with the solution surface. Since PFAS concentrate close to the interface, degradation occurs by ionizing PFOS molecules present in solution reacting with M⁺ at the interface to generate an unstable perfluoroalkyl radical ($C_8F_{17}SO_3\bullet$), which dissociates to form fluorinated radical ($C_8F_{17}\bullet$) and sulfur trioxide (SO_3). The formed $C_8F_{17}\bullet$ and SO_3 reacts with water to yield an unstable alcohol, $C_8F_{17}OH$, and sulfuric acid (H_2SO_4). The unstable $C_8F_{17}OH$ undergoes hydrolysis to form a carboxyl compound with a shorter carbon and fluorine chain length (PFOA). The resulting carboxyl compound is assumed to interact with M⁺ in a similar manner, to yield a fluorinated radical that undergoing a series of reactions to form shorter chain length PFAS. The study by Yasuoka et al. (2011) reported that the effect of hydrated electrons on PFAS decomposition was not significant.

Contrarily, Stratton et al. (2017) reported that hydrated electrons are important species in the degradation of PFAS, in addition to plasma argon ions and high-energy free electrons. The authors hypothesized that since PFAS (PFOA) adsorbs on the plasma–liquid interface, their hydrophobic tail protrudes into the plasma phase thus reacting with two abundant reactive species (argon ions and high-energy free electrons) present in the plasma interior (Thagard et al. 2017; Stratton et al. 2017). The authors also suggested that thermal decomposition might have played a role in the degradation.

12.6.2 Supporting Research

There have been several attempts to utilize plasma to transform PFAS (Yasuoka et al. 2010, 2011; Takeuchi et al. 2013; Matsuya et al. 2014; Hayashi et al. 2015; Stratton et al. 2017). Of these studies, a laminar jet with bubbling (LJB) reactor (Figure 12.5) was found to be the most promising (Stratton et al. 2017). It was used to degrade PFOA and PFOS in prepared solutions and in groundwater. In the case of PFOA, a 30 min plasma treatment of a 1.4 L aqueous solution removed 90% of the initial PFOA concentration of 20 µM with an input power of 76.5 W (high-removal-rate case) or 25% using 4.1 W (high-efficiency case) (Figure 12.6). Continuing the treatment for

Oxidation and Reduction Approaches for Treatment of PFAS 287

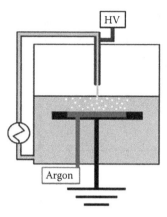

FIGURE 12.5 LJB reactor. The reactor was operated in semibatch mode, with liquid recirculating at 1.4 L/min. The headspace was purged with argon at 3.9 L/min through a submerged diffuser. Plasma was generated at the tip of the high-voltage (HV) electrode by charging the load capacitor by a high-voltage power supply and discharging the stored electrical pulse via a rotary spark gap (Reprinted and adapted with permission from Stratton, G. R. et al., *Environ. Sci. Technol.*, 51(3), 1643–1648, 2017. Copyright 2017, American Chemical Society.)

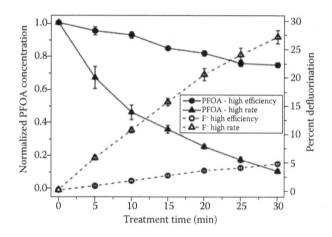

FIGURE 12.6 Normalized PFOA concentration and defluorination profiles for the LJB reactor configured for high treatment efficiency and high treatment rate. (Reprinted and adapted with permission from Stratton, G. R. et al., *Environ. Sci. Technol.*, 51(3), 1643–1648, 2017. Copyright 2017, American Chemical Society.)

an additional 40 min removed PFOA to below detection limits, significantly below USEPA drinking water guidance levels of 70 ppt.

The high-efficiency process was further used to treat groundwater containing PFOA and several co-contaminants, including PFOS, TCE, and PCE (the total TOC concentration was 0.67 mg/L), demonstrating that the process was not significantly affected by co-contaminants and that the process was capable of rapidly degrading PFOS (Figure 12.7).

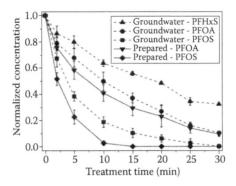

FIGURE 12.7 Normalized concentration profiles for PFAS in the contaminated groundwater and prepared solution. Groundwater experiments were conducted using unmodified samples from the effluent of an air stripper within the former NAWC Warminster groundwater treatment plant in Warminster, Pennsylvania. The initial PFOA concentration in groundwater was 0.0014 mg/L, and that for PFOS was 0.00035 mg/L. Initial concentrations of PFOA and PFOS in prepared solutions were 0.0018 and 0.00014 mg/L, respectively (Reprinted with permission from Stratton, G. R. et al., *Environ. Sci. Technol.*, 51(3), 1643–1648, 2017. Copyright 2017, American Chemical Society.)

A preliminary investigation conducted to quantify major by-products produced during treatment of a mixture of PFOA and PFOS confirmed the presence of shorter-chain PFCAs, including PFHpA, PFHxA, and perfluoropentanoic acid (PFPnA), as shown in Figure 12.8. The concentrations of these products account for only about 10% of the degraded PFOA and PFOS, which is much lower than for oxidation-based processes, where shorter-chain perfluoroalkyl substances (PFAAs) account for most of the degraded PFOA (85%–95% for activated persulfate) (Chen and Zhang

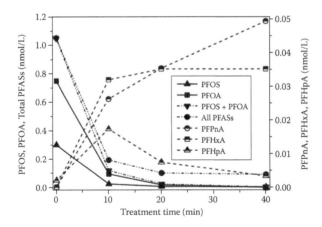

FIGURE 12.8 Concentration profiles showing the reduction in concentrations of PFOA and PFOS, and a corresponding increase in concentrations of PFHpA, PFHxA, and PFPnA. (Reprinted with permission from Stratton, G. R. et al., *Environ. Sci. Technol.*, 51(3), 1643–1648, 2017. Copyright 2017, American Chemical Society.)

2006; Stratton et al. 2017). However, qualitative evidence of volatile F⁻-containing by-products, such as perfluorohexane, hexadecafluorooctyl fluorosulfate, and hexadecafluoroheptane, among other compounds, has been observed in the reactor headspace samples (unpublished data).

Experiments designed to elucidate why plasma is so effective at degrading PFAS suggest that e^-_{aq} produced by the plasma treatment are responsible for degrading PFOA, which is in agreement with past PFOA transformation studies (Park et al. 2009; Song et al. 2013; Stratton et al. 2017). However, in addition to e^-_{aq}, high-energy electrons and argon ions present in the plasma interior have also been suggested as the reactive species responsible for initiating primary reactions with the PFOA in gas phase discharge reactors (Mededovic et al. 2008; Thagard et al. 2016). Because PFOA is a surfactant and thus readily adsorbs to the plasma–water interface (Stratton et al. 2015), reactions with gaseous plasma species are assumed to involve the hydrophobic portion of the PFOA molecule protruding into the plasma (Campbell et al. 2009).

The high-rate and high-efficiency cases presented in Figure 12.6 attained high PFOA removal and mineralization efficiencies compared with leading alternative treatment technologies (Table 12.6). In fact, the high-efficiency plasma-based water treatment process achieved a defluorination efficiency about 30 times greater than that of activated persulfate, 10 times greater than that of sonolysis, and 15% greater than that of electrochemical treatment.

Yasuoka et al. (2011) investigated the decomposition of PFOA and PFOS using DC plasma generated within gas bubbles (argon, neon, helium, and oxygen) in solution. For PFOS degradation with an initial concentration of 50 mg/L and solution volume of 50 mL, the defluorination ratio and decomposition ratio of sulfate was 54.8 and 57%, respectively, after 4 h of treatment. The highest fluoride concentration was measured in the oxygen plasma, and the lowest in the helium plasma. The energy efficiency for the generation of fluoride ions varied with the type of the bubbled gas and was the highest for argon and lowest for helium. The authors suggested that the high-energy efficiency displayed by the argon plasma was because it extended largely along the inner surface of bubbles and had the largest expansion, while the helium plasma extended vertically and had the smallest interfacial area among the gases used. Takeuchi et al. (2013) examined the decomposition of an initial concentration of PFOA of 156 µM using plasma generated inside oxygen bubbles. They reported complete degradation after 2.5 h. In another study, Takeuchi et al. (2013) investigated the degradation of a mixture of PFHpA and PFOA. The observed rate constants decreased to about 80% of those in individual solutions. In the same study, mass balance of PFOA (initial concentration = 115 µM) decomposition was conducted. A mass balance of 93% for carbon and 95% for fluorine was achieved when 34% of PFOA was decomposed after 30 min treatment time was reported. Fluoride ions were identified as the dominant by-product in the solution.

Hayashi et al. (2015) reported a complete degradation of PFOA and PFOS with initial concentrations of 41.4 and 60 mg/L, respectively, with a DC plasma within oxygen bubbles. Complete decomposition was achieved for PFOA after 3 h treatment time and for PFOA after 8 h for a treatment solution volume of 20 mL. For PFOA, the authors reported 94.5% defluorination after 3 h of treatment with an input power of 32 W. Smaller chain PFCAs were identified as the main reaction by-products.

TABLE 12.6
Performance Indicators and Corresponding Experimental Parameters for High-Rate Plasma Water Treatment (PWT), High-Efficiency PWT, Sonolysis, UV-Activated Persulfate, Electrochemical Treatment, and an Alternative Direct Current (DC) Plasma Treatment

Treatment	$[PFOA]_0$ (µM)	PD^c (W/L)	k_{obs} (min^{-1})	$\dfrac{k_{obs}}{PD}\left(10^{-4}\cdot\dfrac{min^{-1}}{W/L}\right)^d$	$\dfrac{F_{50}}{t_{50}}\left(\dfrac{\%}{min}\right)^e$	$\dfrac{F_{50}/t_{50}}{PD}\left(10^{-2}\cdot\dfrac{\%/min}{W/L}\right)^f$
High-rate PWT	20	54.6^b	0.074	14	2.08	3.8
High-efficiency PWT	20	2.90^b	0.012	41	0.31	11
Sonolysis	20	250	0.018	0.72	2.47	0.99
UV-activated persulfate	50^a	23	0.012	5.2	0.09	0.38
Electrochemical treatment	0.031^a	5.0	0.0057	11	0.47	9.5
DC plasma in O$_2$ bubbles	100	1550	0.030	0.20	4.4	0.28

Source: Data for UV-activated persulfate and sonolysis are from Vecitis et al. (2009). Data for electrochemical treatment is for a current density of 10 mA/cm^2 and are from Schaefer et al. (2015). Data for DC plasma in bubbles are from Yasuoka et al. (2011). Reprinted with permission from Stratton, G. R. et al., *Environ. Sci. Technol.*, 51(3), 1643–1648, 2017. Copyright 2017, American Chemical Society.

Note: k_{obs} is the observed first-order removal rate constant. t_{50} is the time to reduce PFOA by 50%. F_{50} is the percentage of fluorine removed from the degraded PFOA and is detected as fluoride ions.

[a] Performance may be sensitive to initial PFOA concentration; thus, comparisons are approximate.
[b] Input power includes power requirements for gas pump and plasma generation.
[c] Power density (PD = input power/treated volume).
[d] Removal efficiency (= k_{obs}/PD).
[e] Rate of mineralization.
[f] Mineralization efficiency.

For PFOS and the same solution volume, the authors reported 70% defluorination ratio and 75% mass balance of sulfur. The reaction by-products identified were also smaller PFCAs. Hayashi et al. (2015) also suggested the possibility of forming perfluoroheptanesulfonic acid (PFHpS) because of their inability to obtain a complete mass balance of fluorine and sulfur.

12.7 IMPLICATIONS

While there is a need for more knowledge on the remediation of PFAS-contaminated water, particularly in a field setting, laboratory-scale studies have proven to be a valuable tool to develop, validate, and improve treatment approaches for PFAS-impacted groundwater. Table 12.7 provides a summary of the remediation techniques that have been discussed in this chapter based on their effectiveness and technical feasibility.

Activated persulfate: Activated persulfate has proven effective for PFOA but not for the destruction of PFOS. Generally, higher doses of persulfate, high temperature, or creating alkaline pH conditions is necessary to achieve an effective degradation of PFOA (Lee et al. 2012b). The presence of environmental matrices such as carbonate and chloride can reduce the efficiency of the activated persulfate system (Lee et al. 2012a).

Electrochemical oxidation: Bench-scale studies have shown success in the degradation and defluorination of both PFOA and PFOS. The implementation and wider adoption of this technique in environmental and industrial applications are still limited due to challenges associated with energy consumption (Schaefer et al. 2015), the cost associated with building effective electrodes (Panizza and Cerisola 2009), and the formation of toxic by-products (Schaefer et al. 2015; Trautmann et al. 2015). Designing and

TABLE 12.7
PFAS Remediation Techniques Based on Effectiveness and Technical Feasibility

Technology	Effective?		Technically Feasible?	
	PFOA	**PFOS**	*In Situ*	*Ex Situ*
Activated persulfate	Yes	No	Yes	Yes
Electrochemical oxidation	Yes	Yes	No	Yes
Sonolysis	Yes	Yes	No	Yes
Reduction processes	Yes	Yes (some)	Yes (some)	Yes
Plasma	Yes	Yes	No	Yes

Note: These approaches have resulted in the removal of PFOA and/or PFOS in laboratory evaluations. While the removal of contaminants was confirmed, not all studies confirmed the actual destruction of contaminants through fluoride mass balance or other means. Treatment can modify the sorption or partitioning of contaminants, which can result in removal from the aqueous phase, but not destruction.

developing inexpensive electrodes with long-lasting electrocatalytic activity may help curb the cost associated with building electrodes (Anglada et al. 2009). Also, further research is warranted for the development of electrodes that inhibit the formation of toxic by-products and for the investigation of optimum conditions that avoid the production of toxic by-products. Finally, further studies mimicking the environmental conditions found in field applications, such as low PFAS concentration, low conductivity, and the presence of NOM and inorganic ions, should be conducted to investigate their effect on electrochemical oxidation efficiency before electrochemical oxidation can be promoted for full-scale remediation in the field (Niu et al. 2016).

Ultrasonication: Sonochemical processes are efficient for the treatment of PFAS contaminated water in the laboratory, as studies have reported mineralization to SO_4^{2-}, CO_2, CO, and F^- after decarboxylation and desulfonation (Vecitis et al. 2008). However, scaling up this treatment technique is challenging. Designing an effective ultrasound reactor has proven to be one of the most challenging aspects for the application of sonolysis in water treatment due to challenges in the optimization of operating conditions and parameters such as frequency and power (Rayaroth et al. 2016). These parameters influence degradation rates and energy efficiency (Campbell and Hoffman 2015). The issue of significant energy requirement needs to be addressed prior to field implementation. During sonolysis, part of the energy input is lost during heat production and acoustic cavitation (Adewuyi 2001), resulting in reduced sonochemical efficiency. The presence of background organics, like humic substances, and inorganics, such as bicarbonate and sulfate, may cause a reduction in the sonochemical degradation rates of PFAS (Cheng et al. 2009). Coupling ultrasound with other treatment techniques might help reduce energy demand and increase degradation of PFAS (Lin et al. 2015; Lee et al. 2016).

Reduction processes: Reductive processes have been proven feasible for the degradation of most PFAS; however, they can require extreme operating conditions, such as high temperature, high reductant dosage, and high solution pH, and they also can be energy-intensive. Most laboratory studies have been conducted using simulated water. Research should be conducted using natural water matrices to evaluate the impact of background interferences on degradation rates.

Plasma technology: Plasma-based water treatment effectively and efficiently transforms PFOA and PFOS in both synthetic water and groundwater. Because chemical reactivity takes place primarily at the plasma–liquid interface, plasma-based water treatment appears to be less sensitive than most other treatment processes to the presence of organic and inorganic co-contaminants. This lack of sensitivity to co-contaminants coupled with high PFAS removal and defluorination efficiencies makes plasma-based water treatment a promising technology for the remediation of PFAS-contaminated water (Stratton et al. 2017). Further work to confirm treatment by-products is warranted.

12.8 CONCLUSIONS

This chapter focused on the oxidative and reductive processes for the treatment of PFAS in the aqueous phase and their associated reaction mechanisms. The processes described show that an in-depth knowledge and understanding of the behavior, fate, and transport of PFAS in the environment and during treatment is needed to design effective removal processes for these contaminants. Because numerous laboratory treatability studies have been successful for treating at least some PFAS, performing the research necessary to scale these processes for field implementation is needed to determine if they will be cost-effective and technically feasible for remediating contaminated groundwater. There is also a need to consider coupling these treatment processes with other available treatment techniques (including those mentioned in this chapter) in treatment trains, as their synergistic effects may improve the removal efficiency and help offset many of the drawbacks associated with individual processes in order to reduce oxidant dosage, treatment time, and energy consumption, as well as to increase sustainability by reducing or eliminating toxic by-product formation.

REFERENCES

Adewuyi, Y. G. (2001). Sonochemistry: Environmental science and engineering applications. *Ind. Eng. Chem. Res.*, 40(22), 4681–4715.

Anglada, A., Urtiaga, A., and Ortiz, I. (2009). Contributions of electrochemical oxidation to waste-water treatment: Fundamentals and review of applications. *J. Chem. Technol. Biotechnol.*, 84(12), 1747–1755.

Antonin, V. S., Santos, M. C., Garcia-Segura, S., and Brillas, E. (2015). Electrochemical incineration of the antibiotic ciprofloxacin in sulfate medium and synthetic urine matrix. *Water Res.*, 83, 31–41.

Arvaniti, O. S., Hwang, Y., Andersen, H. R., Nikolaos, T. S., and Athanasios, S. S. (2014). Removal of perfluorinated compounds from water using nanoscale zero-valent iron. Singapore International Water Week.

Arvaniti, O. S., Hwang, Y., Andersen, H. R., Stasinakis, A. S., Thomaidis, N. S., and Aloupi, M. (2015). Reductive degradation of perfluorinated compounds in water using Mg-aminoclay coated nanoscale zero valent iron. *Chem. Eng. J.*, 262, 133–139.

Azizi, O., Hubler, D., Schrader, G., Farrell, J., and Chaplin, B. P. (2011). Mechanism of perchlorate formation on boron-doped diamond film anodes. *Environ. Sci. Technol.*, 45(24), 10582–10590.

Backe, W. J., Day, T. C., and Field, J. A. (2013). Zwitterionic, cationic, and anionic fluorinated chemicals in aqueous film forming foam formulations and groundwater from US military bases by nonaqueous large-volume injection HPLC-MS/MS. *Environ. Sci. Technol.*, 47(10), 5226–5234.

Bagastyo, A. Y., Batstone, D. J., Kristiana, I., Gernjak, W., Joll, C., and Radjenovic, J. (2012). Electrochemical oxidation of reverse osmosis concentrate on boron-doped diamond anodes at circumneutral and acidic pH. *Water Res.*, 46(18), 6104–6112.

Bennedsen, L. R., Muff, J., and Søgaard, E. G. (2012). Influence of chloride and carbonates on the reactivity of activated persulfate. *Chemosphere*, 86(11), 1092–1097.

Bergmann, M. H., and Rollin, J. (2007). Product and by-product formation in laboratory studies on disinfection electrolysis of water using boron-doped diamond anodes. *Catal. Today*, 124(3), 198–203.

Bergmann, M. H., Rollin, J., and Iourtchouk, T. (2009). The occurrence of perchlorate during drinking water electrolysis using BDD anodes. *Electrochim. Acta*, 54(7), 2102–2107.

Block, P. A., Brown, R. A., and Robinson, D. (2004). Novel activation technologies for sodium persulfate in-situ chemical oxidation. In *Proceedings of the Fourth International Conference on the Remediation of Chlorinated and Recalcitrant Compounds*. Columbus, OH: Battelle Press, 24–27.

Buck, R. C., Franklin, J., Berger, U., Conder, J. M., Cousins, I. T., De Voogt, P., Jensen, A. A., Kannan, K., Mabury, S. A., and van Leeuwen, S. P. (2011). Perfluoroalkyl and polyfluoroalkyl substances in the environment: Terminology, classification, and origins. *Integr. Environ. Assess. Manag.*, 7(4), 513–541.

Buck, R. C., Murphy, P. M., and Pabon, M. (2012). Chemistry, properties, and uses of commercial fluorinated surfactants. In *Polyfluorinated Chemicals and Transformation Products*, ed. T. P. Knepper and F. T. Lange. Berlin: Springer, 1–24.

Bruggeman, P., and Locke, B. R. (2013). Assessment of potential applications of plasma with liquid water. In *Low Temperature Plasma Technology: Methods and Applications*, ed. X. L. P. Chu. Boca Raton, FL: Taylor & Francis, 368–369.

Campbell, T. Y., Vecitis, C. D., Mader, B. T., and Hoffmann, M. R. (2009). Perfluorinated surfactant chain-length effects on sonochemical kinetics. *J. Phys. Chem. A*, 113(36), 9834–9842.

Campbell, T., and Hoffmann, M. R. (2015). Sonochemical degradation of perfluorinated surfactants: Power and multiple frequency effects. *Sep. Purif. Technol.*, 156, 1019–1027.

Carter, K. E., and Farrell, J. (2008). Oxidative destruction of perfluorooctane sulfonate using boron-doped diamond film electrodes. *Environ. Sci. Technol.*, 42(16), 6111–6115.

Carter, K. E., and Farrell, J. (2009). Electrochemical oxidation of trichloroethylene using boron-doped diamond film electrodes. *Environ. Sci. Technol.*, 43(21), 8350–8354.

Chang, M. C., and Kang, H. Y. (2009). Remediation of pyrene-contaminated soil by synthesized nanoscale zero-valent iron particles. *J. Environ. Sci. Health A*, 44(6), 576–582.

Chaplin, B. P. (2014). Critical review of electrochemical advanced oxidation processes for water treatment applications. *Environ. Sci. Process. Impacts*, 16(6), 1182–1203.

Chen, J., and Zhang, P. (2006). Photodegradation of perfluorooctanoic acid in water under irradiation of 254 nm and 185 nm light by use of persulfate. *Water Sci. Technol.*, 54(11–12), 317–325.

Chen, J., Qian, Y., Liu, H., and Huang, T. (2016). Oxidative degradation of diclofenac by thermally activated persulfate: Implication for ISCO. *Environ. Sci. Pollut. Res.*, 23(4), 3824–3833.

Cheng, J., Vecitis, C. D., Park, H., Mader, B. T., and Hoffmann, M. R. (2008). Sonochemical degradation of perfluorooctane sulfonate (PFOS) and perfluorooctanoate (PFOA) in landfill groundwater: Environmental matrix effects. *Environ. Sci. Technol.*, 44(21), 8057–8063.

Cheng, J., Vecitis, C. D., Park, H., Mader, B. T., and Hoffmann, M. R. (2009). Sonochemical degradation of perfluorooctane sulfonate (PFOS) and perfluorooctanoate (PFOA) in groundwater: Kinetic effects of matrix inorganics. *Environ. Sci. Technol.*, 44(1), 445–450.

Chowdhury, P., and Viraraghavan, T. (2009). Sonochemical degradation of chlorinated organic compounds, phenolic compounds and organic dyes—A review. *Sci. Total Environ.*, 407(8), 2474–2492.

Costanza, J., Otaño, G., Callaghan, J., and Pennell, K. D. (2010). PCE oxidation by sodium persulfate in the presence of solids. *Environ. Sci. Technol.*, 44(24), 9445–9450.

Crimi, M. L., and Taylor, J. (2007). Experimental evaluation of catalyzed hydrogen peroxide and sodium persulfate for destruction of BTEX contaminants. *Soil Sed. Contam.*, 16(1), 29–45.

Donaghue, A., and Chaplin, B. P. (2013). Effect of select organic compounds on perchlorate formation at boron-doped diamond film anodes. *Environ. Sci. Technol.*, 47(21), 12391–12399.

Elliott, D. W., Lien, H., and Zhang, W. (2008). Zerovalent iron nanoparticles for treatment of ground water contaminated by hexachlorocyclohexanes. *J. Environ. Qual.*, 37(6), 2192–2201.

Even-Ezra, I., Mizrahi, A., Gerrity, D., Snyder, S., Salveson, A., and Lahav, O. (2009). Application of a novel plasma-based advanced oxidation process for efficient and cost-effective destruction of refractory organics in tertiary effluents and contaminated groundwater. *Desalination Water Treat.*, 11(1–3), 236–244.

Fan, Y., Ji, Y., Kong, D., Lu, J., and Zhou, Q. (2015). Kinetic and mechanistic investigations of the degradation of sulfamethazine in heat-activated persulfate oxidation process. *J. Hazard. Mater.*, 300, 39–47.

Faouzi, A. M., Nasr, B., and Abdellatif, G. (2007). Electrochemical degradation of anthraquinone dye Alizarin Red S by anodic oxidation on boron-doped diamond. *Dyes Pigments*, 73(1), 86–89.

Fordham, J., and Williams, H. L. (1951). The persulfate-iron (II) initiator system for free radical polymerizations. *J. Am. Chem. Soc.*, 73(10), 4855–4859.

Franco, D. V., Da Silva, L. M., and Jardim, W. F. (2009). Reduction of hexavalent chromium in soil and ground water using zero-valent iron under batch and semi-batch conditions. *Water Air Soil Pollut.*, 197(1–4), 49–60.

Fridman, A., and Kennedy, L. A. (2004). *Plasma Physics and Engineering.* New York: Taylor & Francis.

Fu, R., Yang, Y., Xu, Z., Zhang, X., Guo, X., and Bi, D. (2015). The removal of chromium (VI) and lead (II) from groundwater using sepiolite-supported nanoscale zero-valent iron (S-NZVI). *Chemosphere*, 138, 726–734.

Furman, O. S., Teel, A. L., and Watts, R. J. (2010). Mechanism of base activation of persulfate. *Environ. Sci. Technol.*, 44(16), 6423–6428.

Furuta, M., Yamaguchi, M., Tsukamoto, T., Yim, B., Stavarache, C., Hasiba, K., and Maeda, Y. (2004). Inactivation of *Escherichia coli* by ultrasonic irradiation. *Ultrason. Sonochem.*, 11(2), 57–60.

Grymonpre, D. R., Finney, W. C., and Locke, B. R. (1999). Aqueous-phase pulsed streamer corona reactor using suspended activated carbon particles for phenol oxidation: Model-data comparison. *Chem. Eng. Sci.*, 54(15), 3095–3105.

Guan, X., Sun, Y., Qin, H., Li, J., Lo, I. M., He, D., and Dong, H. (2015). The limitations of applying zero-valent iron technology in contaminants sequestration and the corresponding countermeasures: The development in zero-valent iron technology in the last two decades (1994–2014). *Water Res.*, 75, 224–248.

Hao, F., Guo, W., Wang, A., Leng, Y., and Li, H. (2014). Intensification of sonochemical degradation of ammonium perfluorooctanoate by persulfate oxidant. *Ultrason. Sonochem.*, 21(2), 554–558.

Hawley, E. L., Deeb, R. A., Kavanaugh, M. C., and Jacobs, J. A. (2004). Treatment technologies for chromium (VI). In *Chromium (VI) Handbook*, eds. J. Guertin, J. A. Jacobs, and C. P. Avakian. Boca Raton, FL: CRC Press, 275–309.

Hayashi, R., Obo, H., Takeuchi, N., and Yasuoka, K. (2015). Decomposition of perfluorinated compounds in water by DC plasma within oxygen bubbles. *Electric. Eng. Japan*, 190(3), 9–16.

Herzke, D., Schlabach, M., and Mariussen, E. (2007). Literature survey of polyfluorinated organic compounds, phosphor containing flame retardants, 3-nitrobenzanthrone, organic tin compounds, platinum and silver. Norway: Universita NILU (Norwegian Institute for Air Research).

Holzer, F., and Locke, B. R. (2008). Multistage gas-liquid electrical discharge column reactor for advanced oxidation processes. *Ind. Eng. Chem. Res.*, 47(7), 2203–2212.

Hori, H., Nagaoka, Y., Yamamoto, A., Sano, T., Yamashita, N., Taniyasu, S., Kutsuna, S., Osaka, I., and Arakawa, R. (2006). Efficient decomposition of environmentally persistent perfluorooctanesulfonate and related fluorochemicals using zerovalent iron in subcritical water. *Environ. Sci. Technol.*, 40(3), 1049–1054.

Hori, H., Murayama, M., Inoue, N., Ishida, K., and Kutsuna, S. (2010). Efficient mineralization of hydroperfluorocarboxylic acids with persulfate in hot water. *Catal. Today*, 151(1), 131–136.

Huang, K., Zhao, Z., Hoag, G. E., Dahmani, A., and Block, P. A. (2005). Degradation of volatile organic compounds with thermally activated persulfate oxidation. *Chemosphere*, 61(4), 551–560.

Huling, S. G., Ko, S., Park, S., and Kan, E. (2011). Persulfate oxidation of MTBE- and chloroform-spent granular activated carbon. *J. Hazard. Mater.*, 192(3), 1484–1490.

Ji, Y., Ferronato, C., Salvador, A., Yang, X., and Chovelon, J. (2014). Degradation of ciprofloxacin and sulfamethoxazole by ferrous-activated persulfate: Implications for remediation of groundwater contaminated by antibiotics. *Sci. Total Environ.*, 472, 800–808.

Ji, Y., Dong, C., Kong, D., Lu, J., and Zhou, Q. (2015). Heat-activated persulfate oxidation of atrazine: Implications for remediation of groundwater contaminated by herbicides. *Chem. Eng. J.*, 263, 45–54.

Jiang, B., Zheng, J., Qiu, S., Wu, M., Zhang, Q., Yan, Z., and Xue, Q. (2014). Review on electrical discharge plasma technology for wastewater remediation. *Chem. Eng. J.*, 236, 348–368.

Joshi, R. P., and Thagard, S. M. (2013). Streamer-like electrical discharges in water: Part II. Environmental applications. *Plasma Chem. Plasma Process.*, 33(1), 17–49.

Kanthale, P., Ashokkumar, M., and Grieser, F. (2008). Sonoluminescence, sonochemistry (H2O2 yield) and bubble dynamics: Frequency and power effects. *Ultrason. Sonochem.*, 15(2), 143–150.

Khan, S., He, X., Khan, H. M., Boccelli, D., and Dionysiou, D. D. (2016). Efficient degradation of lindane in aqueous solution by iron (II) and/or UV activated peroxymonosulfate. *J. Photochem. Photobiol. A*, 316, 37–43.

Kolthoff, I., and Miller, I. (1951). The chemistry of persulfate. I. The kinetics and mechanism of the decomposition of the persulfate ion in aqueous medium. *J. Am. Chem. Soc.*, 73(7), 3055–3059.

Lacina, P., Dvorak, V., Vodickova, E., Barson, P., Kalivoda, J., and Goold, S. (2015). The application of nano-sized zero-valent iron for in-situ remediation of chlorinated ethylenes in groundwater: A field case study. *Water Environ. Res.*, 87(4), 326–333.

Lee, K., Lee, Y., Yoon, J., Kamala-Kannan, S., Park, S., and Oh, B. (2009). Assessment of zero-valent iron as a permeable reactive barrier for long-term removal of arsenic compounds from synthetic water. *Environ. Technol.*, 30(13), 1425–1434.

Lee, Y., Lo, S., Chiueh, P., and Chang, D. (2009). Efficient decomposition of perfluorocarboxylic acids in aqueous solution using microwave-induced persulfate. *Water Res.*, 43(11), 2811–2816.

Lee, Y., Lo, S., Chiueh, P., Liou, Y., and Chen, M. (2010). Microwave-hydrothermal decomposition of perfluorooctanoic acid in water by iron-activated persulfate oxidation. *Water Res.*, 44(3), 886–892.

Lee, Y., Lo, S., Kuo, J., and Hsieh, C. (2012a). Decomposition of perfluorooctanoic acid by microwave-activated persulfate: Effects of temperature, pH, and chloride ions. *Front. Environ. Sci. Eng.*, 6(1), 17–25.

Lee, Y., Lo, S., Kuo, J., and Lin, Y. (2012b). Persulfate oxidation of perfluorooctanoic acid under the temperatures of 20–40 C. *Chem. Eng. J.*, 198, 27–32.

Lee, Y., Chen, M., Huang, C., Kuo, J., and Lo, S. (2016). Efficient sonochemical degradation of perfluorooctanoic acid using periodate. *Ultrason. Sonochem.*, 31, 499–505.

Li, S., Wang, W., Yan, W., and Zhang, W. (2014). Nanoscale zero-valent iron (nZVI) for the treatment of concentrated Cu (II) wastewater: A field demonstration. *Environ. Sci. Process. Impacts*, 16(3), 524–533.

Li, X. Q., Elliott, D. W., and Zhang, W. X. (2006). Zero-valent iron nanoparticles for abatement of environmental pollutants: Materials and engineering aspects. *Critical Reviews in Solid State and Materials Sciences*, 31(4), 111–122.

Liang, C., Bruell, C. J., Marley, M. C., and Sperry, K. L. (2004a). Persulfate oxidation for in-situ remediation of TCE. I. Activated by ferrous ion with and without a persulfate–thiosulfate redox couple. *Chemosphere*, 55(9), 1213–1223.

Liang, C., Bruell, C. J., Marley, M. C., and Sperry, K. L. (2004b). Persulfate oxidation for in-situ remediation of TCE. II. Activated by chelated ferrous ion. *Chemosphere*, 55(9), 1225–1233.

Liang, C., Huang, C., Mohanty, N., Lu, C., and Kurakalva, R. M. (2007). Hydroxypropyl-β-cyclodextrin-mediated iron-activated persulfate oxidation of trichloroethylene and tetrachloroethylene. *Ind. Eng. Chem. Res.*, 46(20), 6466–6479.

Liang, C., and Lai, M. (2008). Trichloroethylene degradation by zero valent iron activated persulfate oxidation. *Environ. Eng. Sci.*, 25(7), 1071–1078.

Liang, C., Guo, Y., Chien, Y., and Wu, Y. (2010). Oxidative degradation of MTBE by pyrite-activated persulfate: Proposed reaction pathways. *Ind. Eng. Chem. Res.*, 49(18), 8858–8864.

Liao, Z., and Farrell, J. (2009). Electrochemical oxidation of perfluorobutane sulfonate using boron-doped diamond film electrodes. *J. Appl. Electrochem.*, 39(10), 1993–1999.

Lin, H., Niu, J., Ding, S., and Zhang, L. (2012). Electrochemical degradation of perfluorooctanoic acid (PFOA) by Ti/SnO2–Sb, Ti/SnO2–Sb/PbO2 and Ti/SnO2–Sb/MnO2 anodes. *Water Res.*, 46(7), 2281–2289.

Lin, J., Lo, S., Hu, C., Lee, Y., and Kuo, J. (2015). Enhanced sonochemical degradation of perfluorooctanoic acid by sulfate ions. *Ultrason. Sonochem.*, 22, 542–547.

Lin, J., Hu, C., and Lo, S. (2016). Effect of surfactants on the degradation of perfluorooctanoic acid (PFOA) by ultrasonic (US) treatment. *Ultrason. Sonochem.*, 28, 130–135.

Lin, Y., Liang, C., and Yu, C. (2016). Trichloroethylene degradation by various forms of iron activated persulfate oxidation with or without the assistance of ascorbic acid. *Ind. Eng. Chem. Res.*, 55(8), 2302–2308.

Liu, C., Higgins, C., Wang, F., and Shih, K. (2012a). Effect of temperature on oxidative transformation of perfluorooctanoic acid (PFOA) by persulfate activation in water. *Sep. Purif. Technol.*, 91, 46–51.

Liu, C., Shih, K., Sun, C., and Wang, F. (2012b). Oxidative degradation of propachlor by ferrous and copper ion activated persulfate. *Sci. Total Environ.*, 416, 507–512.

Liu, X., Zhong, J., Fang, L., Wang, L., Ye, M., Shao, Y., and Zhang, T. (2016). Trichloroacetic acid reduction by an advanced reduction process based on carboxyl anion radical. *Chem. Eng. J.*, 303, 56–63.

Locke, B., Sato, M., Sunka, P., Hoffmann, M., and Chang, J. (2006). Electrohydraulic discharge and nonthermal plasma for water treatment. *Ind. Eng. Chem. Res.*, 45(3), 882–905.

Locke, B. R., Lukes, P., and Brisset, J. L. (2012). Elementary chemical and physical phenomena in electrical discharge plasma in gas-liquid environments and in liquids. In *Plasma Chemistry and Catalysis in Gases and Liquids*, ed. M. M. V. I. Parvulescu and P. Lukes. Weinheim. Wiley-VCH Verlag.

Locke, B. R., and Thagard, S. M. (2017). Electrical discharge plasma for water treatment. In *Advanced Oxidation Processes for Water Treatment: Fundamentals and Applications*, ed. M. I. Stefan. London: IWA, 493–521.

Lukes, P., Appleton, A. T., and Locke, B. R. (2004). Hydrogen peroxide and ozone formation in hybrid gas-liquid electrical discharge reactors. *IEEE Trans. Ind. Appl.*, 40(1), 60–67.

Lukes, P., and Locke, B. R. (2005). Plasmachemical oxidation processes in a hybrid gas-liquid electrical discharge reactor. *J. Phys. D Appl. Phys.*, 38(22), 4074–4081.

Lukes, P., Locke, B. R., and Brisset, J. L. (2012). Aqueous-phase chemistry of electrical discharge plasma in water and in gas-liquid environments. *Plasma Chemistry and Catalysis in Gases and Liquids*, ed. M. M. V. I. Parvulescu and P. Lukes. Weinheim: Wiley-VCH Verlag.

Matsuya, Y., Takeuchi, N., and Yasuoka, K. (2014). Relationship between reaction rate of perfluorocarboxylic acid decomposition at a plasma–liquid interface and adsorbed amount. *Electric. Eng. Japan*, 188(2), 1–8.

Mededovic, S., Takashima, K., and Mizuno, A. (2008). Decolorization of indigo carmine dye by spark discharge in water. *Int. J. Plasma Environ. Sci. Technol.*, 2, 56–66.

Moody, C. A., and Field, J. A. (2000). Perfluorinated surfactants and the environmental implications of their use in fire-fighting foams. *Environ. Sci. Technol.*, 34(18), 3864–3870.

Moreau, M., Orange, N., and Feuilloley, M. G. J. (2008). Non-thermal plasma technologies: New tools for bio-decontamination. *Biotechnol. Adv.*, 26(6), 610–617.

Moriwaki, H., Takagi, Y., Tanaka, M., Tsuruho, K., Okitsu, K., and Maeda, Y. (2005). Sonochemical decomposition of perfluorooctane sulfonate and perfluorooctanoic acid. *Environ. Sci. Technol.*, 39(9), 3388–3392.

Murugananthan, M., Yoshihara, S., Rakuma, T., Uehara, N., and Shirakashi, T. (2007). Electrochemical degradation of 17β-estradiol (E2) at boron-doped diamond (Si/BDD) thin film electrode. *Electrochim. Acta*, 52(9), 3242–3249.

Němeček, J., Lhotský, O., and Cajthaml, T. (2014). Nanoscale zero-valent iron application for in-situ reduction of hexavalent chromium and its effects on indigenous microorganism populations. *Sci. Total Environ.*, 485, 739–747.

Niu, J., Lin, H., Xu, J., Wu, H., and Li, Y. (2012). Electrochemical mineralization of perfluorocarboxylic acids (PFCAs) by Ce-doped modified porous nanocrystalline PbO_2 film electrode. *Environ. Sci. Technol.*, 46(18), 10191–10198.

Niu, J., Lin, H., Gong, C., and Sun, X. (2013). Theoretical and experimental insights into the electrochemical mineralization mechanism of perfluorooctanoic acid. *Environ. Sci. Technol.*, 47(24), 14341–14349.

Niu, J., Li, Y., Shang, E., Xu, Z., and Liu, J. (2016). Electrochemical oxidation of perfluorinated compounds in water. *Chemosphere*, 146, 526–538.

Ochiai, T., Iizuka, Y., Nakata, K., Murakami, T., Tryk, D. A., Fujishima, A., Koide, Y., and Morito, Y. (2011). Efficient electrochemical decomposition of perfluorocarboxylic acids by the use of a boron-doped diamond electrode. *Diamond Relat. Mater.*, 20(2), 64–67.

Ochoa-Herrera, V., Sierra-Alvarez, R., Somogyi, A., Jacobsen, N. E., Wysocki, V. H., and Field, J. A. (2008). Reductive defluorination of perfluorooctane sulfonate. *Environ. Sci. Technol.*, 42(9), 3260–3264.

Oh, S., Kim, H., Park, J., Park, H., and Yoon, C. (2009). Oxidation of polyvinyl alcohol by persulfate activated with heat, Fe2, and zero-valent iron. *J. Hazard. Mater.*, 168(1), 346–351.

Panizza, M., and Cerisola, G. (2009). Direct and mediated anodic oxidation of organic pollutants. *Chem. Rev.*, 109(12), 6541–6569.

Pardo, F., Rosas, J. M., Santos, A., and Romero, A. (2015). Remediation of a biodiesel blend-contaminated soil with activated persulfate by different sources of iron. *Water Air Soil Pollut.*, 226(2), 17.

Park, H., Vecitis, C. D., Cheng, J., Choi, W., Mader, B. T., and Hoffmann, M. R. (2009). Reductive defluorination of aqueous perfluorinated alkyl surfactants: Effects of ionic headgroup and chain length. *J. Phys. Chem. A*, 113(4), 690–696.

Park, H., Vecitis, C. D., Cheng, J., Dalleska, N. F., Mader, B. T., and Hoffmann, M. R. (2011). Reductive degradation of perfluoroalkyl compounds with aquated electrons generated from iodide photolysis at 254 nm. *Photochem. Photobiol. Sci.*, 10(12), 1945–1953.

Park, S., Lee, L. S., Medina, V. F., Zull, A., and Waisner, S. (2016). Heat-activated persulfate oxidation of PFOA, 6:2 fluorotelomer sulfonate, and PFOS under conditions suitable for in-situ groundwater remediation. *Chemosphere*, 145, 376–383.

Petri, B. G., Thomson, N. R., and Urynowicz, M. A. (2011a). Fundamentals of ISCO using permanganate. In *In-Situ Chemical Oxidation for Groundwater Remediation*. New York: Springer, 89–146.

Petri, B. G., Watts, R. J., Tsitonaki, A., Crimi, M., Thomson, N. R., and Teel, A. L. (2011b). Fundamentals of ISCO using persulfate. In *In-Situ Chemical Oxidation for Groundwater Remediation*. New York: Springer, 147–191.

Phenrat, T., Saleh, N., Sirk, K., Tilton, R. D., and Lowry, G. V. (2007). Aggregation and sedimentation of aqueous nanoscale zerovalent iron dispersions. *Environ. Sci. Technol.*, 41(1), 284–290.

Phillips, D. H., Nooten, T. V., Bastiaens, L., Russell, M., Dickson, K., Plant, S., Ahad, J., Newton, T., Elliot, T., and Kalin, R. (2010). Ten year performance evaluation of a field-scale zero-valent iron permeable reactive barrier installed to remediate trichloroethene contaminated groundwater. *Environ. Sci. Technol.*, 44(10), 3861–3869.

Qi, C., Liu, X., Lin, C., Zhang, X., Ma, J., Tan, H., and Ye, W. (2014). Degradation of sulfamethoxazole by microwave-activated persulfate: Kinetics, mechanism and acute toxicity. *Chem. Eng. J.*, 249, 6–14.

Qu, Y., Zhang, C., Li, F., Chen, J., and Zhou, Q. (2010). Photo-reductive defluorination of perfluorooctanoic acid in water. *Water Res.*, 44(9), 2939–2947.

Qu, Y., Zhang, C., Chen, P., Zhou, Q., and Zhang, W. (2014). Effect of initial solution pH on photo-induced reductive decomposition of perfluorooctanoic acid. *Chemosphere*, 107, 218–223.

Radjenovic, J., Bagastyo, A., Rozendal, R. A., Mu, Y., Keller, J., and Rabaey, K. (2011). Electrochemical oxidation of trace organic contaminants in reverse osmosis concentrate using RuO2/IrO2-coated titanium anodes. *Water Res.*, 45(4), 1579–1586.

Radjenovic, J., and Sedlak, D. L. (2015). Challenges and opportunities for electrochemical processes as next-generation technologies for the treatment of contaminated water. *Environ. Sci. Technol.*, 49(19), 11292–11302.

Rastogi, A., Al-Abed, S. R., and Dionysiou, D. D. (2009). Effect of inorganic, synthetic and naturally occurring chelating agents on Fe (II) mediated advanced oxidation of chlorophenols. *Water Res.*, 43(3), 684–694.

Rayaroth, M. P., Aravind, U. K., and Aravindakumar, C. T. (2016). Degradation of pharmaceuticals by ultrasound-based advanced oxidation process. *Environ. Chem. Lett.*, 14(3), 259–290.

Rodriguez, S., Vasquez, L., Costa, D., Romero, A., and Santos, A. (2014). Oxidation of Orange G by persulfate activated by Fe (II), Fe (III) and zero valent iron (ZVI). *Chemosphere*, 101, 86–92.

Rodriguez-Freire, L., Balachandran, R., Sierra-Alvarez, R., and Keswani, M. (2015). Effect of sound frequency and initial concentration on the sonochemical degradation of perfluorooctane sulfonate (PFOS). *J. Hazard. Mater.*, 300, 662–669.

Ruma, Lukes, P., Aoki, N., Spetlikova, E., Hosseini, S. H. R., Sakugawa, T., and Akiyama, H. (2013). Effects of pulse frequency of input power on the physical and chemical properties of pulsed streamer discharge plasmas in water. *J. Phys. D Appl. Phys.*, 46(12).

Sahni, M., Finney, W. C., and Locke, B. R. (2005). Degradation of aqueous phase polychlorinated biphenyls (PCB) using pulsed corona discharges. *J. Adv. Oxidation Technol.*, 8(1), 105–111.

Schaefer, C. E., Andaya, C., Urtiaga, A., McKenzie, E. R., and Higgins, C. P. (2015). Electrochemical treatment of perfluorooctanoic acid (PFOA) and perfluorooctane sulfonic acid (PFOS) in groundwater impacted by aqueous film forming foams (AFFFs). *J. Hazard. Mater.*, 295, 170–175.

Sharma, A., Locke, B., Arce, P., and Finney, W. (1993). A preliminary study of pulsed streamer corona discharge for the degradation of phenol in aqueous solutions. *Hazard. Waste Hazard. Mater.*, 10(2), 209–219.

Siegrist, R. L., Crimi, M., and Simpkin, T. J. (2011). *In-Situ Chemical Oxidation for Groundwater Remediation.* Berlin: Springer Science & Business Media.

Singh, U. C., and Venkatarao, K. (1976). Decomposition of peroxodisulphate in aqueous alkaline solution. *J. Inorg. Nucl. Chem.*, 38(3), 541–543.

Song, S., Fan, J., He, Z., Zhan, L., Liu, Z., Chen, J., and Xu, X. (2010). Electrochemical degradation of azo dye CI Reactive Red 195 by anodic oxidation on Ti/SnO2–Sb/PbO2 electrodes. *Electrochim. Acta*, 55(11), 3606–3613.

Song, Z., Tang, H., Wang, N., and Zhu, L. (2013). Reductive defluorination of perfluorooctanoic acid by hydrated electrons in a sulfite-mediated UV photochemical system. *J. Hazard. Mater.*, 262, 332–338.

Sopaj, F., Rodrigo, M. A., Oturan, N., Podvorica, F. I., Pinson, J., and Oturan, M. A. (2015). Influence of the anode materials on the electrochemical oxidation efficiency. Application to oxidative degradation of the pharmaceutical amoxicillin. *Chem. Eng. J.*, 262, 286–294.

Stefaniuk, M., Oleszczuk, P., and Ok, Y. S. (2016). Review on nano zerovalent iron (nZVI): From synthesis to environmental applications. *Chem. Eng. J.*, 287, 618–632.

Stratton, G. R., Bellona, C. L., Dai, F., Holsen, T. M., and Thagard, S. M. (2015). Plasma-based water treatment: Conception and application of a new general principle for reactor design. *Chem. Eng. J.*, 273, 543–550.

Stratton, G. R., Dai, F., Bellona, C. L., Holsen, T. M., Dickenson, E. R. V., and Mededovic Thagard, S. (2017). Plasma-based water treatment: Efficient transformation of perfluoroalkyl substances in prepared solutions and contaminated groundwater. *Environ. Sci. Technol.*, 51(3), 1643–1648.

Sugiarto, A. T., Ito, S., Ohshima, T., Sato, M., and Skalny, J. D. (2003). Oxidative decoloration of dyes by pulsed discharge plasma in water. *J. Electrostat.*, 58(1–2), 135–145.

Sun, H., Wang, L., Zhang, R., Sui, J., and Xu, G. (2006). Treatment of groundwater polluted by arsenic compounds by zero valent iron. *J. Hazard. Mater.*, 129(1), 297–303.

Sun, J., Lu, H., Lin, H., Du, L., Huang, W., Li, H., and Cui, T. (2012). Electrochemical oxidation of aqueous phenol at low concentration using Ti/BDD electrode. *Sep. Purif. Technol.*, 88, 116–120.

Sunka, P., Babický, V., Clupek, M., Lukes, P., Simek, M., Schmidt, J., and Cernak, M. (1999). Generation of chemically active species by electrical discharges in water. *Plasma Sources Sci. Technol.*, 8(2), 258.

Takeuchi, N., Kitagawa, Y., Kosugi, A., Tachibana, K., Obo, H., and Yasuoka, K. (2013). Plasma–liquid interfacial reaction in decomposition of perfluoro surfactants. *J. Phys. D Appl. Phys.*, 47(4), 045203.

Tan, C., Gao, N., Deng, Y., An, N., and Deng, J. (2012). Heat-activated persulfate oxidation of diuron in water. *Chem. Eng. J.*, 203, 294–300.

Thagard, S. M., Stratton, G. R., Dai, F., Bellona, C. L., Holsen, T. M., Bohl, D. G., Paek, E., and Dickenson, E. R. (2016). Plasma-based water treatment: Development of a general mechanistic model to estimate the treatability of different types of contaminants. *J. Phys. D Appl. Phys.*, 50(1), 014003.

Trautmann, A., Schell, H., Schmidt, K., Mangold, K., and Tiehm, A. (2015). Electrochemical degradation of perfluoroalkyl and polyfluoroalkyl substances (PFAS) in groundwater. *Water Sci. Technol.*, 71(10), 1569–1575.

Truex, M. J., Macbeth, T., Vermeul, V. R., Fritz, B. G., Mendoza, D. P., Mackley, R. D., Wietsma, T. W., Sandberg, G., Powell, T., and Powers, J. (2011). Demonstration of combined zero-valent iron and electrical resistance heating for in-situ trichloroethene remediation. *Environ. Sci. Technol.*, 45(12), 5346–5351.

Tsitonaki, A., Petri, B., Crimi, M., Mosbæk, H., Siegrist, R. L., and Bjerg, P. L. (2010). In-situ chemical oxidation of contaminated soil and groundwater using persulfate: A review. *Crit. Rev. Environ. Sci. Technol.*, 40(1), 55–91.

USEPA (U.S. Environmental Protection Agency). (2009). Provisional health advisories for perfluorooctanoic acid (PFOA) and perfluorooctane sulfonate (PFOS). Washington, DC: USEPA, Office of Water. https://www.epa.gov/sites/production/files/2015-09/documents/pfoa-pfos-provisional.pdf (accessed April 2016).

USEPA (U.S. Environmental Protection Agency). (2016a). Drinking water health advisories for PFOA and PFOS. https://www.epa.gov/ground-water-and-drinking-water/drinking-water-health-advisories-pfoa-and-pfos (accessed May 2016).

USEPA (U.S. Environmental Protection Agency). (2016b). Public water systems, disinfection byproducts, and the use of monochloramine. https://www.epa.gov/dwreginfo/public-water-systems-disinfection-byproducts-and-use-monochloramine (accessed February 2017).

van Veldhuizen, E. M., and Rutgers, W. R. (2002). Pulsed positive corona streamer propagation and branching. *J. Phys. D Appl. Phys.*, 35(17), 2169–2179.

Vecitis, C. D., Park, H., Cheng, J., Mader, B. T., and Hoffmann, M. R. (2008). Kinetics and mechanism of the sonolytic conversion of the aqueous perfluorinated surfactants, perfluorooctanoate (PFOA), and perfluorooctane sulfonate (PFOS) into inorganic products. *J. Phys. Chem. A*, 112(18), 4261–4270.

Vecitis, C., Park, H., Cheng, J., Mader, B., and Hoffmann, M. (2009). Treatment technologies for aqueous perfluorooctanesulfonate (PFOS) and perfluorooctanoate (PFOA). *Front. Environ. Sci. Eng. China*, 3(2), 129–151.

Vellanki, B. P., Batchelor, B., and Abdel-Wahab, A. (2013). Advanced reduction processes: A new class of treatment processes. *Environ. Eng. Sci.*, 30(5), 264–271.

Vicente, F., Santos, A., Romero, A., and Rodriguez, S. (2011). Kinetic study of diuron oxidation and mineralization by persulphate: Effects of temperature, oxidant concentration and iron dosage method. *Chem. Eng. J.*, 170(1), 127–135.

Wang, L., Jiang, X. Z., and Liu, Y. J. (2008). Degradation of bisphenol A and formation of hydrogen peroxide induced by glow discharge plasma in aqueous solutions. *J. Hazard Mater.*, 154(1–3), 1106–1114.

Xiao, H., Lv, B., Zhao, G., Wang, Y., Li, M., and Li, D. (2011). Hydrothermally enhanced electrochemical oxidation of high concentration refractory perfluorooctanoic acid. *J. Phys. Chem. A*, 115(47), 13836–13841.

Yang, S., Cheng, J., Sun, J., Hu, Y., and Liang, X. (2013). Defluorination of aqueous perfluorooctanesulfonate by activated persulfate oxidation. *PloS One*, 8(10), e74877.

Yang, X., Bai, M., and Han, F. (2009). Treatment of phenol wastewater using hydroxyl radical produced by micro-gap discharge plasma technique. *Water Environ. Res.*, 81(4), 450–455.

Yang, Y., Cho, Y. I., and Fridman, A. (2012). *Plasma Discharge in Liquid: Water Treatment and Applications*. Boca Raton, FL: CRC Press.

Yano, T., Shimomura, N., Uchiyama, I., Fukawa, F., Teranishi, K., and Akiyama, H. (2009). Decolorization of indigo carmine solution using nanosecond pulsed power. *IEEE Trans. Dielectr. Electr. Insul.*, 16(4), 1081–1087.

Yasuoka, K., Sasaki, K., Hayashi, R., Kosugi, A., and Takeuchi, N. (2010). Degradation of perfluoro compounds and F-recovery in water using discharge plasmas generated within gas bubbles. *Int. J. Plasma Environ. Sci. Technol.* 4(2), 113–117.

Yasuoka, K., Sasaki, K., and Hayashi, R. (2011). An energy-efficient process for decomposing perfluorooctanoic and perfluorooctane sulfonic acids using dc plasmas generated within gas bubbles. *Plasma Sources Sci. Technol.*, 20(3), 034009.

Yin, P., Hu, Z., Song, X., Liu, J., and Lin, N. (2016). Activated persulfate oxidation of perfluorooctanoic acid (PFOA) in groundwater under acidic conditions. *Int. J. Environ. Res. Public Health*, 13(6), 602.

Zhang, B., Zhang, Y., Teng, Y., and Fan, M. (2015). Sulfate radical and its application in decontamination technologies. *Crit. Rev. Environ. Sci. Technol.*, 45(16), 1756–1800.

Zhang, C., Qu, Y., Zhao, X., and Zhou, Q. (2015). Photoinduced reductive decomposition of perfluorooctanoic acid in water: Effect of temperature and ionic strength. *Clean (Weinh)*, 43(2), 223–228.

Zhang, Y., Xie, X., Huang, S., and Liang, H. (2014). Effect of chelating agent on oxidation rate of aniline in ferrous ion activated persulfate system at neutral pH. *J. Central South Univ.*, 21 1441–1447.

Zhao, B., Lv, M., and Zhou, L. (2012). Photocatalytic degradation of perfluorooctanoic acid with β-Ga2O3 in anoxic aqueous solution. *J. Environ. Sci.*, 24(4), 774–780.

Zhao, H., Gao, J., Zhao, G., Fan, J., Wang, Y., and Wang, Y. (2013). Fabrication of novel SnO 2-Sb/carbon aerogel electrode for ultrasonic electrochemical oxidation of perfluorooctanoate with high catalytic efficiency. *Appl. Catal. B*, 136, 278–286.

Zhao, L., Hou, H., Fujii, A., Hosomi, M., and Li, F. (2014). Degradation of 1,4-dioxane in water with heat- and Fe2-activated persulfate oxidation. *Environ. Sci. Pollut. Res.*, 21(12), 7457–7465.

Zhou, L., Zheng, W., Ji, Y., Zhang, J., Zeng, C., Zhang, Y., Wang, Q., and Yang, X. (2013). Ferrous-activated persulfate oxidation of arsenic (III) and diuron in aquatic system. *J. Hazard. Mater.*, 263, 422–430.

Zhou, M., Chi, M., Wang, H., and Jin, T. (2012). Anode modification by electrochemical oxidation: A new practical method to improve the performance of microbial fuel cells. *Biochem. Eng. J.*, 60 151–155.

Zhuo, Q., Deng, S., Yang, B., Huang, J., and Yu, G. (2011). Efficient electrochemical oxidation of perfluorooctanoate using a Ti/SnO2-Sb-Bi anode. *Environ. Sci. Technol.*, 45(7), 2973–2979.

Zhuo, Q., Deng, S., Yang, B., Huang, J., Wang, B., Zhang, T., and Yu, G. (2012). Degradation of perfluorinated compounds on a boron-doped diamond electrode. *Electrochim. Acta*, 77, 17–22.

Zhuo, Q., Li, X., Yan, F., Yang, B., Deng, S., Huang, J., and Yu, G. (2014). Electrochemical oxidation of 1H, 1H, 2H, 2H-perfluorooctane sulfonic acid (6: 2 FTS) on DSA electrode: Operating parameters and mechanism. *J. Environ. Sci.*, 26(8), 1733–1739.

13 Reactivation of Spent Activated Carbon Used for PFAS Adsorption

John Matthis and Stephanie Carr

CONTENTS

13.1 Introduction 303
13.2 Basics of Reactivation and a Carbon Reactivation System 305
13.3 The Chemistry behind Carbon Reactivation 306
    13.3.1 Drying of Spent Carbon 307
    13.3.2 Devolatilization of Spent Carbon 308
    13.3.3 Pyrolyzation of Spent Carbon 309
    13.3.4 Carbon or Steam Gasification 309
13.4 Some Rules of Carbon Reactivation That Should Be Considered When Evaluating a Spent Carbon and Its Potential for Reactivation 311
13.5 Reactivation Equipment and Furnaces 312
13.6 Pooled versus Custom Reactivation Options 313
    13.6.1 Regional or Off-Site Carbon Supplier–Owned Reactivation Facility for Pooled Reactivation 315
    13.6.2 Regional Carbon Supplier–Owned Reactivation Facility for Custom or Segregated Reactivation 315
    13.6.3 Customer On-Site Reactivation Facility 316
13.7 Activated Carbon Tests Used for Monitoring and Controlling Reactivation Product Quality 316
13.8 Effectiveness of Reactivation in the Destruction of PFAS Compounds 318
13.9 Properties of Activated Carbon That Can Change with Reactivation 318
13.10 Spent Carbon Profiling and Reactivation 321
13.11 Possible Issues with a Spent Carbon That Would Result in a Poor Candidate for Reactivation 322
13.12 Reactivation as a Sustainable and Environmentally Responsible Approach 323

13.1 INTRODUCTION

The reactivation of activated carbon is a well-established, high-temperature process for the thermal destruction of adsorbed chemicals, after which the reactivated carbon can be reused. The desorbed chemical constituents are thermally destroyed in the process, and the reactivation of spent carbon containing perfluorooctane sulfonate (PFOS), perfluorooctanoic acid (PFOA), and other poly- and perfluorinated alkyl substances (PFAS) has been practiced for more than 10 years. Thus, to understand the reactivation of spent activated carbon with PFAS, one must understand the process of reactivation.

Granular activated carbon (GAC) is a technology commonly used to remove organic species (volatile organic compound [VOC], total organic carbon [TOC], petroleum hydrocarbons, and PFAS) from a variety of water sources (groundwater, surface water, and wastewater). Activated carbon's service life is dependent on the capacity for components that require removal by a specific application and the treatment objective of the system. Once activated carbon's capacity becomes exhausted or is no longer meeting the required treatment objective of the system, the carbon is said to be "spent" or "exhausted." At this time, the spent carbon must be removed from the adsorption vessel and be replaced with a fresh carbon so that the system can continue to meet its treatment objective.

Once the spent carbon is removed from the vessel, there are three options that are typically available for dealing with the spent carbon: (1) disposal via landfill, (2) disposal via an incinerator (for destruction of the carbon and adsorbates or PFAS), or (3) reactivation of the spent carbon for reuse. Disposal options 1 and 2 can incur long-term liability or high costs. Thus, reactivation is an attractive option when it can be performed.

The term *reactivation* commonly refers to the use of a high-temperature thermal treatment process (approximately 1800°F) to completely remove all adsorbates on the spent carbon and either desorb, devolatilize, dissociate, or char the adsorbate in place in the carbon structure itself. The desorbed chemical constituents (whether desorbed completely or in pieces due to the dissociation of the chemical bonds) are thermally destroyed in the reactivation process, by which the spent activated carbon enters a furnace or kiln where temperatures are increased to approximately 1800°F. In addition to the adsorbed species being removed or destroyed in the reactivation process, there are also losses to the carbon structure (carbon atoms are removed) in this process or during the gasification steps. The removal of adsorbed species results in gaining back adsorptive capacity of the material; however, the loss of carbon structure/atoms results in a carbon pore structure that is different from the virgin carbon.

The term *regeneration*, which may also be familiar to the reader, is the restoration of a portion (not all) of the adsorptive capacity so that the carbon product can be used in another adsorption cycle. Methods include a low-temperature thermal (<400°F) procedure (using steam, air, nitrogen, or hot gas) or chemical extraction or solvent stripping (such as a pH swing to desorb some of the adsorbate from the spent carbon). It can be performed on site *in situ*. The additional adsorption cycles will have less adsorption capacity than fresh carbon. Regenerated carbon maintains the original carbon skeleton structure and can contain some to most of the original adsorbates. High-temperature thermal reactivation is the preferred method for reuse of activated carbon, since with efficient reactivation, most of the adsorptive capacity can be restored.

Carbon reactivation is usually performed at a centralized reactivation facility (some applications, such as sugar decolorization, may generate significant amounts of spent carbon to economically warrant an on-site carbon reactivation facility, but that is not usually the case with water treatment applications). This off-site reactivation facility can be (1) a facility that handles both industrial and potable spent carbons, permitted for both hazardous and nonhazardous spent carbon, or (2) a reactivation facility dedicated to custom-food-grade or drinking water spent carbon whereby the incoming spent carbon is segregated by customer, reactivated, and returned to the same customer or application.

In off-site carbon reactivation, the spent carbon is properly transported to a reactivation facility once the necessary waste profile and representative spent carbon sample have been tested or profiled and approved by the reactivation facility. The carbon is then reactivated for reuse by others (can be referred to as "pooled reactivation") or by the same customer (typically referred to as "custom reactivation").

Thermal reactivation is an attractive option from an environmental and cost standpoint. It uses fewer resources to reactivate a spent carbon than activate virgin carbon from a raw material source, reduces one's carbon footprint (produces only 20% of the greenhouse gases compared with virgin activated carbon production), minimizes waste, and can reduce operating costs through the continued use of a reactivated carbon versus a virgin activated carbon.

Commercial carbon reactivation services have been offered and conducted for more than 40 years for a wide variety of applications and adsorbates. Reactivation of spent carbons containing adsorbed PFAS (specifically PFOA or PFOS) has been performed commercially for more than 10 years.

13.2 BASICS OF REACTIVATION AND A CARBON REACTIVATION SYSTEM

As mentioned above, reactivation uses a high-temperature thermal process whereby adsorbed organic compounds (adsorbates) are destroyed and the GAC's adsorptive capacity is restored. Types of thermal reactivation hardware include multiple hearth furnace, rotary kiln (direct fired and indirectly fired), infrared belt furnace, and fluidized bed furnaces. A general overview of a carbon reactivation system is shown in Figure 13.1.

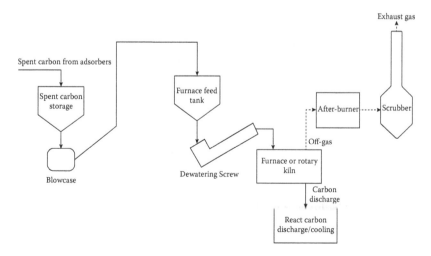

FIGURE 13.1 Generic layout for the typical reactivation process.

A reactivation system will typically require spent carbon storage; a transport mechanism to a furnace feed tank and/or dewatering screw; the multiple hearth furnace or kiln; an afterburner; a furnace or kiln off-gas treatment, such as a chemical scrubber and/or particulate capture (baghouse); cooling of the reactivated carbon discharge; final product or carbon screening; and reactivated carbon storage. Pumps and/or blowcases will also be needed to convey the various carbon or carbon–water slurries throughout the process.

There are two primary types of reactivation furnaces that are commercially employed as the mainstay of a reactivation system: a multiple hearth furnace or a rotary kiln. More will be discussed about these two types of furnaces in the next section.

13.3 THE CHEMISTRY BEHIND CARBON REACTIVATION

During the reactivation process, the spent carbon will not only have the adsorbates removed and destroyed, but also, some of the carbon structure will be altered, changing some of the material properties compared with the original carbon. It is important to realize that the spent carbon quality (adsorbates, inorganics, and metals present on the spent carbon), as well as the original carbon's base structure, will determine the quality and properties of the reactivated carbon.

The chemistry of carbon reactivation and what happens to the adsorbates and carbon structure during each phase are discussed next. This would apply to any spent carbon containing any type of adsorbate: hydrocarbons, volatile organics, and chlorinated, brominated, or fluorinated compounds.

The four general steps in carbon reactivation (Figure 13.2) are

1. Low-temperature pretreatment: Drying of carbon to remove the water remaining in the carbon pores (212°F)
2. Physical processes and reactions: Thermal devolatilization and desorption of weakly held adsorbates (212°F–1000°F)
3. High-temperature carbon condensation reactions: High-temperature pyrolysis or calcination, in which adsorbates that are not removed in step 2 (thermal devolatilization and desorption) are converted to char (1000°F–1450°F)
4. High-temperature carbon gas or solid reactions: Chemical reactions for carbon gasification with water vapor, carbon dioxide, or oxygen to gasify (partially remove) char formed in the step 3 and regain some adsorptive capacity (1450°F–1800°F)

During a reactivation cycle, the carbon skeleton and adsorption pore volume are partially displaced by char and reduced. However, the sum of the transport and adsorption pore structure remains unchanged in terms of percentage of total granule, measured on a volume basis, during reactivation cycles. This is discussed in greater detail in later segments.

Reactivation of Spent Activated Carbon Used for PFAS Adsorption

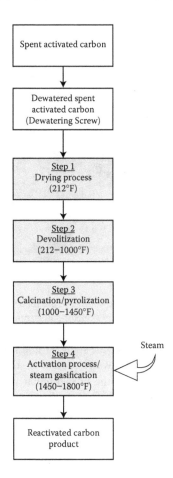

FIGURE 13.2 Steps of reactivating a spent carbon. The gray boxes indicate the steps taking place inside of a furnace.

13.3.1 Drying of Spent Carbon

One commonly used rule of thumb is referred to as the "40/40/20 rule," where 40% of a carbon bed is intergranular voids, 40% is intragranular pores (adsorptive and transport pores), and 20% is the carbon skeleton, by volume. Figure 13.3 illustrates the 40/40/20 rule for a carbon bed, as well as the breakdown of the types of pores (adsorption and transport) and the carbon skeleton for an individual carbon granule. When spent carbon is fed into a reactivation furnace, it is classified as either "wet" or "dewatered." In the case where the carbon is fed wet, it is fed as a carbon–water slurry and the water content is roughly 80% by volume (40% water in the intergranular voids and 40% intragranular pores). In this scenario, a dewatering screw is used prior to the reactivation furnace or kiln to remove the free water present in the intergranular

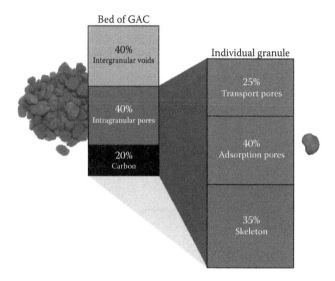

FIGURE 13.3 Graphical depiction of the 40/40/20 rule for a GAC bed (left) and the individual carbon granule breakdown for a carbon skeleton, adsorption pores, and transport pores.

voids (roughly 40% by volume). Once the spent carbon has been dewatered, the carbon undergoes a drying stage as the water content is still roughly 40% by volume, and there is still water remaining in the intragranular pores. To remove this water content, the first one-third to one-half of a react furnace is typically used for drying the carbon, and is the first step in reactivation, as shown in Figure 13.2. Temperatures during the drying stage in the furnace reach 100°C (212°F), simultaneously evaporating moisture and removing the more volatile organics from the granular carbon.

13.3.2 Devolatilization of Spent Carbon

The devolatilization process is desorption of volatile organics that were present as adsorbates on the spent carbon. This is accomplished, in order of preference, by (1) vaporizing low boiling or thermally stable adsorbates, (2) thermally breaking chemical bonds and desorbing smaller molecules, or (3) pyrolizing adsorbates that were not devolatilized by the previous two mechanisms.

The devolatilization process is a function of the individual adsorbates' stabilities and boiling points. Because the adsorption forces of activated carbon act like a pressure cooker, this step typically requires temperatures 400°F above the boiling point of an adsorbate to achieve desorption. When the species has a low boiling point or high stability, species will desorb from the pores completely (species such as alkanes [<C10], aromatics [<C4 side chains], and aliphatic alcohols [<C10]). When the species are high boiling or thermally unstable adsorbates, a different mechanism of desorption will occur whereby the adsorbate bonds break and the resulting fragments desorb (typically observed at temperatures <1000°F). Examples of these types of species are alkanes (>C10), aromatics (>C4 side chains), aliphatic alcohols (>C10), and organic halides.

13.3.3 Pyrolyzation of Spent Carbon

The pyrolyzation process is the thermal destruction of adsorbates. What does not desorb during devolatilization forms char in the adsorbing regions of the carbon structure. Char formation means a loss of adsorption volume; it does not possess any adsorption properties. Char can also be further condensed through thermal calcinations to form a purer and denser form of carbon and restore some adsorptive properties. Whether a species is desorbed, devolatilized, or charred will be adsorbate specific.

13.3.4 Carbon or Steam Gasification

At 1450°F–1800°F, the char is burned along with a small amount of the original carbon. Both the char and the graphitic plates within the activated carbon granule are able to undergo gasification reactions with water vapor, carbon dioxide, and oxygen present in the kiln or furnace atmosphere. It is important to note that the gasification reactions will increase the adsorption pore volume of the activated carbon product, and the pore structure will be intrinsically different and typically less than the adsorption pore volume of the product pre-reactivation.

Figures 13.4 and 13.5 show water–carbon dioxide–oxygen gasification reactions that occur with the carbon plates and structure whereby a carbon atom is removed from the carbon platelet. The activated carbon structure is composed of randomly

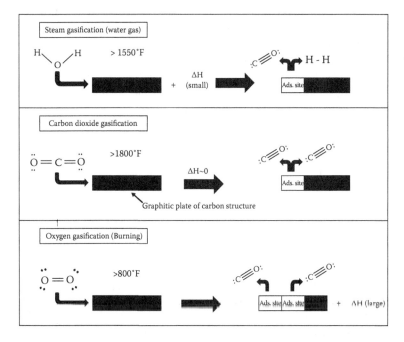

FIGURE 13.4 The three gasification reactions during the reactivation process that can occur with both the carbon structure and the char that is formed in the pores. Ads. = adsorption.

Solid/Gas Reactions	
$C_{(s)} + H_2O_{(g)} \rightarrow CO_{(g)} + H_{2(g)}$	Endothermic
$C_{(s)} + CO_{2(g)} \rightarrow 2\,CO_{(g)}$	Endothermic (3× slower than H_2O)
$C_{(s)} + \frac{1}{2}O_{2(g)} \rightarrow CO_{2(g)}$	Exothermic
$C_{(s)} + O_{2(g)} \rightarrow CO_{2(g)}$	Exothermic
Gas/Gas Reactions	
$CO_{(g)} + \frac{1}{2}O_{2(g)} \rightarrow CO_{2(g)}$	Exothermic
$H_{2(g)} + \frac{1}{2}O_{2(g)} \rightarrow H_2O_{(g)}$	Exothermic

FIGURE 13.5 The gasification reactions and their enthalpy values (ΔH), where exothermic reactions are $\Delta H < 0$ and endothermic values are $\Delta H > 0$.

arranged (amorphous) graphitic plates, commonly referred to as "platelets." The removal of the carbon atom from the platelet leaves a "spot" that is then available for an adsorption site. While the gasification of the carbon structure and platelets can increase adsorption pore volume, it should be balanced and minimized as much as possible as losses to the carbon skeleton can reduce the hardness and abrasion of the product (discussed in future sections), as well as decrease the number of reactivation cycles a product can undergo. Thus, the atmospheric conditions of the furnace should be carefully controlled.

As mentioned above, whatever is not devolatilized or desorbed out of the carbon structure is converted to char in the adsorption pores. Since char has no adsorption characteristics, its presence reduces the adsorption pore volume for a reactivated material. In order to restore some of the adsorptive pore volume lost to char formation, the same three gasification reactions (water vapor, carbon dioxide, and oxygen) discussed in Figure 13.4 for the carbon platelet gasification also occur with the char. It is important to reemphasize that the gasification reactions, while restoring some adsorptive capacity by reducing the char, also react with the original carbon structure at the edges of the graphitic plates and dislocations from the carbon plates. Figure 13.6 shows the order of attack on the carbon structure from these gasification reactions.

In summary, reactivation is a process that restores a spent or used activated carbon product's adsorptive capacity so that the product can be reused. This is accomplished by the thermal devolatilization and desorption of weakly held adsorbates and pieces of contaminants whose bonds are broken and desorbed at elevated temperatures. This devolatilization restores a fraction of the adsorptive pore volume of the original activated carbon product. The adsorbates that are not removed are converted to char within the adsorptive pore space, which has no adsorptive capacity. To restore a larger percentage of adsorptive pore volume, gasification reactions are performed that remove the char that has formed in these pores. These gasification reactions must be balanced carefully, as they will also react with the carbon structure.

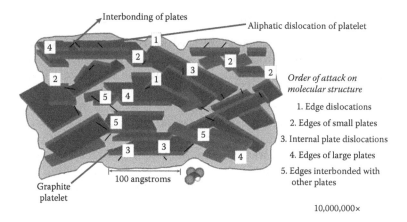

FIGURE 13.6 Order of carbon atom removal on graphite plate, creating greater adsorption capacity during gasification.

13.4 SOME RULES OF CARBON REACTIVATION THAT SHOULD BE CONSIDERED WHEN EVALUATING A SPENT CARBON AND ITS POTENTIAL FOR REACTIVATION

Rule 1: The characteristics of a spent carbon predetermine the best available reactivated carbon product that can be achieved. A poor spent activated carbon structure and ash and char levels will limit the best obtainable reactivated product.

Ash is defined as the noncombustible mineral matter that is contained in activated carbon and is the residue that remains after the combustion of a carbonaceous material. The measurement of ash is performed under specified conditions and is normally defined on a weight percent basis.

Char is defined as the resulting residue of organic matter of adsorbates that are not devolatilized or vaporized from the surface of activated carbon at high temperatures.

Rule 2: High-ash (such as calcium and other inorganics and metals) adsorbates on spent carbons make high-ash reactivated carbons or cause other ash-related problems (e.g., leaching of inorganic species).

Rule 3: The spent carbon cannot be reactivated if it cannot be fed to the furnace.

Rule 4: Balance char buildup with acceptable performance decay. Minimize carbon graphite plate gasification, as a shortening or loss of graphite plates will result in a loss of adsorptive capacity.

Rule 5: Only gasify the carbon plate structure when justified by actual performance decay in an application. Do not gasify the plate structure just to achieve an arbitrary activity specification.

Rule 6: Treat reactivated carbons as differentiated products. A reactivated carbon with virgin density and activity is not equal to a virgin carbon structure. The pore structure has been altered in the reactivation process due to the formation of char and gasification reactions as addressed above.

13.5 REACTIVATION EQUIPMENT AND FURNACES

The reactivation process equipment will need to include auxiliary equipment in addition to the reactivation furnace or kiln. At the onset, spent carbon is transferred from an adsorber or a delivery container (bulk truckload or supersaks if spent carbon is sent to an off-site regional reactivation facility by the customer) to a storage tank or elevated furnace feed tank. If dewatering is required, the spent carbon is then metered into a furnace feed dewatering screw, an inclined screw conveyor that serves the dual purpose of draining slurry water from the carbon and providing a water seal for the top of the furnace. Drained but still wet (40% by volume, 50% by weight moisture), the spent carbon flows into the furnace and proceeds downward through the hearths, seeing increasingly higher-temperature reactivation zones in the furnace.

Six to nine hearths are used in carbon reactivation furnaces that consist of a cylindrical refractory-lined steel shell containing the hearths, one above the other, and a central rotating shaft that drives rabble arms across the hearths (Figure 13.7a and b). Carbon entering the top hearth is raked toward the center and down a central hole to the center of the next hearth, from where the rabble teeth move it outward again in a spiral path toward the rim. The flow pattern is repeated through subsequent pairs of

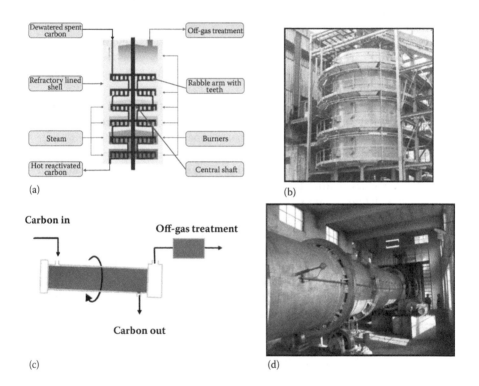

FIGURE 13.7 Illustrations and photographs of both styles of furnaces: (a and b) multiple hearth and (c and d) rotary kiln.

hearths, until the reactivated carbon is discharged from the bottom hearth. Burners using natural gas or fuel oil are mounted on the shell so as to direct their flames tangentially into the space above the hearths. Steam for controlling the reactivation stage is added through ports above the bottom several hearths.

A rotary kiln is a refractory-lined steel cylinder closed at the ends by stationary hoods and mounted so as to slope downward from inlet to outlet (Figure 13.7c and d). This type of kiln passes carbon granules either cocurrent with or countercurrent to combustion gases and steam. The kiln rotates with wet spent carbon entering onto flights through a feed screw or chute. The reactivated carbon discharges from the lower end of the kiln. The atmosphere inside the kiln is controlled by burners plus steam addition. Similar treatment of the off-gases is used for reactivation furnaces and kilns.

Reactivated carbon exits the furnace and is dried through one of two methods: (1) entering a water quench tank or (2) being cooled by contact with air. A water quench tank is typically used by on-site reactivation facilities, as the reactivated carbon can be transferred as a water and carbon slurry from the kiln or furnace. This method also has the added benefits of removing the fines present via the quench tank overflow water and provides a bottom seal for the furnace. Dry cooling, or cooling the carbon with air, is typically used at off-site reactivation centers, as the carbon can be rescreened and packaged into bags or supersaks. This cooling method is typically used by off-site reactivation centers where the carbon is being shipped to other locations since wet carbon will weigh roughly twice as much as dry carbon (see Figure 13.3).

Reactivation furnaces or kilns should be maintained under negative pressure, ensuring that there will be no leaks to the outside environment. In most reactivation facilities, an afterburner (destroys any unburned organics), chemical scrubber (removal of acid gases), and baghouse are also installed for destroying organics and removing residual particles from the furnace off-gases. The reactivation facility should be permitted and required to meet certain local, state, and federal regulations and requirements.

Afterburners and thermal oxidizers combust desorbed organic compounds. Dry scrubbers and spray dryer units are used to remove acid gases from the furnace or kiln off-gas stream. A baghouse (dust collector) will remove particulate matter from the afterburner and spray dryer as well as furnace dust. For some operations, a wet venturi scrubber is used for particulate control.

Any wastewater or solid waste generated in the process may require treatment and discharge and/or waste disposal in accordance with the classification of the waste and plant discharge permits.

13.6 POOLED VERSUS CUSTOM REACTIVATION OPTIONS

There are three typical reactivation scenarios for spent carbons. Regional or off-site carbon supplier–owned reactivation facilities are used for pooled reactivation or custom or segregated reactivation (Figures 13.8 and 13.9). Customers may also use on-site reactivation facilities.

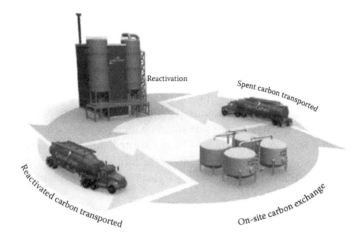

FIGURE 13.8 Example of a custom reactivation cycle.

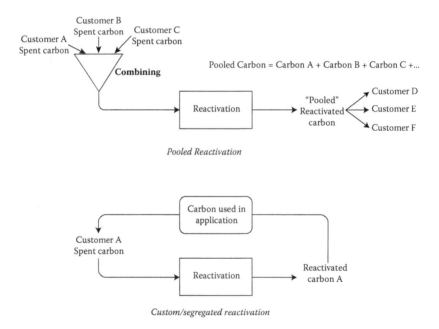

FIGURE 13.9 Visual representation of pooled reactivation versus custom reactivation.

13.6.1 REGIONAL OR OFF-SITE CARBON SUPPLIER–OWNED REACTIVATION FACILITY FOR POOLED REACTIVATION

More prevalent for spent carbon from water treatment applications is the use of an off-site reactivation facility that services many types of applications, sites, and customers. These reactivation centers are available worldwide and typically produce an industrial-grade reactivated carbon. Pooled reactivated carbons are approved spent carbons from many applications (potable, food grade, industrial process, remediation, etc.) and are all received at one reactivation facility and combined into a singular "pool" of carbon. It should be noted that this pool is an assortment of carbons from a wide range of applications, typically with separate pools for liquid phase and vapor phase carbons. In addition, note that a non-food-grade or industrial reactivation facility can also be a Resource Conservation and Recovery Act (RCRA)–approved facility. The pool of spent carbons is fed to a reactivation furnace or kiln. This yields a reactivated grade of carbon that can be used for many nonpotable or non-food-grade applications.

These sites are usually considered "industrial reactivation sites," meaning that the reactivated carbons will be used for industrial or non-food-grade applications. These reactivation centers, if RCRA hazardous permitted, can then handle both RCRA hazardous and non-RCRA-classified spent carbons from any application. The spent carbons received are combined, fed through a reactivation furnace or kiln, and produce a consistent quality and "age" of reactivated carbon product. This industrial reactivated carbon can then be used by appropriate customers or applications. Since the spent carbons received may be from previous react-grade carbon or from virgin carbons that have been used once by a customer, and some reactivated carbons shipped from the reactivation pool do not get returned for reactivation, the overall life cycle or age of the reactivated carbon pool remains relatively constant.

A particular customer or application can be supplied any type carbon that is required (virgin or reactivated) and still return their spent carbon for reactivation pending reactivation approval.

Since vapor phase (commonly 3 or 4 mm pellets, 4×6 and 4×10 mesh sizes) activated carbon and liquid or water phase (commonly 8×30 and 12×40 mesh sizes) activated carbons are different particle size ranges, typically there would be separate vapor phase and liquid phase reactivated carbon product pools. Both coal-based and coconut-based activated carbons can be reactivated in similar manners and reused successfully. They may be reactivated in the same pool or be provided as separate reactivation products, depending on the supplier.

13.6.2 REGIONAL CARBON SUPPLIER–OWNED REACTIVATION FACILITY FOR CUSTOM OR SEGREGATED REACTIVATION

There is also a need in the food and drinking water market segments for reactivation of a specific customer's spent carbon and the return of their same reactivated carbon for reuse in the same application. The amount of spent carbon generated may not justify on-site reactivation, or the customer may not desire to run a carbon reactivation facility, making an off-site custom or segregated food-grade or potable reactivation process a favorable option. Custom food-grade or potable water carbon reactivation is conducted at a dedicated food-grade facility where the process should be tailored to meet the requirements of the American Water Works Association and National Sanitary Foundation (NSF)/ANSI Standard 61. The customer's spent carbon will be segregated and traceable from the time it leaves the specific customer's location, through the receiving, storage, and reactivation at the reactivation facility, and until it is delivered back to the same customer or site (Figure 13.8).

13.6.3 CUSTOMER ON-SITE REACTIVATION FACILITY

If an application will exhaust a sufficient amount of carbon per day (10,000 lb spent carbon per day as an estimated minimum), logistics and costs may justify an on-site reactivation facility. This will involve spent and react carbon handling systems, a reactivation furnace or kiln, off-gas treatment permits, and manpower to operate and oversee the reactivation operation. On-site reactivation is not typically used in water treatment applications and is more common in sugar or sweetener GAC applications.

The reactivation process equipment will be essentially the same for both pooled reactivation and custom reactivation. Depending on the nature of the spent carbon and how much adsorbate it contains, as well as the species, furnace or kiln operating conditions may be different, but the overall flow diagram given in Figure 13.1 applies to both pooled and custom reactivation.

13.7 ACTIVATED CARBON TESTS USED FOR MONITORING AND CONTROLLING REACTIVATION PRODUCT QUALITY

To ensure the quality and consistency of a virgin or reactivated carbon, physical and adsorptive properties are specified. Certain industry standard test methods or ASTM methods are used for this purpose. Although the carbon reactivation process will result in a reactivated carbon that has a different structure than virgin carbon, the desired end goal is to produce a carbon of sufficient adsorptive quality to make reactivation cost-effective.

There are three industry-wide parameters typically used and accepted as reactivated carbon specifications:

1. An adsorption capacity or activity parameter (usually iodine number or I_2#)
2. Apparent density (AD)
3. A screen or mesh size (typically an over and under mesh size, maximum percentage value)

Iodine number (ASTM D4607): The iodine number is a measurement of the amount of iodine (concentration of 0.02 N) adsorbed by 1 g of carbon. The iodine number is reported as milligrams of iodine per gram of activated carbon. Iodine is very strongly adsorbed and will occupy all available adsorption sites over every adsorption energy level. It is understood as an indicator of total pore volume. In a given application, the adsorbate of interest (e.g., PFAS) may only be able to adsorb in a portion of these sites.

Apparent density (ASTM D2854): AD is a physical property defined as the mass per unit volume of a granular material under specified conditions and dense packing. The AD includes the carbon skeleton volume plus the pore and void volumes. The units are typically grams of activated carbon per cubic centimeter.

Mesh size (ASTM D2852): The measurement of the particle size of GAC determined by the US Sieve Series. This is generally represented by a two-sieve size designation (e.g., 8 × 30 or 12 × 40), representing particles passing through the first sieve size but being retained by the second sieve size. The particle size distribution within a mesh series is typically given in the specification of a particular carbon.

When a carbon has been saturated with adsorbates (spent), this will be reflected in an increase in the AD of the carbon. Ideally, in a carbon reactivation process the goal is to remove the adsorbates and restore the carbon to a usable form. Thus, the AD of a reactivated product should be similar to the AD of the carbon that was initially in the vessel (virgin or previously reactivated product). In addition to a quality specification for the reactivated product, the AD is a simple and quick in-house test method to monitor the reactivation process and adjust reactivation furnace or kiln conditions as needed. Targeting a reactivated carbon's AD to be close to the virgin or original carbon's AD will ensure that the adsorbates are removed while minimizing char formation and loss of carbon adsorptive structure in the reactivation process. If the AD of the reactivated carbon is too high, it can indicate that (1) the spent carbon adsorbate loading has changed in nature or increased and the current furnace conditions are no longer adequate, or (2) the adsorbed species are not being adequately removed by the current furnace conditions (i.e., furnace temperatures, atmosphere of steam and/or O_2 or air, and residence time). If the AD is too low, that can indicate overactivation and loss of activated carbon structure along with the adsorbate. This can lead to poor adsorption capacity or performance as well as a loss of too much carbon skeleton. Such a loss can result in fewer reactivation cycles before the carbon granular "falls apart" from losing enough skeleton.

Although the $I_2\#$ is a measurement of total adsorption volume over all sizes of adsorption pores, it is an industry standard test and is typically used for the measurement of spent and reactivated carbon. A reactivated carbon product would ideally have an $I_2\#$ that was near the virgin or fresh carbon specification. In reality, perhaps an $I_2\#$ 100–200 mg/g less than the corresponding virgin or fresh carbon value is more realistic. The $I_2\#$ of the reactivated carbon will depend on the reactivation conditions as well as the spent carbon's $I_2\#$ value. Typically, an increase of more than 300 $I_2\#$ points from the spent carbon $I_2\#$ value is not advisable, as this often indicate overreactivation and a significant loss of carbon skeleton structure.

During the reactivation process, there can be physical carbon attrition due to transport, the dewatering screw, gasification, or the particle's life cycle (the point where a carbon granule disintegrates). This can result in fines that are not removed in the overall process. Therefore, a reactivated carbon that is indirectly cooled is usually screened after it exits the reactivation furnace or kiln. It will typically be screened to match the over- and underspecifications of the virgin or fresh carbon used. As an example, if a virgin or fresh carbon specification had 5.0 weight percent maximum on 12 US mesh and 4.0 weight percent maximum smaller than 40 US mesh, the reactivated carbon for this product on a custom reactivation basis would be screened and have similar mesh size specifications.

13.8 EFFECTIVENESS OF REACTIVATION IN THE DESTRUCTION OF PFAS COMPOUNDS

As discussed in this chapter, the goal of reactivation is to restore the adsorptive capacity of GAC utilizing a high temperature thermal process that devolatilizes, dissociates, and/or chars adsorbed material in the carbon pores. This process destroys the adsorbed compounds, eliminates liability, and allows the activated carbon to be reused in appropriate applications. This process is different from regeneration, where adsorbates are partially removed and the adsorption pore volume is restored to a fraction of the original levels. In applications where destruction of adsorbates is required, regeneration processes are not acceptable. While both terms are often used interchangeably, it is critical to differentiate these processes.

In order to demonstrate the effectiveness of reactivation in the destruction of adsorbed material, such as PFAS compounds, Calgon Carbon, a subsidiary of Kuraray Co., Ltd., conducted leachability studies on reactivated reagglomerated bituminous coal-based GAC (FILTRASORB) previously used in a PFAS removal application. The carbon used in this trial treated approximately 310 million gallons of drinking water with influent PFAS concentrations ranging from 0.5 to 1 ppb (µg/L). This spent carbon was reactivated in a full-scale custom reactivation facility operated by Calgon Carbon Corporation and the reactivation conditions were as outlined in the above reactivation section.

The spent carbon was tested and approved for reactivation via Calgon Carbon's Carbon Acceptance program. This determined that the spent samples were able to be safely and effectively reactivated, as described in Section 13.10.

In order to determine if there were leachable PFAS compounds, the reactivated carbon underwent the NSF42 leaching protocol, a well-accepted and robust method for determining extractable material from activated carbon. The reactivated carbon sample was loaded into a column, was backwashed with approximately 8 bed volumes of NSF42 leaching protocol water (50 ppm total dissolved solids [TDSs], 0.5 ppm Cl-, and pH 6.75), was allowed to sit for 24 hours in 1 BV of water, and then sampled. These steps were repeated two additional times and the three leachate samples were composited for analysis. All samples were analyzed per the EPA 537 method and the results of the leaching test are reported in Table 13.1. These results are shown in Table 13.1. Calgon Carbon Corporation also attempted to utilize a thermal destruction technique on the reactivated carbon to examine fluoride present before and after reactivation, but, because the detection limit of this technique is so high, the data generated was inconclusive and not meaningful. The fluoride present on the spent carbon was at the detection limit prior to reactivation and therefore not included in Table 13.1.

13.9 PROPERTIES OF ACTIVATED CARBON THAT CAN CHANGE WITH REACTIVATION

We have mentioned previously that the reactivation process removes not only adsorbates from the spent carbon but also some carbon atoms or carbon structure, creating a slightly different activated carbon product with restored adsorptive capacity.

TABLE 13.1
EPA Method 537 Results on Leachate from an NSF42 Extraction on a Reactivated GAC Used in a PFAS Application

		PFAS Customer-Plant React
PFBA	ppt	1.9
PFPeA	ppt	<0.43
PFHxA	ppt	<0.51
PFHpA	ppt	<0.22
PFOA	ppt	<0.75
PFNA	ppt	<0.24
PFDA	ppt	<0.27
PFUnA	ppt	<0.97
PFDoA	ppt	<.049
PFTriDA	ppt	<1.1
PFTeA	ppt	<0.26
PFBS	ppt	<.18
PFHxS	ppt	0.23[a,b]
PFHpS	ppt	<0.17
PFOS	ppt	<0.48
PFDS	ppt	<0.28

Note: All leachate from the reactivated reagglomerated bituminous carbon was at or below the quantifiable limit of EPA 537 method, indicating that reactivation was effective in destroying the adsorbed PFAS compounds.

[a] Compound was found in blank.
[b] Result is less than the RL but greater than or equal to the MDL and the concentration is an approximate value.

In addition to the I_2#, AD, and mesh size, some other properties that can be altered in a reactivation cycle are

1. The mean particle diameter (MPD): Even with screening to meet the under and over mesh size of the original carbon, a reactivated carbon can have 99%–95% of the original carbon's MPD.
2. Particle size distribution: Reactivated carbon will be slightly wider.
3. Particle shape: Reactivated carbon will have fewer sharp edges as a result of physical attrition.
4. With each reactivation cycle for a given carbon granule, the volume percent char and percent transport pores of that particle will increase as the volume percent adsorption pores and carbon skeleton decrease. As shown in Table 13.2, the volume percentage of transport pores will increase with the

TABLE 13.2
Volume Percent of Activated Carbon Granule as a Function of the Reactivation Cycle

Component	Virgin	First Cycle	Second Cycle	Fifth Cycle
1. Transport pores	22%	23%	24%	30%
2. Adsorption pores	36%	35%	34%	28%
3. Carbon skeleton	34%	30%	25%	14%
4. Ash	8%	8%	8%	8%
5. Char	0%	4%	9%	20%
Total	100%	100%	100%	100%
Sum of adsorption pores + transport pores	58%	58%	58%	58%
Sum of carbon skeleton (graphitic plates) + char formed	34%	34%	34%	34%

number of reactivation cycles, while the volume percentage of adsorption pores will decrease. The sum of transport pores and adsorption pores will remain unchanged since the gasification reactions convert adsorption areas to transport structure as the graphitic plates within the carbon structure are shortened slightly each reactivation cycle.

If a carbon reactivation is performed properly, the react carbon will have good reactivation results. Characteristics of good results include

1. Slight increase in ash content
2. I_2# within 100–200 mg/g of the original carbon
3. AD of the reactivated carbon relatively similar to that of the original carbon
4. Screens under and over the specified mesh size relatively similar to those of the original carbon used

If a carbon reactivation is performed poorly, the react carbon will be of lower quality and lead to poor adsorption performance or pressure drop problems (too many fines) as characterized by the following:

1. Hardness and abrasion will drop. The hardness number (ASTM D3802) is a measurement of the resistance of a granular carbon to the degradation action of steel balls in a Ro-Tap machine. This number is calculated by using the weight of granular carbon retained on a particular sieve after the carbon has been in vigorous contact with the steel balls. The abrasion number (AWWA-B604) is obtained from a test performed on a particulate material to define the resistance of the particles to degrade on handling. It is calculated by contacting a sample with steel balls in a Ro-Tap machine and determining the ratio of the final to the original MPD.
2. AD (g/cc) will be too low or be too high.

3. $I_2\#$ will be too low or too high.
4. Too many fines will result in pressure drop problems.

In a pooled reactivated product, there is a continuous infusion of spent carbons that result from one adsorption cycle of virgin carbon, as well as spent carbon of previously reactivated carbon. With some particles being lost or degraded structurally after a finite number of react cycles, the overall quality of a reactivated pool reaches equilibrium and remains essentially constant. In custom reactivation, there is typically the addition of some amount of virgin carbon to a specific custom reactivated product to make up for reactivation volume or mass losses and/or to bring the react quality back to some predetermined specification.

Since there can be overall physical carbon losses of typically 5%–10% per reactivation cycle, custom reactivation will require, at a minimum, virgin carbon makeup addition to compensate for these physical losses to supply the customer the same weight or volume of reactivated carbon back for their use. Virgin makeup also maintains the overall quality through multiple reactivation cycles and typically is the same carbon type as initially used in the application system. A higher amount of virgin carbon makeup may be desired on a case-by-case basis if a higher quality of reactivated–virgin blend (usually reported by $I_2\#$) is required for a given application.

The original carbon base material can influence not only the amount of losses per cycle but also the quality of the reactivated carbon. The density, particle shape, hardness or abrasion number, and manufacturing process of the original carbon are important. Some GACs will reactivate more effectively than others and thus require less virgin makeup. Reagglomerated bituminous coal activated carbons typically require less makeup than direct activated (coal or coconut carbons) or lignite carbons. A reagglomerated bituminous coal activated carbon has superior physical durability than direct activated coal, coconut, or lignite products, which results in less attrition and fewer losses in each reactivation cycle.

13.10 SPENT CARBON PROFILING AND REACTIVATION

Before a specific spent carbon can be reactivated (pooled or custom reactivation) at a reactivation center, there is usually an acceptance procedure that a reactivation provider will require. It typically involves completion of a waste and adsorbate profile document by the customer and testing of a representative spent carbon sample to ensure that the spent carbon can be handled and reactivated safely and yield a reusable product. Once this acceptance procedure is completed, spent carbon from the given application or site can be shipped to the reactivation service provider. Depending on the spent carbon classification (nonhazardous or RCRA hazardous) and whether it will be for potable custom reactivation, the receiving reactivation facility will have the necessary permits and certifications to handle the spent carbon.

In most instances and for most applications, including spent carbons that have adsorbed PFAS, the spent carbon can be approved and reactivated at an appropriate reactivation facility. The classification of the spent carbon (RCRA hazardous or non-RCRA hazardous) is determined by the generator of the spent carbon and should be handled accordingly.

There can be some instances or applications where the spent carbon quality or characteristics would make it unacceptable for reactivation, such as

1. If certain contaminants are adsorbed and present that are not permitted in a given react facility or under their permits (e.g., polychlorinated biphenyls [PCBs] and dioxins)
2. If the amount of volatile halides (chlorides, fluorides, and bromides) present on the carbon is higher than can be handled by the off-gas treatment system at the reactivation facility
3. If the carbon contains certain metals above a guideline for pool reactivated carbon
4. If the spent carbon is ignitable

Typical modes of spent carbon packaging, depending on its classification, may include bulk dump trucks, bulk pneumatic trailers, roll-off boxes, supersaks, and drums.

13.11 POSSIBLE ISSUES WITH A SPENT CARBON THAT WOULD RESULT IN A POOR CANDIDATE FOR REACTIVATION

There may be cases where spent carbon generated from an application will contain inorganics, metals, solids, or another property that would make it unacceptable as good-quality feed to a reactivation furnace or kiln, or would result in a reactivated carbon quality that is not usable by the same customer (custom reactivation) or by others (pooled reactivation). Some of these spent carbon characteristics that need to be considered and can result in an unacceptable reactivated carbon or damage to the reactivation equipment are

1. High content of calcium and sodium salt species on or in the spent carbon can lead to furnace hearth failure by attacking the brick or linings in the reactivation furnace, in addition to forming slag and linkers (mass of incombustible matter fused together with no adsorption properties) in the furnace or kiln.
2. Metals such as iron and magnesium can agglomerate and harden the carbon and result in a non-free-flowing react product. Metals may also remain on the reactivated carbon for the next adsorption cycle or reuse.
3. Hardness as $CaCO_3$ can agglomerate and harden the carbon, cause processing or mechanical problems in a furnace or kiln, and produce a poor-quality reactivated carbon.
4. Oil and grease (O&G) or excess hydrocarbons, if free or emulsified, can (a) coat the carbon, making it nonwettable and non–free flowing, preventing it from being able to be fed to the react furnace; and (b) result in high adsorbate or hydrocarbon loading, which makes the carbon ignitable.

13.12 REACTIVATION AS A SUSTAINABLE AND ENVIRONMENTALLY RESPONSIBLE APPROACH

Incineration of any concentrated PFAS waste is required for complete destruction. Spent activated carbon containing adsorbed PFAS compounds can be thermally reactivated, destroying the adsorbed contaminants and allowing the activated carbon to be recycled and reused. When reactivated, carbon can be restored at a lower cost than the manufacture of virgin activated carbon from raw starting materials. Reactivation also avoids the cost and long-term liability associated with disposal and produces only about 20% of the greenhouse gases as does new carbon production.

14 Ion Exchange for PFAS Removal

Steven Woodard, Michael G. Nickelsen, and Marilyn M. Sinnett

CONTENTS

14.1 Properties of Ion Exchange Resins 326
    14.1.1 Physical Properties 326
        14.1.1.1 Structure 326
        14.1.1.2 Functional Groups and Exchange Sites 327
        14.1.1.3 Adsorption Sites 328
    14.1.2 Chemical Properties 329
        14.1.2.1 IEX Capacity 329
        14.1.2.2 Kinetics 329
        14.1.2.3 Selectivity 329
        14.1.2.4 Swelling 330
        14.1.2.5 Stability 330
14.2 Mechanisms of PFAS Removal Using IEX Resins 330
14.3 Importance of Water Quality Analysis 332
    14.3.1 Solids 333
    14.3.2 Natural Organic Matter 333
    14.3.3 Nitrate, Sulfate, and Other Inorganic Anions: Relative Affinity 334
    14.3.4 Iron Fouling 335
    14.3.5 pH 335
14.4 Important Design Parameters 336
    14.4.1 Influent Equalization 336
    14.4.2 Pretreatment Considerations 336
    14.4.3 Empty Bed Contact Time 337
    14.4.4 Vessel Sizing and Geometry 338
    14.4.5 Flow Distribution through the IEX Resin Bed 338
    14.4.6 Face Piping and Valves 338
    14.4.7 Fixed Bed versus Backwashable 339
    14.4.8 Lead–Lag Vessel Arrangement 340
    14.4.9 Single-Use versus Regenerable Systems 341
14.5 IEX Cycles 341
    14.5.1 Loading Cycle 341
    14.5.2 Regeneration Cycle 342
    14.5.3 Rinse Cycle 343
    14.5.4 Recovery and Reuse 343

14.6 Equipment 344
  14.6.1 Primary Treatment Equipment 344
    14.6.1.1 Extraction Well Pumps 344
    14.6.1.2 Equalization Tank 346
    14.6.1.3 System Feed Pumps 346
    14.6.1.4 Pretreatment Equipment 346
    14.6.1.5 IEX Contactor Vessels 346
  14.6.2 Regeneration Equipment 348
    14.6.2.1 Supply and Spent Regenerant Tanks 348
    14.6.2.2 Pumps and Instrumentation 348
    14.6.2.3 Regenerant Recovery Equipment 349
    14.6.2.4 Solvent Recovery Unit (Distiller) 349
    14.6.2.5 Still-Bottoms Handling 349
14.7 Summary 350
References 351

14.1 PROPERTIES OF ION EXCHANGE RESINS

The predominant carboxylic acid and sulfonic acid poly- and perfluorinated alkyl substances (PFAS) are strong acids and therefore almost completely dissociate at common water pH values (i.e., 4 < pH < 10). In the anionic form (carboxylate ion or sulfonate ion), PFAS are susceptible to removal with ion exchange (IEX) resins.

14.1.1 Physical Properties

14.1.1.1 Structure

IEX resins consist of a polymer matrix and introduced functional groups that interact with ions of interest. The most common IEX resin matrix is a copolymer made from styrene and divinylbenzene (DVB) (Figure 14.1), although polyacrylic, phenol-formaldehyde, and polyalkylamine matrices are also available for specialty applications. Standard IEX resins are spherical shaped, with bead diameters ranging from 0.3 to 1.2 mm.

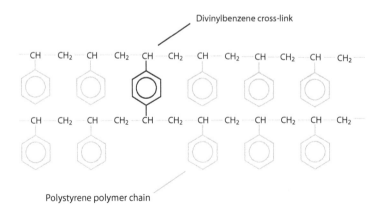

FIGURE 14.1 Structure of styrene–DVB copolymer.

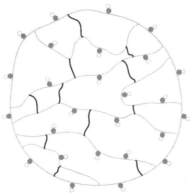

- Polystyrene polymer chain
- Divinylbenzene cross-link
- Fixed ion exchange group, e.g., sulfonic acid, $-SO_3^-$, for cation IEX or quarternary ammonium, $-EN^+$, for anion IEX
- Exchangeable counterion, e.g., hydrogen ion, H^+, for cation IEX or chloride ion, $Cl-$, for anion IEX

FIGURE 14.2 Structural model of a microporous IEX resin.

IEX resins are classified as either cationic (resins that remove positively charged anions, such as Na^+ and Ca^{2+}) or anionic (resins that remove negatively charged anions, such as SO_4^{2-}, NO_3^-, and PFAS). Cation IEX resins are further subdivided into strongly acidic cation (SAC) (sulfonic acid) and weakly acidic cation (WAC) (carboxylic acid) active functional groups. Anion IEX resins are similarly subdivided into strongly basic anionic (SBA) (quaternary ammonium functional groups) and weakly basic anionic (WBA) (primary, secondary, or tertiary amine groups).

A typical IEX resin bead is a complex three-dimensional structure of cross-linked elastic polystyrene chains with DVB cross-linking of irregular lengths and intervals (Figure 14.2).

The structure and porosity of an IEX resin are determined principally by the conditions of polymerization of the backbone polymer. Porosity determines the size of the molecule or ion that may enter a specific structure and its rate of diffusion and exchange. There also is a strong interrelationship between the equilibrium properties of swelling and ionic selectivity.

For example, in IEX resins with a styrene–DVB copolymer backbone, porosity is inversely related to the DVB cross-linking. The amount of DVB cross-linking in a typical SBA IEX resin ranges from a low of 5% to a high of 12% and can be tailored to optimize the removal of specific molecules or anions. Below 5% DVB, the resin is too soft and susceptible to oxidative attack; above 12%, the resin is too dense (molecules or anions cannot enter the pore structure) and too hard. Beads frequently shatter from osmotic pressures experienced during IEX (swelling and shrinkage).

14.1.1.2 Functional Groups and Exchange Sites

Since most PFAS are strong acids (i.e., very low pKa values) and therefore occur predominantly in anionic form at common water pH values, the most applicable IEXs for their removal are SBA IEX resins. SBA IEX resins are prepared by chloromethylation of the benzene ring of polystyrene, followed by replacement of the chlorine

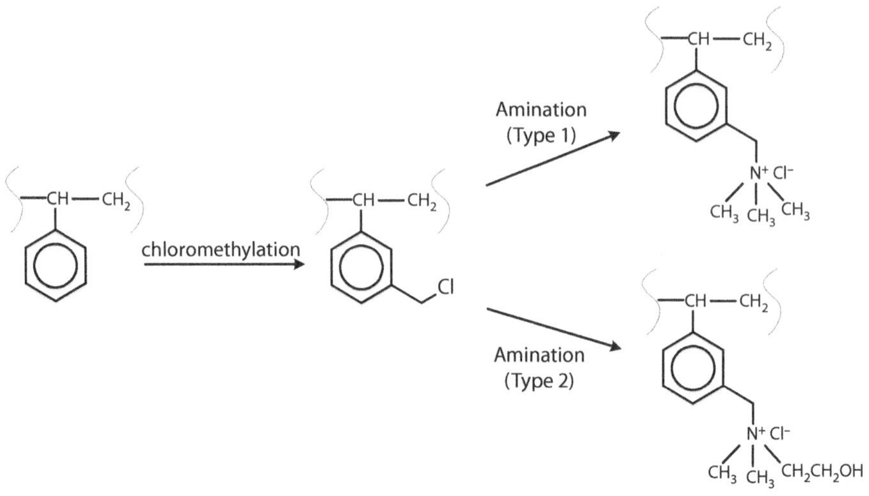

FIGURE 14.3 Functionalization of styrene–DVB copolymer—SBA Type 1 and SBA Type 2 IEX resins.

in the chloromethylated group with a quaternary ammonium group (Figure 14.3). Strong base anion resins are classed as Type 1 and Type 2. The Type 1 functional group is a quaternized trimethylamine and is the most strongly basic functional group available. Type 1 SBA IEX resins have the highest affinity for weak acids. Type 2 functionality is obtained by the reaction of the styrene–DVB copolymer with dimethylethanolamine. This quaternary amine has lower basicity than that of the Type 1 resin, yet it is high enough to remove the weak acid anions for most applications. In general, Type 2 resins have less chemical stability than their Type 1 cousins. Type 1 resins are favored for high-temperature applications (Owens, 1995).

14.1.1.3 Adsorption Sites

Depending on the percentage of DVB cross-linking, porosity structure (micro-, meso-, and macropores), and functional group, the IEX resin can exhibit some adsorptive properties akin to nonfunctionalized styrene–DVB polymeric-based adsorbents. The primary driving forces for adsorption are the hydrophobic (water-disliking) character of the molecule to be adsorbed (sorbate) and/or the high affinity of the sorbate for the resin (sorbent). Non-IEX-based sorption results from the net effect of the combined interactions of these two driving forces. The affinity of the sorbate for the sorbent can result from physical or chemical mechanisms.

Physical mechanisms include dipole–dipole interactions and van der Waals interactions. Dipole moments are caused by a net separation of positive and negative charges derived from the position of atoms and electrons within the molecule. When dipoles from two molecules (e.g., sorbate and sorbent) are near each other, they tend to orient their charges to lower their combined free energy; the negative poles of one molecule tend to approach the positive pole of another, and vice versa. This realignment results in a net attraction between the two molecules.

Ion Exchange for PFAS Removal

When two neutral molecules with no permanent dipoles approach each other, a weak polarization is induced in each because of slight variations in each molecule's charge distribution. The slight variation in charge distribution generates a weak attraction between the two molecules. This attractive phenomenon is known as the van der Waals force. Van der Waals interactions are generally weaker than dipole–dipole interactions.

Chemical mechanisms, or chemisorption, is a type of adsorption involving a chemical reaction between the sorbate and the surface of the sorbent. This strong attraction or reaction creates new chemical bonds. Resins that employ chemisorption for constituent removal are single use and disposable because they are very difficult or impossible to regenerate.

14.1.2 Chemical Properties

14.1.2.1 IEX Capacity

IEX capacity is expressed as either total capacity or operating capacity. Total capacity is based on the total number of fixed sites available for exchange, whereas operating capacity is a measure of the useful capacity observed during site-specific operating conditions (i.e., site water, temperature, flow rate, etc.). Some factors affecting operating capacity will be described in more detail in Section 14.3.

Total capacity is expressed on a dry weight, wet weight, or wet volume basis. Total capacity is a measure of the number of available IEX sites and is typically expressed as equivalents of ionic charge per mass or volume of resin.

Operating capacity is a measure of the useful performance obtained during dynamic loading conditions (flow conditions in a column or vessel containing the IEX resin). Operating capacity is dependent on the total capacity of the resin, the completeness of conversion or regeneration, the complexity or composition of the matrix being treated, the flow rate through the column or vessel, temperature, particle size, and flow distribution through the resin bed (i.e., plug flow or uneven flow distribution).

14.1.2.2 Kinetics

Kinetics are the speed or rate at which IEX or adsorption takes place. The rate of IEX or adsorption is directly related to the diffusion of the sorbate through the film of solution that is in close contact with the resins and diffusion within the resin particle. Film diffusion is the rate-limiting step at low sorbate concentrations, whereas particle diffusion (diffusion through the bulk medium) is the rate-limiting step at high sorbate concentrations. Resin bead particle size is also a rate-determining factor; uniform particle size (i.e., relatively small) yields enhanced kinetic behavior, while nonuniform particle size yields slower kinetic behavior due to the presence of kinetically slower, larger resin beads.

14.1.2.3 Selectivity

For an IEX resin to be effective, there must exist a difference in the affinity for the counterion on the resin in its as-received form (e.g., typically the chloride form for an

TABLE 14.1
Selectivity Coefficients of Various Anions (Compared with the Hydroxyl Ion) on Functionalized Styrene–DVB Type 1 and Type 2 Anion Exchange Resins

Anion	Type 1	Type 2
OH^-	1.0	1.0
SO_4^{2-}	85	15
ClO_3^-	74	12
NO_3^-	65	8
CN^-	28	3
Cl^-	22	2.3

SBA IEX resin) and the ion to be removed from solution. That is, the resin must have a higher affinity for the ion in solution than the ion already on the resin.

Several resin manufacturers have developed selectivity coefficients of various ions for both cation and anion IEX resins. These selectivity coefficients have been normalized to a value of 1.0 for hydrogen ion (H^+) for cation IEX resins and hydroxide ion (OH^-) for anion IEX resins. Table 14.1 shows examples of selectivity coefficients for Type 1 and Type 2 SBA IEX resins.

14.1.2.4 Swelling
Swelling of IEX resins results from hydration of the fixed IEX group. Resins with a lower percentage of cross-linking experience more swelling from hydration. IEX resin volume also changes when converting from the as-received form, typically the chloride ion form for SBA IEX resins, to the PFAS-exchanged form. Allowance must be made in system design to accommodate resin swelling and shrinkage.

14.1.2.5 Stability
IEX resins are readily and rapidly attacked by strong oxidizing agents, such as nitric acid. More common oxidants, such as peroxide, oxygen, or chlorine, react more slowly, but the rate and severity of oxidation can be induced catalytically by the presence of metal ions, such as iron, copper, titanium, and manganese. Cation IEX resins are principally attacked at the polymer backbone, whereas anion IEX resins are principally attacked at the functional group, both of which can lead to total loss of capacity.

From a thermal stability perspective, cation IEX resins are more stable and can handle temperatures as high as 150°C, whereas anion IEX resins have a temperature limitation of approximately 60°C. Both cation and anion IEX resin stability is influenced by pH; anion IEX resins are more stable at lower pH, whereas cation IEX resins are more stable at elevated pH.

14.2 MECHANISMS OF PFAS REMOVAL USING IEX RESINS
Anion exchange resins are highly effective at removing PFAS from water because of the multiple removal methods involved. The molecular structure of most PFAS

compounds can be broken into two functional units: the hydrophobic, nonionic "tail," consisting of the fluorinated carbon chain, and the anionic "head," having a negative charge. Figure 14.4 illustrates this structure.

Anion exchange resins are essentially adsorbents with IEX functionality. The resin beads are composed of neutral copolymers (plastics) that have positively charged exchange sites. Figure 14.5 shows the basic structure of an anion exchange resin with its neutral, hydrophobic backbone and one of its positively charged exchange sites. Anion exchange resins tend to be effective at removing PFAS from water because they take advantage of the unique properties of both the resin and the perfluorinated contaminants. The hydrophobic carbon-fluorine tail of the PFAS adsorbs to the hydrophobic backbone on the resin, and the negatively charged head of the contaminant is attracted to the positively charged IEX site on the resin. Depending on the specific properties of both the resin and the PFAS molecule, this dual mechanism of removal can be highly effective, and certain anion exchange resins have very high removal capacity for PFAS (Nickelsen and Woodard, 2017).

While this dual mechanism of PFAS removal can be highly effective at removing PFAS from water, it also makes resin regeneration and reuse challenging. A brine solution can be used to effectively desorb the anionic head of the molecule from the resin IEX site, but the hydrophobic carbon-fluorine tail tends to stay adsorbed to the resin backbone. Similarly, an organic solvent, like methanol or ethanol, can be used to effectively desorb the hydrophobic tail, but then the anionic head of the PFAS stays attached to the resin IEX site. Research to date has demonstrated that effective regeneration techniques must address both mechanisms of attraction. For example, certain surfactants that have both nonionic and anionic properties have shown promise for use in regenerant solutions (Nickelsen and Woodard, 2017). Solutions combining organic solvents and sodium chloride have produced the most successful results to date (Deng et al., 2010; Chularueangaksorn et al., 2013; Nickelsen and Woodard, 2017). Other research has focused on using combinations of ammonium

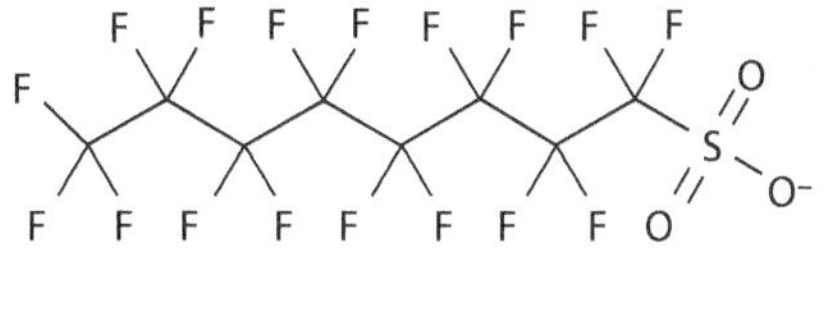

FIGURE 14.4 Structure of PFOS molecule.

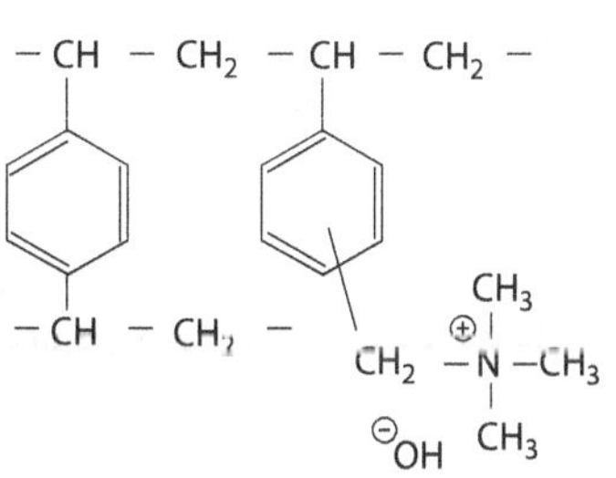

FIGURE 14.5 Molecular structure of a typical anion exchange resin.

salts, including ammonium hydroxide and ammonium chloride (Conte et al., 2015). Du et al. (2014) disclosed a need to further treat the resulting waste regenerant solution to concentrate the PFAS and minimize the volume of waste. Chapter 21 is a case study that describes isotherm, bench-scale, and pilot-scale work conducted at the former Pease Air Force Base, including the process that was used to narrow the field from eight candidate resins to the single best performer (taking into account both the effectiveness of PFAS removal and the ability of the resin to be regenerated).

It should be noted that although the majority of PFAS compounds have an anionic charge, certain PFAS molecules are cationic and others are zwitterionic, having both a positive and a negative electrical charge. This opens the door for future research in the rapidly changing field of PFAS removal from water. IEX resins hold significant promise given the wide variety of resins, both cationic and anionic in nature.

In summary, the PFAS removal effectiveness of individual resins varies based on multiple factors, which will be explored in more detail in Section 14.3 (Deng et al., 2010; Dudley, 2012; Appleman et al., 2014). For example, the PFAS capacity of some resins is highly dependent on the inorganic content of the water, including both the type and concentration of ions in solution. Other resins are less sensitive to inorganic ions, as these resins have a substantially higher affinity for the PFAS molecules than for the inorganic ions. Given the resin specificity and highly variable nature of PFAS-contaminated water, bench-scale column testing is recommended to determine the most effective resin for a specific application.

14.3 IMPORTANCE OF WATER QUALITY ANALYSIS

For proper sizing of an IEX treatment system, the processing flow rate, quality of water being treated, required quality of treated water, and capital and operation and maintenance (O&M) costs must all be considered. This section focuses on the importance of the quality of the water to be treated, specifically the types and concentrations of nontarget constituents present in the water, as their presence has direct implications on system design and capital and O&M costs.

The capacity of an IEX resin is expressed in equivalents per liter (eq/L) of resin or equivalents per kilogram (eq/kg) of resin, where an equivalent is defined as the number of moles of a given ion in solution multiplied by the valence state of the ion in solution. For example, if 1 mol of nitrate ion and 1 mol of sulfate ion were present in a solution, there would exist 1 eq of nitrate (nitrate ion has a valence state of 1) and 2 eqs of sulfate (sulfate ion has a valence state of 2). Carboxylate and sulfonate PFAS compounds each have an ionic valence state of 1.

The major ions typically found in natural groundwater are shown in Figure 14.6 (Dardel and Arden, 2001). Natural organic matter (NOM) is included under the anion group because NOM is a complex mixture of cross-linked organic polymers containing phenolic and carboxylic acid functional groups, and at typical groundwater pH levels, the carboxylic acid functional groups are ionized to carboxylates.

In addition to the major ions shown in Figure 14.6, additional ions, such as iron (Fe^{2+}/Fe^{3+}), other transition metals, transition metal complexes, and silicon dioxide, can impact the performance of IEX resins. The discussion in this section, however, is limited to how water quality parameters can impact the performance of SBA IEX

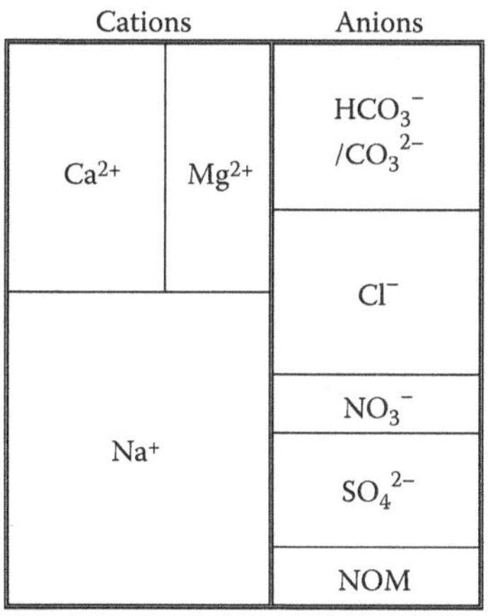

Ion Exchange for PFAS Removal

Cations		Anions
Ca²⁺	Mg²⁺	HCO_3^- / CO_3^{2-}
Na⁺		Cl^-
		NO_3^-
		SO_4^{2-}
		NOM

FIGURE 14.6 Composition of water.

resins for the removal of PFAS compounds. It is important to understand the impact of water quality on IEX resin effectiveness and to perform a detailed water characterization prior to design.

14.3.1 SOLIDS

Solids, whether suspended or dissolved, can prevent the efficient exchange of ions by blocking access to IEX sites. Prefiltration of feedwater is often accomplished with small-pore (<10 μm) filtration media, deep-bed sand filtration, or multimedia filtration for the removal of particulate matter. In some cases, pretreatment with a strong acid cation resin for the removal of the cation portion of total dissolved solids (TDS) may be necessary for effective and/or economical PFAS treatment. For some applications, pretreatment with a cation exchange resin will enhance anion exchange resin performance because the cation exchange resin prevents precipitation of insoluble metal hydroxides and shifts NOM to a protonated, nonionic form.

14.3.2 NATURAL ORGANIC MATTER

The heterogeneous mixture of organic compounds in natural waters, more commonly known as NOM, can reduce the number of resin sites available for PFAS removal, either by direct competition for adsorption sites or by pore blockage. In general, NOM is an ill-defined mixture of naturogenic organic compounds arising from plants and animals and their decomposition products. Two major organic constituents of NOM are humic acid (Figure 14.7) and lignin. Both are large organic molecules that contain a relatively complex mixture of phenol and carboxylic acid groups. The presence of phenolate and carboxylate groups means that the mixture behaves functionally as a dibasic or tribasic acid, and at common groundwater pH values, NOM is negatively charged and is therefore readily captured by anion IEX resins. Pretreatment vial physical filtration or pH adjustment (via acid addition or

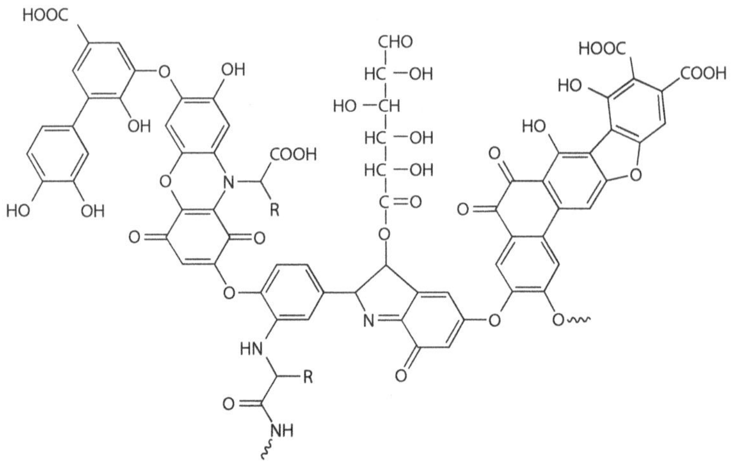

FIGURE 14.7 Generalized structure of humic acid.

use of a cation IEX resin) may be required for effective and/or economical PFAS treatment.

14.3.3 NITRATE, SULFATE, AND OTHER INORGANIC ANIONS: RELATIVE AFFINITY

The effectiveness of SBA IEX resins is highly dependent on the water quality analysis. Specifically, the constituents to be removed, the treatment objective, and what is present in the water could adversely affect IEX performance. Ion selectivity is the primary driving force governing appropriate resin selection. Selectivity is determined by the ionic charge of the specific ion and the resin type. Typically, IEX resins have a higher selectivity for higher-valence state ions (2+, 2−, 3−, etc.) than for singly charged ions. As discussed in Section 14.1.2.3, the general affinity of an SBA resin for anions in solution follows the general trend shown in Table 14.2.

In waters with moderate to high alkalinity, pH adjustment to acidic levels or use of an SBA IEX resin in the chloride form will minimize the impact of competition from bicarbonate anion (HCO_3^-). However, even if selectivity is low, IEX can

TABLE 14.2
SBA Resin Anion Selectivity

Anion	Affinity
SO_4^{2-}	Higher
NO_3^-	
Cl^-	↓
HCO_3^-	
$HSiO_3^-$	
OH^-	Lower

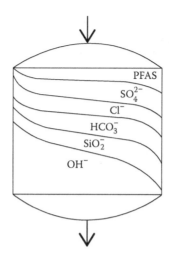

FIGURE 14.8 SBA selectivity profile.

still take place on a mass action basis. That is, if the resin is clean and free from the contaminant of interest, the resin will still remove a modest percentage of that contaminant.

During typical SBA IEX treatment, more attractive ions are removed first (i.e., SO_4^{2-}) and concentrate at the inlet side of the IEX bed. As more ions pass through the bed, ions with less affinity are displaced and gradually move downgradient through the resin bed. This is known as the "chromatography effect" (anions with lower affinity are displaced with ions of higher affinity). Based on observed ion elution profiles, PFAS constituents have a much higher selectivity profile than even sulfate ions on SBA IEX resins (Figure 14.8), because of the combined IEX and adsorption mechanisms described in Section 14.2.

14.3.4 Iron Fouling

Iron fouling is one of the predominant causes of IEX failure. Most waters contain some level of soluble iron in the reduced (Fe^{2+}) state. To minimize the potential for iron fouling, iron must either be kept in the reduced state by preventing aeration of the process water or be removed by precipitation and filtration.

Aeration induces oxidation of ferrous iron (Fe^{2+}) to ferric iron (Fe^{3+}), resulting in precipitation of ferric hydroxide, carbonates, and complexes, which clog resin beads, preventing IEX. If aeration cannot be avoided, another approach is to intentionally oxidize the iron using aeration or by adding a chemical oxidant. This promotes the precipitation of insoluble ferric iron, which is then removed via sedimentation and/or filtration prior to the IEX resin.

14.3.5 pH

In general, anion exchange resin performance increases as pH decreases due to less interference from metal hydroxide precipitates and NOM. Once the pH drops to less than 3, however, the efficiency of short-chain PFAS compound removal begins to diminish since these compounds begin to shift from their ionic to neutral, protonated forms. At a pH value less than 1, PFAS removal is significantly impacted and pH adjustment is required.

14.4 IMPORTANT DESIGN PARAMETERS

There are several key parameters to consider when designing an IEX resin treatment system for PFAS removal. These include, but are not limited to, influent equalization, pretreatment, empty bed contact time (EBCT), vessel sizing and geometry, flow distribution through the resin bed, face piping and valves, fixed-bed versus backwashable systems, vessel arrangement, and whether to use single-use or regenerable IEX media. Each of these important parameters is discussed below.

14.4.1 Influent Equalization

Vessel sizing is based mainly on EBCT, so it is important to maintain a relatively steady influent flow rate to the IEX resin system to minimize occurrences of low-level leakage. Groundwater extraction systems often employ multiple extraction wells, and influent flow rates can fluctuate substantially. Equalization is generally recommended to dampen these variations. This topic is covered in more detail in Section 14.6, which discusses major equipment used in the IEX treatment and regeneration processes.

14.4.2 Pretreatment Considerations

Pretreatment for fouling control should be considered for all PFAS IEX resin systems. Important water quality considerations have been covered in Section 14.3, including solids, iron and manganese, NOM, inorganic anions, and pH control. Bench and/or pilot testing is typically recommended to better understand and provide design criteria for pretreatment systems. When fixed-bed IEX vessels are used, suspended solids must be removed upstream of the IEX media to avoid solids accumulation in the IEX vessels. As discussed in Section 14.3.4, when metals (i.e., iron, manganese, etc.) are present in the dissolved form, the treatment train can be designed to keep solids in the dissolved state (i.e., avoid oxidation or aeration) and minimize the need for pretreatment. Sequestering agents and/or pH adjustment may be incorporated to aid in keeping these metals in the dissolved state.

Common pretreatment methods include

- Particulate filtration (multimedia, catalytic media, sand filters, bag filters, cartridge filters)
- Oxidation using aeration and/or chemical agents
- Granular activated carbon (GAC)
- Sequestering agents (examples include Redux 601, a chelating dispersant blend; Redux RDB, a reducing dispersant blend; and CARUSQUEST 101, a phosphonic acid)

Ion Exchange for PFAS Removal

FIGURE 14.9 Catalytic media filtration system for pretreatment.

FIGURE 14.10 Plate and frame filter press.

- pH adjustment
- IEX resins for softening, total organic carbon (TOC) removal, and so forth

Backwash, air scour, and solids handling may also be required depending on the selected pretreatment approach. Figure 14.9 shows a four-vessel catalytic media filtration system for removing suspended solids, iron, and manganese. Figure 14.10 shows a plate and frame filter press for dewatering the solids generated by backwashing the pretreatment filters.

14.4.3 Empty Bed Contact Time

IEX resin systems have faster kinetics and higher capacities than GAC systems for PFAS removal. This allows the use of shorter EBCTs. Whereas GAC systems are usually designed in the 10- to 20-minute EBCT range, a resin system can be sized for EBCTs as low as 2 minutes. This provides advantages in terms of footprint and capital cost. Final design EBCT selection is based on numerous factors, including footprint, regeneration requirements, capital cost, and O&M costs.

14.4.4 Vessel Sizing and Geometry

IEX contactor vessels are sized based on process flow rate and the minimum EBCT necessary to balance achievement of treatment objectives and O&M costs. Using a height-to-diameter ratio of 1.5 to 2.0 is preferred for resin system vessels when used for adsorption. Lower height-to-diameter ratios can be and often are used for IEX vessels. However, minimum bed depth, velocities, flow distribution, spatial considerations, and material cost must be considered in the design at lower ratios.

Height-to-diameter ratios greater than 2.0 yield better treatment capacity because the hydraulic conditions are closer to plug flow, but higher ratios also cause increased head lead loss. Pump size and energy use increase as differential pressure increases across the resin vessel(s). These issues are discussed in greater detail in Section 14.6.

Cylindrical pressure vessels are normally used to contain the resin. The tanks can be made of steel or fiberglass-reinforced plastic (FRP) construction, and tank operating pressures normally range from 10 to 50 psig. In addition to vessel sizing, the design engineer must consider inlet and outlet connections, resin fill and removal connections, vent connections, and instrumentation ports when selecting the appropriate vessel.

14.4.5 Flow Distribution through the IEX Resin Bed

Treatment vessel internals should be designed to maximize plug flow conditions and minimize short circuiting by distributing process water and regenerant solution uniformly through the resin bed. This is accomplished by the use of flow distributors and collectors. To maximize resin bed capacity, it is critical that the distributors or collectors are properly sized and installed to control flow and flow velocity. Distributor or collector slot width should be small enough to keep resin particles from escaping while providing sufficient open area to handle the design flow range and maintaining enough back pressure to uniformly distribute the flow throughout the entire cross section of the vessel (Owens, 1995).

There are various types of flow distributors, including perforated plates outfitted with screens, filter caps, sintered metal plates, and lateral and/or hub-and-spoke assemblies. Figures 14.11 and 14.12 show examples of a hub-and-spoke design and a lateral-type distributor or collector design, respectively.

14.4.6 Face Piping and Valves

An IEX resin system for PFAS removal typically consists of multiple vessels connected with face piping to allow different directions and volumes of flow, including forward flow (loading cycle), regeneration, and backwash (if applicable). It is desirable to design these systems to maintain forward-flow treatment while one or more vessels is offline for regeneration and/or backwash. The valves that control these various flows can be either manual or automatic, depending on the project needs and

Ion Exchange for PFAS Removal

FIGURE 14.11 Hub and spoke distributor or collector.

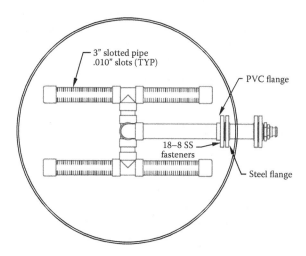

FIGURE 14.12 Lateral-type distributor or collector. (Used with permission from Carbon Service & Equipment.)

system requirements. Valve choice is important to prevent leakage and associated cross-contamination from regenerant or backwash water.

14.4.7 Fixed Bed versus Backwashable

Although the design engineer should strongly consider providing the ability to backwash the resin vessels, there are some notable exceptions. The primary advantage of backwashable beds is the ability to relieve excess pressure drop caused by the buildup of solids. The disadvantage is that backwashing disturbs the mass transfer zone, which can cause premature leakage and reduced treatment capacity. Disruption of the mass transfer zone also makes it more difficult to effectively regenerate the resin using countercurrent flow, and therefore requires more regenerant solution to completely desorb the PFAS from the resin bed.

It is generally safer to include backwash capability, but the decision is situation specific, depending on influent water quality, fouling potential, and the effectiveness of pretreatment, including solids filtration and iron and manganese removal. If a backwash system is included, then the design should include

- Added tankage for clean and dirty backwash water
- Pumps, piping, valves, and solids processing
- Freeboard in the vessel to allow for media expansion
- Air scour capability
- Appropriate pipe sizes, including the distributors and collectors, to handle the potentially higher flow rates associated with backwash

Fixed-bed system designs, on the other hand, should provide some freeboard to account for potential resin swelling. Countercurrent operation is preferred for these systems to maintain the highest PFAS concentrations at the influent end of the bed. This helps maintain relatively clean conditions at the discharge end and reduces the potential for "smearing" the mass transfer zone. The overall goal of countercurrent operation is to minimize the volume of regenerant solution required, which helps reduce the volume of waste that must ultimately be disposed and the associated O&M costs.

14.4.8 LEAD–LAG VESSEL ARRANGEMENT

IEX resin treatment systems are typically designed with multiple vessels in series. Vessel arrangements include lead–lag and lead–lag–polish configurations (Figure 14.13).

Having multiple vessels in series accommodates longer loading cycles and more complete resin use before a regeneration cycle must be initiated. The higher capital cost associated with multiple vessels in series is offset, to varying degrees, by the reduced O&M costs and the enhanced ability to maintain compliance (i.e., better

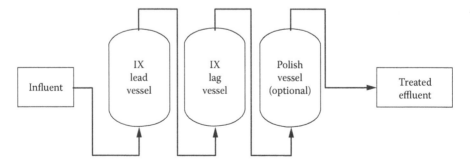

FIGURE 14.13 Typical IEX vessel configuration.

Ion Exchange for PFAS Removal   341

treatment performance). It also allows continuous operation while one or more vessels are regenerated or backwashed. The decision to add a third (polish) vessel is largely dependent on influent PFAS concentrations and effluent permit limits. Again, the trade-off between capital and O&M costs needs to be considered, as additional vessels increase capital cost and power costs to pump water through the beds, but can reduce the required regeneration frequency.

14.4.9 SINGLE-USE VERSUS REGENERABLE SYSTEMS

The design engineer must decide whether to incorporate a resin regeneration system or simply go with single-use resin. The inclusion of a regeneration system is typically based on payback. If the contaminated groundwater is from a source area, such as a fire training area, then the resin will likely require regeneration with such frequency that the O&M savings would more than offset the additional capital expense. More dilute applications, including plume containment and drinking water treatment, may have low enough PFAS concentrations that a resin regeneration system does not make financial sense. In these cases, the IEX resin systems are still normally more cost-effective than GAC because the resin's treatment capacity is much higher (i.e., smaller vessels and longer time between media change-outs). The resulting waste resin can be either landfilled or thermally destroyed via incineration.

14.5 IEX CYCLES

Regenerable IEX treatment systems are operated using three primary operating cycles: loading, regeneration, and rinse. When using a single-use IEX media, the regeneration cycle is replaced with media change-out. This chapter describes the loading cycles for regenerable IEX media.

14.5.1 LOADING CYCLE

In the loading cycle, contaminated water is pumped through the IEX media vessels, typically in a lead–lag or lead–lag–polish configuration, with treated water exiting the bed. The loading cycle is the longest stage in the typical IEX cycle, normally consisting of 80%–95% of the overall cycle duration, with the regeneration and rinse cycles comprising the remaining 5%–20%. This breakdown differs somewhat when treating PFAS, since the influent concentrations are normally much lower (i.e., ppb and ppt levels instead of ppm) and the resins have a higher selectivity for PFAS than for inorganic ions. The PFAS loading cycles are typically in the 90%–99% range, with the regeneration and rinse cycles comprising the remaining 1%–10%. The lead vessel is subject to the highest contaminant loading during the loading cycle, which concludes when a regeneration trigger is reached. The trigger can be a predetermined amount of time, contaminant breakthrough, differential pressure, or a combination of these factors. The regeneration trigger is influenced by several factors, including (1) the contaminant breakthrough curve, (2) vessel configuration, (3) effluent limitations, (4) differential pressure, and (5) owner or operator preference. These factors are expanded on in Table 14.3.

TABLE 14.3
Factors Influencing IEX Regeneration or Change-Out

Factor	Description/Considerations
Contaminant breakthrough curve	The contaminant breakthrough curve will be developed during bench and/or pilot testing. In general, the steeper the breakthrough curve, the lower the allowable breakthrough percentage prior to triggering a regeneration.
Vessel configuration	Adding a polish vessel (lead–lag–polish) provides a safety factor that can allow for an increased or extended regeneration trigger with reduced risk of effluent breakthrough.
Effluent limitations	If contaminant effluent limits are near the analytical detection limit, regeneration may be triggered by timing before any breakthrough is observed at the lead bed effluent.
Differential pressure	If excessive differential pressure is observed, a regeneration or backwash may be triggered prior to achieving contaminant breakthrough.
Owner preference and risk tolerance	This may take into account any factor, or combination of factors, listed above. For example, operators with variable influent concentrations or dynamic contaminant loading scenarios that result in less predictable breakthrough times may elect regeneration at predefined times, regardless of effluent breakthrough.

Data collected during bench and/or pilot testing will provide guidelines for loading cycle duration for full-scale systems. Once the full-scale system is operational, these parameters are refined during start-up and the initial operating period, for optimal system efficiency and risk minimization.

14.5.2 REGENERATION CYCLE

When a regeneration is initiated, the lead vessel is taken offline and the lag vessel becomes the new lead. The vessel designated for regeneration (either in place or off-site) is typically drained prior to starting the regeneration to minimize dilution of the regenerant. The water drained from the regeneration vessel can be recycled to the system influent and metered into the forward-flow stream. Next, regenerant is pumped through the vessel, typically in a countercurrent arrangement (discussed further in Section 14.4). A block flow diagram for one patent-pending IEX regeneration process is shown in Figure 14.14.

Empirical data suggest that a constant, low, single-pass flow-through of regenerant solution achieves optimal regeneration results. The optimum regenerant solution, concentration, flow rate, and volume used during regeneration to desorb PFAS from the IEX media and restore capacity are best approximated through bench testing, followed by optimization work at full scale. After the desired amount of regenerant has passed through the media, the remaining regenerant left over in the vessel is then drained to recapture the maximum amount of regenerant and minimize the amount of regenerant lost during the rinse cycle. Spent regenerant is collected in a tank or drum(s) and stored for additional processing to recycle the regenerant. Simple distillation, adsorption, IEX, filtration, and/or membrane processes can be used to effectively recover and reuse a substantial portion (greater than 98%) of the regenerant. The resulting PFAS

Ion Exchange for PFAS Removal

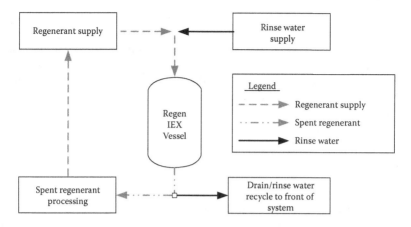

FIGURE 14.14 IEX regeneration block flow diagram.

waste stream can be converted to solid waste through a process called superloading, which minimizes the volume of waste to be landfilled or incinerated. This results in significant overall process savings, helping to make regenerable IEX resin systems an attractive, affordable, and sustainable alternative to GAC for PFAS removal.

14.5.3 Rinse Cycle

After the regenerant has been drained from the vessel, the IEX resin is flushed with clean water until the rinse water exiting the bed is below target concentrations (typically below effluent discharge limitations). The volume of rinse water required is best determined through testing during system start-up. Processed rinse water can either be collected in a holding tank or metered directly into the influent flow stream. The rinse water will initially contain a small amount of residual regenerant, more concentrated at the beginning of the rinse cycle and increasingly diluted throughout the rinse cycle. Testing must be performed during system start-up to understand the concentrations of regenerant and PFAS in the rinse water since rinse water will be a component of the overall system mass balance. Incorporating a rinse water holding tank provides the advantage of equalizing the rinse water, thereby eliminating a potential spike of spent regenerant concentration in the influent flow stream.

At the conclusion of the rinse cycle, the regenerated vessel is placed back into service in the lag vessel position. Having multiple vessels provides the opportunity to maintain forward flow while one vessel is offline for regeneration. This is an advantage compared with systems where the forward flow is interrupted for multiple hours while media change-out takes place.

14.5.4 Recovery and Reuse

Solvent-based spent regenerant can be recycled for reuse in the next regeneration by using a patent-pending distillation and superloading process (Nickelsen and Woodard, 2017). First, the spent regenerant is transferred to a solvent recovery unit. Solvent is recovered by the distillation process, given its low boiling point relative to water, and transferred to the regenerant supply tank for reuse in the next regeneration cycle. The leftover still bottoms (i.e., brine and concentrated PFAS) from the distiller are either further concentrated by evaporation or pumped through superloader vessels, transferring the bulk of the PFAS onto the superloaded media. Superloading is the process by which a relatively small volume of highly concentrated PFAS solution is passed slowly through a relatively small volume of media (either GAC or IEX resin). The long EBCT approaches equilibrium (isotherm) conditions, thereby maximizing PFAS mass transfer onto the media and minimizing the amount of solid waste requiring disposal or incineration.

14.6 EQUIPMENT

IEX resin treatment for PFAS removal can involve either single-use or regenerable resin systems. The type of resin system selected depends on multiple factors, including influent PFAS concentrations, treatment system objectives, characteristics of the water treated, and presence of co-contaminants. Figure 14.15 shows a generic process flow diagram of a groundwater pump-and-treat system using regenerable IEX treatment.

Some guidance and considerations for the treatment system components are provided in this section, which is broken into two parts: primary treatment equipment and regeneration system equipment.

In general, the primary treatment components include

- Extraction well pumps
- Influent equalization tank
- System feed pump(s)
- Pretreatment equipment
- IEX contactor vessels

Regeneration equipment for IEX media regenerated using a solvent–brine mixture typically includes the following

- Regenerant supply tank
- Spent regenerant tank
- Solvent recovery unit (distiller)
- Cooling equipment (chiller, fan cooler, or cooling tower)
- Still-bottoms tank
- Superloader contactor vessels
- Pumps and instrumentation

14.6.1 Primary Treatment Equipment

14.6.1.1 Extraction Well Pumps

Extraction well pumps are typically controlled based on flow rate or water level in the extraction well. Extracted groundwater is either pumped to an equalization tank or sent directly to the groundwater treatment system. If an equalization tank is not used, steady, consistent flow is recommended for consistent treatment performance.

Ion Exchange for PFAS Removal

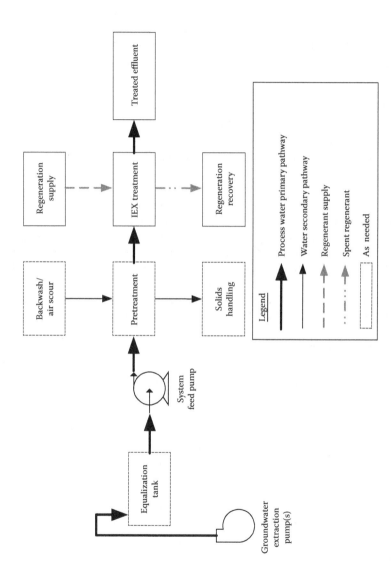

FIGURE 14.15 Typical pump-and-treat system using IEX.

Therefore, when equalization tanks are not used, extraction pumps with variable-frequency drives (VFDs) are recommended to maintain steady flow.

14.6.1.2 Equalization Tank

An equalization tank provides a collection and equalization point for extracted groundwater. If iron is present in the reduced state, the design engineer may choose to consider bypassing the equalization tank to avoid aeration and subsequent oxidation or precipitation. Keeping the iron in its reduced, soluble ferrous form and taking advantage of the zero-headspace IEX vessels will potentially reduce pretreatment equipment and solid waste generation, as discussed in Sections 14.3.4 and 14.4.2. If the extraction well pumps are configured for on–off operation (e.g., on at high level, off at low level) or surge flow (i.e., air-driven pumps), an equalization tank helps to balance flow and avoid surging through the contactor vessels.

Equalization tanks are typically plastic or steel. When steel tanks are used, consideration is given to the corrosiveness of the groundwater being treated.

14.6.1.3 System Feed Pumps

System feed pumps are selected to provide steady, continuous flow at the desired rate with sufficient discharge head to move the feedwater through the treatment system. A VFD is recommended to control pump speed based on the water level in the equalization tank to provide steady flow and minimize cycling and surging. The design engineer should consider overall system losses (including pretreatment equipment, IEX vessels, effluent discharge, and interconnecting pipe and fittings), as well as allowable differential pressure buildup across pretreatment and IEX equipment. Duplex system feed pumps are typically used to provide redundancy and minimize downtime.

14.6.1.4 Pretreatment Equipment

The design engineer must understand the site-specific constituents in groundwater and corresponding effluent limitations prior to selecting a pretreatment approach. In addition, bench and/or pilot testing is recommended to confirm that pretreatment is sufficiently protective of the IEX media. Typical constituents that may foul or limit the capacity of an IEX bed include suspended solids, iron, manganese, nitrate, and high concentrations of NOM as indicated by TOC levels. Pretreatment to control and/or remove these constituents was discussed in more detail in Section 14.4.2.

14.6.1.5 IEX Contactor Vessels

IEX contactor vessels are sized to achieve target contaminant reduction and adequate EBCT. EBCTs for IEX treatment system vessels are typically lower than those using GAC, resulting in a smaller system footprint, as discussed in Section 14.4.3. Materials of construction for contactor vessels are selected for compatibility with groundwater and regenerant and compliance with applicable regulations. Vessels are designed to be fully draining to maximize effectiveness of media regeneration. Using greater height-to-diameter ratios offers treatment advantages, as discussed in Section 14.4; however, this must be balanced with differential pressure (head loss), material cost, and practical spatial considerations. Figure 14.16 shows a three-vessel IEX resin treatment system with associated face piping and valves.

Ion Exchange for PFAS Removal

If differential pressure increases across IEX contactor vessels due to solids buildup and is not restored after regeneration, treatment with a reducing agent or Iron-Out® may be considered. However, a better approach is to improve pretreatment to minimize suspended solids reaching the IEX contactor vessels.

IEX systems can be installed in shipping containers to provide a compact, mobile, efficient, cost-effective PFAS treatment solution. Figure 14.17 is a photo of the outside of a shipping container housing a treatment system, and Figure 14.18 shows a plan view representation of IEX resin vessels inside the box.

FIGURE 14.16 Three-vessel IEX resin treatment system.

FIGURE 14.17 Shipping container housing groundwater treatment system.

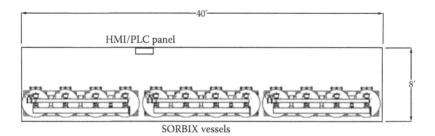

FIGURE 14.18 Plan view of IEX resin vessels inside storage container. HMI = Human–Machine Interface; PLC = programmable logic controller.

14.6.2 REGENERATION EQUIPMENT

Depending on the influent PFAS concentrations and life cycle cost analysis, it may be desirable to install a resin regeneration system. Regeneration systems can be in place or at a centralized regeneration facility where vessels are transported to the facility for regeneration and then returned to service. Regeneration systems provide the capability to periodically desorb PFAS compounds from the IEX resin, facilitating reuse of the resin without removing it from the vessels. Equipment considerations for regeneration system components are provided in this section. When using flammable liquid for regenerant, additional regulations apply, including NFPA 30: "Flammable and Combustible Liquids Code." A code review and area classification should be performed in the early stages of design.

14.6.2.1 Supply and Spent Regenerant Tanks

Supply and spent regenerant solution (regenerant) holding tanks are sized to contain an adequate volume of regenerant to sufficiently restore the IEX media capacity, plus freeboard and safety factor. Bench or pilot testing may be used to verify this volume. Regenerant tank materials of construction are selected for compatibility with all components of the regenerant blend and compliance with regulations. One effective, patent-pending regenerant blend includes flammable liquids and brine. Therefore, tanks must meet applicable codes for storage of flammable liquids, and consideration must be given to materials of construction to minimize corrosion.

14.6.2.2 Pumps and Instrumentation

Regenerant transfer pumps should be self-priming and are selected for material compatibility and area classification. Instrumentation is used in the regeneration cycle for safety and process control and may include some or all the following: flow meters, pressure and temperature gauges or transmitters, conductivity sensors, level transmitters and switches, and lower explosive limit (LEL) sensors or controllers. Instruments are selected for compatibility with fluids and area classification.

14.6.2.3 Regenerant Recovery Equipment

It is often economically and environmentally favorable to recover and reuse regenerant solution, especially since the process is relatively straightforward. Spent regenerant, stored in the spent regenerant tank, is recovered in stages and returned to the regenerant supply tank. Regenerant recovery equipment is sized to ensure that all stages of the recovery process can be completed prior to the next regeneration. For example, if regenerations are expected to occur on a monthly basis, the regenerant recovery process must be completed in less than a month.

14.6.2.4 Solvent Recovery Unit (Distiller)

When solvents are used to regenerate the IEX resin, the first stage of regenerant recovery is solvent recovery, which is typically achieved through distillation. A typical distiller is shown in Figure 14.19.

The distillers are sized such that the total volume of spent regenerant is processed in the first part of the regeneration cycle, allowing adequate time for still-bottoms recovery. The distiller produces two effluent streams: distillate (i.e., recovered solvent) and still bottoms. The vast majority of contaminants are concentrated in the still bottoms.

14.6.2.5 Still-Bottoms Handling

Still bottoms generated from the distillation process are collected in a holding tank and are typically processed through superloader vessels prior to being transferred to the regenerant supply tank. The still-bottoms holding tank is sized to contain the total volume of still bottoms for a regeneration cycle plus margin and freeboard. Materials of construction of the still-bottoms holding tank are selected for compatibility with still-bottoms constituents and temperature considerations.

FIGURE 14.19 Typical distiller (center) shown with supply stocking tank (left), still-bottoms drum (bottom center), and distillate stocking tank (right). (Photo courtesy of NexGenEnviro Systems, Inc.)

Still bottoms are pumped from the still-bottoms holding tank through superloaders at a slow, steady flow rate to maximize superloader performance. Superloading is the final step in the three-step PFAS ultraconcentration process: IEX, distillation, and superloading. Millions of gallons of contaminated groundwater can be concentrated into less than 100 gallons (13 cubic feet) of solid waste.

14.7 SUMMARY

Certain anion exchange resins have properties that lend themselves well to high-capacity removal of PFAS from water. These resins are essentially adsorbents with IEX functionality and take advantage of the unique properties of both the resin and the perfluorinated contaminants: the hydrophobic carbon-fluorine tail of the PFAS adsorbs to the hydrophobic backbone on the resin, and the negatively charged head of the contaminant is attracted to the positively charged exchange site on the resin. This dual-removal mechanism provides higher capacities and faster removal kinetics than GAC, resulting in a treatment system with a smaller footprint and lower life cycle costs. A pilot-scale case study highlighting these advantages is provided in Chapter 21.

Another significant advantage of IEX resin is its ability to be regenerated in vessel. Proprietary regeneration methods have been developed that are capable of restoring resin capacity to near-virgin conditions, making resin a more sustainable treatment solution than GAC. Distillation is used to recover and reuse the majority of regenerant solution, and the concentrated PFAS in the still bottoms can be further concentrated into a solid waste using superloading.

In addition to being more sustainable than GAC, IEX resin systems have the potential to share a central resin regeneration system. For example, multiple groundwater treatment systems are often required to address the PFAS contamination at a single military installation. Using IEX resin treatment provides the opportunity to use one central regeneration facility to serve these multiple treatment systems by simply transporting resin vessels to and from the central regeneration system. IEX resin systems, including the regeneration components, also tend to be compact, relatively lightweight systems that can be installed inside shipping containers. The result is a compact, rapidly deployable, mobile, cost-effective PFAS treatment process.

Although the majority of PFAS compounds have an anionic charge, certain PFAS molecules are cationic, and others are zwitterionic, having both a positive and a negative electrical charge. This opens the door for future research in the rapidly changing field of PFAS removal from water. IEX resins hold significant promise, given the wide variety of resins, both cationic and anionic, in nature. The use of IEX resins for PFAS removal is a rapidly developing field as new resins and regeneration methods are being developed to address the expanding market for PFAS remediation.

REFERENCES

Appleman, T.D., C.P. Higgins, O. Quinones, B.J. Vanderford, C. Kolstad, J.C. Zeigler-Holady, and E.R.V. Dickenson. Treatment of poly- and perfluoroalkly substances in U.S. full-scale treatment systems. *Water Research* 51, 246–255 (2014).

Chularueangaksorn, P., S. Tanaka, S. Fujii, and C. Kunacheva. Regeneration and reusability of anion exchange resin used in perfluorooctane sulfonate removal by batch experiments. *Journal of Applied Polymer Science* 10(1002), 884–890 (2013).

Conte, L., L. Falletti, A. Zaggia, and M. Milan. Polyfluorinated organic micropollutants removal from water by ion exchange and adsorption. *Chemical Engineering Transactions*, 43 (2015).

Dardel, F. De, and T.V. Arden. Ion exchangers. In *Ullmann's Encyclopedia of Industrial Chemistry*. 6th ed. Weinheim: Wiley-VCH Verlag GmbH (2001).

Deng, S., Y. Yu, J. Huang, and G. Yu. Removal of perfluorooctane sulfonate from wastewater by anion exchange resins: Effects of resin properties and solution chemistry. *Water Research* 44, 5188–5195 (2010).

Du, Z., S. Deng, Y. Bei, Q. Huang, B. Wang, J. Huang, and G. Yu. Adsorption behavior and mechanism of perfluorinated compounds on various adsorbents—A review. *Journal of Hazardous Materials* 274, 443–454 (2014).

Dudley, L.M.B. Removal of perfluorinated compounds by powdered activated carbon, superfine powdered activated carbon, and anion exchange resins. Master's thesis, North Carolina State University (2012).

Nickelsen, M., and Woodard, S. A Sustainable System and Method for Removing and Concentrating Per- and Polyfluoroalkyl Substances (PFAS) from Water. U.S. Patent Application Serial No. 15/477,350, April 2017.

Owens, D.L. *Practical Principles of Ion Exchange Water Treatment*. Littleton, CO: Tall Oaks Publishing (1995).

15 Occurrence of Select Perfluoroalkyl Substances at US Air Force Aqueous Film-Forming Foam Release Sites Other than Fire Training Areas
Field Validation of Critical Fate and Transport Properties

R. Hunter Anderson, G. Cornell Long,
Ronald C. Porter, and Janet K. Anderson

CONTENTS

15.1 Introduction	354
15.2 Methods	356
15.2.1 Fieldwork	356
15.2.2 Analytical Analysis	357
15.2.3 Statistical Analysis	360
15.3 Results and Discussion	361
15.3.1 General Summary	361
15.3.2 Discriminant Analysis	364
15.3.3 Categorical Analyses	368
15.3.4 Precursor Biotransformation	368
15.4 Conclusions	369
Acknowledgments	370
References	370

15.1 INTRODUCTION

Poly- and perfluoroalkyl substances (PFAS) are a class of synthetic fluorinated hydrocarbons that have been used in many industrial and consumer products since the 1950s. In the early 1970s, municipalities, the hydrocarbon processing industry, and the US military began using PFAS-based aqueous film-forming foam (AFFF) to efficiently extinguish hydrocarbon-based fires (Moody and Field 2000). During fire training, equipment maintenance, and emergency response, AFFF was released directly to the environment, and it is suggested that decades of AFFF use is a significant source of environmental PFAS (Moody and Field 1999; Moody et al. 2003). PFAS, in particular long-chain perfluoroalkyl acids (PFAAs), are under increased scrutiny from the regulatory community because they are environmentally persistent and globally distributed, bioaccumulate, and have demonstrated some toxicity in laboratory animals, resulting in concerns about human and ecological exposure (USEPA 2009; OECD 2013). The science surrounding PFAS-mediated health and environmental effects is still evolving.

AFFF was developed in the late 1960s by 3M and the US Navy specifically to efficiently extinguish hydrocarbon-based fires. Specific US Department of Defense (USDOD) military specification regulation requires AFFF to conform to specific performance and quality control standards, such as extinguishing time, reignition potential, and surface tension, as well as acute LC_{50} and biological and chemical oxygen demand criteria (MILSPEC 1992). PFAS are the critical active component of AFFF; see Buck et al. (2011) for a comprehensive overview of PFAS chemistry and nomenclature. The use of fluorinated surfactant components within AFFF is a requirement for the USDOD and the US Federal Aviation Administration (FAA). To date, alternative foams that do not contain PFAS have been developed, but performance relative to AFFF is questionable.

Since the initial military specification requirements for AFFF, there have been numerous companies that have manufactured and supplied AFFF to the USDOD (Place and Field 2012). The exact composition of each AFFF formulation is proprietary, but they are all known to be a complex mixture of fluorinated surfactants. The PFAS within AFFF can be synthesized by either electrochemical fluorination or telomerization processes (Kissa 1994; Buck et al. 2011). The AFFF originally sold by 3M contained PFAS synthesized by electrochemical fluorination and therefore contained fully fluorinated perfluoroalkyl sulfonic acids (PFSA), such as perfluorooctane sulfonate (PFOS) and other CF_2 homologues, as well as various perfluoroalkyl sulfonamides and their derivatives (Buck et al. 2011; Backe et al. 2013). AFFF formulations synthesized via telomerization (all other manufacturers), however, contain structurally distinct PFAS; the carbon chains are not fully fluorinated, and instead have homologues of varying C_2F_4 units and are known to contain a highly diverse suite of fluorotelomers (Buck et al. 2011; Backe et al. 2013). The fluorotelomers have been shown to exclusively degrade to perfluorooctoanoic acid (PFOA) and other perfluoroalkyl carboxylic acids (PFCAs) in microcosm and computational studies (Wallington et al. 2006; Wang et al. 2011; Weiner et al. 2013; Jackson et al. 2013). Conversely, perfluoroalkyl sulfonamides and their derivatives can degrade to PFOS and other PFSAs (Houtz et al. 2013; Avendaño and Liu 2015). Importantly, these "precursor" compounds ultimately result in the formation of specific PFAAs (either PFCAs or PFSAs) *in situ* depending on the applicable source of PFAS released to the environment; note that traditionally precursors have been defined as any PFAS that results in the production of a PFCA or PFSA with ≥7 or ≥6 perfluoroalkyl carbons, respectively (OECD 2013). Efforts to reverse engineer the chemical composition of AFFF stocks and elucidate all degradation pathways, as well as to account for the entire mass balance of PFAS in environmental samples, are ongoing (e.g., Houtz et al. 2013; Barzen-Hanson and Field 2015).

In addition to uncertainty about the original product formulations and degradation pathways, studies reporting the occurrence and distribution of PFAS in environmental samples from AFFF-impacted sites are still somewhat scarce. Further, most studies (to date) have mostly focused on a single site and thus lack a comparison of sites with diverse release history and environmental conditions (Awad et al. 2011; D'Agostino and Mabury 2013; Filipovic et al. 2015). Specifically, the occurrence of select PFAS in groundwater as a result of historic AFFF use was first identified at several fire training areas (FTAs) at USDOD facilities (Levine et al. 1997; Moody and Field 1999; Schultz et al. 2004), highlighting the occurrence of PFAS in groundwater decades after release. Consequently, the US Air Force (USAF) Civil Engineer Center's Emerging Issues and Emerging Contaminants Program has since performed sampling at numerous FTAs and confirmed that at all FTAs operable since 1970 that used AFFF for training purposes, select PFAS in soil and groundwater can still be identified (data not published). Detections of PFAS at FTAs coincide with (1) the operational time frame of PFAS-based AFFF use at USDOD facilities and (2) the fact that older FTAs were often not lined and were not constructed to prevent infiltration or runoff of AFFF and combustion products. However, FTAs represent only one type of AFFF release location within the USDOD. A significant data gap exists regarding the magnitude of PFAS associated with other AFFF release sites (e.g., emergency response locations, AFFF lagoons, hangar-related AFFF storage tanks and pipelines, and fire station testing and maintenance areas).

Therefore, the purpose of this investigation was to evaluate select PFAS occurrence at a diverse group of non-FTA sites on active USAF installations with historic AFFF use of varying magnitude. Concentrations of 15 PFAAs and perfluorooctane sulfonamide (PFOSA), an important PFOS precursor (as suggested by Gebbink et al. [2009] and Tomy et al. [2004]), were measured for several hundred samples among multiple media (i.e., surface soil, subsurface soil, sediment, surface water, and groundwater) collected from 40 AFFF-impacted sites across 10 installations between March and September 2014, representing one of the most comprehensive datasets on environmental PFAS occurrence to date. Specific objectives addressed herein are to (1) report the detection frequency and environmentally relevant concentrations for a suite of PFAS, (2) determine whether the various PFAS are statistically distinguishable among environmental media, and (3) critically evaluate empirical evidence of precursor biotransformation. Results are presented within the context of validating research efforts to date and inform future studies related to PFAS in environmental media.

15.2 METHODS

15.2.1 FIELDWORK

A total of 10 active USAF installations were selected for investigation throughout the continental United States, including Alaska. At each installation, potential investigation sites were considered a candidate if there was known historic AFFF release. Candidate historic releases occurred after 1970 but before approximately 1990 (see introduction) such that the observed PFAS profile at each site was recognized to reflect the net effect of several decades' worth of all applicable environmental processes. Typically, AFFF is used in either a 3% or 6% aqueous solution by volume and is potentially released to the environment during training, emergency response, or maintenance and testing at those concentrations. There have also been releases of neat product (premix concentrate) due to equipment failures, vehicle (fire truck) accidents, and human error. Reported AFFF release volumes ranged from less than 1 gallon in a single event to hundreds of gallons over a period of decades. FTAs were specifically excluded from this evaluation in order to focus the investigation on other sites where relatively smaller volumes of AFFF were released. Candidate sites were ranked in terms of logistical criteria (e.g., access and availability of monitoring wells), and the top four sites at each installation were strategically selected for limited investigation. Although previous remedial activities for co-occurring contaminants were not specifically controlled for in the site selection process, active remedies had not been applied at any of the sites ultimately selected for evaluation. Thus, the effect of remediation-induced alterations to the PFAS composition observed by other researchers (McGuire et al. 2014; McKenzie et al. 2015) likely does not confound these results to a significant extent. Selected sites were categorized into three groups according to the assumed volume of AFFF release: low volume (emergency response locations), usually a single AFFF release; medium volume (hangars and buildings), one to five AFFF releases; and high volume (testing and maintenance), multiple releases in the same location over a period of years. The "testing and maintenance" category refers to regular maintenance and equipment performance testing of emergency vehicles and performance testing of the AFFF solution. Approximately 10 samples were collected from each site distributed among soil (surface and unsaturated subsurface), groundwater, sediment, and surface water. Groundwater samples were collected from a combination of existing monitoring wells and temporary monitoring wells installed with direct push technology (DPT). DPT employs a small drill rig to "push" small diameter rods and tools into the subsurface for investigative purposes; applications include soil and groundwater sampling, geophysical sensing, and soil gas sampling. Samples from existing wells were collected at the top of the well screen. Groundwater samples from temporary wells were collected at the water table interface. Surface soil (0–1 ft below ground surface [bgs]) and sediment samples (0–1 ft below top of the sediment) were collected directly into sample containers. Surface water samples were colocated with each sediment sample. Sediment and surface water samples were collected at locations where the conceptual site model (CSM) indicated a hydraulic connection, either through surface water flow (overland sheet flow) or where groundwater discharged to surface waters. Sample locations

TABLE 15.1
Sample Sizes (N_{Total}) of All Environmental Samples Collected by Site Classification

Site Classification	AFFF Release	N_{Sites}	Matrix				
			Surface Soil	Subsurface Soil	Sediment	Surface Water	Groundwater
Emergency response	Low	5	12	17	3	2	24
Hangars and buildings	Medium	27	56	64	35	32	100
Testing and maintenance	High	8	32	31	2	2	25

included engineered storm water channels, engineered AFFF ponds, and natural streams. Approximately three subsurface soil samples were collected at intervals from each DPT boring between the top of the water table and the 0–1 ft bgs sample. At two installations subsurface geology prevented the use of DPT; only surface soils and groundwater samples from existing monitoring wells were collected. Each water sample consisted of a minimum of 250 mL collected into a 1 L high-density polyethylene (HDPE) container. Each soil or sediment sample consisted of a minimum of 10 g collected into a 250 mL HDPE container. Field duplicate samples and matrix spike–matrix spike duplicate samples were collected at a rate of one field duplicate per 20 samples in all media. One field blank and one equipment blank were collected per sampling event. All known or suspected PFAS-containing materials were avoided during sample collection, handling, and transport. Table 15.1 summarizes all samples collected.

15.2.2 ANALYTICAL ANALYSIS

All PFAS analytes selected for evaluation are reported in Table 15.2 and include the standard, commercially available suite of PFAS, including PFOSA, an important PFOS precursor. Other relevant precursors—the fluorotelomer sulfonates (FTSs), such as 6:2 FTS and 8:2 FTS—were not analyzed; nor were the fluorotelomers identified by Place and Field (2012). Further, the shorter-chain-length (<C4) PFSAs recently identified by Barzen-Hanson and Field (2015) were not targeted for analysis either. Therefore, the total PFAS profile for each sampled location was obviously not resolved, and thus reported results likely underestimate the total mass of PFAS.

Analysis of all samples was conducted through a combination of matrix-specific preparatory methods employing solid-phase extraction for aqueous samples and liquid extraction of solid samples, followed by liquid chromatography and detection with tandem mass spectrometry following USEPA Method 537 for drinking water asmodified by TestAmerica's proprietary standard operating procedures

TABLE 15.2
List of Analyzed PFAS with Chemical Abstract Services Registry Number (CASRN)

Analyte	Acronym	CASRN
Perfluorobutanesulfonic acid	PFBS	375-73-5
Perfluorobutanoic acid	PFBA	375-22-4
Perfluoropentanoic acid	PFPA	2706-90-3
Perfluorohexanesulfonic acid	PFHxS	355-46-4
Perfluorohexanoic acid	PFHxA	307-24-4
Perfluoroheptanoic acid	PFHpA	375-85-9
Perfluorooctanesulfonic acid	PFOS	1763-23-1
Perfluorooctanoic acid	PFOA	335-67-1
Perfluorooctanesulfonamide	PFOSA	754-91-6
Perfluorononanoic acid	PFNA	375-95-1
Perfluorodecanesulfonate	PFDS	67906-42-7
Perfluorodecanoic acid	PFDA	335-76-2
Perfluoroundecanoic acid	PFUnA	2058-94-8
Perfluorododecanoic acid	PFDoA	307-55-1
Perfluorotridecanoic acid	PFTriA	72629-94-8
Perfluorotetradecanoic acid	PFTeA	376-06-7

(DV-LC-0012 and DV-LC-0019) for aqueous and solid samples, respectively. For all compounds except PFOSA, water samples were prepared for analysis using solid-phase extraction employing a reversed-phase, weak anion exchange mixed-mode sorbent. Compounds were eluted from the cartridge with an ammonium hydroxide–methanol solution. For PFOSA, a silica-based bonded phase was used as the sorbent and methanol as the elutant. For soils, all samples were mixed with sodium hydroxide, followed by the addition of methanol. The soil–solvent mixture was then sonicated, tumbled, and adjusted to pH <2. The extracts were centrifuged, concentrated, solvent exchanged, cleaned, and reduced to a final volume of 1 mL. Extract cleanup was accomplished using one of several techniques, including solid-phase extraction, temperature-modified phase separation, or graphitized carbon. ^{13}C- or ^{18}O-labeled PFAS were used as isotope dilution standards. Target analytes without a corresponding labeled analog were quantified using the internal standard technique, using the most similar internal standard in terms of carbon chain length. All data reported herein were validated against the quality control and quality assurance parameters reported in Table 15.3. Median reporting limits (RLs) for each PFAS in each environmental matrix are presented in Table 15.4.

TABLE 15.3
Quality Control and Quality Assurance Parameters

Data Quality Indicators	Measurement Performance Criteria	QC Sample and/or Activity Used to Access Measurement Performance	QC Sample Assesses Error for Sampling (S), Analytical (A), or Both (S&A)
All Solid-Phase Samples			
Precision and within-batch accuracy	Laboratory recovery limits, refer to DV-LC-0012 Example: 50%–150%	LCS and/or SRM	A
Precision	RPD ≤50%	Field duplicates	S&A
Accuracy/bias	Refer to DV-LC-0012 Example: 50%–150% recovery, RPD ≤30%	MS/MSD	A
Sensitivity	Results must be less than half the RL	Method and field blanks	S&A
Completeness	Number of samples collected and analyzed = 95% of that specified in work plan	Comparison of actual samples collected in the field and analyzed in the lab with that specified in the work plan	S&A
All Aqueous-Phase Samples			
Precision and within-batch accuracy	Laboratory recovery limits, refer to DV-LC-0019 Example: 50%–150%	LCS and/or SRM	A
Precision	RPD ≤50%	Field duplicates	S&A
Accuracy/bias	Refer to DV-LC-0019 Example: 50%–150% recovery, RPD ≤30%	MS/MSD	A
Sensitivity	Results must be less than half the RL	Method and field blanks	S&A
Completeness	Number of samples collected and analyzed = 95% of that specified in the work plan	Comparison of actual samples collected in the field and analyzed in the lab with that specified in the work plan	S&A

Abbreviations: LCS = laboratory control sample; MS = matrix spike; MSD = matrix spike duplicate; QC = quality control; RPD = relative percent difference; SRM = standard reference manual.

TABLE 15.4
Median Reporting Limits for All 16 PFAS Measured by Matrix

	Matrix				
PFAS	Surface Soil	Subsurface Soil	Sediment	Surface Water	Groundwater
PFBA	0.12	0.13	0.21	0.010	0.010
PFBS	0.15	0.15	0.24	0.008	0.008
PFPA	0.25	0.26	0.41	0.011	0.011
PFHxA	0.16	0.16	0.26	0.003	0.003
PFHxS	0.29	0.31	0.48	0.007	0.007
PFHpA	0.12	0.13	0.21	0.013	0.013
PFOA	0.24	0.25	0.40	0.010	0.010
PFOS	0.15	0.15	0.24	0.013	0.014
PFOSA	0.10	0.11	0.17	0.006	0.006
PFNA	0.23	0.24	0.38	0.017	0.018
PFDA	0.28	0.30	0.46	0.008	0.008
PFDS	0.31	0.33	0.52	0.009	0.009
PFUnA	0.33	0.35	0.55	0.007	0.007
PFDoA	0.59	0.62	0.98	0.014	0.015
PFTriA	0.33	0.35	0.55	0.017	0.018
PFTeA	0.72	0.75	1.2	0.014	0.015

Note: All soil and sediment values are reported in micrograms per kilogram and all water samples are reported in micrograms per liter.

15.2.3 STATISTICAL ANALYSIS

Linear discriminant analysis was used to evaluate intermedia variability as a function of all 16 PFAS analyzed. For this purpose, surface soil and subsurface soil samples were evaluated indiscriminately. All units were standardized to reflect parts per billion (ppb) for the respective media, and nondetects were substituted with one-half the RL (see Table 15.4 for median RLs). All data were subsequently $\log_{10}$ transformed to satisfy the normality assumption for error estimation. The standardized canonical structure (defined for each PFAS as the correlation coefficient between the predicted values of the linear discriminant function [i.e., the canonical scores] and the actual values) was used to determine the relative order of PFAS that defined differences in the chemical signature among environmental media. Canonical variables (abbreviated herein as CAN) are orthogonal (i.e., contain nonoverlapping pieces of information) and were thus evaluated independently. In general, discriminant analysis is applicable when multiple quantitative response variables belong to two or more levels of a classification variable and some linear function is of interest in terms of quantitatively describing the interclass variation. The resulting CAN variables must subsequently be interpreted within the context of the separation observed among the applicable variable space (CAN1, CAN2, etc.). In the current context, the quantitative response variables are the measured concentrations of the various PFAS and the classification variable is the environmental matrix (i.e., soil, sediment, surface water, and groundwater). Thus, the linear discriminant function quantifies intermedia variability as a function of all 16 PFAS analyzed. Discriminant analysis was performed using Proc DISCRIM in 64-bit SAS® version 9.4 for Windows and considered significant at $p \leq 0.05$.

Categorical data analysis methods were used to evaluate differences between surface and subsurface soil samples, specifically to test the hypothesis that sorption is more likely to occur for longer-chain PFAS, resulting in inverted depth profiles that reflect limited transport. All units were standardized and nondetects were substituted with one-half of the reporting limit. Only sites with both surface and subsurface soil samples were evaluated. Sites were dichotomized for each PFAS according to whether mean surface (0–1 ft bgs) concentrations exceeded mean subsurface concentrations (>1 ft bgs). All PFAS were also dichotomized according to carbon chain length, where C6 and less were considered short-chain PFAS and all C7 and above were considered long-chain PFAS, consistent with the description by Buck et al. (2011). Logistic regression was applied to evaluate chain length dependence on the probability of observing inverted depth profiles (reflecting limited transport), whereas a 2×2 cross-classification analysis (i.e., contingency table) was applied to evaluate the more general categorical association. Logistic regression and the cross-classification analyses were performed using Proc LOGISTIC and Proc FREQ, respectively, in 64-bit SAS version 9.4 for Windows and considered significant at $p \leq 0.05$.

Two-way analysis of variance (ANOVA) was used to evaluate differences in mean PFAS concentrations among the various site classifications that reflect increasing volumes of AFFF release. The inherent assumption is made that at higher-volume release sites, greater potential exists for precursor biotransformation and (if true) would result in relatively higher concentrations of select PFAS, primarily since the candidate AFFF release sites have all undergone many years of weathering. Therefore, a departure in parallelism (i.e., significant interaction between the site classification variable and the PFAS variable) was specifically tested. However, because discriminant analysis resulted in significant differences in the PFAS signature among environmental media (as well as the fact that the absolute concentrations are inconsequential), ANOVA was performed on media-normalized concentrations to avoid confounding. Log_{10}-transformed data were normalized by subtracting the media-specific mean from each observation and dividing by the media-specific standard deviation (i.e., standard normal distributions were generated). Consistent with the discriminant analysis, surface soil and subsurface soil samples were evaluated indiscriminately, all units were standardized to reflect ppb for the respective media, and nondetects were substituted with one-half of the reporting limit. ANOVA was performed with the site as a random variable using Proc Mixed in 64-bit SAS version 9.4 for Windows and considered significant at $p \leq 0.05$.

15.3 RESULTS AND DISCUSSION

15.3.1 GENERAL SUMMARY

Summary statistics for all 16 PFAS are presented by matrix in Table 15.5. Across all sites and media, PFOS was the predominant PFAS detected, followed by PFHxS.

TABLE 15.5
Summary Statistics[a] for All 16 PFAS Measured[b] by Matrix

		Matrix				
PFAS	Parameter	Surface Soil	Subsurface Soil	Sediment	Surface Water	Groundwater
PFBA	DF	38.46%	29.81%	24.24%	84.00%	85.51%
	Median	1.00	0.960	1.70	0.076	0.180
	Maximum	31.0	14.0	140	110	64.0
PFBS	DF	35.16%	34.62%	39.39%	80.00%	78.26%
	Median	0.775	1.30	0.710	0.106	0.200
	Maximum	52.0	79.0	340	317	110
PFPA	DF	53.85%	45.19%	45.45%	92.00%	87.68%
	Median	1.20	0.960	1.70	0.230	0.530
	Maximum	30.0	50.0	210	133	66.0
PFHxA	DF	70.33%	65.38%	63.64%	96.00%	94.20%
	Median	1.75	1.04	1.70	0.320	0.820
	Maximum	51.0	140	710	292	120
PFHxS	DF	76.92%	59.62%	72.73%	88.00%	94.93%
	Median	5.70	4.40	9.10	0.710	0.870
	Maximum	1300	520	2700	815	290
PFHpA	DF	59.34%	45.19%	48.48%	84.00%	85.51%
	Median	0.705	0.660	1.07	0.099	0.235
	Maximum	11.4	17.0	130	57.0	75.0
PFOA	DF	79.12%	48.08%	66.67%	88.00%	89.86%
	Median	1.45	1.55	2.45	0.382	0.405
	Maximum	58.0	140	950	210	250
PFOSA	DF	64.84%	29.81%	75.76%	52.00%	48.55%
	Median	1.20	0.470	1.30	0.014	0.032
	Maximum	620	160	380	15.0	12.0
PFOS	DF	98.90%	78.85%	93.94%	96.00%	84.06%
	Median	52.5	11.5	31.0	2.17	4.22
	Maximum	9700	1700	190000	8970	4300
PFNA	DF	71.43%	14.42%	12.12%	36.00%	46.38%
	Median	1.30	1.50	1.10	0.096	0.105
	Maximum	23.0	6.49	59.0	10.0	3.00
PFDA	DF	67.03%	12.50%	48.48%	52.00%	34.78%
	Median	0.980	1.40	1.90	0.067	0.023
	Maximum	15.0	9.40	59.0	3.20	1.80
PFDS	DF	48.35%	11.54%	33.33%	8.00%	20.29%
	Median	3.70	3.55	2.00	17.8	0.125
	Maximum	265	56.0	2200	35.6	2.00
PFUnA	DF	45.05	9.62	24.24	20.00	8.70
	Median	0.798	1.15	1.60	0.021	0.025
	Maximum	10.0	2.00	14.0	0.210	0.086

(*Continued*)

TABLE 15.5 (CONTINUED)
Summary Statistics[a] for All 16 PFAS Measured[b] by Matrix

		Matrix				
PFAS	Parameter	Surface Soil	Subsurface Soil	Sediment	Surface Water	Groundwater
PFDoA	DF	21.98%	6.73%	45.45%	20.00%	4.35%
	Median	1.95	2.40	2.80	0.058	0.022
	Maximum	18.0	5.10	84.0	0.071	0.062
PFTriA	DF	15.38%	13.46%	24.24%	0.00%	1.45%
	Median	0.665	1.90	1.65	na	0.019
	Maximum	6.40	4.70	29.0	na	0.019
PFTeA	DF	10.99%	6.73%	15.15%	0.00%	1.45%
	Median	1.10	3.40	1.66	a	0.021
	Maximum	4.70	5.40	4.16	na	0.027

Abbreviations: DF = detection frequency; NA = not applicable.
[a] Median values are reported using only detected concentrations.
[b] All soil and sediment values are reported in micrograms per kilogram and all water samples are reported in micrograms per liter.

While PFOA was frequently detected in all media, concentrations were generally much lower. This profile is consistent with the findings of previous investigations of AFFF-impacted groundwater at FTAs associated with US military installations, wherein PFOS and PFHxS are the prominent PFAS detected, followed by PFOA (Moody et al. 2003; Barzen-Hanson and Field 2015). Although additional accounts of such a wide suite of PFAS compounds at various AFFF release locations are lacking, detection frequencies and concentrations of PFAS herein are consistent with AFFF formulations and environmental release. Detection frequencies for most PFAS were similar at the high-volume (i.e., testing and maintenance) and medium-volume (i.e., hangars and buildings) sites but were considerably lower at the low-volume (i.e., emergency response) sites, where only a one-time release of AFFF occurred (data not shown). Interestingly, however, PFOA detection frequencies were similar among all three site classifications, although concentrations varied with the estimated volume of AFFF release, suggesting a trace-level background source of PFOA, potentially atmospheric deposition (Wallington et al. 2006; Kim and Kannan 2007). In general, the range in concentrations for PFOA and PFOS in groundwater is similar to that in concentrations previously reported at FTAs (Moody and Field 1999, 2000; Moody et al. 2003; Schultz et al. 2004); however, this is the first study to assess a variety of AFFF-impacted sites to assess detection frequencies of these 16 PFAS. A high degree of variability was observed in measured concentrations within site classifications and simply reflects site-specific conditions and installation-specific operational practices.

Although not definitive as an independent line of evidence (primarily because the FTSs were not evaluated), the predominance of PFOS and PFHxS (both PFSAs) in all media from all site classifications suggests that 3M AFFF is at least a significant contributing source of PFAS at these sites, if not the predominant source (discussed further below). However, while PFSAs dominate the overall multivariate PFAS signature among the various matrices, at least some degree of telomer-based AFFF contamination may be evident given the sporadic occurrence of select PFCAs (i.e., PFNA) that are not present in past 3M AFFF formulations (Backe et al. 2013); the extent to which biotransformation of the various precursor compounds in 3M AFFF results in the accumulation of PFCAs, however, remains uncertain. Nevertheless, multivariate PFAS chemical signatures may also be applicable to source apportionment studies given the seeming ubiquity of PFAS sources reported to date. For example, contrasting PFAS chemical signatures from different sources have been demonstrated (So et al. 2007). Further, PFCAs have been shown as the dominant PFAS signature in multiple studies where samples were collected in urban areas impacted by diffuse non–point PFAS sources (e.g., Loganathan et al. 2007; Nguyen et al. 2012). Quantitative multivariate comparisons of PFAS profiles along groundwater or surface water flow paths may prove useful for distinguishing some AFFF releases from other (i.e., non-AFFF) sources, such as landfill leachate and urban runoff.

15.3.2 Discriminant Analysis

Linear discriminant analysis resulted in three significant canonical variables with an overall 28.2% classification error. However, the first canonical variable (CAN1) accounted for almost all of the pooled covariance (98.4%), whereas CAN2 and CAN3 accounted for 1.26% and 0.340%, respectively. Therefore, only CAN1 and CAN2 are reported. Figure 15.1 illustrates the orthogonal canonical scores for CAN1

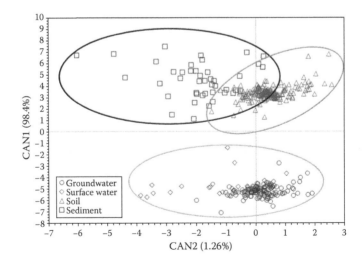

FIGURE 15.1 Canonical scores from linear discriminant analysis. Includes all data presented in Table 15.2. Surface and subsurface soil results were evaluated collectively.

and CAN2. Readily apparent is the complete vertical separation (CAN1) between all solid-phase data (soil and sediment) and all aqueous-phase data (groundwater and surface water), and the partial horizontal (CAN2) separation between the soil and sediment data. Thus, although statistically significant signatures are distinguishable between soil and sediment, the overwhelmingly predominant pattern evident in these data defines phase-dependent (i.e., solid-phase vs. aqueous-phase) differences in PFAS concentration and composition.

The standardized canonical structure for CAN1 is presented in Figure 15.2. All coefficients are positive values reflecting higher mean (log-transformed) soil and sediment concentrations (ppb equivalence) relative to surface water and groundwater (given the direction of the vertical separation observed in Figure 15.1). The magnitude of the coefficients quantifies that relative differences among the PFAS are proportional to their difference. These results, however, are to some extent confounded by differences in the sensitivities among those detected by the analytical method for water samples relative to soil and sediment (Table 15.4) because nondetects were substituted with one-half of the RL but are considered trivial given that multiple PFAS were detected in almost every single sample regardless of media (see Table 15.5 for detection frequencies). So, PFAS with higher coefficients were interpreted as empirical evidence of a greater solid-phase affinity than PFAS with smaller coefficients. Although no obvious pattern was evident among the PFAS with different functional groups, almost complete concordance was observed between the coefficients and the total carbon chain length (Figure 15.2). Further, excluding PFDoA and PFTeA, the rank order of the coefficients was not significantly correlated with

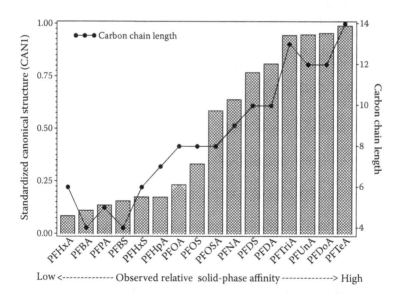

FIGURE 15.2 Standardized canonical structure of the CAN1 variable presented in Figure 15.1. The standardized canonical structure is defined for each PFAS as the correlation coefficient between the predicted values of the linear discriminant function (i.e., the canonical scores presented in Figure 15.1) and the actual values.

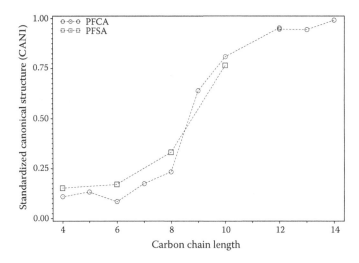

FIGURE 15.3 Standardized canonical structure of the CAN1 variable presented in Figure 15.1 versus the total carbon chain length for both PFCAs and PFSAs.

differences in mean RLs between the corresponding analytical methods (data not shown). Therefore, these results conclusively illustrate carbon chain length dependence on PFAS transport. An alternative way of looking at these same results is to simply plot the CAN1 coefficients against carbon chain length for both the PFCAs and PFSAs; this is presented in Figure 15.3. Subtle differences in the observed sorption behavior between PFCAs and PFSAs were apparent, but only among the PFAS with smaller chain lengths (<C8), where moderately greater sorption was observed among the PFSAs. Also, among all the PFAS there was an obvious inflection point between C8 and C10. Overall, these results suggest that the effect of carbon chain length on PFAS sorption behavior was nonlinear and functional group-mediated differences are secondary relative to the effect of total carbon chain length. It should be noted that in the present context, the total carbon chain length and the fully fluorinated chain length differed proportionally (n – 1) among all 16 compounds, and thus did not affect the results. To date, sorption-mediated differences due to PFAS chain length have only been field validated qualitatively in other (albeit relatively limited) reports of regional PFAS occurrence (Ahrens et al. 2009; Murakami et al. 2009; Gellrich et al. 2012).

The standardized canonical structure for CAN2 is presented in Figure 15.4. All coefficients except PFNA were negative, reflecting higher mean log-transformed sediment concentrations relative to soil (given the direction of the horizontal separation observed in Figure 15.1). Although the maximum observed concentration occurred in sediment (Table 15.2), the median and mean of the log-transformed values for PFNA were higher in soil, indicating inherent log-normality, and reflected an upward shift in the low to midrange of observed concentrations relative to sediment (Table 15.2). This is particularly noteworthy because the mean RL for sediment

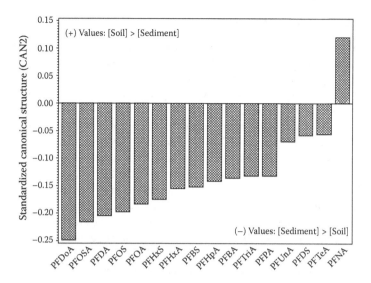

FIGURE 15.4 Standardized canonical structure of the CAN2 variable presented in Figure 15.1. The standardized canonical structure is defined for each PFAS as the correlation coefficient between the predicted values of the linear discriminant function (i.e., the canonical scores presented in Figure 15.1) and the actual values.

was 2.76-fold higher than that for soil (Table 15.4). Thus, nondetects substituted as one-half of the RL would tend to bias results higher for sediment. Although CAN2 only accounted for 1.26% of the total covariance, these results are highly significant ($p < 0.0001$).

There are potentially many different explanations for the observed pattern in CAN2. Primarily, the observed statistically significant discrimination is largely driven by the general differences in concentrations rather than composition; among the 15 PFAS observed with higher mean log-transformed sediment concentrations (Figure 15.4), no pattern was evident among the different functional groups or carbon chain lengths. In general, the observed higher sediment concentrations reflected the greater sorption capacity of sediment (f_{OC}) because all samples were collected from known source areas and are consistent with other investigations (Ahrens 2011; Yang et al. 2011; Zareitalabad et al. 2013; Ahrens and Bundschuh 2014). Nevertheless, the anomalous pattern observed for PFNA is intriguing. Because sediments are oxygen limited relative to typical aerobic soil, reductive defluorination as discussed by Park et al. (2009) could be operative at relatively low concentrations but is purely speculative; indirect accounts of the redox sensitivity of PFNA from wastewater treatment plants with anaerobic digestion have not demonstrated the disappearance of PFNA (Loganathan et al. 2007; Shivakoti et al. 2010). An alternative explanation is simply a greater biotransformation rate (aerobic) of telomer-based PFNA precursors in soil relative to sediment that was not observed among the other PFCAs.

15.3.3 CATEGORICAL ANALYSES

Additional analyses were applied specifically to evaluate differences in PFAS concentrations between surface and subsurface soil samples as a function of carbon chain length. Logistic regression did not yield any significant results and was not improved by also evaluating differences between functional groups. However, the categorical association between long-chain PFAS and cases where mean surface soil concentrations exceeded mean subsurface soil concentrations was significant ($p = 0.027$) with an odds ratio of 2.61, illustrating a strong association. So, while a categorical association was observed consistent with the results of the discriminant analysis (Figure 15.2) and previous investigation of the PFAS depth profile in soil as a result of municipal biosolids application (Sepulvado et al. 2011), excess variability prevented a quantitative relationship. Obviously, carbon chain length dependence reflected partitioning to the various sorption sites within the amorphous soil organic matter (SOM) matrix. Notwithstanding the synergistic effect of co-contaminants on PFAA sorption to soil (Guelfo and Higgins 2013) and potential pH-dependent ionic interactions (Higgins and Luthy 2006; Ahrens et al. 2009), solid-phase PFAS sorption (at least those evaluated, in particular the PFAAs) was therefore at least a function of both the carbon chain length and the SOM content and composition (Gellrich et al. 2012; Guelfo and Higgins 2013). Unfortunately, however, soil physiochemical properties were not analyzed as part of this field effort, nor were co-contaminants. A more comprehensive dataset including these parameters would most likely account in some capacity for observed intersite heterogeneity in PFAS depth profiles that are not accounted for by chain length alone.

15.3.4 PRECURSOR BIOTRANSFORMATION

ANOVA resulted in a significant ($p < 0.0001$) interaction between the site classification variable and the PFAS variable. The interaction plot is presented in Figure 15.5. Increasing trends among mean concentrations from low-volume release sites (i.e., emergency response) to high-volume release sites (i.e., testing and maintenance) were expected due to increases in cumulative AFFF release volumes, but of particular interest was the relative trend among the various PFAS. The relative increase was notably greater for PFOS, followed by PFHxS. PFOS concentrations increased by 1.06 standard normal units, followed by PFHxS, which increased by 0.650 units. PFOA, however, only increased by 0.353 units, which was well within the range of the other PFAS. Because only PFSAs were observed with anomalously higher trends (as well as relatively higher concentrations—discussed above), 3M AFFF was most likely the dominant source of PFAS at the sites evaluated; telomer-based AFFFs neither contain nor result in the *in situ* formation of PFOS or PFHxS (Wang et al. 2011; Weiner et al. 2013). The predominance of telomer-based AFFFs should eventually result in significant production of PFCAs *in situ* (Wang et al. 2011; Weiner et al. 2013). A fundamental uncertainty, however, is the biotransformation rates under ambient field conditions. So, an alternative conclusion may be that telomer-based precursors at the sites evaluated simply have yet to undergo biotransformation to PFCAs, although previous investigations of precursor prevalence from past AFFF

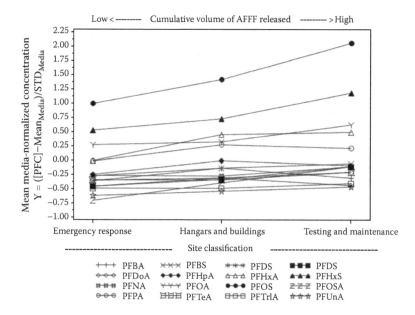

FIGURE 15.5 Interaction plot illustrating a departure in parallelism (i.e., significant interaction between the site classification variable and the PFAS variable). All sites were classified according to the estimated volume of AFFF release, which consisted of emergency response, hangars and buildings, and testing and maintenance sites corresponding to low, medium, and high cumulative release volumes, respectively.

releases have clearly shown evidence of biotransformation (Houtz et al. 2013). Based on the mean difference in the normalized concentrations among all PFAS except PFOS and PFHxS (assumed to reflect differences in the AFFF release volume), approximately 40% of the PFOS and 36% of the PFHxS observed at the high-volume release sites appear to have originated *in situ*. Collectively, these results reflect the dominance of PFSA precursor biotransformation relative to the carboxylic analogs at the sites evaluated, although future investigations should at least include evaluation of the FTSs for comparison to PFAAs. However, PFOSA was not observed with a decreasing trend and thus cannot be confirmed as a relevant precursor notwithstanding that (1) stoichiometric degradation is rarely observed with field data, and (2) relevant precursors and reaction pathways are still the subject of intensive research (Houtz et al. 2013). Overall, these results demonstrate that PFOS (and to a slightly lesser extent PFHxS) was formed *in situ* with a magnitude dependent on the cumulative volume of AFFF released.

15.4 CONCLUSIONS

In summary, multivariate data analyses suggest that 3M AFFF is a significant contributing source of PFAS at the sites evaluated. Differences have been observed in detection frequencies and observed concentrations as a result of AFFF release volume.

Furthermore, field validation of two fundamental conclusions of laboratory-based research to date has been provided. Nonlinear phase-dependent (i.e., solid-phase vs. aqueous-phase) differences in the PFAS chemical signature as a function of carbon chain length and *in situ* PFOS and PFHxS formation (40% and 36%, respectively), presumably due to precursor biotransformation, have been empirically demonstrated. Additional investigation, however, is needed to validate mechanistic assumptions of PFAS transport that account for site-specific physiochemical properties. Future investigations should at least also include an evaluation of the FTSs for comparison to PFAAs and quantification of the total oxidizable precursors (Houtz et al. 2013). In general, this investigation demonstrates the complexity of issues practitioners are faced with in accurately characterizing the nature and extent of PFAS at AFFF-impacted sites.

ACKNOWLEDGMENTS

The fieldwork for this effort was funded under contract number W912HN-12-D-0021 to SES Construction and Fuel Services LLC. Special thanks are given to Cassandra Bergstedt (Noblis) and Michael Bruckner (Noblis) for transcribing the data.

REFERENCES

Ahrens, L., Yamashita, N., Yeung, L. W., Taniyasu, S., Horii, Y., Lam, P. K., & Ebinghaus, R. (2009). Partitioning behavior of per-and polyfluoroalkyl compounds between pore water and sediment in two sediment cores from Tokyo Bay, Japan. *Environmental Science & Technology*, *43*(18), 6969–6975.

Ahrens, L. (2011). Polyfluoroalkyl compounds in the aquatic environment: A review of their occurrence and fate. *Journal of Environmental Monitoring*, *13*(1), 20–31.

Ahrens, L., & Bundschuh, M. (2014). Fate and effects of poly- and perfluoroalkyl substances in the aquatic environment: A review. *Environmental Toxicology and Chemistry*, *33*(9), 1921–1929.

Avendaño, S. M., & J. Liu. (2015). Production of PFOS from aerobic soil biotransformation of two perfluoroalkyl sulfonamide derivatives. *Chemosphere*, *119*, 1084–1090.

Awad, E., Zhang, X., Bhavsar, S. P., Petro, S., Crozier, P. W., Reiner, E. J. et al. (2011). Long-term environmental fate of perfluorinated compounds after accidental release at Toronto airport. *Environmental Science & Technology*, *45*(19), 8081–8089.

Backe, W. J., Day, T. C., & Field, J. A. (2013). Zwitterionic, cationic, and anionic fluorinated chemicals in aqueous film forming foam formulations and groundwater from US military bases by nonaqueous large-volume injection HPLC-MS/MS. *Environmental Science & Technology*, *47*(10), 5226–5234.

Barzen-Hanson, K. A., & Field, J. A. (2015). Discovery and implications of C2 and C3 perfluoroalkyl sulfonates in aqueous film-forming foams and groundwater. *Environmental Science & Technology Letters*, *2*(4), 95–99.

Buck, R. C., Franklin, J., Berger, U., Conder, J. M., Cousins, I. T., de Voogt, P. et al. (2011). Perfluoroalkyl and polyfluoroalkyl substances in the environment: Terminology, classification, and origins. *Integrated Environmental Assessment and Management*, *7*(4), 513–541.

D'Agostino, L. A., & Mabury, S. A. (2013). Identification of novel fluorinated surfactants in aqueous film forming foams and commercial surfactant concentrates. *Environmental Science & Technology*, *48*(1), 121–129.

Filipovic, M., Woldegiorgis, A., Norström, K., Bibi, M., Lindberg, M., & Österås, A. H. (2015). Historical usage of aqueous film forming foam: A case study of the widespread distribution of perfluoroalkyl acids from a military airport to groundwater, lakes, soils and fish. *Chemosphere, 129,* 39–45.

Gebbink, W. A., Hebert, C. E., & Letcher, R. J. (2009). Perfluorinated carboxylates and sulfonates and precursor compounds in herring gull eggs from colonies spanning the Laurentian Great Lakes of North America. *Environmental Science & Technology, 43*(19), 7443–7449.

Gellrich, V., Stahl, T., & Knepper, T. P. (2012). Behavior of perfluorinated compounds in soils during leaching experiments. *Chemosphere, 87*(9), 1052–1056.

Guelfo, J. L., & Higgins, C. P. (2013). Subsurface transport potential of perfluoroalkyl acids at aqueous film-forming foam (AFFF)-impacted sites. *Environmental Science & Technology, 47*(9), 4164–4171.

Higgins, C. P., & Luthy, R. G. (2006). Sorption of perfluorinated surfactants on sediments. *Environmental Science & Technology, 40*(23), 7251–7256.

Houtz, E. F., Higgins, C. P., Field, J. A., & Sedlak, D. L. (2013). Persistence of perfluoroalkyl acid precursors in AFFF-impacted groundwater and soil. *Environmental Science & Technology, 47*(15), 8187–8195.

Jackson, D. A., Wallington, T. J., & Mabury, S. A. (2013). Atmospheric oxidation of polyfluorinated amides: Historical source of perfluorinated carboxylic acids to the environment. *Environmental Science & Technology, 47*(9), 4317–4324.

Kim, S. K., & Kannan, K. (2007). Perfluorinated acids in air, rain, snow, surface runoff, and lakes: Relative importance of pathways to contamination of urban lakes. *Environmental Science & Technology,* 41(24), 8328–8334.

Kissa, E. (1994). *Fluorinated Surfactants: Synthesis, Properties, Applications.* New York: M. Dekker.

Levine, A. D., Libelo, E. L., Bugna, G., Shelley, T., Mayfield, H., & Stauffer, T. B. (1997). Biogeochemical assessment of natural attenuation of JP-4-contaminated ground water in the presence of fluorinated surfactants. *Science of the Total Environment, 208*(3), 179–195.

Loganathan, B. G., Sajwan, K. S., Sinclair, E., Kumar, K. S., & Kannan, K. (2007). Perfluoroalkyl sulfonates and perfluorocarboxylates in two wastewater treatment facilities in Kentucky and Georgia. *Water Research, 41*(20), 4611–4620.

McGuire, M. E., Schaefer, C., Richards, T., Backe, W. J., Field, J. A., Houtz, E. et al. (2014). Evidence of remediation-induced alteration of subsurface poly- and perfluoroalkyl substance distribution at a former firefighter training area. *Environmental Science & Technology, 48*(12), 6644–6652.

McKenzie, E. R., Siegrist, R. L., McCray, J. E., & Higgins, C. P. (2015). Effects of chemical oxidants on perfluoroalkyl acid transport in one-dimensional porous media columns. *Environmental Science & Technology, 49*(3), 1681–1689.

MILSPEC. (1992). MIL-F-24385F: Fire extinguishing agent, aqueous film-forming foam (AFFF) Liquid concentrate, for fresh and sea water, 7. U.S. Naval Research Laboratory.

Moody, C. A., & Field, J. A. (1999). Determination of perfluorocarboxylates in groundwater impacted by fire-fighting activity. *Environmental Science & Technology, 33*(16), 2800–2806.

Moody, C. A., & Field, J. A. (2000). Perfluorinated surfactants and the environmental implications of their use in fire-fighting foams. *Environmental Science & Technology, 34*(18), 3864–3870.

Moody, C. A., Hebert, G. N., Strauss, S. H., & Field, J. A. (2003). Occurrence and persistence of perfluorooctanesulfonate and other perfluorinated surfactants in groundwater at a fire-training area at Wurtsmith Air Force Base, Michigan, USA. *Journal of Environmental Monitoring: JEM, 5*(2), 341.

Murakami, M., Kuroda, K., Sato, N., Fukushi, T., Takizawa, S., & Takada, H. (2009). Groundwater pollution by perfluorinated surfactants in Tokyo. *Environmental Science & Technology, 43*(10), 3480–3486.

Nguyen, V. T., Gin, K. Y. H., Reinhard, M., & Liu, C. (2012). Occurrence, fate, and fluxes of perfluorochemicals (PFCs) in an urban catchment: Marina Reservoir, Singapore. *Water Science and Technology*, 66(11), 2439.

OECD (Organisation for Economic Co-operation and Development). (2013). OECD/UNEP Global PFC Group. Synthesis paper on per- and polyfluorinated chemicals (PFCs). Environment, Health and Safety, Environment Directorate, OECD. https://www.google.com/url?sa=t&rct=j&q=&esrc=s&source=web&cd=1&ved=0ahUKEwjltpmR0YzaAhVCU98KHYl1A9IQFggnMAA&url=https%3A%2F%2Fwww.oecd.org%2Fenv%2Fehs%2Frisk-management%2FPFC_FINAL-Web.pdf&usg=AOvVaw2MlNyxaGylUGBigePvn8sa (Accessed 15 December 2015).

Park, H., Vecitis, C. D., Cheng, J., Choi, W., Mader, B. T., & Hoffmann, M. R. (2009). Reductive defluorination of aqueous perfluorinated alkyl surfactants: Effects of ionic headgroup and chain length. *Journal of Physical Chemistry A*, 113(4), 690–696.

Place, B. J., & Field, J. A. (2012). Identification of novel fluorochemicals in aqueous film-forming foams used by the US military. *Environmental Science & Technology*, 46(13), 7120–7127.

Schultz, M. M., Barofsky, D. F., & Field, J. A. (2004). Quantitative determination of fluorotelomer sulfonates in groundwater by LC MS/MS. *Environmental Science & Technology*, 38(6), 1828–1835.

Sepulvado, J. G., Blaine, A. C., Hundal, L. S., & Higgins, C. P. (2011). Occurrence and fate of perfluorochemicals in soil following the land application of municipal biosolids. *Environmental Science & Technology*, 45(19), 8106–8112.

Shivakoti, B. R., Tanaka, S., Fujii, S., Kunacheva, C., Boontanon, S. K., Musirat, C. et al. (2010). Occurrences and behavior of perfluorinated compounds (PFCs) in several wastewater treatment plants (WWTPs) in Japan and Thailand. *Journal of Environmental Monitoring*, 12(6), 1255–1264.

So, M. K., Miyake, Y., Yeung, W. Y., Ho, Y. M., Taniyasu, S., Rostkowski, P. et al. (2007). Perfluorinated compounds in the Pearl River and Yangtze River of China. *Chemosphere*, 68(11), 2085–2095.

Tomy, G. T., Tittlemier, S. A., Palace, V. P., Budakowski, W. R., Braekevelt, E., Brinkworth, L., & Friesen, K. (2004). Biotransformation of N-ethyl perfluorooctanesulfonamide by rainbow trout (*Onchorhynchus mykiss*) liver microsomes. *Environmental Science & Technology*, 38(3), 758–762.

USEPA (U.S. Environmental Protection Agency). (2009). Long-chain perfluorinated chemicals (PFCs) action plan. December 30. https://www.epa.gov/assessing-and-managing-chemicals-under-tsca/long-chain-perfluorinated-pfcs-action-plan (Accessed 15 December 2015).

Wallington, T. J., Hurley, M. D., Xia, J. D. J. W., Wuebbles, D. J., Sillman, S., Ito, A. et al. (2006). Formation of $C_7F_{15}COOH$ (PFOA) and other perfluorocarboxylic acids during the atmospheric oxidation of 8:2 fluorotelomer alcohol. *Environmental Science & Technology*, 40(3), 924–930.

Wang, N., Liu, J., Buck, R. C., Korzeniowski, S. H., Wolstenholme, B. W., Folsom, P. W., & Sulecki, L. M. (2011). 6:2 fluorotelomer sulfonate aerobic biotransformation in activated sludge of waste water treatment plants. *Chemosphere*, 82(6), 853–858.

Weiner, B., Yeung, L. W., Marchington, E. B., D'Agostino, L. A., & Mabury, S. A. (2013). Organic fluorine content in aqueous film forming foams (AFFFs) and biodegradation of the foam component 6:2 fluorotelomermercaptoalkylamido sulfonate (6:2 FTSAS). *Environmental Chemistry*, 10(6), 486–493.

Yang, L., Zhu, L., & Liu, Z. (2011). Occurrence and partition of perfluorinated compounds in water and sediment from Liao River and Taihu Lake, China. *Chemosphere*, 83(6), 806–814.

Zareitalabad, P., Siemens, J., Hamer, M., & Amelung, W. (2013). Perfluorooctanoic acid (PFOA) and perfluorooctanesulfonic acid (PFOS) in surface waters, sediments, soils and wastewater—A review on concentrations and distribution coefficients. *Chemosphere*, 91(6), 725–732.

16 A Preliminary Treatment Train Study

Removal of Perfluorinated Compounds from Postemergency Wastewater by Advanced Oxidation Process and Granular Activated Carbon Adsorption

Sean M. Dyson, Christopher T. Schmidt, and John E. Stubbs

CONTENTS

16.1 Introduction 373
16.2 Statement of the Problem 374
16.3 Investigation of a Treatment Train Approach for AFFF-Contaminated Natural Water 375
16.4 Treatment Train Process of AFFF-Contaminated Water: Preliminary Findings 377
    16.4.1 TOC Removal by UV/H_2O_2 AOP 377
    16.4.2 Effects of AOP Pretreatment on GAC Capacity 378
16.5 Summary 379
References 380

16.1 INTRODUCTION

An incident aboard the US Navy's USS *Forrestal* in 1967 highlighted the need to field an improved firefighting foam to fight hydrocarbon-based fires. As a response to the incident, the Department of Defense (DoD) fielded their Military Specification (MilSpec) version of aqueous film-forming foam (AFFF) developed by 3M in 1963 (Sheinson et al., 2015). Legacy MilSpec AFFF consists of long-chain perfluorinated alkyl substances (PFAS) (SERDP, 2015). These PFAS contain perfluorooctanoic acid (PFOA) or perfluorooctane sulfonic acid (PFOS), which have degradation half-lives of 92 and 41 years, respectively (USEPA, 2012). AFFFs used at military training and real-world crash sites have been successful because PFAS are made up of C–F bonds, which is the strongest covalent bond type there is (O'Hagan, 2008). The C–F bond makes PFAS thermally inert, and PFAS have hydrophobic and lipophobic tendencies, making PFAS highly enduring in the environment (Zhao et al., 2016).

PFAS are not only used in AFFF; they are also commonly used in industrial products, such as nonstick, stain-resistant, food contact materials; fast-food wrappers; french fry paperboards; and outdoor apparel for their waterproofing abilities (Schaider et al., 2017). Decades-long use of long-chain AFFF poses the potential to contaminate surface and groundwater drinking systems (USEPA, 2016). PFAS have been shown to have some toxicity impact in laboratory animal testing, leading to concerns regarding human and environmental exposures (Anderson et al., 2016).

In May 2016, the US Environmental Protection Agency (USEPA) lowered its lifetime PFAS drinking water advisory limits from 400 and 200 parts per trillion (ppt) for PFOA and PFOS, respectively, to 70 ppt for the sum of PFOA and PFOS. This lower limit was based on the best available peer-reviewed studies indicating that exposure to PFOA and PFOS over certain levels may result in adverse health effects, including developmental effects to fetuses during pregnancy or to breastfed infants (e.g., low birth weight, accelerated puberty, and skeletal variations), cancer (e.g., testicular, kidney), liver effects (e.g., tissue damage), immune effects (e.g., antibody production and immunity), thyroid effects, and other effects (e.g., cholesterol changes) (USEPA, 2016). This new limit compelled Wright-Patterson Air Force Base in Ohio to shut down two of its groundwater wells in 2016 that had confirmed levels of PFOS as high as 110 ppt (88th Air Base Wing Public Affairs, 2016). PFAS contamination is not limited to DOD sources. Resident drinking water sources in Belmont, Michigan, tested for combined PFOA and PFOS levels as high as 37,800 ppt (Ellison, 2017). Ellison reports that a local tannery facility—which used PFAS chemicals for waterproofing leather—dumped its industrial waste at an adjacent landfill to the resident's property and is suspected of being the source of PFAS contamination.

16.2 STATEMENT OF THE PROBLEM

PFAS have been found worldwide in surface waters and groundwater aquifers (Zhang et al., 2016). Research has shown that granular activated carbon (GAC), acting as a sorbent, works to purify and remove PFAS from drinking water (Zhang et al., 2016). There is a body of research showing how certain types of GAC work more efficiently than others at removing target contaminants. Benchtop flow through research using rapid small-scale column tests (RSSCTs) has shown that Calgon Filtrasorb® 600 (F600) GAC is more efficient at removing PFAS than some other GAC types (Schmidt, 2017).

Schmidt's research also showed a 30-fold increase in total organic carbon (TOC) concentration with the introduction of AFFF, which resulted in fouling and reduced capacity of the GAC. Schmidt suggested that compounds within the TOC may have a higher affinity for adsorption, leaving fewer sites for PFAS to adsorb onto the GAC.

A Preliminary Treatment Train Study

In Schmidt's research, source water with a starting TOC concentration of 3.3 mg/L ended up with 99.1 mg/L TOC upon the addition of 3M FC-203CF Light Water™ AFFF, targeting a concentration of approximately 4 mg/L. Based on this result, further research was warranted to determine if the removal capacity by F600 GAC may be improved if TOC concentrations are reduced first. For instance, research has shown that the removal of PFAS by GAC is more effective when the dissolved organic matter (DOM) concentration is low (Appleman et al., 2013). Because DOM content is correlated with TOC content, a pretreatment step that removes TOC is expected to increase GAC capacity by reducing fouling on the adsorption sites.

Advanced oxidation process (AOP) is one method shown to be effective in reducing TOC concentrations. However, to date, no studies have been conducted using AOP as initial treatment of PFAS-contaminated water, followed by further treatment with GAC. Research has shown that a UV/hydrogen peroxide (H_2O_2) AOP treatment can achieve >60% TOC removal from natural water when operated under optimal conditions (Rezaee et al., 2014). Independent variables, such as H_2O_2 concentration, pH, and contact time, are all found to affect TOC removal efficiency (Rezaee et al., 2014). The Rezaee et al. (2014) model limits the initial TOC concentration to 10 mg/L.

UV/H_2O_2 AOP has shown negligible degradation of PFOA (Phillips et al., 2016). PFOS degradation fares better with up to 50% removal when using AOP (Schroder et al., 2010), but generally the fluorine bonds in PFAS resist oxidation. Therefore, oxidation is not an effective removal treatment for PFAS (Appleman et al., 2014), especially for PFOS and PFOA. These results suggest that a treatment train process, combining an initial UV/H_2O_2 AOP to remove TOC with subsequent PFAS adsorption by GAC, might be viable to investigate.

16.3 INVESTIGATION OF A TREATMENT TRAIN APPROACH FOR AFFF-CONTAMINATED NATURAL WATER

To investigate the treatment of AFFF containing PFAS, including PFOS and PFOA, and a number of co-contaminants, a treatment train that involves an AOP followed by filtration through GAC was utilized. This research included studies with much higher initial levels of TOC than summarized above, but which were realistic concentrations following AFFF use.

The specific AOP process used was UV combined with hydrogen peroxide. The UV/H_2O_2 AOP setup is shown below in Figure 16.1, and the process flow diagram is shown in Figure 16.2. Each test utilized 5 gallons of natural groundwater sourced from Idaho National Laboratory (INL), contained in a 10-gallon vessel and pumped through a 20-inch sediment prefilter before going through a VIQUA UV MAX Model D4 254 nm mercury lamp with a disinfection dose of 30 mJ/cm². Hydrogen peroxide was added in various amounts to test the concentration effects. UV contact times were also varied.

Experimental design parameters for the GAC adsorbent step included the type of adsorbent, empty bed contact time (EBCT), and whether to put the beds in series or parallel (Jarvie et al., 2005). The performance of GAC adsorbent was analyzed by RSSCTs. This study used 80 × 200 sieved F600 GAC (US standard sieve size),

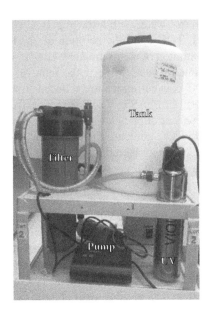

FIGURE 16.1 UV/H_2O_2 AOP setup.

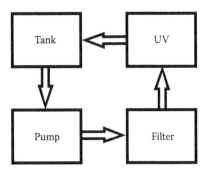

FIGURE 16.2 UV/H_2O_2 process flow diagram.

an EBCT of 9 minutes, and constant diffusivity design. Treated GAC effluent was analyzed for concentrations of PFOA and PFOS by a method described by Schmidt (2017) using fluorine-free ultra-high-performance liquid chromatography (UPLC)–tandem mass spectrometry (MS-MS) equipment. The RSSCT setup is shown below in Figure 16.3.

The objective of this research was to develop a treatment train process to more effectively and efficiently remove PFAS from an impacted groundwater source. A comparison between the treatment train method (AOP followed by GAC) with GAC treatment alone was made to determine:

1. Whether UV/H_2O_2 AOP reduced TOC in AFFF-contaminated groundwater
2. Whether pretreatment with AOP affected PFAS GAC adsorption capacity

A Preliminary Treatment Train Study

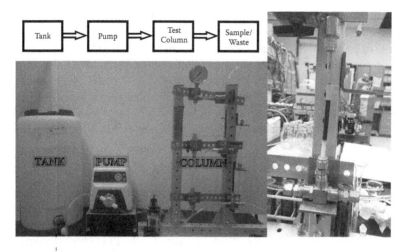

FIGURE 16.3 RSSCT process flow diagram (top left), RSSCT setup (bottom left), and close-up of a test column filled with GAC supported by approximately 1 inch of glass wool (right).

16.4 TREATMENT TRAIN PROCESS OF AFFF-CONTAMINATED WATER: PRELIMINARY FINDINGS

16.4.1 TOC Removal by UV/H_2O_2 AOP

To answer the first research question, an AOP experiment using a 500 mg/L H_2O_2 concentration was conducted. Figure 16.4 shows that the 500 mg/L H_2O_2 concentration reduced the TOC concentration by 98%. A control test with no H_2O_2 added was also conducted to ensure that UV contact time alone would not degrade TOC. While the control test results showed a total TOC reduction of 16% after 8 hours of treatment, most of the TOC removal (11%) took place within the first hour. However, the rate of reduction beyond the first hour was too slow to be applicable in a real-world treatment plant.

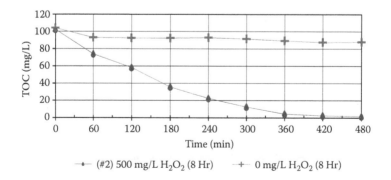

FIGURE 16.4 TOC degradation of INL water with AFFF using 500 mg/L H_2O_2 in a UV/H_2O_2 AOP.

One disadvantage of pretreating the PFAS-contaminated water with an AOP appears to be an increased concentration of both PFOA and PFOS, the constituents that were to be removed. Increased concentrations may be attributable to other PFAS chains in the matrix of the source water that reacted in the AOP. One path may be precursor PFAS chains forming into the larger PFOA and PFOS compounds. Another potential path may be the larger PFAS chains degrading into PFOA and PFOS. Further research is needed to determine this phenomenon. More discussion on PFAS precursor transformation is included in Chapter 3 by Chandramouli.

16.4.2 Effects of AOP Pretreatment on GAC Capacity

To answer the second research question, adsorption of PFAS from AOP pretreated water was investigated via RSSCTs. AOP experiments are shown in Figure 16.5. Included in Figure 16.5 is the TOC breakthrough curve for a test of water without AOP pretreatment (Schmidt, 2017). Schmidt's data for the PFOS breakthrough curve (square markers) are plotted in Figure 16.5 for comparison. The RSSCT PFOS results for the 500 mg/L H_2O_2 (teardrop markers) 8-hour AOP test indicate an increase in the capacity to approximately 4000 bed-volumes before 5% breakthrough.

These results, shown in Figure 16.5, demonstrate that a pretreatment step of TOC removal by AOP does result in increased GAC adsorption capacity. Even though the correlation seems to be clear, it is still premature to conclude with complete confidence that the increased GAC capacity for PFAS removal is indeed due to decreased fouling from TOC. The total TOC concentration does not always reflect TOC adsorption onto GAC; that is, a decreased total TOC concentration does not necessarily mean decreased adsorption. Depending on the nature of the TOC, TOC adsorption

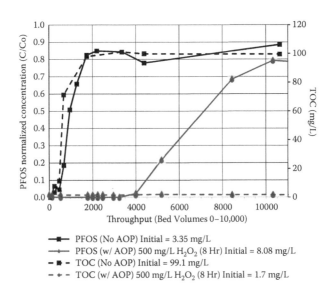

- PFOS (No AOP) Initial = 3.35 mg/L
- PFOS (w/ AOP) 500 mg/L H_2O_2 (8 Hr) Initial = 8.08 mg/L
- TOC (No AOP) Initial = 99.1 mg/L
- TOC (w/ AOP) 500 mg/L H_2O_2 (8 Hr) Initial = 1.7 mg/L

FIGURE 16.5 PFOS RSSCT results for 9-minute EBCT with and without pre-AOP treatment.

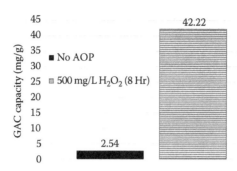

FIGURE 16.6 Calgon Filtrasorb F600 GAC capacities for PFOS with no pre-AOP treatment and at a concentration of 500 mg/L H_2O_2 with UV/H_2O_2 AOP pretreatment.

could be very different at the same total concentration or even change with time as compounds within the TOC undergo changes that affect adsorption. Future studies with more specific TOC characterization technologies may help answer these questions.

The calculated GAC capacities for PFOS with no pretreatment and with AOP pretreatment are shown in Figure 16.6. GAC capacities were calculated by using the treated volume of water in liters (L) for the sample result prior to the first nonzero, or essentially the point immediately before breakthrough. The GAC capacity formula is shown in Equation 16.1, where C_0 = influent concentration (mg/L), C_e = effluent concentration (mg/L), and C_c = concentration of activated carbon (g/L) (Clark, 2009). A C_e concentration of 0 mg/L was used.

$$q_e = \frac{mass\ adsorbate\ (mg)}{mass\ adsorbent\ (g)} = \frac{(C_0 - C_e) \cdot V}{C_c \cdot V} \tag{16.1}$$

In Figure 16.6, the PFOS 500 mg/L H_2O_2 (striped bar) 8-hour AOP pretreatment experiment shows that GAC capacity increased by a factor of 16 versus the test run with no AOP pretreatment (solid bar).

16.5 SUMMARY

A treatment train process was investigated to determine whether UV/H_2O_2 AOP reduced TOC in AFFF-contaminated groundwater and whether pretreatment with AOP affected GAC adsorption capacity for PFAS. Results indicated using UV/H_2O_2 AOP to pretreat impacted groundwater contaminated with AFFF can reduce TOC from an initial concentration of 99.1 mg/L to less than 2 mg/L. This was accomplished by using a 500 mg/L H_2O_2 concentration and operating the UV/H_2O_2 AOP for 8 hours. This effective pretreatment process improved GAC capacity for PFOS from 2.5 mg/g without pretreatment up to ≥42 mg/g. This increase could correspond to significant operational savings and reduction in logistical complexities of PFAS treatment.

Additional research is recommended to determine the optimal H_2O_2 concentration and UV contact time during the AOP process. An optimized AOP pretreatment process, coupled with post-GAC treatment, has the potential to scale up to full-scale treatment plants to remediate PFAS contamination from affected drinking water sources with greater efficiency than currently employed methods.

REFERENCES

88th Air Base Wing Public Affairs. (2016, May 20). Base issues drinking water advisory. Retrieved November 14, 2017, from http://www.wpafb.af.mil/News/Article-Display/Article/818472/base-issues-drinking-water-advisory/.

Anderson, R. H., Long, G. C., Porter, R. C., & Anderson, J. K. (2016). Occurrence of select perfluoroalkyl substances at U.S. Air Force aqueous film-forming foam release sites other than fire-training areas: Field-validation of critical fate and transport properties. *Chemosphere*, *150*, 678–685. https://doi.org/10.1016/j.chemosphere.2016.01.014.

Appleman, T. D., Dickenson, E. R. V., Bellona, C., & Higgins, C. P. (2013). Nanofiltration and granular activated carbon treatment of perfluoroalkyl acids. *Journal of Hazardous Materials*, *260*, 740–746. https://doi.org/10.1016/j.jhazmat.2013.06.033.

Appleman, T. D., Higgins, C. P., Quiñones, O., Vanderford, B. J., Kolstad, C., Zeigler-Holady, J. C., & Dickenson, E. R. V. (2014). Treatment of poly- and perfluoroalkyl substances in U.S. full-scale water treatment systems. *Water Research*, *51*, 246–255. https://doi.org/10.1016/j.watres.2013.10.067.

Clark, M. M. (2009). Adsorption, partitioning and interfaces. In *Transport Modeling for Environmental Engineers and Scientists*, pp. 135–174. Second Edition. John Wiley & Sons, INC.

Ellison, G. (2017). Tannery chemicals are 540 times above EPA level in Belmont well. Retrieved December 7, 2017, from http://www.mlive.com/news/grand-rapids/index.ssf/2017/09/wolverine_house_street_well.html.

Jarvie, M. E., Hand, D. W., Bhuvendralingam, S., Crittenden, J. C., & Hokanson, D. R. (2005). Simulating the performance of fixed-bed granular activated carbon adsorbers: Removal of synthetic organic chemicals in the presence of background organic matter. *Water Research*, *39*(11), 2407–2421. https://doi.org/10.1016/j.watres.2005.04.023.

O'Hagan, D. (2008). Understanding organofluorine chemistry. An introduction to the C-F bond. *Chemical Society Reviews*, *37*(2), 308–319.

Phillips, R., Magnuson, M., James, R., Benotti, & Benotti, M. (2016). Comparing advanced oxidation processes for treatment of heavily contaminated water. Presented at 2016 Annual Conference and Exposition, Chicago, IL.

Rezaee, R., Maleki, A., Jafari, A., Mazloomi, S., Zandsalimi, Y., & Mahvi, A. H. (2014). Application of response surface methodology for optimization of natural organic matter degradation by UV/H2O2 advanced oxidation process. *Journal of Environmental Health Science and Engineering*, *12*(1), 67. https://doi.org/10.1186/2052-336X-12-67.

Schaider, L. A., Balan, S. A., Blum, A., Andrews, D. Q., Strynar, M. J., Dickinson, M. E. et al. (2017). Fluorinated compounds in U.S. fast food packaging. *Environmental Science & Technology Letters*, *4*(3), 105–111. https://doi.org/10.1021/acs.estlett.6b00435.

Schmidt, C. T. (2017). Adsorption of perfluorinated compounds from post-emergency response wastewater. Master's thesis. Air Force Institute of Technology.

Schroder, H. Fr., Jose, H. J., Gebhardt, W., Moreira, R. F. P. M., & Pinnekamp, J. (2010). Biological wastewater treatment followed by physicochemical treatment for the removal of fluorinated surfactants. *Water Science and Technology*, *61*(12), 3208–3215.

SERDP (Strategic Environmental Research and Development Program). (2015). Fluorine-free aqueous film forming foam. SERDP WPSON-17-01.

Sheinson, R. S., Williams, B. A., Green, C., Fleming, J. W., Anleitner, R., Ayers, S., Maranghides, A., and Barylski, D. (2015). The future of aqueous film forming foam (AFFF): Performance parameters and requirements. Retrieved from https://www.nist.gov/sites/default/files/documents/el/fire_research/R0201327.pdf.

USEPA. (2012, May). Emerging contaminants—Perfluorooctane sulfonate (PFOS) and perfluorooctanoic acid (PFOA). Retrieved from https://nepis.epa.gov/Exe/ZyNET.exe/P100LTG6.TXT?ZyActionD=ZyDocument&Client=EPA&Index=2011+Thru+2015&Docs=&Query=&Time=&EndTime=&SearchMethod=1&TocRestrict=n&Toc=&TocEntry=&QField=&QFieldYear=&QFieldMonth=&QFieldDay=&IntQFieldOp=0&ExtQFieldOp=0&XmlQuery=&File=D%3A%5Czyfiles%5CIndex%20Data%5C11thru15%5CTxt%5C00000014%5CP100LTG6.txt&User=ANONYMOUS&Password=anonymous&SortMethod=h%7C-&MaximumDocuments=1&FuzzyDegree=0&ImageQuality=r75g8/r75g8/x150y150g16/i425&Display=hpfr&DefSeekPage=x&SearchBack=ZyActionL&Back=ZyActionS&BackDesc=Results%20page&MaximumPages=1&ZyEntry=1&SeekPage=x&ZyPURL.

USEPA. (2016, May). Drinking Water Health Advisory for Perfluorooctanoic acid (PFOA). Retrieved from https://www.epa.gov/sites/production/files/2016-05/documents/pfoa_health_advisory_final-plain.pdf.

USEPA. (2016, November). Fact sheet PFOA & PFOS drinking water health advisories. Retrieved from https://www.epa.gov/sites/production/files/2016-06/documents/drinkingwaterhealthadvisories_pfoa_pfos_updated_5.31.16.pdf.

Zhang, D., Luo, Q., Gao, B., Chiang, S. Y. D., Woodward, D., & Huang, Q. (2016). Sorption of perfluorooctanoic acid, perfluorooctane sulfonate and perfluoroheptanoic acid on granular activated carbon. *Chemosphere, 144*, 2336–2342. https://doi.org/10.1016/j.chemosphere.2015.10.124.

Zhao, P., Xia, X., Dong, J., Xia, N., Jiang, X., Li, Y., & Zhu, Y. (2016). Short- and long-chain perfluoroalkyl substances in the water, suspended particulate matter, and surface sediment of a turbid river. *Science of the Total Environment, 568*, 57–65. https://doi.org/10.1016/j.scitotenv.2016.05.221.

17 Remediation of PFAS-Contaminated Soil

Konstantin Volchek, Yuan Yao, and Carl E. Brown

CONTENTS

17.1 Introduction 383
17.2 PFAS Contamination in Soil 384
    17.2.1 Fate of PFAS in Soil 384
    17.2.2 PFAS Bioaccumulation Factors in Soil... 387
17.3 Current Options for the Remediation of PFAS-Contaminated Soil 388
    17.3.1 Nondestructive Methods 389
        17.3.1.1 Removal 389
        17.3.1.2 Immobilization of PFAS in Soil 394
    17.3.2 Destructive Methods... 396
        17.3.2.1 Vitrification and Incineration... 396
        17.3.2.2 *In Situ* Chemical Oxidation 397
17.4 Conclusions... 398
References... 399

17.1 INTRODUCTION

The world of per- and polyfluoroalkyl substances (PFAS) is full of many useful and unique industrial and consumer products. Some of them are tailored for specific applications, such as aqueous film-forming foams (AFFFs), against hydrocarbon-fueled fires. Used as intended, PFAS provide great benefits to society. However, when released to the soil environment, a number of problems can result. These include human and environmental toxicity, movement through soil to water, and uptake from soil to biota.

Environmental site remediation is required when contaminants are present in concentrations above acceptable limits. Depending on specific site conditions, remediation may involve the treatment of soil, sediments, groundwater, surface water, and other affected media. As contaminants may migrate from one medium to another, more than one medium normally requires treatment. For example, soil and groundwater remediation are often carried out together to avoid recontamination.

There has been more work done and more information available on the remediation of PFAS-contaminated water compared with soil since adsorption, a common treatment technique, can effectively remove most PFAS from aqueous systems. Soil remediation typically poses a greater challenge due to complex and often strong interactions between PFAS and the soil matrix. These interactions, along with a high chemical stability of PFAS molecules, make most "conventional" soil remediation methods ineffective (Vecitis et al. 2009; Yao et al. 2013). Having said this, there are some established techniques and new promising methods that can be used to deal with the soil. This chapter discusses PFAS contamination in soil as well as potentially applicable remedial options.

17.2 PFAS CONTAMINATION IN SOIL

17.2.1 FATE OF PFAS IN SOIL

The two main paths of PFAS entry into soil are industrial waste sites and the application of AFFFs at military bases, airports, and municipal firefighting training areas (FFTAs) (Houtz et al. 2013). Although PFAS presence in soil may not pose an immediate threat, soil treatment is essential for long-term remediation. The dangers associated with PFAS in soil include the potential for leaching into water systems and entry into the food chain through bioaccumulation in plants or soil-dwelling animals, such as worms or insects.

Leaching of PFAS from soil to groundwater is of particular concern. Once PFAS have become dissolved in water, their potential for migration greatly increases (Nunes et al. 2011). Furthermore, PFAS may seep into sources of drinking water. Wilhelm et al. (2008) performed a case study following the accidental application of PFAS to nearly 1000 agricultural land sites in Europe. PFAS reportedly leached into nearby rivers, leading to perfluorooctanoic acid (PFOA) and perfluorooctanesulfonic acid (PFOS) levels of up to 150 µg/L. The contaminated water has subsequently been restored to suitable PFAS levels through activated carbon filtration.

Higgins and Luthy (2006) analyzed the sorption of perfluorinated surfactants on sediments and were able to observe several trends. First, the main property affecting the strength of PFAS sorption is the organic content of the soil. Higher organic content generally favors sorption of PFAS. Other parameters observed leading to lower perfluorinated surfactant mobility were increasing Ca^{2+} concentration and decreasing pH. This study reported that Ca^{2+} ions may act to reduce the charge present on organic matter, which would in turn decrease the repulsion of anionic surfactants, such as perfluoroalkyl sulfonates (PFSAs) and perfluoroalkyl carboxylates (PFCAs). You et al. (2010) also observed an increased sorption due to a greater salinity; however, they justified this trend using the salting-out effect, which would increase the order between water molecules and reduce the solubility of PFOS in aqueous media. Higgins and Luthy (2006) observed that sorption is affected by PFAS chain length and the presence of the sulfate moiety. Sorption is increased with each additional CF_2 moiety, and the interaction is consistently stronger in PFSAs than in their PFCA analogs.

The behavior of PFAS in soils during leaching experiments was reported by Gellrich and others (Gellrich and Knepper 2012; Gellrich et al. 2012). The 2-year laboratory study simulated the effect that precipitation would have on the transport of PFAS from soil to groundwater using flow-through columns 60 cm in length. It was observed that short-chain PFAS with alkyl chains less than seven carbons were much more mobile than PFAS with longer alkyl chain lengths. The leaching behavior was dependent on the organic content of the soil sample. With an organic content of 2%, PFOS was eluted after 70 weeks, compared with perfluorobutyric acid (PFBA) and perfluorobutanesulfonic acid (PFBS), which began to elute after just 7 weeks. When the organic content of soil was increased to 7% and 14%, PFOS was not observed to elute at all within the 2-year period. These results indicate that soil is a significant environmental sink for PFOS. Without remediation, the soil may continue to be a long-term source of PFOS leachate for other environmental receptors.

Sludge effluent from wastewater treatment plants, also referred to as biosolids, can be composed of anything that enters the processing plant. The makeup of wastewater sludge is dependent on the location of the treatment plant and its proximity to communities or industry. Approximately 50% of biosolids generated in the United States are subsequently applied to agricultural land as fertilizer (Lindstrom et al. 2014). The land application of biosolids in the United States is controlled by Part 503 of the US Code of Federal Regulations (USGPO 2014), "Standards for the Use or Disposal of Sewage Sludge." The regulations control the level of metals and pathogens present in sludge destined for land application, but organic pollutants such as PFAS remain unmeasured and unregulated. Direct application of biosolids to agricultural fields may hence be a significant source of PFAS soil pollution and potentially a dietary source for consumers of the agricultural products produced on the land.

PFCAs can enter the wastewater system through a variety of industrial sources. Direct sources include the manufacturing of PFCAs, fluoropolymers, and AFFFs. Indirect sources include the breakdown of fluorotelomer-based products and perfluorooctyl sulfonyl fluoride (POSF)–based products in the environment (Prevedouros et al. 2006). Several studies have attempted to quantify the degree of PFAS presence in wastewater sludge. Higgins et al. (2005) sampled digested sludge from eight wastewater treatment plants scattered around the United States receiving at least 50% domestic waste. The concentration of total PFAS in the domestic sludge ranged from 73 to 3390 ng/g, with total PFCA concentrations between 5 and 152 ng/g and total PFSA concentrations between 55 and 3370 ng/g. The highest concentration of PFOS was measured at 2610 ng/g from a wastewater treatment plant receiving 90% domestic waste and only 10% light industrial waste. This sample was, however, the only sample taken from 1998, before the PFOS production phaseout in 2002. A more recent sample taken from the same plant had a total PFAS concentration of 474 ng/g and a PFOS concentration of only 167 ng/g. If the average concentration is adjusted to exclude the sample from 1998, the total PFAS concentration drops to 436 ng/g, and 124 ng/g for PFOS alone. This evidence suggests a trend of decreasing PFOS levels entering the environment through treatment plant sludge following PFOS production phaseout.

Sun et al. (2011) conducted a similar study with digested sewage sludge from 20 wastewater treatment plants in Switzerland, where biosolids application to agricultural fields is currently forbidden. The total concentration of PFAS in the sludge ranged from 28 to 637 ng/g, while the total concentration of PFOS ranged from 15 to 600 ng/g. The predominant PFAS in every sample was PFOS, and higher levels of PFOS had no correlation with higher levels of PFCAs within the same sample. Archived sludge samples were also screened from 1993 and 2002 accompanying the samples taken in 2008 to assess the effect of the 2002 production phaseout.

Interestingly, no clear pattern of decline was observed for PFOS or PFOA concentrations over the years, with some wastewater treatment plants even measuring their highest concentrations in 2008.

A review by Jensen et al. (2012) compiled additional PFAS concentrations in wastewater sludge from Europe. Nordic countries had relatively low concentrations in the range of 0.6 to 15.2 ng/g total PFAS (Kallenborn et al. 2004). Levels of PFAS contamination in Swedish sludge ranged from 0.6 to 23.9 ng/g and 1.6 to 54.8 ng/g for PFOA and PFOS, respectively (Haglund and Olofsson 2009).

Sepulvado et al. (2011) analyzed PFAS concentration in agricultural fields that had received applications of municipal biosolids. The amount of PFAS in biosolids-amended fields was measurably higher than that in background fields that had never received biosolids application. PFAS were found to within a depth of 120 cm below the soil surface. The concentration of PFOS was up to 483 ng/g in fields that had experienced long-term biosolids application.

Washington et al. (2010) investigated sludge-amended soils in proximity to a wastewater treatment plant handling sewage waste from industries known to work with perfluoroalkylates and their precursors. Sludge from this plant was measured to contain PFOA concentrations up to 1875 ng/g; however, the application of biosolids from this plant was discontinued in 2007, and the concentration of PFAS in the sludge subsequently decreased dramatically. The measured PFAS of highest concentration in the sludge-applied fields was perfluorodecanoic acid (PFDA) (C_{10}) at 989 ng/g. The highest concentrations of PFOS and PFOA were 408 and 312 ng/g, respectively. The concentrations observed in the sludge-amended soils were compared to those in background fields that had never received biosolids application. The concentrations for all measured PFAS in the background fields were either below the limit of quantification or in the picogram per gram range. Evidence from the background fields suggests that PFAS contamination in agricultural soil is likely a direct result of biosolids application.

In addition to the direct sources of PFAS into wastewater treatment plants, a significant amount of precursors exist that can indirectly contribute to the final concentration of PFAS in digested sludge. The Organisation for Economic Co-operation and Development (OECD) has listed 875 chemicals with the potential to degrade to PFAS. For PFOS, these precursors include derivatives and polymers of perfluoroalkyl sulfonyl or sulfonamide compounds, while PFOA precursors have even more diversity, including derivatives and polymers of perfluoroalkyl alcohols, amines, carboxylic acids, esters, ethers, and iodides (OECD 2007).

Several studies (Sinclair and Kannan 2006; Loganathan et al. 2007; Murakami et al. 2009) have reported that in addition to the wastewater treatment plants being inefficient at general PFAS removal, they may actually cause an increase in PFAS, as evidenced by increased concentrations of PFAS in wastewater effluents versus influents. Oil, fuel, and AFFF components, common co-contaminants in PFAS-impacted wastewater, have also been shown to have a negative effect on the activated sludge process (Moody and Field 2000). The increase of PFAS in effluents is theorized to be attributed to the degradation of more complex PFAS precursors during activated sludge treatment. Schultz et al. (2006) sampled 10 wastewater treatment plants, and in 7 of these plants the PFOA concentration increased between 9% and 352%.

Although PFOS often exhibited a decrease in concentration via sludge treatment, it should also be taken into account that the K_d value for PFOS on sludge is more than three times higher than that for PFOA. This indicates a higher affinity for PFOS to become adsorbed to the sludge components, causing final PFOS concentrations to appear lower in effluents (J. Yu et al. 2009). Becker et al. (2008) observed a 20-fold increase in PFOA from influents to effluents. For PFOS, the total increase from inlet to outlet was about threefold. PFAS generated in the treatment process not only ended up in the effluent but also remained in the sludge. Following the treatment, the concentration of PFOA in the sludge increased by 10%. As for PFOS, almost half of that generated during the treatment remained in the sludge.

Liu and Mejia Avendaño (2013) compiled a list of biodegradability studies for perfluoroalkyl acid (PFAA) precursors conducted in conditions including microbial cultures, activated sludge, soil, and sediment. The degradation of 8:2 fluorotelomer alcohol (FTOH) caused an increase of PFOA by 6% in mixed bacterial culture (Wang et al. 2005a), 2.1% in activated sludge (Wang et al. 2005b), and up to ~40% in aerobic soils (Wang et al. 2009). The degradation of perfluorooctane sulfonamidoethanol (EtFOSE) caused an increase of PFOS by 7% in activated sludge (Lange 2000) and 12% in marine sediment (Benskin et al. 2013). Liu and Mejia Avendaño (2013) conclude by proposing that a knowledge gap exists for several important classes of precursors, such as

- Perflurooctane sulfonamide–based side-chain polymers
- Zwitterionic, cationic, and anionic fluoroalkyl surfactants (from AFFFs)
- Fluorotelomer iodides (FTIs)

While the microbial degradation of fluorotelomer and perfluoroalkyl sulfonamide compounds has been well documented (Lange 2000; Rhoads et al. 2008; Frömel and Knepper 2010), the subsequent degradation of PFOS and PFOA is likely minimal or nonexistent in sludge and soils under normal environmental conditions (Schröder et al. 2003; Sáez et al. 2008).

17.2.2 PFAS BIOACCUMULATION FACTORS IN SOIL

PFAS left in soil have the potential to enter the food chain through plants and animals. Several studies exist that quantify the bioaccumulation factors (BAFs) of PFAS in soil. This is calculated as a ratio of the PFAS concentration in biota to the concentration in the soil from which it was extracted.

Stahl et al. (2009) studied the transfer of PFOA and PFOS from artificially contaminated soil to wheat, oat, maize, potato, and grass plants. The amount of PFOS and PFOA uptake was dependent on their original concentration in the soil. Concentrations were greater in the vegetative portion of the plant than in the storage organs. Yoo et al. (2011) conducted a similar study using grass grown in fields contaminated with PFAS via biosolids application. As well as calculating their own grass–soil accumulation factors (GSAFs), they used the previous research by Stahl et al. (2009) to generate GSAFs for comparison. Perfluorohexanoic acid (PFHxA), the shortest PFCA tested in the study, showed the highest GSAF value (mean 3.8), and the results indicated lower transfer potential from soil to grasses for long-chain PFCAs. The GSAF for PFOA was calculated to be around 0.25, while PFOS was lower, at around 0.07.

Beach et al. (2006) conducted an ecotoxicological evaluation of PFOS, including the results of a study by Brignole et al. (2003), to determine the BAFs of seven plant species (i.e., onion, ryegrass, alfalfa, flax, lettuce, soybean, and tomato) under varying soil PFOS concentrations between 3.61 and 278 μg/g (dry weight). The soybean plant had the highest measured BAF at 4.3 within its vegetative tissue. Enhanced BAFs (2–3) were generally observed in the vegetative tissues harvested from the plants grown in the soils with low PFOS concentrations (<11.1 μg/g).

Lasier et al. (2011) studied the transfer of PFAS from soil to an aquatic worm, *Lumbriculus variegatus*, and calculated BAFs. The researchers found that the tendency to bioaccumulate increased with increasing PFCA chain length and the presence of the sulfonate moiety. The average BAFs for PFOA and PFOS were 0.07 and 0.49, respectively. Although limited in number, the observations suggested that PFCAs with fewer than six carbons have little potential to bioaccumulate in oligochaetes; however, perfluorohexane sulfonate (PFHxS) (mean BAF 0.48) and perfluoroheptane sulfonate (PFHpS) (BAF 2.6) appear to have the same or even greater bioaccumulation potential than PFOS.

Das et al. (2015) calculated BAFs based on PFOS for earthworms (*Eisenia fetida*) in soil that had been impacted by AFFFs. The values they calculated ranged from 1.23 to 13.9, and the highest bioaccumulation was observed in soil that contained just 0.8 μg/g PFOS, indicating that bioaccumulation can occur even in soils with less heavy pollution. It was observed that bioaccumulation was more pronounced in soils with lower organic content, perhaps due to weaker sorption of PFOS to the soil.

A similar study on earthworms conducted by the Norwegian Pollution Control Authority (SFT 2008) found the BAF for PFOS to be 2.6, while for PFOA it was 5.9. By comparing the opposing trends for worms (i.e., BAF increases with increasing chain length) and grasses (i.e., BAF decreases with increasing chain length), Yoo et al. (2011) hypothesized that BAFs tend to be higher in worms than in plants because the worms directly ingest PFAS from organic matter, while the plants depend on uptake of dissolved PFAS from soil–water. They pointed out that food chains based on direct ingestion of soils contaminated with PFAS might tend to accumulate longer-chained PFAS. Conversely, the food chains in which rooted plants serve as the basal trophic level would accumulate shorter-chained PFAS.

17.3 CURRENT OPTIONS FOR THE REMEDIATION OF PFAS-CONTAMINATED SOIL

All remediation methods can be divided into two categories: destructive and nondestructive. Destructive methods involve chemical transformations of PFAS leading to a change (ideally, reduction) in their toxicological characteristics. Most of these methods employ oxidation facilitated by the use of catalysts, heat, ultrasound, and so forth. Nondestructive methods can be further divided into two subcategories: removal and immobilization. Removal can be accomplished by several means, ranging from mechanical soil excavation to flushing the soil and collecting the PFAS-containing

Remediation of PFAS-Contaminated Soil

leachate for further treatment or disposal. Immobilization leaves PFAS within the soil but creates a stronger contaminant–soil bond to the soil matrix, and thus makes them less mobile.

There have been a number of methods reportedly used for the remediation of PFAS-contaminated water; however, much less is known about soil remediation (Törneman 2012). Many of the strategies effective for water treatment cannot be translated to solid matrices, such as soil, because the PFAS molecules are much less accessible. *In situ* technologies for soil remediation that have already been proven ineffective include air sparging, soil–vapor extraction, biodegradation, and hydrolysis (Pancras et al. 2013). As yet, there are no techniques available that allow total desorption of PFAS from soil.

Presently, the cost of PFAS remediation from soil is quite high, so that the remediation of PFAS-contaminated sites is rarely performed (Eschauzier et al. 2012). This section examines the current and developing technologies available for soil remediation, as well as their limitations, most notably the lack of field studies available for emerging technologies. An overview of PFAS treatment technology effectiveness is provided in Table 17.1, and some available cost estimates are summarized in Table 17.2.

17.3.1 Nondestructive Methods

17.3.1.1 Removal

17.3.1.1.1 Excavation and Specialized Landfills and Encapsulation

Contaminated soils can be excavated and moved to areas designed for long-term containment. These areas can be constructed to be surrounded by low-permeability barriers built of materials such as clay caps or synthetic textiles to reduce or eliminate the migration of contaminants (Khan et al. 2004). Similarly, treatment walls can be set up at the original area of contamination or at the designated long-term storage area. Treatment walls can be constructed by digging a trench around the contaminated area and filling it with materials that will treat contaminated groundwater as it passes through the barrier. There are three main types of treatment walls: sorption barriers, precipitation barriers, and degradation barriers. Since PFAS are very nonreactive chemicals, it is difficult to expect their precipitation or degradation from groundwater. However, filling the barrier with a sorbent to capture and hold PFAS could be feasible.

With assistance from the Minnesota Pollution Control Agency (MPCA 2009), 3M (2010) developed an action plan for the treatment of their PFAS-contaminated industrial waste sites. Strategies for soil remediation included removal of PFAS-contaminated soils in the cove leading to the Mississippi River and capping the area with soils heavy in clay content. The plan also included excavation of all soils on site that exceeded the Minnesota industrial soil reference values (SRVs) for PFOS (14 µg/g) and PFOA (13 µg/g), followed by backfilling of the site with uncontaminated soil. The contaminated soil was transported to a long-term containment facility, engineered specifically to hold PFAS. All leachate from the containment site was continuously pumped out and treated with activated carbon. Remediation of the sites was completed in 2010, but soil and groundwater sampling procedures are still

TABLE 17.1
Summary of PFAS Treatment Technology Effectiveness

Method	Highlights	Relevant Papers
Excavation and specialized landfills	• Temporary solution—landfills becoming increasingly unwilling to accept PFOS and PFOA • Potential still exists for soil leaching • Sorption barrier treatment walls have the potential to effectively inhibit the migration of PFAS-contaminated groundwater	Khan et al. (2004) MPCA (2009)
Groundwater pumping and treatment	• GAC treatment of collected groundwater is highly effective (>99%) • Separation of PFAS from soil is slow or does not occur	Paterson et al. (2008) Wilhelm et al. (2008) 3M (2010)
Soil flushing	• SDBS (anionic surfactant) could potentially mobilize PFOS from sediments • Possible adaptation of organic solvent extraction techniques • Requires more study into possible interactions with AFFF co-contaminants • Sc-CO_2 flushing with methanol and HNO_3 demonstrated possible success as a low-cost, environmentally friendly method for PFOS decontamination from solid matrices	Mulligan et al. (2001) Sun et al. (2011) Powley et al. (2005) Llorca et al. (2011) Schröder et al. (2003) Pan et al. (2009) Guelfo and Higgins (2013) Tang et al. (2006) Chen et al. (2012)
Soil washing	• Reduces the volume of contaminated soil requiring decontamination • Exhibited past success with a variety of halogenated organic compounds • Must be combined with another treatment technology	Khan et al. (2004) Chu and Kwan (2003) Chu and Chan (2003) Hasegawa et al. (1997) Pancras et al. (2013)
Immobilization	• CTAB (cationic surfactant) showed effective immobilization of PFOS to sediments • Evidence suggests that activated carbon may be able to immobilize PFAS in soil as well as water • Clay-based adsorbent, MatCARE, is in late commercial stages as a remediation technology involving PFOS immobilization in soil	Pan et al. (2009) Brändli et al. (2008) Hale et al. (2012) Jakob et al. (2012) Zimmerman et al. (2004) Das et al. (2013b)
Vitrification and incineration	• Extreme temperatures (1600°C–2000°C) destroy all organic pollutants causing very few by-products • Requires powerful thermal or electrical energy source • Could be adapted for temperatures suitable for PFAS incineration (>600°C)	Khan et al. (2004) Vecitis et al. (2009) Yamada et al. (2005)
Persulfate oxidation	• Exhibited past success with other POPs for *in situ* remediation of soil • SCISOR technology currently in development to be combined with advanced soil washing; lab tests indicate >99% removal of PFOS from soil • Formation of PFAS degradation products must be investigated	Nadim et al. (2006) Tsitonaki et al. (2010) Watts and Teel (2006) Hori et al. (2008) Lee et al. (2012) Hawley et al. (2012) Pancras et al. (2013)

TABLE 17.2
Cost Data for Presented Soil Remediation Technologies

Technology	Relevant Cost Data
Encapsulation	Dependent on depth of contamination and physical characteristics of site[a]
Groundwater pump-and-treat	$200,000–$900,000 for design and installation cost $1–$100/1000 gal (3785 L) of groundwater[a]
Soil flushing	$25–$250/yd^3 ($33–$327/m^3)[a]
Soil washing (including excavation)	$75–$170/ton (1000 kg) of soil[a]
Immobilization	Dependent on choice of adsorbent:[b] • $5.77/kg Filtrasorb 400 GAC with 0.002 mmol PFOS/g adsorption capacity • $14.60/kg Hydraffin CC8x30 GAC with 0.07 mmol PFOS/g adsorption capacity • $26.00/kg MatCARE with 0.09 mmol PFOS/g adsorption capacity • $88.00/kg Amberlite IRA 400 resin with 0.42 mmol PFOS/g adsorption capacity • $218.00/kg Amberlite XAD4 resin with 1.59 mmol PFOS/g adsorption capacity
Vitrification	$270/yd^3 treated (electrical) ($353/m^3)[a]
Persulfate oxidation	$1.65/lb ($3.64/kg) of sodium persulfate[c] Average total cost (capital plus operating) for one site is around $450,000[c]

Note: Dollar amounts are given in US dollars.
[a] From Khan, F.I. et al., *Journal of Environmental Management*, 71: 95–122, 2004.
[b] From Das, P. et al., *Water, Air, & Soil Pollution*, 224: 1–14, 2013.
[c] From Rosansky, S., and A. Dindal, Cost and Performance Report for Persulfate Treatability Studies, 2010, retrieved May 2014 from http://clu-in.org/download/techfocus/chemox/ISCO-Persulfate-C&P.pdf.

periodically implemented. No information about the size of the site was provided; however, the initial estimated cost for site cleanup, including soil and groundwater, was around $20 million.

17.3.1.1.2 Groundwater Pump-and-Treat

Groundwater treatment for contaminated soil leachate is generally a reactionary measure that has been proven effective for preventing PFAS release to sources of drinking water (Wilhelm 2008; 3M 2010). The theory could, however, be adapted as a precautionary measure to pump contaminated water from the soil before it has a chance to leach into surrounding areas, thereby facilitating direct site remediation.

Paterson et al. (2008) conducted a field study of the effectiveness of groundwater extraction as a remediation technique for PFAS-contaminated soil at an AFFF-impacted site. The field trial involved the installation of an *in situ* vacuum-enhanced multiphase extraction (VEMPE) system at a former FFTA in British Columbia, Canada. Four rotary claw pumps removed groundwater and vapor from the subsurface. The extracted groundwater was first pumped through an oil–water separator and air stripping unit, and then treated in tanks via granular activated carbon (GAC) filtration and released back to the soil. Although the GAC filtration had an efficiency of greater than 99% PFAS removal from the groundwater, it was estimated that only 0.1% of the total PFAS in the soil were removed over a period of 2 years. This demonstration emphasizes a strong affinity of PFAS to the soil. It also suggests that treating groundwater alone, without trying to remediate the soil, is not feasible.

17.3.1.1.3 Soil Flushing

Even if groundwater extraction by itself may not be an effective remediation strategy, the release of PFAS from soil may be enhanced using a flushing agent. Soil flushing is an *in situ* process that involves the injection of a flushing solution into the ground for extracting contaminants. Soil flushing has been used for the removal of persistent organic pollutants (POPs) and heavy metals from soil in the past with apparent success, the main advantage being that large quantities of soil can be treated *in situ*, making excavation and transport unnecessary (Jawitz et al. 2000; Svab et al. 2009; Zheng et al. 2012). The efficiency of this technique was evaluated at the Canadian Forces Base Borden for the removal of trichloroethylene from soil (Mulligan et al. 2001). It was estimated that with the addition of 1% surfactant to water, the remediation of the site would take approximately 21 pore volumes of solution over 4 years. In comparison, a simple pump-and-treat groundwater system was estimated to require 2000 pore volumes and more than 100 times longer time for decontamination.

Surfactants contain both hydrophobic and hydrophilic portions, making them particularly effective soil flushing additives for the mobilization of organic contaminants. That said, many PFAS, including PFOS and PFOA, are themselves surfactants, which can make their behavior more difficult to predict. Other common soil flushing additives include organic and inorganic acids and bases, solvents such as methanol or ethanol, oxidizing/reducing agents, and chelating agents. To make the process cost-effective, the flushing solution should be recyclable and nontoxic. In an ideal scenario, once the contaminant has been flushed out of the soil and collected, the solution could be decontaminated using water treatment technologies such as adsorption, ion exchangers, or filtration, and then reused for subsequent flushes.

Pan et al. (2009) investigated the effect of cationic and anionic surfactants on the sorption and desorption of PFOS on sediments. The cationic surfactant tested was cetyltrimethylammonium bromide (CTAB), while the anionic surfactant was sodium dodecylbenzene sulfonate (SDBS). Batch sorption and desorption tests were conducted with increasing concentrations of surfactant, and the effect was observed as the changing concentration of PFOS in the water, measured through high-performance liquid chromatography (HPLC). CTAB seemed to effectively immobilize PFOS at any surfactant concentration. The researchers justified the observed effect of the cationic surfactant by postulating that CTAB may have adsorbed to the sediment first, thus exposing its hydrophobic tails, which can act as a sink for PFOS. In comparison, the anionic surfactant SDBS had a concentration-dependent effect. At concentrations of SDBS lower than 4.34 mg/L, sorption of PFOS to sediments was increased; however, at concentrations of 21.7 and 43.4 mg/L, sorption of PFOS was significantly decreased. This reduction became more pronounced with increasing concentrations of SDBS.

In terms of soil flushing strategies, SDBS can act as an effective surfactant for the increased mobilization of PFOS from sediments. The results by Pan et al. (2009) should encourage further study into a variety of surfactants and their interactions with a larger suite of PFAS. For example, in a study by Guelfo and Higgins (2013) on the interactions between PFAS and co-contaminants in AFFFs, it was found that sorption was decreased in the presence of anionic surfactant sodium dodecyl sulfate for low concentrations of PFOS, perfluorononanoic acid (PFNA), and PFDA. At the same time, it appeared to have no effect on long-chain PFAAs and actually increased sorption for PFBA, perfluoropentanoic acid (PFPeA), PFHxA, perfluoroheptanoic acid (PFHpA), and PFBS. Other co-contaminants tested were amphoteric surfactant N,N-dimethyldodecylamine, N-oxide, and trichloroethylene as non-aqueous-phase liquid. All PFAAs showed increased sorption or no effect in the presence of the surfactant. The addition of trichloroethylene caused decreased PFOS sorption to soil at low concentrations of PFOS (1 µg/L) but increased sorption at a higher PFOS concentration (500 µg/L). For their experiments with reverse osmosis, Tang et al. (2006) found that an organic solvent, isopropyl alcohol, was capable of increasing the solubility of PFOS but also caused a decrease in the flux of a reverse osmosis membrane. This finding is important as reverse osmosis may be used to treat the leachate generated in soil flushing. The membrane separation process rejects the contaminants and some other dissolved compounds and collects them as a concentrate. AFFF-impacted sites may contain any number of co-contaminants, which means that the latter may affect the effectiveness of leachate treatment.

Soil flushing only transfers the contaminated components of the soil into the produced wastewater (leachate). This aqueous stream must be further treated to remove or destroy the contaminants and ideally reuse the water for the soil flushing again and again. Common second-stage treatments include adsorption, liquid–liquid extraction, membrane separation, advanced oxidation/reduction, and bioremediation (Hasegawa et al. 1997; Khan et al. 2004).

17.3.1.1.4 Soil Washing

The physicochemical principle of soil washing is the same as that for soil flushing— it is an extraction process of PFAS with other contaminants using aqueous solutions. However, the implementation of soil washing is different from that of soil flushing. It involves the excavation of contaminated soils and their *ex situ* treatment. Soil washing involves the separation of coarse components of the soil, such as sand and gravel, from the finer components, such as clay and silt (Khan et al. 2004), and is usually the first step of an overall treatment train. Coarse fractions normally have low contamination levels and do not require further treatment. They can often be used as backfill on the site. Fine fractions are more heavily contaminated, as hydrophobic organic contaminants tend to adhere to smaller soil particles.

This separation process reduces the amount of contaminated soil that requires treatment. Similar to soil flushing, the efficiency of soil washing can be enhanced by the addition of a solvent or surfactant to the wash water. Soil washing has been recognized as effective at removing hydrophobic organic compounds, such as aromatic compounds, polyaromatic hydrocarbons (PAHs), polychlorinated biphenyls (PCBs), herbicides, and pesticides (Chu and Chan 2003; Chu and Kwan 2003).

The wash water and the fine soil fractions produced in the process must be further treated. Second-stage water treatment options are similar to those listed in Section 3.1.1.3. In addition, the following technologies can be used to deal with the fine fractions: landfilling, solid-phase extraction, solidification, incineration, and bioremediation (Hasegawa et al. 1997; Khan et al. 2004).

17.3.1.1.5 Solvent Extraction

The principles of small-scale extraction developed for the analytical application may have the potential for a full-scale application for soil remediation purposes. Analysis of PFAS in soil requires an extraction step before liquid chromatography and mass spectrometry can be performed. Several separation methods exist involving organic solvent extraction. In a method used by Sun et al. (2011) and Powley et al. (2005), sludge or soil samples are flushed three times, with methanol, shaking, sonication, and centrifugation between each flush. In the pressurized solvent extraction procedure used by Llorca et al. (2011), sludge samples were treated to two cycles of methanol at 70°C and 100 bar of pressure. Schröder et al. (2003) experimented with several organic solvents to determine the solvent or combination of solvents with the greatest PFAS extraction potential. Sludge samples were subjected to pressurized solvent extraction at 150°C and 143 bar using solvents, namely, ethyl acetate, dimethylformamide, pyridine, tert-butyl methyl ether, 1,4-dioxane, or tetrahydrofuran. The most effective extraction procedure involved sequential flushing with a mixture of ethyl acetate and dimethylformamide, and then methanol modified with phosphoric acid, although the extraction was almost as effective without dimethylformamide.

Chen et al. (2012) developed a method for supercritical fluid extraction of PFOS and PFOA from solid matrices, using supercritical carbon dioxide (Sc-CO_2), methanol, and nitric acid (HNO_3). This method takes advantage of the readily achievable critical point of carbon dioxide (i.e., temperature of 31.1°C and pressure of 74 bar). Concentrated HNO_3 is used to suppress the polarity of the PFOS and PFOA molecules, therefore increasing their solubility in supercritical carbon dioxide. Methanol, a highly polar solvent, was used to modify supercritical carbon dioxide by increasing its polarity, which also ultimately leads to increased solubility of PFOS and PFOA. The contaminated samples were treated in pressurized cells at 20.3 MPa (203 bar) and 50°C with 16 M HNO_3 under dynamic and static extraction conditions. The extract was collected in vials filled with methanol and prepared for liquid chromatography/mass spectrometry (LC/MS) analysis. Extraction took between 40 and 180 minutes, and the expired gaseous carbon dioxide was released to the atmosphere. The extraction efficiency from sand (after double extraction) was 77% for PFOA and 59% for PFOS. The method was also tested against paper and fabric matrices, with efficiencies of 100% and 80% for PFOA and PFOS, respectively. The researchers postulate that this method would be a rapid, low-cost, environmentally friendly solution to PFOA and PFOS remediation; however, no field tests were reported for this technology to verify this claim.

17.3.1.2 Immobilization of PFAS in Soil

If removal of PFAS from soil is not feasible, an alternative method, such as immobilization, can be used. The immobilizing agent could be an adsorbent, similar or identical to those used in water treatment, or a liquid chemical solution. The idea behind this technique would be to prevent leaching of PFAS from the initial source of pollution, and evidence suggests that it could even reduce the bioavailability of the contaminants in the soil.

Pan et al. (2009) calculated the thermodynamic index of irreversibility (TII) for PFOS to sediments in the presence of cationic surfactant CTAB to quantify the degree of sorption irreversibility caused by CTAB; a value of 0 represents a highly reversible system and 1 represents irreversible sorption. A value of 1 was approached for concentrations of 18.1 and 36.1 mg/L CTAB, indicating its potential use for site immobilization of PFOS. The cationic surfactant could be delivered to the source of pollution using *in situ* percolation or injection. Due to the possibility of biodegradation of the cationic surfactant, this remediation approach may provide just a temporary measure rather than a long-term solution.

Aly (2016) investigated the sorption of six PFAS onto Ottawa sand in the presence of four cationic coagulants: polyaluminum chloride, polyamine, polydiallyldimethyl ammonium chloride (polyDADMAC), and a tannin-based cationic polymer. These coagulants reportedly enhanced the sorption of PFAS onto the sand, with polyDADMAC performing the most effectively. Aly suggested that the addition of cationic coagulants may facilitate the remediation of groundwater. One can further speculate that this method might be useful for soil remediation to immobilize PFAS within the soil matrix. It is, however, an open question whether coagulation-enhanced sorption can be adopted for site remediation. Further studies are required to answer this question.

Brändli et al. (2008) tested whether activated carbon could immobilize PAHs in soil. Powdered activated carbon (PAC) and GAC were mixed into the soil, and the dissolved PAH concentration was measured in a soil–water slurry system. It took 42 days for the GAC-amended soils to reach equilibrium versus 31 days for PAC-amended soils. The PAC was found to be more effective than GAC at reducing the freely dissolved PAH concentration. It was observed that only 2% PAC was needed to reduce the aqueous PAH concentration by 99%.

A follow-up to the above study was conducted by Hale et al. (2012) in order to test the results in field conditions. This was the first *in situ* field study of activated carbon amendment in soil. Three plots were constructed in the soil measuring 25 m^3: one reference area, one mixed with 2% (weight) PAC, and another mixed with 2% (weight) GAC. After 17 months, the free aqueous PAH concentration from the soil was reduced by 93% with PAC and 84% with GAC. In another follow-up study by Jakob et al. (2012), the environmental implications of activated carbon amendment were evaluated, specifically observing the effects on plant growth and BAFs. PAC was found to inhibit the growth of plants and reduced their BAF by an average of 53%, while GAC had a positive effect on plant growth and reduced their BAF by an average of 46%. PAC was toxic to earthworms, as evidenced by their observed significant weight loss. The toxicity of GAC for earthworms was inconclusive. The BAF for earthworms was reduced by an average of 72% and 47% for PAC and GAC, respectively.

Zimmerman et al. (2004) tested the efficiency of using activated carbon in reducing the release of POPs from marine sediments. They found that upon addition of activated carbon to the sediments, the available aqueous concentrations of PCBs and PAHs were reduced by 92% and 84%, respectively. While activated carbon has been shown to be effective at binding PFAS in water (Ochoa-Herrera and Sierra-Alvarez 2008), no studies were found specifically testing whether activated carbon can efficiently immobilize PFAS in soil. Other documented adsorbents for PFAS in water include anion exchange resins (Q. Yu et al. 2009; Senevirathna et al. 2010), chars and ash (Chen et al. 2011), and carbon nanotubes (Li et al. 2011).

A modified clay adsorbent for the immobilization of PFOS in contaminated soils was reported by Das et al. (2013). The adsorbent, a palygorskite-based clay modified with oleylamine, is in the advanced stages of commercialization under the trade name MatCARE™. Soils from four different sites impacted by AFFFs were measured for their initial PFOS concentration and then used for treatability studies, with 10 g of MatCARE being applied for every 100 g of soil. Without treatment with MatCARE, the average PFOS release from the control soils after 1 year at 25°C was 8.14%, and 9.48% at 37°C. With the application of MatCARE, there was no detectable PFOS release after 1 year at 25°C, and only 0.15% at 37°C. The experiment was also performed with spiked soils, with the average PFOS release decreasing from around 18% (control soil) to 0.5% (treated soil). It was found that for most soils, total immobilization of PFOS occurred with the application of 100 g MatCARE/kg of soil. Application of MatCARE has also been reported to be successful in field trials.

Cost comparisons between MatCARE and other common adsorbents were included in the chapter. The cost of MatCARE is listed as $26.0/kg, compared with $14.6/kg for a commercial GAC (Hydraffin CC8x30) and $88.0/kg for an anion exchange resin (Amberlite IRA 400). Although MatCARE is more expensive than GAC, it should also be noted that MatCARE has a greater adsorption capacity at 0.09 mmol PFOS/g, compared with the adsorption capacity for Hydraffin GAC at 0.07 mmol PFOS/g. MatCARE has been successfully used for water remediation, with more than 1 million L of AFFF-contaminated water being reduced to PFOS concentrations of less than 5 ppb (CRC CARE 2013). MatCARE has been advertised to reduce the cost of cleaning an AFFF-contaminated water site from $300,000 to $30,000.

RemBind and RemBind Plus technologies are other sorbent media with the potential to remove organics from impacted sites. As discussed in Chapter 18, RemBind products consist of a proprietary mix of aluminum hydroxide, activated carbon, organic matter, and kaolinite that have demonstrated the capacity to bind to hydrophobic compounds, including total petroleum hydrocarbons (TPHs), PAHs, PCBs, and PFAS. See Chapter 18 for more specifics on this immobilization technique.

17.3.2 Destructive Methods

17.3.2.1 Vitrification and Incineration

Also referred to as the molten glass process, vitrification involves using a powerful energy source to essentially melt soil, causing pyrolysis or immobilization of virtually all contaminants (Khan et al. 2004). Temperatures required for vitrification range between 1600°C and 2000°C. An advantage to this process in the context of PFAS is the lack of by-products generated because all organic contaminants are destroyed. There are several processes available to reach vitrification temperatures: electrical, thermal, and plasma. The electrical process is *in situ* and involves construction of a zone surrounded by graphite electrodes inserted in the ground that pass energy through the soil. The thermal process is *ex situ* and is generally carried out in a rotary kiln. Plasma processes are only necessary when temperatures of up to 5000°C are required.

Electrical or thermal processes could be used to target the combustion of PFAS on site without using more extreme temperatures. Typical municipal incinerators operate at temperatures of around 600°C–1000°C for approximately 2 seconds (Yamada et al. 2005; Vecitis et al. 2009). These temperatures are generally sufficient for the incineration of PFOS and PFOA. Yamada et al. (2005) tested whether fluorotelomer-treated textiles and paper, after being destroyed under municipal incinerator conditions, would form PFOA as a degradation product. There was no PFOA detected in the samples after incineration. Even if PFOA was temporarily formed during incineration, it would also have been destroyed in the process, meaning that degradation of precursors during incineration would not be a significant source of PFOA to the environment.

17.3.2.2 *In Situ* Chemical Oxidation

In situ chemical oxidation using peroxydisulfate ($S_2O_8^{2-}$), often simply referred to as persulfate, is a promising technique for soil remediation. The process has previously been used for the remediation of pollutants such as chlorinated ethenes and benzenes; oxygenates; benzene, toluene, ethylbenzene, and xylene (BTEX); and PAHs from soil (Nadim et al. 2006; Tsitonaki et al. 2010). Persulfate can first be delivered to the soil subsurface in an inactive form, and then activated once contact with the contaminated zone has occurred. An example of a delivery system would be a network of high-pressure injection points, followed by mixing of the soil with a backhoe (Tsitonaki et al. 2010). The activated persulfate radical can be produced through ultraviolet exposure, heat, high pH (alkaline conditions), hydrogen peroxide, and a variety of transition metals (Watts and Teel 2006). Activation by heat can be accomplished using steam injection. Heat can also be incorporated into the soil using six-phase soil heating, which involves the use of electricity to pass current through the soil, resulting in thermal energy production (Heine and Steckler 1999). Six-phase soil heating can be used as a stand-alone soil remediation technique for more volatile compounds, as it encourages their release from the soil matrices. Nadim et al. (2006) used a complex of divalent iron Fe (II) with ethylenediaminetetraacetic acid (EDTA) to activate persulfate for the degradation of PAHs. The addition of a chelating agent effectively kept the Fe (II) in solution even at neutral pH.

Hori et al. (2008) found that persulfate oxidation, activated by hot water, was effective at degrading PFOA to below the detection limit after 6 hours of treatment at 80°C. Formation of CO_2, fluoride ions, and shorter-chain PFCAs was indicative of the degradation. Lee et al. (2012) were able to degrade PFOA in aqueous solution with an efficiency of 80.5% at temperatures as low as 20°C by changing the persulfate dose and pH of the system. Complete degradation was observed at a pH of 2.5 after 72 hours at 40°C and after 215 hours at 30°C.

Hawley et al. (2012) reported the experiments assessing the best method for activation of persulfate oxidation with the goal of groundwater and soil treatment. Activators that were able to degrade PFOS by more than 97.5% included Fenton's reagent, peroxide-activated persulfate, and heat-activated persulfate. Strong reducing agents, such as sodium dithionite and sodium hypophosphate, were also tested, but only partial degradation of PFOS was observed. Follow-up tests were conducted to develop a mixture of common oxidants that could function under less extreme conditions (such as a field setting) to efficiently degrade PFOS and PFOA in groundwater and soil. The method, called smart combination *in situ* oxidation/reduction (SCISOR), reportedly reduced the amount of PFOS in soil by 60% after one contact phase; however, there was no information provided on the formation of degradation products that were most likely generated during the process.

The idea of using advanced soil washing in tandem with SCISOR was presented by Pancras et al. (2013), with the results of lab tests indicating that the technique was able to remove >99% of PFOS from contaminated soil. At the same time, Place and Field (2012) expressed concern regarding the use of *in situ* chemical oxidation for the remediation of AFFF-impacted sites, as advanced oxidation techniques have been known to encourage PFSA and PFCA formation from the more complex precursors present in AFFFs.

17.4 CONCLUSIONS

The analysis of published literature on the remediation of PFAS-contaminated soil reveals a significant gap between what are currently available as commercial technologies and what can be effectively used to remediate PFAS-contaminated sites. Two main reasons for this are as follows:

1. PFAS are emerging contaminants. Not enough efforts have been made to develop and/or optimize the technologies specifically for PFAS.
2. PFAS are difficult to treat due to their high chemical stability and a broad variety of physicochemical properties within the PFAS family. This, in many instances, forces the site owners to resort to excavation and landfilling, which is "low-tech" and merely removes bulk soil from one location to another for storage.

On a positive side, there are many methods not specifically developed for PFAS treatment that seem to be applicable to deal with these contaminants, subject to technology optimization and successful demonstration. For example, technologies such as soil washing and soil flushing have been successfully used on a large scale for a number of contaminants; however, their application for PFAS has been limited to bench-scale studies so that their economic efficiency is yet to be assessed. On the other hand, pump-and-treat of the contaminated groundwater is a well-developed approach to soil remediation, and this includes PFAS contamination cases. This approach, however, would not be effective in cases of low-level PFAS contamination, as the remediation would take a very long time.

In terms of treatment effectiveness, it was reported only for PFOS and/or PFOA in vast majority of studies. While these two compounds are often present in soil at much higher levels than other PFAS, it is known that AFFFs are composed of hundreds of diverse fluorinated surfactants. It is therefore desirable to know their fate, as well to fully assess the overall treatment effectiveness for the entire suite of PFAS.

Some emerging remediation methods, such as immobilization and *in situ* persulfate oxidation, show promise and may become commercial technologies. In particular, complete mineralization of PFAS in the case of oxidation would be highly desirable—it would mean that PFAS no longer exist and do not require contaminant storage, monitoring, and other long-term commitments.

As a final statement, the demand for effective remedial solutions for PFAS-contaminated soil is very high. Fortunately, there are existing and emerging methods that have the potential to become commercial technologies. The technology development and transition to end users can be accelerated through relevant applied research (e.g., evaluation of the treatment effectiveness for both long- and short-chain PFAS, identification of effective soil flushing and washing solutions, and investigation of degradation by-products in the persulfate oxidation process), as well as through successful field demonstrations of best-performing technologies.

REFERENCES

3M Company. Construction Completion Report—D1 and D2 Areas—Cottage Grove Site, Cottage Grove, Minnesota. 2010. Retrieved May 2014 from http://www.pca.state.mn.us/index.php/view-document.html?gid=14280.

Aly, Y.H. Enhanced Sorption of Perfluoro-Alkyl Substances (PFAS) onto Ottawa Sand. Master thesis, University of Minnesota, 2016.

Beach, S.A., J.L. Newsted, K. Coady, and J.P. Giesy. Ecotoxicological Evaluation of Perfluorooctanesulfonate (PFOS). *Reviews of Environmental Contamination and Toxicology*, 186: 133–174, 2006.

Becker, A.M., S. Gerstmann, and H. Frank. Perfluorooctane Surfactants in Waste Waters, the Major Source of River Pollution. *Chemosphere*, 72: 115–121, 2008.

Benskin, J.P., M.G. Ikonomou, F.A. Gobas, T.H. Begley, M.B. Woudneh, and J.R. Cosgrove. Biodegradation of N-Ethyl Perfluorooctane Sulfonamido Ethanol (EtFOSE) and EtFOSE-Based Phosphate Diester (SAmPAP Diester) in Marine Sediments. *Environmental Science & Technology*, 47: 1381–1389, 2013.

Brändli, R.C., T. Hartnik, T. Henriksen, and G. Cornelissen. Sorption of Native Polyaromatic Hydrocarbons (PAH) to Black Carbon and Amended Activated Carbon in Soil. *Chemosphere*, 73: 1805–1810, 2008.

Brignole, A., J. Porch, H. Krueger, and R. Van Hoven. PFOS: A Toxicity Test to Determine the Effects of the Test Substance on Seedling Emergence of Seven Species of Plants. In Toxicity to Terrestrial Plants. EPA Docket AR226-1369. Easton, MD: Wildlife International, 2003.

Chen, H., W. Liao, B. Wu, H. Nian, K. Chiu, and H. Yak. Removing Perfluorooctane Sulfonate and Perfluorooctanoic Acid from Solid Matrices, Paper, Fabrics, and Sand by Mineral Acid Suppression and Supercritical Carbon Dioxide Extraction. *Chemosphere*, 89: 179–184, 2012.

Chen, X., X. Xia, X. Wang, J. Qiao, and H. Chen. A Comparative Study on Sorption of Perfluorooctane Sulfonate (PFOS) by Chars, Ash and Carbon Nanotubes. *Chemosphere*, 83: 1313–1319, 2011.

Chu, W., and K. Chan. The Mechanism of the Surfactant-Aided Soil Washing System for Hydrophobic and Partial Hydrophobic Organics. *Science and the Total Environment*, 307: 83–92, 2003.

Chu, W., and C. Kwan. Remediation of Contaminated Soil by a Solvent/Surfactant System. *Chemosphere*, 53: 9–15, 2003.

CRC CARE (Cooperative Research Centre for Contamination Assessment and Remediation of the Environment). Fighting Fire-Fighting Foam. 2013. Retrieved May 2014 from http://www.crccare.com/case-study/fighting-fire-fighting-foam.

Das, P., M. Megharaj, and R. Naidu. Perfluorooctane Sulfonate Release Pattern from Soils of Fire Training Areas in Australia and Its Bioaccumulation Potential in the Earthworm *Eisenia fetida*. *Environmental Science & Pollution Research*, 22: 8902–8910, 2015.

Das, P., V. Kambala, M. Mallavarapu, and R. Naidu. Remediation of Perfluorooctane Sulfonate in Contaminated Soils by Modified Clay Adsorbent—A Risk-Based Approach. *Water, Air, & Soil Pollution*, 224: 1–14, 2013.

Eschauzier, C., P. de Voogt, H. Brauch, and F.T. Lange. Polyfluorinated Chemicals in European Surface Waters, Ground-and Drinking Waters. In Knepper, T.P., and F.T. Lange (eds.), *The Handbook of Environmental Chemistry*, Berlin: Springer-Verlag, 2012, Vol. 17, pp. 73–102.

Frömel, T., and T.P. Knepper. Biodegradation of Fluorinated Alkyl Substances. *Reviews of Environmental Contamination and Toxicology*, 208: 161–177, 2010.

Gellrich, V., and T.P. Knepper. Sorption and Leaching Behavior of Perfluorinated Compounds in Soil. In Knepper, T.P., and F.T. Lange (eds.), *Polyfluorinated Chemicals and Transformation Products*. Berlin: Springer-Verlag, 2012, Vol. 17, pp. 63–72.

Gellrich, V., T. Stahl, and T. Knepper. Behavior of Perfluorinated Compounds in Soils during Leaching Experiments. *Chemosphere*, 87: 1052–1056, 2012.

Guelfo, J.L., and C.P. Higgins. Subsurface Transport Potential of Perfluoroalkyl Acids at Aqueous Film-Forming Foam (AFFF)-Impacted Sites. *Environmental Science & Technology*, 47: 4164–4171, 2013.

Haglund, P., and U. Olofsson. Miljöövervakning av Slam-Redovisning av Resultat från 2009 års Provtagning. Report to the Swedish EPA (Naturvårdsverket), 2009.

Hale, S.E., M. Elmquist, R. Brändli, T. Hartnik, L. Jakob, T. Henriksen, D. Werner, and G. Cornelissen. Activated Carbon Amendment to Sequester PAHs in Contaminated Soil: A Lysimeter Field Trial. *Chemosphere*, 87: 177–184, 2012.

Hasegawa, M.A., D.A. Sabatini, and J.H. Harwell. Liquid-Liquid Extraction for Surfactant-Contaminant Separation and Surfactant Reuse. *Journal of Environmental Engineering*, 123: 691–697, 1997.

Hawley, E.L., T. Pancras, and J. Burdick. Remediation Technologies for Perfluorinated Compounds (PFCs), Including Perfluorooctane Sulfonate (PFOS) and Perfluorooctanoic Acid (PFOA). ARCADIS White Paper, 2012.

Heine, K.S., and D.J. Steckler. Augmenting In-Situ Remediation by Soil Vapor Extraction with Six-Phase Soil Heating. *Remediation Journal*, 9: 65–72, 1999.

Higgins, C.P., J.A. Field, C.S. Criddle, and R.G. Luthy. Quantitative Determination of Perfluorochemicals in Sediments and Domestic Sludge. *Environmental Science & Technology*, 39: 3946–3956, 2005.

Higgins, C.P., and R.G. Luthy. Sorption of Perfluorinated Surfactants on Sediments. *Environmental Science & Technology*, 40: 7251–7256, 2006.

Hori, H., Y. Nagaoka, M. Murayama, and S. Kutsuna. Efficient Decomposition of Perfluorocarboxylic Acids and Alternative Fluorochemical Surfactants in Hot Water. *Environmental Science & Technology*, 42: 7438–7443, 2008.

Houtz, E.F., C.P. Higgins, J.A. Field, and D.L. Sedlak. Persistence of Perfluoroalkyl Acid Precursors in AFFF-Impacted Groundwater and Soil. *Environmental Science & Technology*, 47: 8187–8195, 2013.

Jakob, L., T. Hartnik, T. Henriksen, M. Elmquist, R.C. Brändli, S.E. Hale, and G. Cornelissen. PAH-Sequestration Capacity of Granular and Powder Activated Carbon Amendments in Soil, and Their Effects on Earthworms and Plants. *Chemosphere*, 88: 699–705, 2012.

Jawitz, J.W., R.K. Sillan, M.D. Annable, P.S.C. Rao, and K. Warner. In-Situ Alcohol Flushing of a DNAPL Source Zone at a Dry Cleaner Site. *Environmental Science & Technology*, 34: 3722–3729, 2000.

Jensen, J., S. Toft Ingvertsen, and J. Magid. Risk Evaluation of Five Groups of Persistent Organic Contaminants in Sewage Sludge. Environmental Project No. 1406. Danish Ministry of the Environment, 2012.

Kallenborn, R., U. Berger, U. Järnberg, M. Dam, O. Glesne, B. Hedlund, J. Hirvi, A. Lundgren, B.B. Mogensen, and A.S. Sigurdsson. Perfluorinated Alkylated Substances (PFAS) in the Nordic Environment. Nordic Council of Ministers, Copenhagen, 2004.

Khan, F.I., T. Husain, and R. Hejazi. An Overview and Analysis of Site Remediation Technologies. *Journal of Environmental Management*, 71: 95–122, 2004.

Lange, C. The Aerobic Biodegradation of N-EtFOSE Alcohol by the Microbial Activity Present in Municipal Wastewater Treatment Sludge. For 3M Company, 2000.

Lasier, P.J., J.W. Washington, S.M. Hassan, and T.M. Jenkins. Perfluorinated Chemicals in Surface Waters and Sediments from Northwest Georgia, USA, and Their Bioaccumulation in *Lumbriculus variegatus*. *Environmental Toxicology & Chemistry*, 30: 2194–2201, 2011.

Lee, Y., S. Lo, J. Kuo, and Y. Lin. Persulfate Oxidation of Perfluorooctanoic Acid under the Temperatures of 20–40°C. *Chemical Engineering Journal*, 198: 27–32, 2012.

Li, X., S. Chen, X. Quan, and Y. Zhang. Enhanced Adsorption of PFOA and PFOS on Multiwalled Carbon Nanotubes under Electrochemical Assistance. *Environmental Science & Technology*, 45: 8498–8505, 2011.

Lindstrom, A.B., M.J. Strynar, L. McMillan, and R. McMahen. Measurement of Perfluorinated Compounds in Matrices That Are Important for Environmental Distribution and Human Exposure. Federal Contaminated Sites Action Plan at Management of Perfluorinated Compounds at Federal Contaminated Sites, Ottawa, ON, 2014.

Liu, J., and S. Mejia Avendaño. Microbial Degradation of Polyfluoroalkyl Chemicals in the Environment: A Review. *Environment International*, 61: 98–114, 2013.

Llorca, M., M. Farré, Y. Picó, and D. Barceló. Analysis of Perfluorinated Compounds in Sewage Sludge by Pressurized Solvent Extraction Followed by Liquid Chromatography-Mass Spectroscopy. *Journal of Chromatography A*, 1218: 4840–4846, 2011.

Loganathan, B.G., K.S. Sajwan, E. Sinclair, K. Senthil Kumar, and K. Kannan. Perfluoroalkyl Sulfonates and Perfluorocarboxylates in Two Wastewater Treatment Facilities in Kentucky and Georgia. *Water Research*, 41: 4611–4620, 2007.

Moody, C.A., and J.A. Field. Perfluorinated Surfactants and the Environmental Implications of Their Use in Fire-Fighting Foams. *Environmental Science & Technology*, 34: 3864–3870, 2000.

MPCA (Minnesota Pollution Control Agency). 3M Cottage Grove Site: Proposed Cleanup Plan for PFCs. 2009. Retrieved May 2014 from http://www.pca.state.mn.us/index.php/view-document.html?gid=2895.

Mulligan, C., R. Yong, and B. Gibbs. Surfactant-Enhanced Remediation of Contaminated Soil: A Review. *Engineering Geology*, 60, 371–380, 2001.

Murakami, M., H. Shinohara, and H. Takada. Evaluation of Wastewater and Street Runoff as Sources of Perfluorinated Surfactants (PFSs). *Chemosphere*, 74: 487–493, 2009.

Nadim, F., K. Huang, and A.M. Dahmani. Remediation of Soil and Groundwater Contaminated with PAH Using Heat and Fe (II)-EDTA Catalyzed Persulfate Oxidation. *Water, Air, & Soil Pollution: Focus*, 6: 227–232, 2006.

Nunes, L.M., Y. Zhu, T. Stigter, J.P. Monteiro, and M. Teixeira. Environmental Impacts on Soil and Groundwater at Airports: Origin, Contaminants of Concern and Environmental Risks. *Journal of Environmental Monitoring*, 13: 3026–3039, 2011.

Ochoa-Herrera, V., and R. Sierra-Alvarez. Removal of Perfluorinated Surfactants by Sorption onto Granular Activated Carbon, Zeolite and Sludge. *Chemosphere*, 72: 1588–1593, 2008.

Organisation for Economic Co-operation and Development (OECD). Lists of PFOS, PFAS, PFOA, PFCA, Related Compounds and Chemicals That May Degrade to PFCA. Joint Meeting of the Chemicals Committee and the Working Party on Chemicals, Pesticides and Biotechnology, Environment Directorate Series on Risk Management No. 21, 2007.

Pan, G., C. Jia, D. Zhao, C. You, H. Chen, and G. Jiang. Effect of Cationic and Anionic Surfactants on the Sorption and Desorption of Perfluorooctane Sulfonate (PFOS) on Natural Sediments. *Environmental Pollution*, 157: 325–330, 2009.

Pancras, T., W. Plaisier, and A. Barbier. Challenges of PFOS Remediation. Presented at the Proceedings of the AquaConSoil Conference, Barcelona, 2013.

Paterson, L., T.S. Kennedy, and D. Sweeney. Remediation of Perfluorinated Alkyl Compounds at a Former Fire Fighting Training Area. Presented at the Proceedings of the Remediation Technologies Symposium, Banff, AB, 2008.

Place, B.J., and J.A. Field. Identification of Novel Fluorochemicals in Aqueous Film-Forming Foams Used by the US Military. *Environmental Science & Technology*, 46: 7120–7127, 2012.

Powley, C.R., S.W. George, T.W. Ryan, and R.C. Buck. Matrix Effect-Free Analytical Methods for Determination of Perfluorinated Carboxylic Acids in Environmental Matrixes. *Analytical Chemistry*, 77: 6353–6358, 2005.

Prevedouros, K., I.T. Cousins, R.C. Buck, and S.H. Korzeniowski. Sources, Fate and Transport of Perfluorocarboxylates. *Environmental Science & Technology*, 40: 32–44, 2006.

Rhoads, K.R., E.M. Janssen, R.G. Luthy, and C.S. Criddle. Aerobic Biotransformation and Fate of N-Ethyl Perfluorooctane Sulfonamidoethanol (N-EtFOSE) in Activated Sludge. *Environmental Science & Technology*, 42: 2873–2878, 2008.

Rosansky, S., and A. Dindal. Cost and Performance Report for Persulfate Treatability Studies. 2010. Retrieved May 2014 from http://clu-in.org/download/techfocus/chemox /ISCO-Persulfate-C&P.pdf.

Sáez, M., P. de Voogt, and J.R. Parsons. Persistence of Perfluoroalkylated Substances in Closed Bottle Tests with Municipal Sewage Sludge. *Environmental Science & Pollution Research*, 15: 472–477, 2008.

Schröder, H.F. Determination of Fluorinated Surfactants and Their Metabolites in Sewage Sludge Samples by Liquid Chromatography with Mass Spectrometry and Tandem Mass Spectrometry after Pressurised Liquid Extraction and Separation on Fluorine-Modified Reversed-Phase Sorbents. *Journal of Chromatography A*, 1020: 131–151, 2003.

Schultz, M.M., D.F. Barofsky, and J.A. Field. Quantitative Determination of Fluorinated Alkyl Substances by Large-Volume-Injection Liquid Chromatography Tandem Mass Spectrometry Characterization of Municipal Wastewaters. *Environmental Science & Technology*, 40: 289–295, 2006.

Senevirathna, S., S. Tanaka, S. Fujii, C. Kunacheva, H. Harada, B. Shivakoti, and R. Okamoto. A Comparative Study of Adsorption of Perfluorooctane Sulfonate (PFOS) onto Granular Activated Carbon, Ion-Exchange Polymers and Non-Ion-Exchange Polymers. *Chemosphere*, 80: 647–651, 2010.

Sepulvado, J.G., A.C. Blaine, L.S. Hundal, and C.P. Higgins. Occurrence and Fate of Perfluorochemicals in Soil Following the Land Application of Municipal Biosolids. *Environmental Science & Technology*, 45: 8106–8112, 2011.

SFT (Norwegian Pollution Control Authority). Screening of Polyfluorinated Organic Compounds at Four Fire Training Facilities in Norway. STF Report No. TA-2444. STF, 2008.

Sinclair, E., and K. Kannan. Mass Loading and Fate of Perfluoroalkyl Surfactants in Wastewater Treatment Plants. *Environmental Science & Technology*, 40: 1408–1414, 2006.

Stahl, T., J. Heyn, H. Thiele, J. Hüther, K. Failing, S. Georgii, and H. Brunn. Carryover of Perfluorooctanoic Acid (PFOA) and Perfluorooctane Sulfonate (PFOS) from Soil to Plants. *Archives of Environmental Contamination and Toxicology*, 57: 289–298, 2009.

Sun, H., A.C. Gerecke, W. Giger, and A.C. Alder. Long-Chain Perfluorinated Chemicals in Digested Sewage Sludges in Switzerland. *Environmental Pollution*, 159: 654–662, 2011.

Svab, M., M. Kubal, M. Müllerova, and R. Raschman. Soil Flushing by Surfactant Solution: Pilot-Scale Demonstration of Complete Technology. *Journal of Hazardous Materials*, 163: 410–417, 2009.

Tang, C.Y., Q.S. Fu, A. Robertson, C.S. Criddle, and J.O. Leckie. Use of Reverse Osmosis Membranes to Remove Perfluorooctane Sulfonate (PFOS) from Semiconductor Wastewater. *Environmental Science & Technology*, 40: 7343–7349, 2006.

Törneman, N. Remedial Methods and Strategies for PFCs. Presented at the 4th Joint Nordic Meeting on Remediation of Contaminated Sites (NORDROCS 2012), Oslo, Norway, 2012. http://nordrocs.org/wp-content/uploads/2012/09/Session-VI-torsdag-1-Torneman-short-paper.pdf.

Tsitonaki, A., B. Petri, M. Crimi, H. Mosbæk, R.L. Siegrist, and P.L. Bjerg. In Situ Chemical Oxidation of Contaminated Soil and Groundwater Using Persulfate: A Review. *Critical Reviews in Environmental Science & Technology*, 40: 55–91, 2010.

USGPO (U.S. Government Publishing Office). Electronic Code of Federal Regulations. Title 40: Protection of Environment, Part 503—Standards for the Use or Disposal of Sewage Sludge. 2014. Retrieved May 2014 from http://www.ecfr.gov/cgi-bin/retrieveECFR?gp=&SID=fc4dfb52783c2e7b63a4b7e918a2e189&mc=true&n=pt40.30.503&r=PART&ty=HTML.

Vecitis, C.D., H. Park, J. Cheng, B.T. Mader, and M.R. Hoffmann. Treatment Technologies for Aqueous Perfluorooctanesulfonate (PFOS) and Perfluorooctanoate (PFOA). *Frontiers of Environmental Science & Engineering in China*, 3: 129–151, 2009.

Wang, N., B. Szostek, R.C. Buck, P.W. Folsom, L.M. Sulecki, V. Capka, W.R. Berti, and J.T. Gannon. Fluorotelomer Alcohol Biodegradation Direct Evidence That Perfluorinated Carbon Chains Breakdown. *Environmental Science & Technology*, 39: 7516–7528, 2005a.

Wang, N., B. Szostek, P.W. Folsom, L.M. Sulecki, V. Capka, R.C. Buck, W.R. Berti, and J.T. Gannon. Aerobic Biotransformation of 14C-Labeled 8-2 Telomer B Alcohol by Activated Sludge from a Domestic Sewage Treatment Plant. *Environmental Science & Technology*, 39: 531–538, 2005b.

Wang, N., B. Szostek, R.C. Buck, P.W. Folsom, L.M. Sulecki, and J.T. Gannon. 8-2 Fluorotelomer Alcohol Aerobic Soil Biodegradation: Pathways, Metabolites, and Metabolite Yields. *Chemosphere*, 75: 1089–1096, 2009.

Washington, J.W., H. Yoo, J.J. Ellington, T.M. Jenkins, and E.L. Libelo, Concentrations, Distribution, and Persistence of Perfluoroalkylates in Sludge-Applied Soils near Decatur, Alabama, USA. *Environmental Science & Technology*, 44: 8390–8396, 2010.

Watts, R.J., and A.L. Teel. Treatment of Contaminated Soils and Groundwater Using ISCO. *Practice Periodical of Hazardous, Toxic, and Radioactive Waste Management*, 10: 2–9, 2006.

Wilhelm, M., M. Kraft, K. Rauchfuss, and J. Hölzer. Assessment and Management of the First German Case of a Contamination with Perfluorinated Compounds (PFC) in the Region Sauerland, North Rhine-Westphalia. *Journal of Toxicology and Environmental Health A*, 71: 725–733, 2008.

Yamada, T., P.H. Taylor, R.C. Buck, M.A. Kaiser, and R.J. Giraud. Thermal Degradation of Fluorotelomer Treated Articles and Related Materials. *Chemosphere*, 61: 974–984, 2005.

Yao, Y., K. Volchek, C.E. Brown, P. Lambert, L. Gamble, K. Kitagawa, J. Anglesey, A. Dugas, M. Punt, and D. Velicogna. Remedial Options for PFC Contaminated Sites: A Review. In *Proceedings of the 36th AMOP Technical Seminar on Environmental Contamination and Response*, Halifax, Canada, 2013, pp. 654–672.

Yoo, H., J.W. Washington, T.M. Jenkins, and J.J. Ellington. Quantitative Determination of Perfluorochemicals and Fluorotelomer Alcohols in Plants from Biosolid-Amended Fields Using LC/MS/MS and GC/MS. *Environmental Science & Technology*, 45: 7985–7990, 2011.

You, C., C. Jia, and G. Pan. Effect of Salinity and Sediment Characteristics on the Sorption and Desorption of Perfluorooctane Sulfonate at Sediment-Water Interface. *Environmental Pollution*, 158: 1343–1347, 2010.

Yu, J., J. Hu, S. Tanaka, and S. Fujii. Perfluorooctane Sulfonate (PFOS) and Perfluorooctanoic Acid (PFOA) in Sewage Treatment Plants. *Water Research*, 43: 2399–2408, 2009.

Yu, Q., R. Zhang, S. Deng, J. Huang, and G. Yu. Sorption of Perfluorooctane Sulfonate and Perfluorooctanoate on Activated Carbons and Resin: Kinetic and Isotherm Study. *Water Research*, 43: 1150–1158, 2009.

Zheng, G., A. Selvam, and J.W. Wong. Enhanced Solubilization and Desorption of Organochlorine Pesticides (OCPs) from Soil by Oil-Swollen Micelles Formed with a Nonionic Surfactant. *Environmental Science & Technology*, 46: 12062–12068, 2012.

Zimmerman, J.R., U. Ghosh, R.N. Millward, T.S. Bridges, and R.G. Luthy. Addition of Carbon Sorbents to Reduce PCB and PAH Bioavailability in Marine Sediments: Physicochemical Tests. *Environmental Science & Technology*, 38: 5458–5464, 2004.

18 Soil and Groundwater PFAS Remediation
Two Technology Examples

David F. Alden, John Archibald,
Gary M. Birk, and Richard J. Stewart

CONTENTS

18.1 Introduction 405
18.2 PerfluorAd.. 405
18.3 RemBind 408
    18.3.1 Sorption of PFAS from Contaminated Water to Activated
           Carbon-Based Materials 409
    18.3.2 Binding PFAS in Soil.. 410
18.4 Conclusions... 414
References.. 414

18.1 INTRODUCTION

Per- and polyfluoroalkyl substances (PFAS) are extraordinarily stable and difficult to volatilize. Today's soil and groundwater remediation methods primarily include soil incineration, excavation to landfill, and groundwater extraction with PFAS adsorption onto activated carbon or resins.

Two promising and potentially synergistic solutions have recently been brought to the market: PerfluorAd and RemBind. The former is a surfactant-based technology that targets PFAS in water, while the latter is a mixture of compounds that capitalize on numerous binding mechanisms, addressing a variety of PFAS molecular structures in soils. This chapter discusses each of these technologies for soil and groundwater remediation. Since these techniques do not include the destruction of the PFAS molecules, regulatory agencies may require disposal for the precipitates and sorptive materials, albeit at greatly reduced volumes.

18.2 PERFLUORAD

In 2008, Cornelsen Umwelttechnologie GmbH, based in Essen, Germany, in collaboration with the Fraunhofer Institute for Environmental, Safety, and Energy Technology (UMSICHT), started developing a water purification process specifically aimed at removing PFAS. Their research and development has led to three patent applications. This innovative technology for purifying water contaminated with

PFAS (designated the PerfluorAd procedure) has reached a mature and marketable stage of development.

PerfluorAd agent is a liquid ingredient tailored to address dissolved PFAS in water. Waters impacted with PFAS are dosed with PerfluorAd, which flocculates with PFAS to form insoluble flake structures that precipitate. A particulate filtration process separates the newly formed adducts from the water (Figure 18.1). The process can remove more than 97% of PFAS, as exemplified in the case study below, and is therefore commonly paired with a downstream granular activated carbon (GAC) filter as a final polishing measure. As the PerfluorAd procedure removes such a large percentage of PFAS, GAC loadings, usually the highest cost consumable in PFAS treatment, drastically decrease, thus reducing overall treatment costs.

As opposed to generating chemical PFAS destruction reactions, the PerfluorAd agent is metered into a stirred reactor in order to react with PFAS and form flocs. This in turn reduces the risk of generating transformation by-products. That said, the eventual precipitate disposal may involve a PFAS destruction method.

Since PerfluorAd stirred reactors only target PFAS, additional treatment steps may be implemented to minimize downstream GAC requirements, particularly in the case of complex or mixed-contaminant waters. In the case of waters contaminated with industrial wastes or aqueous film-forming foam (AFFF), for example, a series of stirred reactors could instigate a process in which a variety of added reagents, such as flocculants, defoamers, or powdered adsorbents, address total dissolved solids (TDSs), froth, pH, and other issues. Other pretreatment steps common to groundwater remediation include the removal of iron, manganese, volatiles, and other constituents.

PerfluorAd was used to treat PFAS-impacted groundwater at the Nuremberg Airport in southeastern Germany near a groundwater source for drinking water. The local government evaluated the PerfluorAd process in a pilot test conducted between September and November 2014. In this case study, the groundwater contained high levels of PFAS (averaging 357 µg/L); petroleum hydrocarbons above the C5 range; ethylbenzene, xylene, and traces of polycyclic aromatic hydrocarbons (PAHs); and chlorinated solvents such as cis-1,2-dichloroethene, trichloroethylene, and vinyl chloride. As PerfluorAd provided the most cost-effective solution among the various technologies piloted, the city went on to commission a full-scale treatment system in September 2015.

Once the remotely controlled, full-scale PerfluorAd system was installed and optimized at a 9 gallon/minute flow rate (2 m^3/h), it attained PFAS removal efficiencies

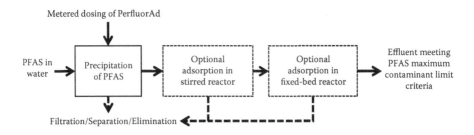

FIGURE 18.1 Principle of the PerfluorAd procedure.

of 86.7%–98.1% prior to GAC polishing, as shown in Figure 18.2. With PFAS influent concentrations ranging from 390 to 480 μg/L, the system's remediation target was both a maximum PFOS concentration of 0.23 μg/L and 0.3 μg/L for the sum of perfluorooctane sulfonic acid (PFOS), perfluorooctanoic acid (PFOA), and perfluorohexanesulfonic acid (PFHxS) (Figure 18.3). By evaluating post-GAC PFAS breakthrough under field operations, the contractor did not require a GAC change-out until 18 months later.

The PerfluorAd procedure is economically and environmentally attractive because the amount of active ingredient used, as well as the amount of waste generated, is relatively low. Incineration, a relatively expensive procedure, is the predominantly known method that destroys PFAS in waste, that is, floc and GAC. A PerfluorAd treatment system thus reduces waste and subsequent incineration costs. For instance, the observed floc production rate was 0.42 pounds per thousand gallons of treated water, with a total PFAS influent concentration of 500 μg/L, which is equivalent to about a 55-gallon drum of waste per million gallons of water treated. By comparison, the GAC-only treatment prior to using PerfluorAd had a design loading capacity rate of 0.03% by weight, which generated 30 times more waste (Cornelsen 2014).

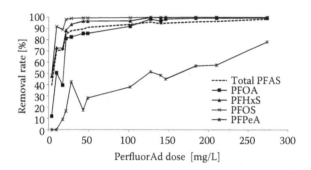

FIGURE 18.2 Nuremberg field pilot test (2014). PFAS removal efficiencies are plotted as a function of PerfluorAd liquid used during a 2 m³/h influent continuous flow rate.

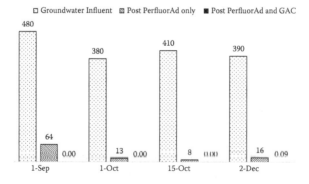

FIGURE 18.3 Nuremberg pilot test total PFAS sampling results. Concentrations in are in micrograms per liter.

18.3 REMBIND

The Commonwealth Scientific and Industrial Research Organization (CSIRO) in Australia, in collaboration with Ziltek Pty Ltd., developed and patented an adsorption reagent called RemBind. It consists of a coarse powder that, upon contact, binds organic chemical contaminants found in soil and water. The product constituents include aluminum hydroxide, activated carbon, organic matter, and kaolinite, resulting in a high-surface-area media capable of binding hydrophobic chemicals, as well as positively and negatively charged molecules capable of adsorbing ionic chemicals. Remediation practitioners have used RemBind to address PAHs, total petroleum hydrocarbons (TPHs), PFAS, polychlorinated biphenyls (PCBs), trichloroethylene, and some amphoteric metals, such as arsenic and chromium. Current research using infrared and x-ray spectroscopy aims to confirm and better understand the proposed binding mechanisms of RemBind ingredients.

Aluminum hydroxide ($Al(OH)_3$) is naturally found in gibbsite and is the predominant ingredient in RemBind. Its basic structure consists of stacked sheets of linked octahedrons of aluminum hydroxide. These are in turn composed of (+)3 charged aluminum ions bonded to six (–)1 charged hydroxides coordinated octahedrally. The amorphous or noncrystalline nature of aluminum hydroxide provides a large surface area in which additional interactions with chemicals are possible. A point zero charge (PZC) value of 9.1 (Sverjensky and Fukushi 2006) indicates that the mineral is positively charged in most soils, allowing it to bind anionic contaminants through electrostatic interactions. This property is particularly enhanced at low pH conditions as hydroxyl groups (^-OH) protonate to become $-OHH^+$.

RemBind also contains activated carbon, which has a large hydrophobic area. It binds molecules through hydrophobic interactions such as Van der Waals forces and other physicochemical bonds involving the presence of both micro- and macropore types within its structure. Studies featuring granular and powdered activated carbon with laboratory-spiked and environmentally observed concentrations of selected PFAS show that although the method is somewhat effective for the larger PFAS compounds, the carbon has limitations for the shorter-chain molecules.

The presence of kaolinite in RemBind allows sorption of cationic PFAS. Kaolinite is an aluminosilicate clay mineral comprising one octahedral sheet of aluminum-oxygen linked to one tetrahedral sheet of silica-oxygen bonds. It contains net a negative charge due to substitution of lower-valence atoms into the aluminum octahedral position (such as Mg^{+2}) or the Si tetrahedral positions (e.g., Al^{+3}). At low pH, it may also contain some positive charges located at the unpaired oxygen atoms at the edge of the crystal lattice, but these would be small in magnitude when compared with the large positive charge afforded by the aluminum hydroxide in RemBind.

The organic matter component of RemBind contains a highly carbonaceous surface capable of hydrophobic interactions with molecules, as occurs with activated carbon. The difference relies on the material's negatively charged surface due to the presence of organic acids, particularly carboxylate and phenolic groups. RemBind thus provides additional binding sites for cationic constituents of concern.

The predominant PFAS compounds in industrial formulations are anionic due to their acidic functional groups. Common examples are PFOS, PFOA, and the fluorotelomer sulfonates (6:2 and 8:2 FtS). The aluminum hydroxide component of RemBind forms electrostatic interactions with the anionic PFAS molecules, binding the functional head of the PFAS molecules. The activated carbon and organic matter components of RemBind bind to the lipophilic tail of PFAS molecules through hydrophobic interactions and Van der Waals forces.

Some PFAS compounds are cationic due to the presence of amine groups or zwitterionic (containing both positive and negative charges) due to the presence of two functional groups, as in the case of perfluorosulfamide amino carboxylate (C4–C8 PFSaAm). The kaolinite and organic matter components of RemBind give it the ability to bind a portion of these positively charged compounds through electrostatic interactions, regardless of the C–F backbone chain length. The activated carbon component binds the PFAS cations through hydrophobic interactions, as it does for longer-chain PFAS anions.

Most of the longer-chain PFAS molecules ($n \geq 5$, where n = the number of C–F units in the chemical backbone) bind to activated carbon. However, the smaller-chain compounds ($n < 5$) do not bind as effectively to activated carbon because of their greater hydrophilicity (Appleman et al. 2014; Xiao et al. 2017). Furthermore, there are fewer contact points with the carbon surface, and Van der Waals forces are relatively weak. Some of the shorter-chain PFAS compounds, such as PFBA ($n = 3$) and PFBS ($n = 4$), bind poorly to activated carbon. In RemBind, these compounds are bound effectively through electrostatic interactions provided by the aluminum hydroxide.

RemBind's composition creates stronger bonds as the environment becomes acidic. Nevertheless, the binding of anionic compounds may be inadequate in extreme conditions where alkaline-induced chemical oxidation has been employed, or at a site with high levels of calcium carbonate. This is due to the nature of aluminum hydroxide, which exhibits a less positively charged surface as pH approaches 9.1 and becomes neutral or even negatively charged as the pH climbs above 9.1.

18.3.1 SORPTION OF PFAS FROM CONTAMINATED WATER TO ACTIVATED CARBON-BASED MATERIALS

Although several studies have shown that activated carbon can remove PFAS (Ochoa-Herrera and Sierra-Alvarez 2008; Yu and Hu 2011; Rattanaoudom et al. 2012; Yao et al. 2014), gross hydrocarbon content is known to blind this type of media, preventing adsorption. RemBind has been used to remediate both PFAS-laden water (Huttman et al. 2014) and hydrocarbon-contaminated soils and has been reported to maintain its effectiveness. This section describes a case study where an independent water treatment company, BECA, conducted lab-scale trials for an Australian government airport authority in order to assess the locally produced, activated-carbon-based RemBind product as a treatment solution specifically targeting PFAS in water contaminated with hydrocarbons from firefighting training activities as part of aviation rescue (Marquez 2016).

The trials were conducted in two stages, and the first set of experiments served to develop parameters for the second. In Stage 1, the adsorption of PFAS onto

RemBind was determined to be sufficiently fast acting for scaled-up operations with an expected 2-hour mixing period. This first set of experiments used 19 different aliquots with varying volumes of wastewater from the Adelaide hot fire training ground (HFTG), dilution water, RemBind, and various mixing times.

HFTG wastewater was also used as stock sample water for Stage 2, where wastewater aliquots were dosed with RemBind and allowed to mix for 2 hours prior to filtration with a polyethersulfone membrane. The variables tested include RemBind dose, PFAS concentration, and differing amounts of Jet A-1 fuel spikes into the raw stock. Gross hydrocarbons were shown to significantly reduce the adsorption of PFAS onto the RemBind.

Equilibrium concentrations (i.e., the residual solute concentrations after the RemBind dose) were fitted to Freundlich model adsorption isotherms, and results suggested that RemBind had an adsorption capacity of 2560 µg/g for PFOS and 6.76 µg/g for PFOA. As the stock water contained various hydrocarbons, both organic and solids, these capacities were specific to the water used and any use of them for scale-up should allow a contingency factor. The chosen full-scale minimum design adsorption capacity for this project was 20–25 kg of RemBind to reduce PFOS levels in a 20,000 L tank of wastewater with similar characteristics as the stock water to 0.3 µg/L, the safe drinking water limit proposed by the Minnesota Department of Health.

This study led to the conclusion that any process using RemBind as PFAS treatment should include a pretreatment stage to reduce total recoverable hydrocarbons to levels less than 10–15 mg/L. In doing so, RemBind's effectiveness in removing PFAS compounds from raw wastewater was shown at addition rates of less than 0.1% by weight, which in turn has the potential to significantly reduce current full-scale operation waste disposal costs. Field pilot tests are underway using this methodology.

18.3.2 Binding PFAS in Soil

Soil stabilization is a risk mitigation technique that minimizes the potential of hazardous chemicals leaching out of contaminated media. The method is accomplished either by mixing soils with surface adsorption reagents or by adjusting its pH and reduction potential, or by a combination of both. Although the physical nature of the treated material may or may not be changed significantly, the chemical reactions that occur when reagents encounter contaminants render them stable, insoluble, or simply immobile. In addition to organophilic clays, activated carbon has been a useful additive to bond organic materials. Adsorption is a surface phenomenon in which a substance's atoms, ions, or molecules adhere to the surface area of an adsorbent such as activated carbon. Absorption is a different mechanism that also involves chemical uptake, but in this case the molecules actually penetrate or are dissolved by the absorbent. In other words, the concentration of the absorbed molecules is the same in the bulk and on the surface (Agarwal 2016). The sort of chemical interaction that may take place in absorption includes ion exchange involving metal oxyhydroxides, resulting in one chemical being replaced with another. Therefore, a mixture of reagents with several binding mechanisms may provide a robust stabilization agent for various chemical structures within the PFAS family of molecules.

One relevant case study is from an Australian government airport authority responsible for providing environmentally responsible services to the Australian aviation industry, which has in recent years undertaken trials of RemBind as an immobilizing agent for PFAS in impacted soils. Initial laboratory trials of the PFAS-impacted soil from aviation rescue firefighter (ARFF) sites were undertaken by an independent consultancy firm. These trials showed a 99% reduction in leachability or contaminant immobilization levels (Stewart 2017). The airport authority subsequently used this product in its future operations, resulting in more than 700 m^3 of PFAS-impacted soil from one site treated and sent to landfill for disposal.

Another example involves additional investigations in collaboration with the University of Queensland to assess the application of RemBind as an *in situ* treatment for PFAS-impacted soils. In one of the studies, two contaminated Australian airport field soils, referenced as HB and LT, received treatment with 25 wt% by weight RemBind (Table 18.1). This proprietary adsorbent showed effectiveness in reducing PFAS mobility and bioavailability in soils (Braunig et al. 2016). The study investigated whether the treatment and fixation of PFAS in soil will also decrease their bioavailability to worms (*Eisenia fetida*) and wheatgrass (*Elymus scaber*). A standard 28-day worm uptake study was performed to determine the PFAS uptake into worm tissue from untreated and treated soils. Uptake of PFAS from nontreated and treated soils into grasses was measured after a 10-week growing period. Results indicated that the bioavailability of PFAS to earthworms and wheatgrass decreased strongly for all compounds investigated with the application of the adsorbent (Figures 18.4 and 18.5). Uptake into worms is thus largely dominated by the freely available concentration of PFAS measured in leachate water. The study revealed that contaminated soils have a substantially reduced ability to act as an ongoing source, which in turn hinders further spread of PFAS into ground and surface waters.

TABLE 18.1
PFAA Concentrations (ng/g) in Untreated Soil Samples from Sites Referenced as HB and LT

Soil Concentration (ng/g)	HB	LT
Perfluorohexanoic acid (PFHA) C6	45	68
Perfluorooctanoic acid (PFOA) C8	14	55
Perfluorooctanoic sulfonate (PFHxS) C6	123	447
Perfluorooctane sulftonate (PFOS) C8	2,193	13,362

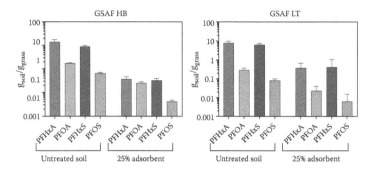

FIGURE 18.4 Grass PFAS accumulation from RemBind-treated and untreated soils, showing a 30-fold decrease in PFOS accumulation for HB soil. Note there is a higher accumulation of molecules with shorter carbon chains.

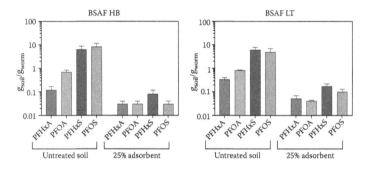

FIGURE 18.5 Earthworm PFAS accumulation from RemBind-treated and untreated soils. RemBind-treated soils (25 wt%) reduced PFAA accumulation in worms. PFHxS and PFOS, the longer-carbon-chain molecules, accumulated more than other PFAS.

RemBind is marketed as a contaminant immobilization agent that can address a source of groundwater PFAS contamination. The technique requires heavy equipment mixing and hydration of contaminated soils with a validated amount of RemBind, with an expected dose in the range of 2%–10% by weight. Validation tests of this *in situ* soil stabilization method via chemical fixation include toxicity characteristic leaching procedure (TCLP), synthetic precipitation leaching procedure, and the multiple extraction procedure (MEP) (US Environmental Protection Agency [USEPA] Method 1320). The MEP leachability test is designed to simulate the leaching that a waste will undergo from repetitive 1000-year acid rain events on an improperly designed sanitary landfill. A study of RemBind-treated soils shows that they could potentially be disposed of in sanitary landfills (Table 18.2) under this criterion. Lath et al. (2017) have produced additional work analyzing the stability of the RemBind binding reaction, particularly to PFOA, across a range of pH values and ionic strengths. Results show a relatively consistent stability across a range of pH values, leading to the conclusion that in addition to electrostatic interactions, ligand exchange may be playing a contaminant binding role.

TABLE 18.2
Summary of PFAS Leachability Characteristics before and after Treatment with RemBind of 14 Soils from Fire Training Grounds in Australia

			PFAS Concentrations in Soil and Soil Leachate[a]							
		RemBind Addition	Before RemBind Treatment			After Treatment				PFOS/ Total
Site	Soil Type	Rate % (w/w)	PFOS mg/kg	PFOS μg/L	PFOA μg/L	PFOS μg/L	PFOA μg/L	PFOS Reduction%	Passed USEPA Method 1320?	PFAS[b] %
1	Silty clay loam	5.0	0.74	34	0.65	0.29	<0.02	99.20	Yes	86
2	Silty clay	7.5	2.24	376	5.6	0.1	<0.02	99.97	Yes	67
3	Clay	5.0	20.9	695	11	1.5	<0.02	99.80	nt	99
4	Clayey silt (organic)	10.0	3.15	38	1.17	1.9	<0.02	95.00	Yes	99
5	Sand	5.0	1.26	1	1	<0.02	<0.02	>98.00	nt	99
6	Heavy clay	5.0	3.01	87	1.54	<0.02	<0.02	>99.98	nt	nt
7	Silty sand	5.0	7.25	190	0.05	0.05	<0.02	99.97	Yes	99
8	Clayey loam	5.0	1.45	62.5	2.7	<0.02	<0.02	>99.97	Yes	98
9	Clay/gravel (spill)	10.0	184	4780	222	3.52	0.21	99.90	Yes	nt
10	Clay/gravel	5.0	1.24	72	0.7	0.1	<0.01	99.90	nt	66
11	Heavy clay	5.0	0.67	36	1	0.1	<0.01	99.70	nt	40
12	Clay	5.0	0.78	43	0.6	0.1	<0.01	99.80	nt	57
13	Silty clay	2.5	nt	120	0.51	0.16	<0.02	99.90	nt	67
14	Silty clay	2.5	nt	184	1.84	0.2	<0.02	99.89	nt	67

Source: Stewart and McFarland, Immobilisation of per- and polyfluorinated alkyl substances (PFAS) in 14 soils from airport sites across Australia, poster presentation, Battelle International Bioremediation Symposium, Miami, FL, 2017.

Note: All treatments passed the New South Wales (NSW) landfill guidelines stipulating a soil leachate criterion of 50 μg/L for PFOS and PFHxS. ASLP = Australian standard leaching procedure; nt = not tested.

[a] As prepared by TCLP or ASLP at pH 5.
[b] Ratio of total PFOS/total PFAS extended suite (20 analytes) run by Australian Laboratory Services.

18.4 CONCLUSIONS

Growing international concern about the pervasiveness and toxicity of PFAS in the environment has led to the development and commercialization of technologies anticipating remediation needs. Two examples, one from a German research institution and another from an Australian one, are described in this chapter. The first is an in-line water treatment step that allows the extension of the lifetimes of GAC filters. Relatively high PFAS concentrations in water (500 µg/L) are addressed by adding a surfactant-based agent, PerfluorAd, that flocculates PFAS and allows their removal via sedimentation or filtration. The Nuremberg case study showed a PerflourAd system capable of reducing PFAS loading on GAC by more than 80%. The Australian-based technology, RemBind, has been specifically designed to address the limitation of inefficient binding of short-chain PFAS (C < 5) by GAC. RemBind uses a combination of aluminum hydroxide and activated carbon to bind with short- and long-chain PFAS through hydrophobic and electrostatic interactions, as well as ligand exchange mechanisms. Independent data suggest that the binding reaction between RemBind and PFAS is stable in the long term under environmentally relevant pH ranges. Mixing it with contaminated soils significantly reduces the bioaccumulation of PFAS in earthworm and plant systems, which may pave the way for the safe reuse, containment, or landfill disposal of RemBind-treated soil containing PFAS.

REFERENCES

Agarwal, S. 2016. Surface chemistry. In *Engineering Chemistry: Fundamentals and Applications*, chap. 12. Cambridge: Cambridge University Press, Technology & Engineering.

Appleman, T. D., Higgins, C. P., Quinones, O., Vanderford, B. J., Kolstad, C., Zeigler-Holady, J. C., and Dickenson, E. R. V. 2014. Treatment of poly- and perfluoroalkyl substances in US full-scale water treatment systems. *Water Research*, 51, 246–255.

Braunig, J., Baduel, C., and Mueller, J. F. 2016. Influence of a commercial adsorbent on the leaching behaviour and bioavailability of selected perfluoroalkyl acids (PFAAs) from soil impacted by AFFFs. Platform presentation, EmCon 2016, 20–23 September, Sydney, Australia.

Cornelsen, M. 2014. Sanierung von PFT-Schadensfälle—Sanierungstechnologien und Sanierungspraxis. http://tu-freiberg.de/sites/default/files/media/professur-fuer-bodenmechanik-6558/2014karlsruhe6.pdf (accessed December 21, 2017).

Huttman, S. et al. December 2014. Laboratory investigations on RemBind sorption capacity. Industry report. Unpublished data.

Lath, S., McLaughlin, M., Navarro, D., Losic, D., and Kumar, A. 2017. Adsorption of PFOA using graphene based materials. Platform presentation, Cleanup Conference, 10–14 September, Melbourne, Australia.

Marquez, N. 2016. A novel adsorption product for the treatment of PFAS in wastewater from airport fire-training grounds. Poster presentation, Battelle International Conference on the Remediation of Chlorinated and Recalcitrant Compounds, 22–26 May, Monterey, CA.

Ochoa-Herrera, V., and Sierra-Alvarez, R. 2008. Removal of perfluorinated surfactants by sorption onto granular activated carbon, zeolite and sludge. *Chemosphere*, 72(10), 1588–1593.

Rattanaoudom, R., Visvanathan, C., and Boontanon, S.K. 2012. Removal of concentrated PFOS and PFOA in synthetic industrial wastewater by powder activated carbon and hydrotalcite. *Journal of Water Sustainability*, 2(4), 245–258.

Stewart, R. 2017. Immobilization and safe disposal of aqueous film forming foam (AFFF) impacted soil in Australia. Platform presentation, Battelle International Bioremediation Symposium, 22–25 May, Miami, FL.

Stewart, R. and McFarland, R. 2017. Immobilisation of per- and polyfluorinated alkyl substances (PFAS) in 14 soils from airport sites across Australia. Poster presentation, Battelle International Bioremediation Symposium, 22–25 May, Miami, FL.

Sverjensky, D. A., and Fukushi, K. 2006. A predictive model (ETLM) for As(III) adsorption and surface speciation on oxides consistent with spectroscopic data. *Geochimica et Cosmochimica Acta*, 70(15), 3778–3802.

Xiao, X., Ulrich, B. A., Chen, B., and Higgins, C. P. 2017. Sorption of poly- and perfluoroalkyl substances (PFASs) relevant to aqueous film-forming foam (AFFF)-impacted groundwater by biochars and activated carbon. *Environmental Science and Technology*, 51, 6342–6351.

Yao, Y., Volchek, K., Brown, C. E., Robinson, A., and Obalm, T. 2014. Comparative study on adsorption of perfluorooctane sulfonate (PFOS) and perfluorooctanoate (PFOA) by different absorbents in water. *Water Science and Technology*, 70(12), 1983–1991.

Yu, J., and Hu, J. 2011. Adsorption of perfluorinated compounds onto activated carbon and activated sludge. *Journal of Environmental Engineering*, 137(10), 945–951.

19 Per- and Polyfluoroalkyl Substances in AFFF-Impacted Soil and Groundwater and Their Treatment Technologies

Bo Wang, Abinash Agrawal, and Marc A. Mills

CONTENTS

19.1 Background 417
19.2 Nomenclature, Classification, and Properties 419
19.3 PFASs in Groundwater and Soil Impacted by AFFF 427
19.4 Treatment Technologies for PFASs 430
    19.4.1 PFAS Removal by Adsorption 430
    19.4.2 PFAS Treatment by Advanced Oxidation Processes 431
    19.4.3 PFAS Treatment by Sonochemical Oxidation 433
    19.4.4 PFAS Treatment by *In Situ* Chemical Reduction 434
    19.4.5 Other Emerging Technologies 434
        19.4.5.1 Electrochemical Treatment 434
        19.4.5.2 Air-Sparged Hydrocyclone 435
        19.4.5.3 Nanomaterial Technology 435
19.5 Gaps and Future Work 437
References 437

19.1 BACKGROUND

In recent years, the presence of per- and polyfluoroalkyl substances (PFASs) in the environment has gained increased attention because they are reported to be widespread, highly persistent, and potentially bioaccumulative, and pose potential adverse effects to humans and wildlife (Moody et al. 2002; Prevedouros et al. 2006; Conder et al. 2008; Paul et al. 2009; Giesy et al. 2010). Since PFASs were manufactured starting in 1949, they have been widely used in numerous industrial, residential, and commercial applications as processing additives in fluoropolymer production and surfactants in consumer products (Buck et al. 2011; Lindstrom et al. 2011). Owing to their strong carbon-fluorine bonds and highly polar character structure,

PFASs exhibit unique physicochemical properties, including thermal and chemical stability, yet they are water and oil repellent and provide low surface tension (Kissa 2001; Kotthoff et al. 2015). PFAS-based products are widely used for nonstick cookware, textile stains, paper and packaging products, surface coatings, paint additives, carpet treatment, and aqueous film-forming foams (AFFFs) used in firefighting (Paul et al. 2009; Buck et al. 2011). Due to their widespread use in household and commercial products, together with their limited natural biodegradation potential, PFASs have been detected in human blood, biota, food, and water (Giesy et al. 2001; Austin et al. 2003; Olsen et al. 2008; Oakes et al. 2010).

PFASs are widely distributed in the aquatic system, including surface water, groundwater, wastewater, and tap water, due to their high solubility in water (Fujii et al. 2007). PFASs have also been reported in remote regions of the world, such as the Arctic and the North Pacific Oceans (Giesy et al. 2001). PFASs are released into the aquatic environment during production, along the supply chains, as a result of designed use, disposal of industrial and consumer products, and transformations in the environment (Ahrens 2011). The sources of PFASs into the aqueous environment can be grouped into the following categories: (1) direct emission from production, industrial or municipal wastewater treatment plants, landfill leachate, and rainfall runoff, and (2) indirect emission of PFASs resulting from transformation of their precursors, such as biotransformation in wildlife or humans (e.g., perfluorooctanoic acid [PFOA] forms from 8:2 fluorotelomer alcohol [8:2 FTOH]) and decomposition in the atmosphere (e.g., perfluoropentanoic acid [PFPeA] forms from perfluorobutane sulfonamidoethanol [NMeFBSE]) (Ahrens 2011; Ahrens and Bundschuh 2014). However, in comparison with the aquatic environment, only a relatively small proportion of PFASs (~5%) are directly released into the atmosphere (Prevedouros et al. 2006; Paul et al. 2009).

Despite recent restrictions and regulations on production, emission, and use of specific PFASs, their homologues and related precursor chemicals are still being produced that may degrade to PFASs and enter into the environment, and more specifically perfluorinated carboxylic acids (PFCAs) (Taniyasu et al. 2005; Prevedouros et al. 2006; Ahrens 2011; Ahrens and Bundschuh 2014). The presence of perfluorodecanoic acid (PFDA) and perfluoroundecanoic acid (PFUnDA) in Arctic ice may indicate atmospheric oxidation as a main mechanism for perfluorinated acid (PFA) formation that results in global transport far from the original sources (Young et al. 2007). Ellis et al. (2004) reported that a homologous series of PFCAs are generated from the degradation of FTOHs in the atmosphere. Further, Cai et al. (2012) reported PFASs from the North Pacific being transported to the Arctic Ocean via long-range transport by atmospheric and oceanic currents.

Originally developed in the 1960s, AFFFs are commercial formulations that include solvents, fluorocarbon-based surfactants, hydrocarbon-based surfactants, and stabilizers utilized for extinguishing hydrocarbon-fueled fires (Moody and Field 2000; Place and Field 2012). Fluorocarbon surfactants are the key components of AFFFs, which help to lower the surface tension of the air–water interface to prevent reignition (Moody and Field 2000). Fluorocarbon surfactants in AFFF formulations vary by manufacturer, production process, and year of production. For example, AFFFs sold by 3M were synthesized mainly by electrochemical fluorination (ECF), while a telomerization process was employed by Ansul and National Foam companies (Place and Field 2012).

AFFFs were often used extensively at military bases, firefighter training facilities, and commercial airports. The US Department of Defense (DOD) was responsible for up to 75% of the US market share (Moody and Field 2000). Repeated use of AFFFs results in the discharge of PFASs to wastewater treatment facilities or directly to surrounding surface water and groundwater without prior treatment. These operations have resulted in groundwater and soil contamination at microgram per liter to milligram per liter concentrations (Moody et al. 2003; Schultz et al. 2004; Houtz et al. 2013). Although the use of AFFFs at a few military firetraining sites was at least partially decommissioned between 1993 and 2000, fluorocarbon-based surfactants are still present in the groundwater at measurable levels after 4–10 years, or even longer, after discontinuing AFFF use (Moody and Field 1999; Moody et al. 2003; Baduel et al. 2015).

Due to their high persistence and bioaccumulative potential, it is important to understand the factors controlling the environmental behavior, partitioning and transport mechanisms, and fate of PFASs in AFFF-impacted soil and groundwater. Thus, this chapter provides a review of (1) the nomenclature and unique properties of PFASs frequently detected in the environment; (2) the occurrence and fate of these PFASs in contaminated soil and groundwater, especially from areas impacted by AFFFs, and their potential interactions with co-contaminants; and (3) treatment technologies for PFAS-contaminated groundwater and soil, especially resulting from firetraining operations. This review will help in a better understanding of the emissions, distribution, and behavior of specific PFASs in AFFF-impacted groundwater and soil or sediment, estimation of the extent of the potential environmental impact, and development of the appropriate remediation strategies for PFASs contained in AFFF agents.

19.2 NOMENCLATURE, CLASSIFICATION, AND PROPERTIES

PFASs comprise a diverse group of fluorinated organic chemicals that are oil and water repelling (hydrophobic group) and include a lipophobic or hydrophilic functional ionic head (Figure 19.1) (Kissa 2001; Arvaniti and Stasinakis 2015). Polyfluoroalkyl substances (PolyFASs) have one or more (not all) hydrogens substituted by fluorines, while perfluoroalkyl substances (PerFASs) are fully fluorinated

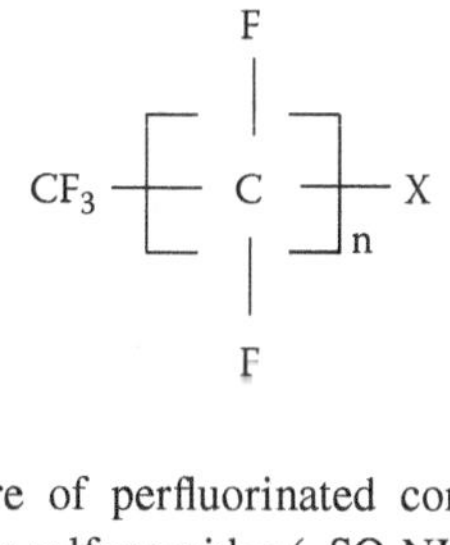

$$CF_3 - \left[\begin{array}{c} F \\ | \\ C \\ | \\ F \end{array} \right]_n - X$$

FIGURE 19.1 General structure of perfluorinated compounds. X represents sulfonate $(-SO_3^-)$, carboxylate ($-COO^-$), or sulfonamides ($-SO_2NH_2$). (Adapted from Arvaniti, O. S., and Stasinakis, A. S., *Science of the Total Environment* 2015, *524*, 81–92.)

with all hydrogens replaced by fluorines (Ahrens 2011; Ahrens and Bundschuh 2014; Arvaniti and Stasinakis 2015). PFASs can also be categorized into nonpolymers and polymers (Buck et al. 2011). Among PFASs, PFOA (F(CF$_2$)$_7$COOH) and perfluorooctane sulfonic acid (PFOS, F(CF$_2$)$_8$SO$_3$H) have received the most attention in recent years, as they are most frequently detected in environmental samples (Table 19.1) (USEPA 2012). PFOA and PFOS belong to PFCAs (F(CF$_2$)$_n$CO2−) and perfluoroalkane sulfonic acids (PFSAs, F(CF$_2$)$_n$SO3−), respectively. The structure and properties of the most studied PFASs in recent literature are summarized in Table 19.2, including nine PFCAs (C$_4$–C$_{12}$ PFCAs), three PFSAs (C$_4$, C$_6$, and C$_8$ PFSAs), and 4:2, 6:2, 8:2, and 10:2 FTOH and isomers, which are the focus of this review.

Among different kinds of PFASs, the fluorinated carbon chain varies in length and can be linear or branched isomers (Ahrens 2011). The Organisation for Economic Co-operation and Development (OECD) defines long-chain PFSAs as those with six or more perfluorinated carbons and long-chain PFCAs as those with seven or more perfluorinated carbons (Buck et al. 2011). Alsmeyer et al. (1994) defined PFAS isomers as linear isomers that have carbon bonded to one or two other carbons with homologues of varying −CH$_2$−, and branched isomers have carbon bonded to more than two carbon atoms with homologues of varying −C$_2$F$_4$−. PFASs normally exist as a mixture of linear and branched isomers in the environmental matrices. In Tokyo Bay, linear isomers contributed 53% of the total PFOA measured in surface water, while branched isomers were enriched (>50%) in PFOS profiles from the Mississippi River (Benskin et al. 2010).

Different patterns of PFASs result from two different manufacturing processes, including ECF and telomerization (Riddell et al. 2009; Benskin et al. 2010). ECF mainly produces mixtures of linear and branched isomers with even- and odd-numbered perfluorinated alkyl compounds (PFACs). In contrast, the telomerizaton

TABLE 19.1
Physical and Chemical Properties of PFOS and PFOA

Properties	PFOS	PFOA
Molecular weight (g/mol)	500.13	414
Molecular formula	C$_8$HF$_{17}$O$_3$S	C$_8$HF$_{15}$O$_2$
Melting point (°C)	>400	45–50
Boiling point (°C)	Not measurable	188–192
Vapor pressure at 20°C (mmHg)	2.48 × 10^{-6}	0.017
Water solubility at 25°C (mg/L)	550–570 (purified), 370 (freshwater), 25 (filtered seawater)	9.5 × 10^3 (purified)
pKa	−3.3	2.80
Organic-carbon partition coefficient (log K$_{oc}$)	2.57 (estimate based on anion and not the salt)	2.06

Source: Adapted from USEPA, Emerging contaminants fact sheet—Perfluorooctane sulfonate (PFOS) and perfluorooctanoic acid (PFOA), in Solid Waste and Emergency Response (5160P), Washington, DC: USEPA, 2012.

TABLE 19.2
Selected Environmentally Relevant Groups of PFASs

Compound Name	Structure	Solubility (mg/L)	Vapor Pressure (Pa)
	Perfluoroalkyl Acids (PFAAs)		
Perfluoroalkyl Carboxylic Acids (PFCAs)			
Perfluorobutanoic acid (PFBA) (C_4)			851 (25°C)
Perfluoropentanoic acid (PFPeA) (C_5)			
Perfluorohexanoic acid (PFHxA) (C_6)			

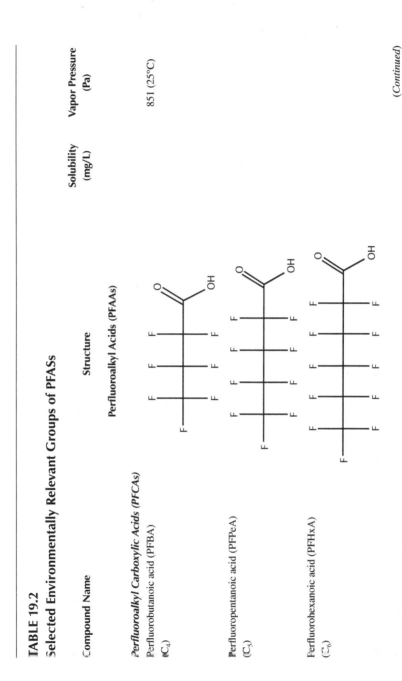

(Continued)

TABLE 19.2 (CONTINUED)
Selected Environmentally Relevant Groups of PFASs

Compound Name	Structure	Solubility (mg/L)	Vapor Pressure (Pa)
Perfluoroheptanoic acid (PFHpA) (C_7)		118,000 (21.6°C)	20.89 (25°C)
Perfluorooctanoic acid (PFOA) (C_8)		4340 (24.1°C)	4.17 (25°C)
Perfluorononanoic acid (PFNA) (C_9)			1.29 (25°C)
Perfluorodecan oic acid (PFDA) (C_{10})		260 (22.4°C)	0.23 (25°C)

(Continued)

TABLE 19.2 (CONTINUED)
Selected Environmentally Relevant Groups of PFASs

Compound Name	Structure	Solubility (mg/L)	Vapor Pressure (Pa)
Perfluoroundecanoic acid (PFUnDA) (C₁₁)		92.3 (22.9°C)	0.10 (25°C)
Perfluorododecanoic acid (PFDoA) (C₁₂)			0.008 (25°C)
Perfluoroalkyl Sulfonic Acids (PFSAs)			
Perfluorobutane sulfonic acid (PFBS) (C₄)		510	
Perfluorohexane sulfonic acid (PFHxS) (C₆)			

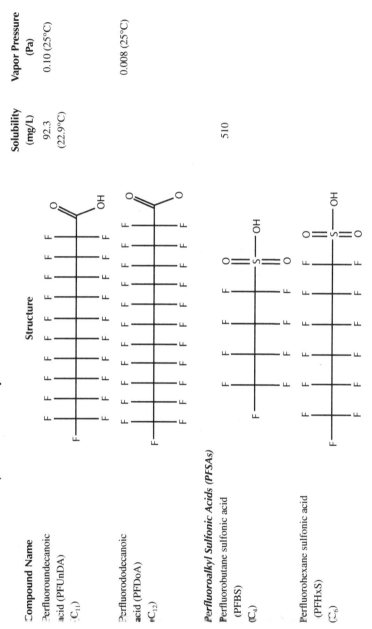

(Continued)

TABLE 19.2 (CONTINUED)
Selected Environmentally Relevant Groups of PFASs

Compound Name	Structure	Solubility (mg/L)	Vapor Pressure (Pa)
Perfluorooctane sulfonic acid (PFOS) (C_8)	[structure: perfluorooctane sulfonic acid]	570	3.31×10^{-4} (25°C)

Precursor Compounds

Fluorotelomer Alcohols (FTOHs)

Compound Name	Structure	Solubility (mg/L)	Vapor Pressure (Pa)
4:2 fluorotelomer alcohol (4:2 FTOH)	[structure: 4:2 FTOH with CH$_2$CH$_2$OH]	974 (22.5°C)	992 (25°C)
6:2 fluorotelomer alcohol (6:2 FTOH)	[structure: 6:2 FTOH with CH$_2$CH$_2$OH]	18.8 (22.5°C)	713 (25°C)

(Continued)

TABLE 19.2 (CONTINUED)
Selected Environmentally Relevant Groups of PFASs

Compound Name	Structure	Solubility (mg/L)	Vapor Pressure (Pa)
8:2 fluorotelomer alcohol (8:2 FTOH)	F-(CF2)8-CH2CH2OH	0.194 (22.3°C)	254 (25°C)
10:2 fluorotelomer alcohol (10:2 FTOH)	F-(CF2)10-CH2CH2OH	0.011	144 (25°C)
Perfluoroalkane Sulfonamides (FASAs)			
Perfluorooctane sulfonamide (FOSA)	F-(CF2)8-S(=O)2-NH2	499.14	2.56

(Continued)

TABLE 19.2 (CONTINUED)
Selected Environmentally Relevant Groups of PFASs

Compound Name	Structure	Solubility (mg/L)	Vapor Pressure (Pa)
N-Alkyl Perfluoroalkane Sulfonamidoethanols (FASEs)			
N-Methyl perfluorooctane sulfonamidoethanol (N-MeFOSE)		0.81 (25°C)	0.70 (25°C)
N-Ethyl perfluorooctane sulfonamidoethanol (N-EtFOSE)		0.89 (25°C)	0.35 (25°C)

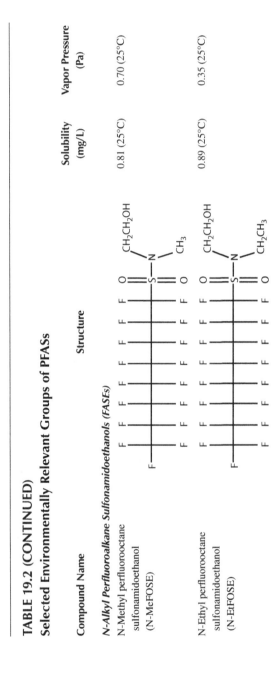

Source: Adapted from Rahman, M. F. et al., *Water Research* 2014, *50*, 318–340.

process produces only linear isomers of odd-numbered PFACs (partially fluorinated carbon chains) (Place and Field 2012). Therefore, the branching patterns in the PFASs of interest may be a reasonable tracer for their manufacturer source and can, in turn, help to determine the best contamination treatment technology (Kissa 1994; Moody and Field 1999).

Besides PolyFASs and PerFASs, PFASs also include precursor compounds, such as FTOHs, perfluoroalkane sulfonamides (FASAs), and perfluoroalkane sulfonamide ethanols (FOSEs), which are now receiving increasing attention due to their potential to degrade to persistent by-products, for example, long-chain PFCAs (Ahrens 2011; Ahrens and Bundschuh 2014). Based on the form in the environment, PFASs can be divided into neutral and ionic forms. The vastly different behaviors of various PFASs in the environment depend on their chain length, functional groups, branched versus linear structure, and so forth. The limited data on calculated pKa values (Table 19.2) indicate that PFCAs and PFSAs are strong acids and normally occur in anionic (negatively charged) form in the environment, while precursor compounds are generally neutral (Rahman et al. 2014).

The physicochemical properties of PFASs include negligible vapor pressure, high water solubility, and moderate sorption to solids (Prevedouros et al. 2006; Ahrens and Bundschuh 2014). These properties suggest that PFASs would be distributed ubiquitously in aquatic environments. Indeed, 40 surface water samples collected from the Tennessee River near a fluorochemical manufacturing facility have reported a range of PFOS (17 to 144 ng/L) and PFOA (<25 to 298 ng/L) concentrations (Moody et al. 2002). Groundwater samples collected from the military firetraining areas were found containing PFOS, PFOA, perfluorohexanoic acid (PFHxA), and perfluorohexane sulfonic acid (PFHxS) with a concentration range from 3 to 120 µg/L (Moody et al. 2003). AFFFs were frequently used in firetraining exercises at Wurtsmith Air Force Base (WAFB), Michigan, during the 1950s through 1993, which reported PFAS-contaminated groundwater and soil or sediment (Moody et al. 2003).

The ability of PFASs to partition into soil (indicated by log K_{oc} values) and their bioaccumulation potential increase with increasing fluorination of carbon chain length (Ahrens et al. 2010; Yeung and Mabury 2013). Short-chain PFSAs and PFCAs are more soluble, whereas long-chain PFSAs and PFCAs tend to bind to particles and have a substantial bioaccumulation potential in various wildlife species in the food web and human populations (Ahrens et al. 2010; Ahrens and Bundschuh 2014). For example, PFOS was found as the dominant PFAS in a subtropical food web in Hong Kong and accounted for 66%, 38%, 49%, and 58% of the total PFAS concentrations in worms, shrimp, fish, and livers of waterbirds, respectively (Loi et al. 2011). Ten known PFASs accounted for 30%–85% of extractable organic fluorine (EOF) in human blood samples from five cities in China (Yeung et al. 2008). The frequent detection of PFASs in a variety of wildlife and humans raises public concerns about their transformation mechanisms in the environment.

19.3 PFASs IN GROUNDWATER AND SOIL IMPACTED BY AFFF

Due to their unique physicochemical properties, the fate of PFASs in the environment is influenced by the interplay of various partitioning, transport, and degradation processes. Two primary pathways that allow PFASs to enter the groundwater are (1) production and use of materials containing PFAS components and (2) wastewater effluent discharge generated from industrial processes that use PFASs. The PFASs released into the soil can also migrate though the subsurface and enter the groundwater. Other sources of PFASs may also include atmospheric contributions and, to a lesser extent, contaminated surface waters that may impact groundwater sources (Loganathan et al. 2007; Ahrens 2011). Airborne PFASs have been receiving growing attention recently since certain PFAS precursors may biodegrade to persistent products such as PFOS and PFOA (Dinglasan et al. 2004). In published studies, certain precursors were shown to be more volatile, which, once released into the air, may become a significant source of more persistent PFASs.

PFAS adsorption to soil is regarded as an important process involving hydrophobic interaction between the perfluorinated carbon chain and organic matter. For example, PFOA and PFOS are mainly transported in the dissolved phase. As a result, the PFAS concentrations of suspended solids in marine environments are greater, because the concentrations of suspended solids are typically low (Ahrens et al. 2011). The sorption affinity of PFASs also increases with the hydrophobicity of the compound (Milinovic et al. 2015).

The concentrations of perfluorocarboxylate surfactants in groundwater collected from former firetraining facilities on Naval Air Station Fallon, Nevada, and Tyndall Air Force Base, Florida, ranged between 124 and 298 µg/L and 540 and 7090 µg/L, respectively (Moody and Field 1999). Another study revealed that total concentrations of PFCAs were lower, ranging from 3 to 120 µg/L at WAFB, Michigan (Moody et al. 2003). While the firetraining activities were discontinued at these three military bases during 1988–1993, PFASs are still present in contaminated groundwater and soil, with concentrations up to 7090 µg/L and detected 500 m downgradient from the site 7–10 years after AFFF use ceased (Moody and Field 1999; Moody et al. 2003). After 2–9 years of an accidental release of AFFFs at the Toronto airport, the concentrations of PFOS in fish and surface waters were 2- to 10-fold greater than those from upstream of the release (Awad et al. 2011). Baduel et al. (2015) reported firefighting training pads with significant concentrations of PFASs as important long-term sources for many decades (half-life, $t_{0.5}$ ~25 years for PFOS).

Due to elevated PFAS levels in soil and groundwater at AFFF-contaminated sites, it is important to understand the environmental behavior of PFASs at these sites. Since short-chain PFASs are more soluble than long-chain PFASs, the depth of their penetration in the soil decreases with the increasing number of carbons in the chain. Baduel et al. (2015) investigated the spatial and vertical distribution of selected PFASs at a firefighting training ground (FTG) in Australia that used AFFF from 1983 to 2010. The concentration of PFASs near the surface (0–0.5 cm deep) of a training pad ranged from 80 ng/g (PFDS) to 33,426 ng/g (PFOS), and the site is expected to remain a major source of PFASs in rainfall-induced runoff in future decades. Similarly, a former firefighting training area at Ellsworth Air Force Base, South Dakota, showed PFAS levels of up to 36,000 µg/kg in the surface soil close to the burn pit area (McGuire et al. 2014). At this site, the vertical profile (0–12 cm deep) indicates that PFOS and PFOA concentrations decrease significantly with depth. In contrast with shorter-chain PFASs (perfluorobutanoic acid [PFBA], PFPA, PFHxA, and perfluorobutane sulfonic acid [PFBS]) measured at nearly constant concentrations through the entire core, the longer-chain PFCAs (9–14 carbons) were detected mainly at the surface (Baduel et al. 2015).

The migration potential of PFASs in groundwater is largely controlled by their interactions with aquifer sediment and is affected by pH and ionic strength. In fact, these are important factors that can influence hydrophobic partitioning (K_d) and affect the sorption of PFASs to soil or sediment (Liu and Lee 2005). The sorption ability of anionic PFAS surfactants (e.g., perfluorocarboxylates, perfluorosulfonates, and perfluoroctyl sulfonamide acetic acids) to sediments can vary with chain length and organic carbon content of the sediment, f_{oc} (Higgins and Luthy 2006; Guelfo and Higgins 2013). The partitioning of PFASs to soil can increase with increasing the ionic strength (e.g., by adding [Ca^{2+}] in solution) and decreasing the pH (Chen et al. 2009; Yu et al. 2009; Wang and Shih 2011) of the groundwater.

Site characterization parameters, such as dissolved organic carbon (DOC), inorganic constituents, and the distribution of co-contaminants, can also affect the occurrence and distribution of PFASs. The mechanisms of perfluoroalkyl acid (PFAAs) sorption to soil at AFFF-impacted sites are influenced by the presence of multiple PFAAs, other nonfluorinated surfactants, and other comingled contaminants, such as fuels and chlorinated solvents (Luthy et al. 1997; Guelfo and Higgins 2013). During firefighting activities, the large amounts of waste fuel components and residual chlorinated hydrocarbons (CHCs) released into the environment can also enter into the subsurface soil and groundwater. These co-contaminants impact the fate and transport of PFASs in the subsurface. Guelfo and Higgins (2013) examined the impact of non–aqueous phase liquids (NAPLs), including chlorinated solvents, fuels, and nonfluorinated AFFF surfactants, on PFAA-contaminated soils that were also amended with trichloroethene (TCE). They discovered that the impact of NAPL depended on the fraction of organic carbon in soil, PFAA chain length, PFAA concentration, and solid phase surface charge. When above the critical separate phase concentration (CSPC), NAPL can serve as an additional sorbent of PFAAs and also block access for PFAA sorption to organic matter in soil with low organic carbon content (f_{oc}). Therefore, the NAPL concentration can influence PFAA adsorption and partitioning in a complex manner. Further, the measurements of aquifer solids and groundwater samples collected from Ellsworth Air Force Base indicated that most of the PFAAs, except for PFOS, were within the groundwater plume containing benzene (McGuire et al. 2014).

AFFF formulations are complex, as they also contain PFAA precursors, in addition to PFAAs (Place and Field 2012; D'Agostino and Mabury 2014). Due to their zwitterionic or cationic nature, PFAA precursors are less soluble than PFAAs and express a stronger sorption to aquifer solids (Place and Field 2012). The sorption of 8:2 FTOH to soils is proportional to the soil f_{oc} (Liu and Lee 2005). Houtz et al. (2013) showed that several classes of AFFF-derived precursors (fluorotelomer and sulfonamide precursors) accounted for 41%–100% of the mass of PFASs in AFFF formulations at a previously contaminated field site at Ellsworth Air Force Base. However, much of the mass of these precursors was not detected in AFFF-impacted environmental samples from the area. This suggests that most of the precursors were degraded into PFAAs, such as PFOAs and PFOSs, which have higher mass fractions in environmental samples than in original AFFF formulation samples. Backe et al. (2013) found similar results in which PFAS profiles from AFFF-impacted groundwater samples were different, depending on the PFAS composition in the original AFFF formulations.

The biodegradation of PFAA precursors is an important source of PFOS and PFOA in the environment (Wang et al. 2005a). Studies have shown the potential of FTOHs ($F(CF_2)nCH_2CH_2OH$, n = 4, 6, and 8) to biodegrade to other precursor compounds, such as $CF_3(CF_2)_6(14)CH=CHCOOH$ (7-3 unsaturated acid or 7-3 U acid) and 8:2 FTUA (Wang et al. 2009), as well as a series of homologous PFAAs in mixed bacterial cultures (Liu et al. 2007), soil (Russell et al. 2008; Liu et al. 2010), landfills (Sepulvado et al. 2011), and aerobic activated sludge from wastewater treatment (Boulanger et al. 2005; Wang et al. 2005b, 2011). Wang et al. (2009) found that 25% of 8:2 FTOH degradation corresponded with 25% PFOA production through aerobic biotransformation in activated sludge.

Microbial populations that can metabolize FTOHs may vary with differences in the environmental matrix. As a result, FTOH undergoes biodegradation by multiple pathways and generates different transformation by-products (Wang et al. 2009). However, FTOH degradation pathways in aerobic soils are similar to those in aerobic sludge and bacterial cultures. In aerobic soils, 8:2 FTOH was converted rapidly to 8:2 fluorotelomer aldehyde (8:2 FTAL) and 8:2 FTUA; following incubation with sassafras soil, the transformation continued to 7:3 U acid and 7:2 sFTOH ($F(CF_2)_7CH(OH)CH_3$), which were then oxidized to generate PFOA as a by-product (Wang et al. 2009). Additionally, Wang et al. (2009) found an intermediate metabolite of 7:2 FT ketone ($F(CF_2)_7COCH_3$) and two other final by-products, 3-OH-7-3 acid ($F(CF_2)_7CHOHCH_2COOH$) and 2H-PFOA ($F(CF_2)_6CFHCOOH$).

19.4 TREATMENT TECHNOLOGIES FOR PFASs

Due to their unique physicochemical properties, the remediation and treatment of PFAS-contaminated water is challenging. Wastewater mass flow studies found similar or higher PFAS concentrations in the effluent than in the influent, presumably from the biodegradation of precursor compounds, indicating that conventional wastewater treatment, including sedimentation, filtration, oxidation, and disinfection, may not be effective for PFAS removal in treated water (Schultz et al. 2006; Sinclair and Kannan 2006). Therefore, other technologies must be pursued for the treatment of aqueous PFASs. These methods can be categorized into conventional technologies (e.g., adsorption, advanced oxidation methods, reduction methods, and sonochemical methods) and emerging treatment methods (e.g., electrochemical, nanomaterials such as nanoscale zerovalent iron [nZVI], and air sparging).

19.4.1 PFAS REMOVAL BY ADSORPTION

Adsorption to activated carbon (AC) is a promising and commonly used technique in PFAS remediation. Previous studies reported that granular activated charcoal (GAC) effectively removes up to 99% of PFOA and PFOS from wastewater effluent and has a maximum PFOA sorption capacity of up to 1100 ng/g from contaminated groundwater (Fujii et al. 2007; Hansen et al. 2010; Yao et al. 2014). Smaller particle diameters and shorter diffusive path lengths give powdered activated charcoal (PAC) faster kinetics than GAC. A commercially available PAC (Filtrasorb 400) reached equilibrium within ~3 to 5 h, whereas the GAC adsorbed PFASs much slower, over ~48 to 168 h (Yu et al. 2009). Although Yu et al. (2009) concluded that PAC was superior to GAC in removing PFOA and PFOS, it has certain limitations. GAC is generally favored for long-term treatment of a constant source of contaminant since it can be regenerated. PAC use, on the other hand, is typically more episodic, as it cannot be regenerated. Long-term use of PAC can lead to seasonal algal or taste and odor issues in treated water. The application of PAC may be preferred for a single or batch operation (Yu et al. 2009).

Field data from full-scale pump-and-treat treatment systems with GAC for PFAS removal from extracted water are limited. An *in situ* remediation system, the vacuum-enhanced multiphase extractor (VEMPE), designed for hydrocarbon removal, was installed at a firefighting training area in the Canadian province of British Columbia (Paterson and Sweeney 2008). The results of the VEMPE system show that only 0.1% of PFASs was recovered in 2 years, indicating that such an approach to cleaning up hydrocarbon contamination may be ineffective for removing PFASs.

The removal of PFASs by adsorption to AC from natural waters is also strongly affected by the presence of dissolved organic matter (DOM). Appleman et al. (2013) and Yu et al. (2011) examined the sorption performance of PFAAs onto PAC and GAC, both with and without DOM; both studies found that the removal of PFAAs from water by adsorption to AC can become inhibited due to the competitive effects of DOM. They also indicated that DOM could cover and block the active adsorption sites on PAC, which led to the rapid breakthrough of PFASs during treatment. Besides DOM, the sorption performance also varied with experimental conditions and different adsorbents. The organic carbon partitioning coefficient (K_{oc}) values of PFASs decrease with decreasing carbon chain length. Therefore, longer-chain PFASs may have greater potential for removal by adsorption to AC (Higgins and Luthy 2006). AC also preferentially adsorbs PFOS compared with PFOA due to the difference in their hydrophobic properties (Yu et al. 2009). Furthermore, in comparison with branched isomers, linear PFAS isomers exhibit greater sorption potential to GAC (Eschauzier et al. 2012).

In addition to GAC, other types of commercially available adsorbents, such as ion exchange resin (Deng et al. 2010), mineral materials (Tang et al. 2010), carbon nanotubes (Chen et al. 2011), alumina (Wang and Shih 2011), and zeolites (Ochoa-Herrera and Sierra-Alvarez 2008), are also used to remove PFASs from extracted groundwater. Although GAC is more cost-effective due to its high adsorption capacity for PFASs, GAC requires frequent regeneration to maintain its high efficiency. Regenerating GAC two to three times per year is recommended to avoid rapid breakthrough of PFOA. Additionally, a subsequent destruction step, such as incineration, may be required for PFASs absorbed on used GAC (Hansen et al. 2010).

19.4.2 PFAS Treatment by Advanced Oxidation Processes

Due to the high strength of the C–F bond and high electronegativity of fluorine, PFASs are often recalcitrant toward chemical oxidation (Moriwaki et al. 2005).

Advanced oxidation processes (AOPs) traditionally utilize strong oxidants for the *in situ* treatment of contaminated sites. Hydrogen peroxide (H_2O_2) is a potent oxidant that can generate hydroxyl radicals (HO·) that react with organic compounds to form carbon-centered radicals (R· or R·–OH) by hydrogen abstraction (Deng and Zhao 2015). PFAS remediation using AOP aims to destroy C–F bonds to form F^- ions, which can combine with Ca^{2+} to form the environmentally friendly, insoluble CaF_2 (Hori et al. 2004). In the presence of O_2, more active agents are formed from radicals to lead to the degradation of the organic chemicals in traditional AOPs. However, the mechanisms of conventional oxygen-based AOPs are not generally effective in PFAS remediation, as there is a lack of hydrogen for abstraction from PFASs (Deng and Zhao 2015).

Recent studies indicate that hydroxyl radicals produced by AOPs using ultraviolet (UV)–H_2O_2 may not be efficient in decomposing PFOA (Hori et al. 2004). Similarly, Schroder and Meesters (2005) reported AOPs by ozone (O_3), O_3/UV, O_3/H_2O_2, and H_2O_2/Fe^{2+} failing to degrade PFOS. However, degradation of PFOA and PFOS observed by O_3 and O_3/H_2O_2 under alkaline conditions can achieve up to 85–100% removal where the treatment is further enhanced with ozone pretreatment at pH 4–5 (A. Lin et al. 2012). Improvement in AOP performance has also been shown by other commonly used oxidation agents, such as activated persulfate $\left(S_2O_8^{2-}\right)$, Fenton's reagent, catalyst (zerovalent metal), and subcritical water, which increased the rate of PFAA oxidation under ozonation (O_3), irradiation (UV light), heat (e.g., microwave), or a combination of these conditions (Lee et al. 2009, 2010; Qu et al. 2010; A. Lin et al. 2012; Jin et al. 2014). Photolytic oxidation methods, namely, direct photolysis (Yamamoto et al. 2007), persulfate photolysis (Hori et al. 2005), photocatalysis (Hori et al. 2004), and microwave- (Lee et al. 2010), and heat-activated persulfate oxidation (Lee et al. 2012), have shown varying degrees of success in PFOA and PFOS degradation.

Several AOP studies using photochemcial oxidation with activated persulfate have shown promising results in PFOS and PFOA treatment. The persulfate ion is a strong oxidant ($E^0 = 2.1$ eV) and can be activated by heat, UV light, elevated pH, and ferrous iron (Fe^{2+}) to form highly oxidizing sulfate radicals ($SO_4 \cdot^-$). These radicals have a higher redox potential ($E^0 = 2.6$ eV) that can destroy persistent organic contaminants (Huang et al. 2005; Tsitonaki et al. 2010; Lee et al. 2012). Unlike hydroxyl radicals added to unsaturated C=C bonds or abstracting H from C–H bonds, sulfate radicals remove electrons directly from organic compounds and produce OH· to enhance the reaction (Equation 19.1) (Vogelpohl and Kim 2004).

$$SO_4^- + H_2O \rightarrow OH + SO_4^2 H^+ \tag{19.1}$$

Hori et al. (2005) showed a complete decomposition of 1.35 mM PFOA with 50 mM of light-activated persulfate $\left(S_2O_8^{2-}\right)$ after 4 h of irradiation with fluoride and carbon dioxide as major by-products, and a small amount of shorter-chain PFAAs. Wang et al. (2010) found similar results on PFDeA decomposition by $S_2O_8^{2-}$ under vacuum ultraviolet (VUV) light irradiation.

Heat- and microwave-assisted hydrothermal methods have also been used to enhance persulfate oxidation performance on PFAS degradation (Hori et al.

2008a; Lee et al. 2010). Traditional thermal treatment conditions require high temperatures, such as subcritical (~300°C) or supercritical (>374°C) water (Huang et al. 2005). However, the presence of persulfate lowers the reaction temperature, making AOP with persulfate more applicable to *in situ* treatments. Hori et al. (2008a) reported that C_5–C_9 PFAAs were efficiently decomposed in hot water at a temperature of 80°C with the addition of $S_2O_8^{2-}$. The high mole percent yields of F⁻ ions (77.5% for PFOA and 82.9%–88.9% for other PFACs) and CO_2 (70.2% for PFOA and 87.7%–100% for other PFAAs) may indicate complete mineralization of PFCAs. The decomposition activity of PFOA can be accelerated by higher temperature, but extremely high temperatures (130°C) may release radical oxidants that consume persulfate. Lee et al. (2009) also found that PFOA was decomposed by persulfate under relatively low temperatures, even at room temperature (27°C). Decomposition of PFOA and other PFAA species (e.g., C_2–C_7 perfluoroalkyl groups) is more active in acidic conditions. Lee et al. (2012) further reported that PFOA was decomposed and mineralized at 20°C–40°C (72–648 h) with a lower initial pH (2.5) or higher persulfate (200 mM) concentration that accelerated the degradation reaction.

19.4.3 PFAS Treatment by Sonochemical Oxidation

Another effective oxidation process, sonochemical oxidation, could achieve complete mineralization of PFOA and PFOS (Cheng et al. 2008). Moriwaki et al. (2005) showed the degradation potential of PFOS and PFOA by ultrasonic irradiation at the bubble–water interface during *in situ* pyrolysis, caused by cavitation bubbles. Ultrasonic irradiation in air for 60 min decomposed 28% of PFOS following pseudo-first-order kinetics. This reaction produced PFOA as an intermediate product, 98% of which was decomposed to *tert*-butyl alcohol. The reaction was accelerated by saturating the reactor atmosphere with argon, where the production of fluoride, sulfate, CO, and CO_2 as by-products was verified by complete mineralization of PFOA or PFOS (Moriwaki et al. 2005). Sonochemical oxidation may also be enhanced by the addition of sulfate to indirectly destroy PFOA by sulfate free radicals (Lin et al. 2015).

Cheng et al. (2008) examined the sonolysis of PFOS and PFOA present in groundwater contaminated with landfill leachate and reported a significant decline in the degradation kinetics of about 60% due to the effect of organic co-constituents. This was especially by the volatile organic compounds (VOCs) that can lower the interfacial tension. However, the DOM at 15 mg/L had only a minor effect on the decline in degradation kinetics, as there was negligible competitive adsorption onto the interface. Nevertheless, the competition at the interface would be significant, with higher concentrations of surfactants (e.g., AFFF) in the environment.

Degradation pathways of PFASs using UV irradiation generally involve the following: (1) oxidative degradation by direct photolysis and (2) reductive degradation of photochemical decomposition (Ochoa-Herrera et al. 2008; Wang et al. 2010). Qu et al. (2010) reported near-complete PFOA defluorination in the aqueous phase with irradiation (254 nm) in the aqueous phase at room temperature and pH 9 under anaerobic conditions. Photoreductive defluorination was mainly attributed to the generation of hydrated electrons $\left(e^{-aq}, E_{aq/e^0} = -2.9V\right)$. Song et al. (2013) reported similar results with near-complete PFOA defluorination by hydrated electrons in a sulfite-mediated UV photochemical system.

19.4.4 PFAS Treatment by *In Situ* Chemical Reduction

Reductive dehalogenation has been widely and extensively utilized on persistent CHC degradation (Matheson and Tratnyek 1994). Reductive treatment using zerovalent iron (ZVI) also has the potential to decompose PFASs. Efficient iron-induced decompositions of PFASs using ZVI in sub- and supercritical water have been observed (Hori et al. 2006, 2008b). The addition of ZVI could facilitate complete PFOS mineralization under high temperature (350°C) and high pressure (250 atm) in 6 h. The reaction occurred on the iron metal surface and produced fluoride ions following pseudo-first-order kinetics. It has been suggested that an increase in the ZVI surface area may accelerate the reaction rate significantly (Hori et al. 2008b). This method has also been effective on shorter-chain (C_2–C_6) PFAAs. In comparison with sonochemical oxidation, Hori et al. (2006) found an absence of undesirable intermediates, such as PFOA and other shorter-chain PFAAs, during treatment by this method. The results further indicated that reductive defluorination by ZVI, followed by a heat treatment, led to PFOS adsorption onto the iron metal (Hori et al. 2006).

A photocatalyst using persulfate activated by ZVI (Fe^0) has also been investigated for PFAS treatment. Lee et al. (2010) found that ZVI amendment with persulfate successfully accelerates the microwave-hydrothermal decomposition rate of PFAAs. PFOA was degraded rapidly up to 67.6% at 90°C after 2 h. Under aerobic or anaerobic conditions, ZVI was oxidized to form Fe^{2+} (Equation 19.2) (Furukawa et al. 2002).

$$Fe^0 + \tfrac{1}{2}O_2 + H_2O \rightarrow Fe^{2+} + 2OH^- \tag{19.2}$$

Fe^{2+} activates $S_2O_8^{2-}$, leading to the formation of sulfate free radicals (Equation 19.3), $SO_4^{\cdot-}$, which are capable of decomposing PFASs (House 1962; Lee et al. 2010).

$$Fe^{2+} + S_2O_8^{2-} \rightarrow Fe^{3+} + SO_4^- + SO_4^{2-} \tag{19.3}$$

19.4.5 Other Emerging Technologies

Most conventional treatment technologies are ineffective or too costly to treat PFAS-contaminated groundwater. There is considerable interest in emerging technologies that have shown promise in treating PFOA and PFOS. As such, the potential for electrochemical treatment, air-sparged hydrocyclone (ASH) technology, and nanomaterials for remediation (Wang et al. 2010; Schaefer et al. 2015) is summarized below.

19.4.5.1 Electrochemical Treatment

A novel SnO_2–Sb/carbon aerogel electrode catalyst has demonstrated efficient removal of PFOA by electrochemical oxidation (Zhao et al. 2013). Coating a catalyst metal on the carbon electrode may further improve the potential for PFAS adsorption and degradation. Schaefer et al. (2015) investigated the potential for the electrochemical treatment of PFOA/PFOS in AFFF-impacted groundwater using a divided electrochemical cell and Ti/RuO$_2$ anode. Several researchers concluded that PFAS degradation using the electrochemical approach occurs on the anode surface via direct electron transfer (H. Lin et al. 2012; Zhao et al. 2014; Schaefer et al. 2015).

19.4.5.2 Air-Sparged Hydrocyclone

This is a treatment technology that has been evaluated for AFFF removal at former firefighting training sites with extensive contamination. The ASH system combines froth flotation principles with the flow characteristics of a hydrocyclone. In an ASH system, AFFF is removed via adsorption to hydrophobic particles or oil droplets that are attached to bubbles formed from pressurized air sheared by the tangential swirl of water. Coagulants and flocculants are also used in this aboveground treatment system. The removal efficiency of AFFF via ASH reached 80%–97% at six US Air Force sites (Van Gils 2003). This system was also efficient in removing oils and grease, petroleum hydrocarbons, and other co-contaminants that accompany AFFF.

The ASH technology is mobile and can be easily relocated on a single semi-truck trailer unit. It has a typical processing capacity of 50 gallons/min and was designed to treat wastewater generated from firetraining activities or ship bilge water containing significant concentrations of AFFF. In comparison with other treatment technologies with equivalent removal efficiencies, ASH is considered more cost-effective for the treatment of water containing significant concentrations of AFFF free product (Van Gils 2003).

19.4.5.3 Nanomaterial Technology

Recently, nZVI has become one of the most innovative technologies for the treatment of contaminated aquifers (O'Carroll et al. 2013). Compared with granular or powder (microscale) ZVI, nZVI shows a significant increase in the reaction kinetics due to a greater specific surface area (Crane and Scott 2012). For instance, Feng and Lim (2005) reported nZVI to have much higher reactivity for carbon tetrachloride (CT) and chloroform (CF) than microscale ZVI. nZVI has also proven effective and efficient for degrading polychlorinated biphenyls (PCBs), CHCs, chlorinated pesticides, and certain inorganic anions, such as nitrate (Wang and Zhang 1997; Choe et al. 2000; Joo and Zhao 2008). Although there have been limited studies on PFAS treatment using nZVI, Arvaniti et al. (2015) reported the removal of selected PFASs from 38% (for PFOA) to 96% (for PFDA) with Mg-aminoclay (MgAC)-coated nZVI at acidic pH (3) and low temperature (20°C). The order of removal efficiency using nZVI was PFOA < perfluorononanoic acid (PFNA) < PFOS ≈ PFDA.

A mass balance experiment demonstrated that PFAS removal by MgAC-coated nZVI involves adsorption to the iron metal surface and reduction via dehydrohalogenation (Arvaniti et al. 2015). It is noteworthy that nZVI mobility in porous media is greatly limited due to its aggregation caused by van der Waals forces and magnetic attraction. The rapid aggregation of nZVI results in a loss of reactive surface area (Phenrat et al. 2007). However, in comparison with commercial ZVI and uncoated nZVI, the MgAC-coated nZVI increases the stability and reactivity for reductive decontamination due to the increased electrostatic repulsion among nZVI particles due to the positively charged MgAC coating (Hwang et al. 2014). When the aminoclay-to-nZVI ratio is 1.0, the material exhibits a highly positively charged surface and a ferromagnetic property (Lee et al. 2014). Therefore, the organo-functional groups of $-(CH_2)_3NH_2$ may form positively charged $-NH_3^+$ ions on the nZVI surface, which strongly adsorbs the negatively charged PFASs and facilitates reductive decomposition and release of fluoride (Hwang et al. 2014; Arvaniti et al. 2015).

Permeable reactive barrier (PRB) is a proven technology for the *in situ* treatment of contaminated groundwater that is more effective than the conventional pump-and-treat technique (Khan et al. 2004; Henderson and Demond 2007). PRBs, filled with a suitable reactive material, such as phosphate, activated charcoal, or ZVI, may be constructed in the subsurface across the flow path of the groundwater plume to intercept and treat dissolved contaminants by precipitation, adsorption, or transformation to harmless products (Puls et al. 1999; Guerin et al. 2002). PRBs do not require aboveground treatment facilities and have low operation and maintenance costs and long-term performance. However, in comparison with the conventional granular ZVI, the injection of nZVI slurry in the subsurface can be employed to create a subsurface reactive zone, analogous to a PRB, for the *in situ* treatment of a groundwater plume. Injected nZVI with good mobility and high reactivity may help in overcoming limitations of depth, site topography, and facility operations for site cleanup. However, care must be taken to overcome the low hydraulic conductivity of nZVI (Kocur, 2015).

Stabilizers and supports, such as MgAC and carboxymethyl cellulose (CMC), may be added to enhance the stability of nanoparticles and facilitate effective remediation (He and Zhao 2007; Arvaniti et al. 2015). The reactivity of nZVI can also be improved in bimetallic preparations by doping it with a secondary, catalyst metal, namely, Pt (0), Ni (0), or Pd (0), to assist in faster degradation (Elliott and Zhang 2001). Such bimetallic ZVI and nZVI reductants (e.g., Fe/Pd, Fe/Pt, Fe/Cu, and Fe/Ni) have shown potential for the treatment of CHCs via reduction dehalogenation (Fennelly and Roberts 1998; Zhang et al. 1998; Xu and Zhang 2000; Lien and Zhang 2001). Cho and Choi (2010) showed 85%, 80%, and 56% removal of TCE, tetrachloroethene (PCE), and 1,1,1-trichloroethane (1,1,1-TCA), respectively, after 120 min using nanoscale Fe/Pd bimetallic particles. Bimetallic ZVI also exhibits faster reaction kinetics and slower deposition of corrosion products on the particles' surface (Fu et al. 2014). nZVI has been demonstrated to be ~50 to 80 times slower in TCE dehalogenation in comparison to nanoscale Fe/Ni (Schrick et al. 2002). Shih et al. (2009) reported that nanoscale Fe/Pd bimetallic particles significantly accelerated hexachlorobenzene dechlorination when compared with nZVI. In addition, Feng and Lim (2005) indicated that the degradation rate constants of CT and CF with nanoscale Fe/Ni particles were two- to eightfold greater than those with nZVI. Despite the success of bimetallic nZVI in treating CHCs, there is a lack of studies on treating PFASs with this method. This approach should be investigated for its suitability for the treatment of PFASs.

In conclusion, the treatment of PFASs in AFFF-impacted groundwater and soil or sediment is challenging in the natural environment. PFASs are generally recalcitrant to most conventional biological and chemical treatments. The constituents of AFFF-impacted groundwater are also complex, involving a wide range of zwitterionic, cationic, and anionic fluorinated chemicals (Backe et al. 2013). The impact of complex mixtures and co-contaminants involved in remediation activities is still under investigation. Most recent studies on PFAS degradation were conducted in laboratory-scale experiments with elevated (spiked) PFAS concentrations (mg/L). In contrast, the AFFF-impacted sites have a wide range of PFAS concentrations, from nanogram per liter to milligram per liter levels (Schultz et al. 2004). Furthermore, the objective of *in situ* remediation is to eliminate PFAS compounds from groundwater or soil or sediment under mild conditions with cost-effective and environmentally friendly approaches. However, some of the emerging technology treatment methods have largely been demonstrated in laboratory settings that consume high energy or require extreme laboratory conditions that are difficult or costly to apply at full-scale field studies.

19.5 GAPS AND FUTURE WORK

PFASs are frequently found in the environment, especially in aquatic systems. Many studies have investigated the behavior of PFASs in wastewater and drinking water treatment plants, but there is limited research identifying the fate and transport of PFASs, especially as a principal component of AFFF-contaminated groundwater and soil. The potential risk to organisms and humans due to the discharge of PFASs contained in AFFF are also not fully understood. Volatile emissions, transformation kinetics, and biodegradation mechanisms of PFASs, in particular short-chain PFASs and precursors, should also be considered for monitoring studies. Quantification methods of isomers with different branching patterns and some airborne PFAS derivatives are still under development, limiting further exploration of their occurrence, fate, and distribution in the environment (Martin et al. 2002).

In the future, long-chain PFASs will continue to be a challenge that may arise from the degradation of precursors and their remobilization into water from contaminated soil. There are fundamental gaps in our understanding about their dominant transport pathways, especially in remote regions. Impacts from the changing environmental conditions during the transport process should also be characterized.

Some advanced treatment processes have efficiently removed aqueous PFASs. Most of these studies were conducted under laboratory conditions at high temperatures or elevated pressures, thus consuming large amounts of energy. However, organic by-products and some volatile compounds produced from the destructive treatment processes have not yet been fully examined. Further studies should investigate and develop these advanced technologies in pilot- and field-scale applications.

REFERENCES

Ahrens, L.; Taniyasu, S.; Yeung, L. W. Y.; Yamashita, N.; Lam, P. K. S.; Ebinghaus, R. Distribution of polyfluoroalkyl compounds in water, suspended particulate matter and sediment from Tokyo Bay, Japan. *Chemosphere* **2010**, *79* (3), 266–272.

Ahrens, L. Polyfluoroalkyl compounds in the aquatic environment: A review of their occurrence and fate. *Journal of Environmental Monitoring* **2011**, *13* (1), 20–31.

Ahrens, L.; Yeung, L. W. Y.; Taniyasu, S.; Lam, P. K. S.; Yamashita, N. Partitioning of perfluorooctanoate (PFOA), perfluorooctane sulfonate (PFOS) and perfluorooctane sulfonamide (PFOSA) between water and sediment. *Chemosphere* **2011**, *85* (5), 731–737.

Ahrens, L.; Bundschuh, M. Fate and effects of poly- and perfluoroalkyl substances in the aquatic environment: A review. *Environmental Toxicology and Chemistry* **2014**, *33* (9), 1921–1929.

Alsmeyer, Y. W.; C. W.; Flynn, R. M.; Moore, G. G. I.; Smeltzer, J. C. Organofluorine chemistry: Principles and commercial applications. In *Electrochemical Fluorination and Its Applications*, Banks, R. E. et al., Eds. New York: Plenum Press, **1994**, p. 121.

Appleman, T. D.; Dickenson, E. R.; Bellona, C.; Higgins, C. P. Nanofiltration and granular activated carbon treatment of perfluoroalkyl acids. *Journal of Hazardous Materials* **2013**, *260*, 740–746.

Arvaniti, O. S.; Stasinakis, A. S. Review on the occurrence, fate and removal of perfluorinated compounds during wastewater treatment. *Science of the Total Environment* **2015**, *524*, 81–92.

Arvaniti, O. S.; Hwang, Y.; Andersen, H. R.; Stasinakis, A. S.; Thomaidis, N. S.; Aloupi, M. Reductive degradation of perfluorinated compounds in water using Mg-aminoclay coated nanoscale zero valent iron. *Chemical Engineering Journal* **2015**, *262*, 133–139.

Austin, M. E.; Kasturi, B. S.; Barber, M.; Kannan, K.; MohanKumar, P. S.; MohanKumar, S. M. J. Neuroendocrine effects of perfluorooctane sulfonate in rats. *Environmental Health Perspectives* **2003**, *111* (12), 1485–1489.

Awad, E.; Zhang, X. M.; Bhavsar, S. P.; Petro, S.; Crozier, P. W.; Reiner, E. J.; Fletcher, R.; Tittemier, S. A.; Braekevelt, E. Long-term environmental fate of perfluorinated compounds after accidental release at Toronto Airport. *Environmental Science & Technology* **2011**, *45* (19), 8081–8089.

Backe, W. J.; Day, T. C.; Field, J. A. Zwitterionic, cationic, and anionic fluorinated chemicals in aqueous film forming foam formulations and groundwater from US military bases by nonaqueous large-volume injection HPLC-MS/MS. *Environmental Science & Technology* **2013**, *47* (10), 5226–5234.

Baduel, C.; Paxman, C. J.; Mueller, J. F. Perfluoroalkyl substances in a firefighting training ground (FTG), distribution and potential future release. *Journal of Hazardous Materials* **2015**, *296*, 46–53.

Benskin, J. P.; Yeung, L. W. Y.; Yamashita, N.; Taniyasu, S.; Lam, P. K. S.; Martin, J. W. Perfluorinated acid isomer profiling in water and quantitative assessment of manufacturing source. *Environmental Science & Technology* **2010**, *44* (23), 9049–9054.

Boulanger, B.; Vargo, J. D.; Schnoor, J. L.; Hornbuckle, K. C. Evaluation of perfluorooctane surfactants in a wastewater treatment system and in a commercial surface protection product. *Environmental Science & Technology* **2005**, *39* (15), 5524–5530.

Buck, R. C.; Franklin, J.; Berger, U.; Conder, J. M.; Cousins, I. T.; de Voogt, P.; Jensen, A. A.; Kannan, K.; Mabury, S. A.; van Leeuwen, S. P. Perfluoroalkyl and polyfluoroalkyl substances in the environment: Terminology, classification, and origins. *Integrated Environmental Assessment and Management* **2011**, *7* (4), 513–541.

Cai, M. H.; Zhao, Z.; Yin, Z. G.; Ahrens, L.; Huang, P.; Cai, M. G.; Yang, H. Z.; He, J. F.; Sturm, R.; Ebinghaus, R.; Xie, Z. Y. Occurrence of perfluoroalkyl compounds in surface waters from the North Pacific to the Arctic Ocean. *Environmental Science & Technology* **2012**, *46* (2), 661–668.

Chen, H.; Chen, S.; Quan, X.; Zhao, Y. Z.; Zhao, H. M. Sorption of perfluorooctane sulfonate (PFOS) on oil and oil-derived black carbon: Influence of solution pH and Ca2+. *Chemosphere* **2009**, *77* (10), 1406–1411.

Chen, X.; Xia, X. H.; Wang, X. L.; Qiao, J. P.; Chen, H. T. A comparative study on sorption of perfluorooctane sulfonate (PFOS) by chars, ash and carbon nanotubes. *Chemosphere* **2011**, *83* (10), 1313–1319.

Cheng, J.; Vecitis, C. D.; Park, H.; Mader, B. T.; Hoffmann, M. R. Sonochemical degradation of perfluorooctane sulfonate (PFOS) and perfluorooctanoate (PFOA) in landfill groundwater: Environmental matrix effects. *Environmental Science & Technology* **2008**, *42* (21), 8057–8063.

Cho, Y.; Choi, S. I. Degradation of PCE, TCE and 1,1,1-TCA by nanosized FePd bimetallic particles under various experimental conditions. *Chemosphere* **2010**, *81* (7), 940–945.

Choe, S.; Chang, Y. Y.; Hwang, K. Y.; Khim, J. Kinetics of reductive denitrification by nanoscale zero-valent iron. *Chemosphere* **2000**, *41* (8), 1307–1311.

Conder, J. M.; Hoke, R. A.; De Wolf, W.; Russell, M. H.; Buck, R. C. Are PFCAs bioaccumulative? A critical review and comparison with regulatory lipophilic compounds. *Environmental Science & Technology* **2008**, *42* (4), 995–1003.

Crane, R. A.; Scott, T. B. Nanoscale zero-valent iron: Future prospects for an emerging water treatment technology. *Journal of Hazardous Materials* **2012**, *211*, 112–125.

D'Agostino, L. A.; Mabury, S. A. Identification of novel fluorinated surfactants in aqueous film forming foams and commercial surfactant concentrates. *Environmental Science & Technology* **2014**, *48* (1), 121–129.

Deng, S. B.; Yu, Q. A.; Huang, J.; Yu, G. Removal of perfluorooctane sulfonate from wastewater by anion exchange resins: Effects of resin properties and solution chemistry. *Water Research* **2010**, *44* (18), 5188–5195.

Deng, Y.; Zhao, R. Advanced oxidation processes (AOPs) in wastewater treatment. *Current Pollution Reports* **2015**, *1* (3), 167–176.

Dinglasan, M. J. A.; Ye, Y.; Edwards, E. A.; Mabury, S. A. Fluorotelomer alcohol biodegradation yields poly- and perfluorinated acids. *Environmental Science & Technology* **2004**, *38* (10), 2857–2864.

Elliott, D. W.; Zhang, W. X. Field assessment of nanoscale biometallic particles for groundwater treatment. *Environmental Science & Technology* **2001**, *35* (24), 4922–4926.

Ellis, D. A.; Martin, J. W.; De Silva, A. O.; Mabury, S. A.; Hurley, M. D.; Andersen, M. P. S.; Wallington, T. J. Degradation of fluorotelomer alcohols: A likely atmospheric source of perfluorinated carboxylic acids. *Environmental Science & Technology* **2004**, *38* (12), 3316–3321.

Eschauzier, C.; Beerendonk, E.; Scholte-Veenendaal, P.; De Voogt, P. Impact of treatment processes on the removal of perfluoroalkyl acids from the drinking water production chain. *Environmental Science & Technology* **2012**, *46* (3), 1708–1715.

Feng, J.; Lim, T. T. Pathways and kinetics of carbon tetrachloride and chloroform reductions by nano-scale Fe and Fe/Ni particles: Comparison with commercial micro-scale Fe and Zn. *Chemosphere* **2005**, *59* (9), 1267–1277.

Fennelly, J. P.; Roberts, A. L. Reaction of 1,1,1-trichloroethane with zero-valent metals and bimetallic reductants. *Environmental Science & Technology* **1998**, *32* (13), 1980–1988.

Fu, F. L.; Dionysiou, D. D.; Liu, H. The use of zero-valent iron for groundwater remediation and wastewater treatment: A review. *Journal of Hazardous Materials* **2014**, *267*, 194–205.

Fujii, S.; Polprasert, C.; Tanaka, S.; Lien, N. P. H.; Qiu, Y. New POPs in the water environment: Distribution, bioaccumulation and treatment of perfluorinated compounds—A review paper. *Journal of Water Supply Research and Technology—Aqua* **2007**, *56* (5), 313–326.

Furukawa, Y.; Kim, J. W.; Watkins, J.; Wilkin, R. T. Formation of ferrihydrite and associated iron corrosion products in permeable reactive barriers of zero-valent iron. *Environmental Science & Technology* **2002**, *36* (24), 5469–5475.

Giesy, J. P.; Kannan, K. Global distribution of perfluorooctane sulfonate in wildlife. *Environmental Science & Technology* **2001**, *35* (7), 1339–1342.

Giesy, J. P.; Naile, J. E.; Khim, J. S.; Jones, P. D.; Newsted, J. L. Aquatic toxicology of perfluorinated chemicals. In *Reviews of Environmental Contamination and Toxicology*, Whitacre, D. M., Ed. New York: Springer, **2010**, Vol. 202, pp. 1–52.

Guelfo, J. L.; Higgins, C. P. Subsurface transport potential of perfluoroalkyl acids at aqueous film-forming foam (AFFF)-impacted sites. *Environmental Science & Technology* **2013**, *47* (9), 4164–4171.

Guerin, T. F.; Horner, S.; McGovern, T.; Davey, B. An application of permeable reactive barrier technology to petroleum hydrocarbon contaminated groundwater. *Water Research* **2002**, *36* (1), 15–24.

Hansen, M. C.; Borresen, M. H.; Schlabach, M.; Cornelissen, G. Sorption of perfluorinated compounds from contaminated water to activated carbon. *Journal of Soils and Sediments* **2010**, *10* (2), 179–185.

He, F.; Zhao, D. Y. Manipulating the size and dispersibility of zerovalent iron nanoparticles by use of carboxymethyl cellulose stabilizers. *Environmental Science & Technology* **2007**, *41* (17), 6216–6221.

Henderson, A. D.; Demond, A. H. Long-term performance of zero-valent iron permeable reactive barriers: A critical review. *Environmental Engineering Science* **2007**, *24* (4), 401–423.

Higgins, C. P.; Luthy, R. G. Sorption of perfluorinated surfactants on sediments. *Environmental Science & Technology* **2006**, *40* (23), 7251–7256.

Hori, H.; Hayakawa, E.; Einaga, H.; Kutsuna, S.; Koike, K.; Ibusuki, T.; Kiatagawa, H.; Arakawa, R. Decomposition of environmentally persistent perfluorooctanoic acid in water by photochemical approaches. *Environmental Science & Technology* **2004**, *38* (22), 6118–6124.

Hori, H.; Yamamoto, A.; Hayakawa, E.; Taniyasu, S.; Yamashita, N.; Kutsuna, S. Efficient decomposition of environmentally persistent perfluorocarboxylic acids by use of persulfate as a photochemical oxidant. *Environmental Science & Technology* **2005**, *39* (7), 2383–2388.

Hori, H.; Nagaoka, Y.; Yamamoto, A.; Sano, T.; Yamashita, N.; Taniyasu, S.; Kutsuna, S.; Osaka, I.; Arakawa, R. Efficient decomposition of environmentally persistent perfluorooctanesulfonate and related fluorochemicals using zerovalent iron in subcritical water. *Environmental Science & Technology* **2006**, *40* (3), 1049–1054.

Hori, H.; Nagaoka, Y.; Murayama, M.; Kutsuna, S. Efficient decomposition of perfluorocarboxylic acids and alternative fluorochemical surfactants in hot water. *Environmental Science & Technology* **2008a**, *42* (19), 7438–7443.

Hori, H.; Nagaoka, Y.; Sano, T.; Kutsuna, S. Iron-induced decomposition of perfluorohexanesulfonate in sub- and supercritical water. *Chemosphere* **2008b**, *70* (5), 800–806.

House, D. A. Kinetics and mechanism of oxidations by peroxydisulfate. *Chemical Reviews* **1962**, *62* (3), 185.

Houtz, E. F.; Higgins, C. P.; Field, J. A.; Sedlak, D. L. Persistence of perfluoroalkyl acid precursors in AFFF-impacted groundwater and soil. *Environmental Science & Technology* **2013**, *47* (15), 8187–8195.

Huang, K. C.; Zhao, Z. Q.; Hoag, G. E.; Dahmani, A.; Block, P. A. Degradation of volatile organic compounds with thermally activated persulfate oxidation. *Chemosphere* **2005**, *61* (4), 551–560.

Hwang, Y.; Lee, Y. C.; Mines, P. D.; Huh, Y. S.; Andersen, H. R. Nanoscale zero-valent iron (nZVI) synthesis in a Mg-aminoclay solution exhibits increased stability and reactivity for reductive decontamination. *Applied Catalysis B—Environmental* **2014**, *147*, 748–755.

Jin, L.; Zhang, P. Y.; Shao, T.; Zhao, S. L. Ferric ion mediated photodecomposition of aqueous perfluorooctane sulfonate (PFOS) under UV irradiation and its mechanism. *Journal of Hazardous Materials* **2014**, *271*, 9–15.

Joo, S. H.; Zhao, D. Destruction of lindane and atrazine using stabilized iron nanoparticles under aerobic and anaerobic conditions: Effects of catalyst and stabilizer. *Chemosphere* **2008**, *70* (3), 418–425.

Khan, F. I.; Husain, T.; Hejazi, R. An overview and analysis of site remediation technologies. *Journal of Environmental Management* **2004**, *71* (2), 95–122.

Kissa, E. *Fluorinated Surfactants: Synthesis, Properties, and Applications.* New York: Marcel Dekker, **1994**.

Kissa, E. *Fluorinated Surfactants and Repellents.* New York: Marcel Dekker, **2001**.

Kocur, C. Field scale application of nanoscale zero valent iron: Mobility, contaminant degradation, and impact on microbial communities. Electronic Thesis and Dissertation Repository, **2015**, 3068.

Kotthoff, M.; Muller, J.; Jurling, H.; Schlummer, M.; Fiedler, D. Perfluoroalkyl and polyfluoroalkyl substances in consumer products. *Environmental Science and Pollution Research* **2015**, *22* (19), 14546–14559.

Lee, Y. C.; Lo, S. L.; Chiueh, P. T.; Chang, D. G. Efficient decomposition of perfluorocarboxylic acids in aqueous solution using microwave-induced persulfate. *Water Research* **2009**, *43* (11), 2811–2816.

Lee, Y. C.; Lo, S. L.; Chiueh, P. T.; Liou, Y. H.; Chen, M. L. Microwave-hydrothermal decomposition of perfluorooctanoic acid in water by iron-activated persulfate oxidation. *Water Research* **2010**, *44* (3), 886–892.

Lee, Y. C.; Lo, S. L.; Kuo, J.; Lin, Y. L. Persulfate oxidation of perfluorooctanoic acid under the temperatures of 20–40 degrees C. *Chemical Engineering Journal* **2012**, *198*, 27–32.

Lee, Y. C.; Lee, K.; Hwang, Y.; Andersen, H. R.; Kim, B.; Lee, S. Y.; Choi, M. H.; Park, J. Y.; Han, Y. K.; Oh, Y. K.; Huh, Y. S. Aminoclay-templated nanoscale zero-valent iron (nZVI) synthesis for efficient harvesting of oleaginous microalga, *Chlorella* sp KR-1. *RSC Advances* **2014**, *4* (8), 4122–4127.

Lien, H. L.; Zhang, W. X. Nanoscale iron particles for complete reduction of chlorinated ethenes. *Colloids and Surfaces A—Physicochemical and Engineering Aspects* **2001**, *191* (1–2), 97–105.

Lin, A. Y. C.; Panchangam, S. C.; Chang, C. Y.; Hong, P. K. A.; Hsueh, H. F. Removal of perfluorooctanoic acid and perfluorooctane sulfonate via ozonation under alkaline condition. *Journal of Hazardous Materials* **2012**, *243*, 272–277.

Lin, H.; Niu, J. F.; Ding, S. Y.; Zhang, L. L. Electrochemical degradation of perfluorooctanoic acid (PFOA) by Ti/SnO2-Sb, Ti/SnO2-Sb/PbO2 and Ti/SnO2-Sb/MnO2 anodes. *Water Research* **2012**, *46* (7), 2281–2289.

Lin, J. C.; Lo, S. L.; Hu, C. Y.; Lee, Y. C.; Kuo, J. Enhanced sonochemical degradation of perfluorooctanoic acid by sulfate ions. *Ultrasonics Sonochemistry* **2015**, *22*, 542–547.

Lindstrom, A. B.; Strynar, M. J.; Libelo, E. L. Polyfluorinated compounds: Past, present, and future. *Environmental Science & Technology* **2011**, *45* (19), 7954–7961.

Liu, J.; Lee, L. S.; Nies, L. F.; Nakatsu, C. H.; Turco, R. F. Biotransformation of 8:2 fluorotelomer alcohol in soil and by soil bacteria isolates. *Environmental Science & Technology* **2007**, *41* (23), 8024–8030.

Liu, J. X.; Lee, L. S. Solubility and sorption by soils of 8:2 fluorotelomer alcohol in water and cosolvent systems. *Environmental Science & Technology* **2005**, *39* (19), 7535–7540.

Liu, J. X.; Wang, N.; Buck, R. C.; Wolstenholme, B. W.; Folsom, P. W.; Sulecki, L. M.; Bellin, C. A. Aerobic biodegradation of C-14 6:2 fluorotelomer alcohol in a flow-through soil incubation system. *Chemosphere* **2010**, *80* (7), 716–723.

Loganathan, B. G.; Sajwan, K. S.; Sinclair, E.; Kumar, K. S.; Kannan, K. Perfluoroalkyl sulfonates and perfluorocarboxylates in two wastewater treatment facilities in Kentucky and Georgia. *Water Research* **2007**, *41* (20), 4611–4620.

Loi, E. I. H.; Yeung, L. W. Y.; Taniyasu, S.; Lam, P. K. S.; Kannan, K.; Yamashita, N. Trophic magnification of poly- and perfluorinated compounds in a subtropical food web. *Environmental Science & Technology* **2011**, *45* (13), 5506–5513.

Luthy, R. G.; Aiken, G. R.; Brusseau, M. L.; Cunningham, S. D.; Gschwend, P. M.; Pignatello, J. J.; Reinhard, M.; Traina, S. J.; Weber, W. J.; Westall, J. C. Sequestration of hydrophobic organic contaminants by geosorbents. *Environmental Science & Technology* **1997**, *31* (12), 3341–3347.

Martin, J. W.; Muir, D. C. G.; Moody, C. A.; Ellis, D. A.; Kwan, W. C.; Solomon, K. R.; Mabury, S. A. Collection of airborne fluorinated organics and analysis by gas chromatography/chemical ionization mass spectrometry. *Analytical Chemistry* **2002**, *74* (3), 584–590.

Matheson, L. J.; Tratnyek, P. G. Reductive dehalogenation of chlorinated methanes by iron metal. *Environmental Science & Technology* **1994**, *28* (12), 2045–2053.

McGuire, M. E.; Schaefer, C.; Richards, T.; Backe, W. J.; Field, J. A.; Houtz, E.; Sedlak, D. L.; Guelfo, J. L.; Wunsch, A.; Higgins, C. P. Evidence of remediation-induced alteration of subsurface poly- and perfluoroalkyl substance distribution at a former firefighter training area. *Environmental Science & Technology* **2014**, *48* (12), 6644–6652.

Milinovic, J.; Lacorte, S.; Vidal, M.; Rigol, A. Sorption behaviour of perfluoroalkyl substances in soils. *Science of the Total Environment* **2015**, *511*, 63–71.

Moody, C. A.; Field, J. A. Determination of perfluorocarboxylates in groundwater impacted by fire-fighting activity. *Environmental Science & Technology* **1999**, *33* (16), 2800–2806.

Moody, C. A.; Field, J. A. Perfluorinated surfactants and the environmental implications of their use in fire-fighting foams. *Environmental Science & Technology* **2000**, *34* (18), 3864–3870.

Moody, C. A.; Martin, J. W.; Kwan, W. C.; Muir, D. C. G.; Mabury, S. C. Monitoring perfluorinated surfactants in biota and surface water samples following an accidental release of fire-fighting foam into Etobicoke Creek. *Environmental Science & Technology* **2002**, *36* (4), 545–551.

Moody, C. A.; Hebert, G. N.; Strauss, S. H.; Field, J. A. Occurrence and persistence of perfluorooctanesulfonate and other perfluorinated surfactants in groundwater at a fire-training area at Wurtsmith Air Force Base, Michigan, USA. *Journal of Environmental Monitoring* **2003**, *5* (2), 341–345.

Moriwaki, H.; Takagi, Y.; Tanaka, M.; Tsuruho, K.; Okitsu, K.; Maeda, Y. Sonochemical decomposition of perfluorooctane sulfonate and perfluorooctanoic acid. *Environmental Science & Technology* **2005**, *39* (9), 3388–3392.

Oakes, K. D.; Benskin, J. P.; Martin, J. W.; Ings, J. S.; Heinrichs, J. Y.; Dixon, D. G.; Servos, M. R. Biomonitoring of perfluorochemicals and toxicity to the downstream fish community of Etobicoke Creek following deployment of aqueous film-forming foam. *Aquatic Toxicology* **2010**, *98* (2), 120–129.

O'Carroll, D.; Sleep, B.; Krol, M.; Boparai, H.; Kocur, C. Nanoscale zero valent iron and bimetallic particles for contaminated site remediation. *Advances in Water Resources* **2013**, *51*, 104–122.

Ochoa-Herrera, V.; Sierra-Alvarez, R. Removal of perfluorinated surfactants by sorption onto granular activated carbon, zeolite and sludge. *Chemosphere* **2008**, *72* (10), 1588–1593.

Ochoa-Herrera, V.; Sierra-Alvarez, R.; Somogyi, A.; Jacobsen, N. E.; Wysocki, V. H.; Field, J. A. Reductive defluorination of perfluorooctane sulfonate. *Environmental Science & Technology* **2008**, *42* (9), 3260–3264.

Olsen, G. W.; Mair, D. C.; Church, T. R.; Ellefson, M. E.; Reagen, W. K.; Boyd, T. M.; Herron, R. M.; et al. Decline in perfluorooctanesulfonate and other polyfluoroalkyl chemicals in American Red Cross adult blood donors, 2000–2006. *Environmental Science & Technology* **2008**, *42* (13), 4989–4995.

Paterson, T. S. K.; Sweeney, D. Remediation of perfluorinated alkyl chemicals at a former fire-fighting training area. Presented at RemTech 2008: Remediation Technologies Symposium, **2008**, paper 34.

Paul, A. G.; Jones, K. C.; Sweetman, A. J. A first global production, emission, and environmental inventory for perfluorooctane sulfonate. *Environmental Science & Technology* **2009**, *43* (2), 386–392.

Phenrat, T.; Saleh, N.; Sirk, K.; Tilton, R. D.; Lowry, G. V. Aggregation and sedimentation of aqueous nanoscale zerovalent iron dispersions. *Environmental Science & Technology* **2007**, *41* (1), 284–290.

Place, B. J.; Field, J. A. Identification of novel fluorochemicals in aqueous film-forming foams used by the US military. *Environmental Science & Technology* **2012**, *46* (13), 7120–7127.

Prevedouros, K.; Cousins, I. T.; Buck, R. C.; Korzeniowski, S. H. Sources, fate and transport of perfluorocarboxylates. *Environmental Science & Technology* **2006**, *40* (1), 32–44.

Puls, R. W.; Paul, C. J.; Powell, R. M. The application of in situ permeable reactive (zerovalent iron) barrier technology for the remediation of chromate-contaminated groundwater: A field test. *Applied Geochemistry* **1999**, *14* (8), 989–1000.

Qu, Y.; Zhang, C. J.; Li, F.; Chen, J.; Zhou, Q. Photo-reductive defluorination of perfluorooctanoic acid in water. *Water Research* **2010**, *44* (9), 2939–2947.

Rahman, M. F.; Peldszus, S.; Anderson, W. B. Behaviour and fate of perfluoroalkyl and polyfluoroalkyl substances (PFAS) in drinking water treatment: A review. *Water Research* **2014**, *50*, 318–340.

Riddell, N.; Arsenault, G.; Benskin, J. P.; Chittim, B.; Martin, J. W.; McAlees, A.; McCrindle, R. Branched perfluorooctane sulfonate isomer qantification and characterization in blood serum samples by HPLC/ESI-MS(/MS). *Environmental Science & Technology* **2009**, *43* (20), 7902–7908.

Russell, M. H.; Berti, W. R.; Szostek, B.; Buck, R. C. Investigation of the biodegradation potential of a fluoroacrylate polymer product in aerobic soils. *Environmental Science & Technology* **2008**, *42* (3), 800–807.

Schaefer, C. E.; Andaya, C.; Urtiaga, A.; McKenzie, E. R.; Higgins, C. P. Electrochemical treatment of perfluorooctanoic acid (PFOA) and perfluorooctane sulfonic acid (PFOS) in groundwater impacted by aqueous film forming foams (AFFF). *Journal of Hazardous Materials* **2015**, *295*, 170–175.

Schrick, B.; Blough, J. L.; Jones, A. D.; Mallouk, T. E. Hydrodechlorination of trichloroethylene to hydrocarbons using bimetallic nickel-iron nanoparticles. *Chemistry of Materials* **2002**, *14* (12), 5140–5147.

Schroder, H. F.; Meesters, R. J. W. Stability of fluorinated surfactants in advanced oxidation processes—A follow up of degradation products using flow injection-mass spectrometry, liquid chromatography-mass spectrometry and liquid chromatography-multiple stage mass spectrometry. *Journal of Chromatography A* **2005**, *1082* (1), 110–119.

Schultz, M. M.; Barofsky, D. F.; Field, J. A. Quantitative determination of fluorotelomer sulfonates in groundwater by LC MS/MS. *Environmental Science & Technology* **2004**, *38* (6), 1828–1835.

Schultz, M. M.; Higgins, C. P.; Huset, C. A.; Luthy, R. G.; Barofsky, D. F.; Field, J. A. Fluorochemical mass flows in a municipal wastewater treatment facility. *Environmental Science & Technology* **2006**, *40* (23), 7350–7357.

Sepulvado, J. G.; Blaine, A. C.; Hundal, L. S.; Higgins, C. P. Occurrence and fate of perfluorochemicals in soil following the land application of municipal biosolids. *Environmental Science & Technology* **2011**, *45* (19), 8106–8112.

Shih, Y. H.; Chen, Y. C.; Chen, M. Y.; Tai, Y. T.; Tso, C. P. Dechlorination of hexachlorobenzene by using nanoscale Fe and nanoscale Pd/Fe bimetallic particles. *Colloids and Surfaces A—Physicochemical and Engineering Aspects* **2009**, *332* (2–3), 84–89.

Sinclair, E.; Kannan, K. Mass loading and fate of perfluoroalkyl surfactants in wastewater treatment plants. *Environmental Science & Technology* **2006**, *40* (5), 1408–1414.

Song, Z.; Tang, H. Q.; Wang, N.; Zhu, L. H. Reductive defluorination of perfluorooctanoic acid by hydrated electrons in a sulfite-mediated UV photochemical system. *Journal of Hazardous Materials* **2013**, *262*, 332–338.

Tang, C. Y.; Fu, Q. S.; Gao, D. W.; Criddle, C. S.; Leckie, J. O. Effect of solution chemistry on the adsorption of perfluorooctane sulfonate onto mineral surfaces. *Water Research* **2010**, *44* (8), 2654–2662.

Taniyasu, S.; Kannan, K.; So, M. K.; Gulkowska, A.; Sinclair, E.; Okazawa, T.; Yamashita, N. Analysis of fluorotelomer alcohols, fluorotelorner acids, and short- and long-chain perfluorinated acids in water and biota. *Journal of Chromatography A* **2005**, *1093* (1–2), 89–97.

Tsitonaki, A.; Petri, B.; Crimi, M.; Mosbaek, H.; Siegrist, R. L.; Bjerg, P. L. In situ chemical oxidation of contaminated soil and groundwater using persulfate: A review. *Critical Reviews in Environmental Science and Technology* **2010**, *40* (1), 55–91.

USEPA (U.S. Environmental Protection Agency). Emerging contaminants fact sheet— Perfluorooctane sulfonate (PFOS) and perfluorooctanoic acid (PFOA). In Solid Waste and Emergency Response (5160P). Washington, DC: USEPA, 2012.

Van Gils, G. Oil/water emulsion and aqueous film forming foam (AFFF) treatment using air-sparged hydrocyclone technology. Environmental Security Technology Certification Program, VA, **2003**. http://www.dtic.mil/dtic/tr/fulltext/u2/a607005.pdf (accessed May 5, 2017).

Vogelpohl, A.; Kim, S. M. Advanced oxidation processes (AOPs) in wastewater treatment. *Journal of Industrial and Engineering Chemistry* **2004**, *10* (1), 33–40.

Wang, B. B.; Cao, M. H.; Tan, Z. J.; Wang, L. L.; Yuan, S. H.; Chen, J. Photochemical decomposition of perfluorodecanoic acid in aqueous solution with VUV light irradiation. *Journal of Hazardous Materials* **2010**, *181* (1–3), 187–192.

Wang, C. B.; Zhang, W. X. Synthesizing nanoscale iron particles for rapid and complete dechlorination of TCE and PCBs. *Environmental Science & Technology* **1997**, *31* (7), 2154–2156.

Wang, F.; Shih, K. M. Adsorption of perfluorooctanesulfonate (PFOS) and perfluorooctanoate (PFOA) on alumina: Influence of solution pH and cations. *Water Research* **2011**, *45* (9), 2925–2930.

Wang, N.; Szostek, B.; Buck, R. C.; Folsom, P. W.; Sulecki, L. M.; Capka, V.; Berti, W. R.; Gannon, J. T. Fluorotelomer alcohol biodegradation—Direct evidence that perfluorinated carbon chains breakdown. *Environmental Science & Technology* **2005a**, *39* (19), 7516–7528.

Wang, N.; Szostek, B.; Folsom, P. W.; Sulecki, L. M.; Capka, V.; Buck, R. C.; Berti, W. R.; Gannon, J. T. Aerobic biotransformation of C-14-labeled 8-2 telomer B alcohol by activated sludge from a domestic sewage treatment plant. *Environmental Science & Technology* **2005b**, *39* (2), 531–538.

Wang, N.; Szostek, B.; Buck, R. C.; Folsom, P. W.; Sulecki, L. M.; Gannon, J. T. 8-2 fluorotelomer alcohol aerobic soil biodegradation: Pathways, metabolites, and metabolite yields. *Chemosphere* **2009**, *75* (8), 1089–1096.

Wang, N.; Liu, J. X.; Buck, R. C.; Korzeniowski, S. H.; Wolstenholme, B. W.; Folsom, P. W.; Sulecki, L. M. 6:2 fluorotelomer sulfonate aerobic biotransformation in activated sludge of waste water treatment plants. *Chemosphere* **2011**, *82* (6), 853–858.

Xu, Y.; Zhang, W. X. Subcolloidal Fe/Ag particles for reductive dehalogenation of chlorinated benzenes. *Industrial & Engineering Chemistry Research* **2000**, *39* (7), 2238–2244.

Yamamoto, T.; Noma, Y.; Sakai, S. I.; Shibata, Y. Photodegradation of perfluorooctane sulfonate by UV irradiation in water and alkaline 2-propanol. *Environmental Science & Technology* **2007**, *41* (16), 5660–5665.

Yao, Y.; Volchek, K.; Brown, C. E.; Robinson, A.; Obal, T. Comparative study on adsorption of perfluorooctane sulfonate (PFOS) and perfluorooctanoate (PFOA) by different adsorbents in water. *Water Science and Technology* **2014**, *70* (12), 1983–1991.

Yeung, L. W. Y.; Miyake, Y.; Taniyasu, S.; Wang, Y.; Yu, H. X.; So, M. K.; Jiang, G. B. et al. Perfluorinated compounds and total and extractable organic fluorine in human blood samples from China. *Environmental Science & Technology* **2008**, *42* (21), 8140–8145.

Yeung, L. W. Y.; Mabury, S. A. Bioconcentration of aqueous film-forming foam (AFFF) in juvenile rainbow trout (*Oncorhyncus mykiss*). *Environmental Science & Technology* **2013**, *47* (21), 12505–12513.

Young, C. J.; Furdui, V. I.; Franklin, J.; Koerner, R. M.; Muir, D. C. G.; Mabury, S. A. Perfluorinated acids in arctic snow: New evidence for atmospheric formation. *Environmental Science & Technology* **2007**, *41* (10), 3455–3461.

Yu, J.; Hu, J. Y. Adsorption of perfluorinated compounds onto activated carbon and activated sludge. *Journal of Environmental Engineering* **2011**, *137* (10), 945–951.

Yu, Q.; Zhang, R. Q.; Deng, S. B.; Huang, J.; Yu, G. Sorption of perfluorooctane sulfonate and perfluorooctanoate on activated carbons and resin: Kinetic and isotherm study. *Water Research* **2009**, *43* (4), 1150–1158.

Zhang, W. X.; Wang, C. B.; Lien, H. L. Treatment of chlorinated organic contaminants with nanoscale bimetallic particles. *Catalysis Today* **1998**, *40* (4), 387–395.

Zhao, H.; Gao, J.; Zhao, G.; Fan, J.; Wang, Y.; Wang, Y. Fabrication of novel SnO2-Sb/carbon aerogel electrode for ultrasonic electrochemical oxidation of perfluorooctanoate with high catalytic efficiency. *Applied Catalysis B: Environmental* **2013**, *136–137*, 278–286.

20 Short-Chain PFAS
Their Sources, Properties, Toxicity, Environmental Fate, and Treatment

*Yuan Yao, Justin Burgess,
Konstantin Volchek, and Carl E. Brown*

CONTENTS

20.1 Introduction 447
20.2 Sources..448
    20.2.1 Waste..448
    20.2.2 Precursor Compounds 449
20.3 Properties..450
    20.3.1 Physical Properties 450
    20.3.2 Kinetics in Organisms 450
    20.3.3 Bioaccumulation 451
20.4 Toxicity 452
20.5 Environmental Fate...454
    20.5.1 Sediment...454
    20.5.2 Soil..456
    20.5.3 Humans and Biota 456
    20.5.4 Precipitation..457
    20.5.5 Arctic Ice and Water 457
20.6 Treatment..458
20.7 Conclusions...460
Acknowledgments..461
References..461

20.1 INTRODUCTION

Several long-chain per- and polyfluoroalkyl substances (PFAS), such as perfluorooctanesulfonic acid (PFOS) and perfluorooctanoic acid (PFOA), have been recognized as persistent, bioaccumulative, and toxic. They have been detected globally in the environment, biota, and humans. This has led to efforts toward the phaseout of long-chain PFAS and the development of a large number of safer alternatives, including short-chain PFAS. Although this book is mainly focused on PFAS contained in aqueous film-forming foams (AFFFs), it should be noted that a variety of short-chain PFAS are being used in consumer products. Kotthoff et al. (2015) analyzed various consumer products and found perfluorobutanoic acid (PFBA), perfluorohexanoic acid (PFHxA), and perfluoroheptanoic acid (PFHpA) in nanosprays, outdoor textiles, carpets, gloves, paper food wrappers, ski wax, and leather. Perfluoropentanoic acid (PFPeA) was found in all of these except for gloves. Perfluorobutanesulfonic acid (PFBS) was found in carpets, ski wax, leather, and awning textiles, while perfluorohexanesulfonic acid (PFHxS) and perfluoroheptanesulfonic acid (PFHpS) were found only in food paper, ski wax, and leather, with PFHpS also in awning textiles. 6:2 fluorotelomer alcohol (6:2 FTOH) was found in cleaning products, nanosprays, outdoor textiles, carpets, gloves, and food paper. Since much of the PFAS research has been focused on PFOS, PFOA, and other long-chain compounds (≥C8), information on short-chain PFAS (<C8) is scattered and scarce. This chapter attempts to summarize the available information on short-chain PFAS.

20.2 SOURCES

20.2.1 WASTE

A major source of PFAS in the environment is industrial and municipal wastes. Short-chain PFAS from these sources are becoming more common since the phasing out of long-chain compounds. High levels of short-chain PFAS were found in the surface water of Tangxun Lake in China (Zhou et al., 2013). PFBA and PFBS were the predominant perfluoroalkyl acids (PFAAs) in the surface water, with average concentrations of 4.77 and 3.66 µg L^{-1}, respectively. The source of the short-chain PFAS was untreated industrial wastewater. The high concentrations of short-chain PFAS demonstrated that the industry in the area was shifting away from long-chain PFAS. A study of the Daling River Basin in China found that short-chain PFAS were being introduced into the river by two nearby fluorochemical industrial parks (Wang P. et al., 2015). The detected highest level of PFBS (2.9 µg L^{-1}) was just below its provisional health-related indication value (HRIV) as safe in drinking water for lifelong exposure in Germany (3 µg L^{-1}) (Wilhelm et al., 2010).

Castiglioni et al. (2015) conducted a study of PFAS in water in Milan, Italy. They found that the largest source of PFAS in the water was industrial wastewater treatment plants (WWTPs), which discharged up to 50 times more than those receiving municipal wastes. Short-chain (C4–C7) perfluoroalkyl carboxylic acids (PFCAs) contributed up to 44% of the total PFAAs. The highest contribution of short-chain PFAS was found to be at an industrial WWTP. The types of influents affected the PFAS profiles in effluents. For example, PFHxA was found to be the most abundant in the effluents from a plant receiving municipal wastes in 2013.

Li et al. (2011) found small concentrations of PFHpA and large concentrations of PFHxA in surface water and sediments in Haihe River, China. Ratios of PFAS varied at different parts of the river, suggesting a variety of PFAS sources. Potential sources of PFAS identified were a nearby industrial center, the degradation of precursor compounds, and agricultural contamination, but further investigation was recommended to determine the specific sources.

PFAS were found at several waste disposal sites in Vietnam (Kim et al., 2013). Short-chain PFAS were detected at a rural dumping site, a municipal dumping site, a municipal wastewater station, a recycling site, and an e-waste recycling site, with the municipal dumping site found to be the most impactful source. The types of PFAS detected were different for each site too, suggesting that each source contributes different profiles of PFAS.

20.2.2 Precursor Compounds

One source of PFAS is precursor compound degradation. These are compounds that are produced and released into the environment and subsequently degrade into PFAS. One such type of precursor compounds is fluorotelomer unsaturated carboxylic acids (FTUCAs) (Anumol et al., 2016). Anumol et al. investigated FTUCA degradation by advanced oxidation processes (AOPs). They found that oxidation would degrade FTUCAs into PFAS with the same carbon chain length, along with some formation of PFAS that were one carbon shorter and a very small amount that were two carbons shorter. Specifically, 6:2 FTUCA degraded into PFHxA and a small concentration of PFPeA, and 8:2 FTUCA was found to degrade into PFOA with a small concentration of PFHpA and minor formation of PFHxA. These findings are important since oxidation methods are currently used in water treatment facilities. This raises concerns that water treatment plants could increase concentrations of PFAS, including short-chain compounds in treated water.

Houtz et al. (2016) investigated wastewater effluent samples from eight WWTPs that discharge to San Francisco Bay, USA. The treatment facilities included both industrial and municipal WWTPs. PFAA precursor compounds accounted for 33%–63% of the total molar concentration of PFAS across all effluent samples. The majority of the precursor compounds were found to produce C6 and shorter PFAS upon total oxidizable precursor (TOP) assay. PFBA, PFPeA, PFHxA, PFHpA, PFBS, PFHxS, and perfluorohexylphosphonic acid (PFHxPA) were detected in the effluent samples. PFHxA had the highest median concentration of all PFAS, including long-chain compounds. Data comparison between 2009 and 2014 showed a significant increase in PFBA and PFHxA concentrations. One of the PFAS sources was localized usage of AFFFs. Some of the plants serviced airports, while some had other customers using AFFFs. PFAS contamination from AFFFs is typically associated with groundwater and accidental releases to surface water. This study suggested that AFFFs might be a primary source of PFAS, particularly short-chain PFAS in WWTP effluent.

Some short-chain PFAS may also come from environmental transformation. For example, trifluoroacetic acid (TFA) can be formed through atmospheric degradation of hydrofluorocarbon (HFC) and hydrochlorofluorocarbon (HCFC). In a study of short- and long-chain PFAS in solid matrices (i.e., sediments, soils, and sludge) collected in Shanghai, China, Li et al. (2010) found that TFA was the major perfluorinated acid (PFA) in most cases and accounted for 22%–90% of the total PFAS (C2–C14). Given that some short-chain PFAS, like TFA, are mildly phytotoxic, the elevated levels of these short-chain PFAS in solid matrices should be investigated further.

20.3 PROPERTIES

There are many large gaps in the research on the properties of PFAS, partly because there are so many different compounds and isomers of those compounds. Most of the research in this area has been focused on PFOS and PFOA, since those were the two most important PFAS. Rayne and Forest (2009) conducted a review of the properties of PFAS. Research on the physical properties of PFAS was found to be lacking, especially for branched isomers of PFAS. They determined that software modeling was not reliable for calculating the physical properties of PFAS.

20.3.1 Physical Properties

ENVIRON International (2014) assessed if 6:2 FTOH, 6:2 fluorotelomer acrylate (6:2 FTAC), 6:2 fluorotelomer methacrylate (6:2 FTMAC), and PFHxA met the criteria for being persistent organic pollutants (POPs). The first three compounds are manufacturing intermediates that degrade into PFHxA. They found that none of them met more than one out of the four criteria for being a POP. None met the criteria for bioaccumulation or toxicity. None of the intermediates met the criteria for persistence since they degrade into PFHxA. However, PFHxA did meet the persistence criteria. 6:2 FTOH met the criteria for long-range transport potential, while PFHxA required further data for this determination. Neither 6:2 FTAC nor 6:2 FTMAC met the transport criteria.

Wang F. et al. (2015) found that humic acid reduces the sorption of PFBS on bohemite. This suggests that the presence of natural organic materials can reduce the ability of PFBS to adsorb onto solids, resulting in greater concentrations in higher organic content waters. Organic matter is very common in natural water systems, so understanding its effects on PFAS, including short-chain compounds, is crucial.

Zhao et al. (2014) investigated PFAS sorption on humic acids and humin from peat soil. They found that PFAS sorption was mainly governed by hydrophobic interactions. PFAS with fewer carbons were found to have lower sorption. pH was also found to have an impact on sorption, with an increase in pH causing decreased sorption. This might be ascribed to the electrostatic interaction and hydrogen bonding at lower pH. Hydrophobic interaction might also be stronger at lower pH due to the aggregation of humic acids.

Sundstrom et al. (2012a) synthesized radioactively labeled PFBS and studied its adsorption onto common laboratory equipment materials. They found that PFBS did not bind to polypropylene, polystyrene, or glass. PFBS was also found to be less soluble in hydrophobic solvents than in water.

20.3.2 Kinetics in Organisms

Sundstrom et al. (2012b) examined the pharmacokinetics of PFHxS in rats, mice, and monkeys. Rats and mice were much faster at eliminating PFHxS than monkeys. Half-lives were found to be on the order of months. Based on comparison with literature, the authors pointed out that the elimination kinetics of PFHxS by these animal species are similar to those of PFOS in male rats, mice, monkeys, and humans.

PFHxS and PFOS have the lowest elimination rates compared with PFBA, PFHxA, PFOA, and PFBS. PFBA and PFHxA appear to be eliminated the most efficiently in the species studied (i.e., mice, rats, monkeys, and humans for PFBA, and rats and monkeys for PFHxA), with serum elimination half-lives of hours to several days.

Olsen et al. (2009) compared the properties of PFBS in rats, monkeys, and humans. Urine seemed to be the major path for excretion, with both rats and monkeys excreting the majority of the dosed PFBS within 24 hours. PFBS excretion through urine was less effective in humans, but PFBS was still completely excreted within 10 days. Elimination was much faster than what had been observed in other studies with PFHxS and PFOS.

Bogdanska et al. (2014) treated mice with radioisotope-labeled PFBS and characterized the tissue distribution of PFBS in mice after 5 days of exposure. After the 5 days, the highest PFBS levels were detected in the liver, gastrointestinal tract, blood, kidney, cartilage, whole bone, lungs, and thyroid gland. The tissue levels increased over the first 3 days and then leveled off between days 3 and 5. Whole bone, liver, blood, skin, and muscle contained approximately 90% of the PFBS recovered in the tissues. This exposure to PFBS resulted in significantly lower tissue levels than did similar exposure to PFOS, as well as in a different pattern of tissue distribution, including lower levels in liver and lungs than in blood.

Chengelis et al. (2009) compared the toxicokinetic behavior of PFHxA and PFBS in cynomolgus monkeys and Sprague-Dawley rats. For both monkeys and rats, equal intravenous dosages of PFHxA and PFBS resulted in lower systemic exposure to PFHxA than PFBS. Serum clearance was more rapid for PFHxA than for PFBS. In monkeys, there were no major gender differences. In rats, it was found that exposure was lower in females than in males. Serum clearance of PFHxA and PFBS was more rapid in females than in males; however, there was no appreciable difference in the extent or rate of urinary elimination between compounds or genders.

Kowalczyk et al. (2013) examined the absorption and excretion of PFAS by dairy cows when fed with contaminated feed. The feed was grown on farmland that was contaminated by PFAS and had PFAS-contaminated fertilizer administered. PFBS had no tendency to accumulate in plasma and was excreted quickly via urine. Very little PFBS was detected in milk, liver, and kidney. All these observations indicated that PFBS did not accumulate in the body of dairy cows. PFHxS had a low tendency to accumulate, lower than that of PFOA, and was found to be excreted via both milk and urine. PFHxS was found in the meat of the cows, but its level significantly declined after 21 days of PFAS-free feeding. Overall, the authors found that elimination was faster for shorter-chain compounds, and shorter compounds had lower accumulation. This study also shows that contaminated fertilizer could potentially be a route of exposure for humans through dairy and meat products from animals fed with feed from contaminated sources.

20.3.3 Bioaccumulation

A study of bioaccumulation of PFAS in *Lumbriculus variegatus* (blackworms) (Lasier et al., 2011) detected higher concentrations of PFHpA, PFBS, PFHxS, and PFHpS in the worm tissue than the surrounding sediments. They found a relationship between increased carbon chain length and increased bioaccumulation. The tendency to bioaccumulate also increased with the presence of the sulfonate moiety. Biota–sediment accumulation factors indicated that short-chain PFCAs with fewer than seven carbons might be environmentally benign alternatives in aquatic ecosystems; however, sulfonates with four to seven carbons might be as likely to bioaccumulate as PFOS.

Hong et al. (2015) studied the bioaccumulation of PFAS in coastal organisms in South Korea. They found that the percent contribution of short-chain PFAS to the total PFAS concentration had been increasing over recent years. Bioaccumulation of different PFAS was found to be species specific. This could be due to different habitats and feeding habits between species, although the main contributing factors remain unknown. Long-chain PFAS had larger bioaccumulation factors (BAFs) than short-chain compounds. Short-chain PFAS were generally below the Toxic Substances Control Act criteria for being bioaccumulative. PFBS and PFHxS were not detected in the coastal organisms in 2008 but were commonly detected in 2010.

A study of PFAS levels in water, sediment, and biota of the Jucar River, Spain (Campo et al., 2016) found that mean concentrations of PFCAs in fish samples were higher than those in water and sediment, indicating possible bioaccumulation. PFPeA in biota was detected at the highest level (max 946 µg kg^{-1}, mean 274 µg kg^{-1}), which is more than 100 times higher than that of PFOS (max 8.13 µg kg^{-1}, mean 2.16 µg kg^{-1}). The estimated BAF (in L kg^{-1}) for PFPeA was 6.63. PFBA was detected in both water (mean 83.1 ng L^{-1}) and sediment (5.85 ng g^{-1}) but not in biota, while PFHxS was detected in water (mean 24.4 ng L^{-1}) and biota (mean 0.63 µg kg^{-1}) but not in sediment. There was no significant correlation found between the trophic-level data and PFAS concentrations.

20.4 TOXICITY

Borg et al. (2013) evaluated available hazard and risk assessments and toxicological data for PFAS congeners. The toxicological end points included hepatotoxicity (hepatocellular hypertrophy, hepatocellular vacuolation, increased liver weight, and liver-to-body ratio) and reproductive toxicity (reduced fetal, perinatal, and neonatal viability; reduced body weight or body weight gain; and litter loss in the dams). They found that long-chain and short-chain PFAS are relatively similar relating to their hepatotoxic and reproductive toxic effect levels based on internal doses. Long-chain PFAS are more potent than short-chain compounds. The hepatotoxicity of PFAS seems to be dependent on the hepatic concentration of the congener and not on its structure, and the difference in potency between congeners likely is due, to some extent, to kinetic differences, with short-chain PFAS being more rapidly excreted than long-chain compounds.

Ding et al. (2012) studied PFAS toxicity to lettuce seeds and green algae. Toxicity for both species was found to decrease with decreasing chain length. The PFBA concentration causing 50% inhibition effect (EC$_{50}$) in lettuce was 2.3 times as large as that of PFOA. This was not true for algae, likely due to the fact that algae are extremely pH sensitive and PFBA was the most acidic PFAS tested.

Eriksen et al. (2010) investigated the potential effects of short- and long-chain PFAS on DNA in HepG2 cells. PFBS and PFHxA were not found to generate reactive oxygen species (ROS) in blood, which may induce oxidative DNA damage. Only the exposure to perfluorononanoic acid (PFNA) caused a modest increase in DNA damage at a cytotoxic concentration level, which was detected as lactate dehydrogenase release into the cell medium.

Gorrochategui et al. (2014) investigated the cytotoxicity of PFAS in human placental choriocarcinoma cell line JEG-3. No significant cytotoxicity was observed for PFBA, PFHxA, PFBS, and PFHxS, while it was for long-chain PFAS, such as PFOA and PFOS. There seemed to be a correlation between chain length and toxicity. Despite the low uptake, PFBS and PFHxS were found to be able to significantly inhibit CYP19 aromatase activity in human placental cells. In addition, a mixture of all the PFAS tested was shown to cause an increase of production of cell lipids in the cellular membrane, especially at low concentrations. These results suggest the ability of PFAS to interact with cellular membranes, possibly inducing the synthesis of lipidome as a defense mechanism of cells.

Hagenaars et al. (2011) evaluated the effects of PFBS and PFBA on zebrafish eggs. Both chemicals only demonstrated toxic effects at the highest concentrations tested. PFBS was found to be more toxic than PFBA, causing malformations in the head and an increased heart rate, while PFBA did not. They both caused tail malformations and an uninflated swim bladder. Eight-carbon PFAS were more toxic than four-carbon PFAS, and to a lesser extent, perfluoroalkyl sulfonic acids (PFSAs) were more toxic than PFCAs.

Ikeda et al. (1985) investigated the induction of peroxisome proliferation in the livers of rats dosed with PFAS. Peroxisomes are organelles used by the cell to break down fatty acids; however, PFAS cannot be metabolized by them. PFBA-tainted feed was found not to increase liver weight, but to increase the activity of enzymes associated with peroxisomes by 42%. A single PFBA injection was not found to induce peroxisomes, which was attributed to PFBA perhaps being excreted more quickly. They also found that the fluorocarbons without a functional group did not induce peroxisomes, suggesting that the functional group is key for this effect.

Lee and Viberg (2013) investigated the effects of a single PFHxS dose on developing mouse pup brains. Changes in the neuroproteins in the neonatal mouse brains were observed within 24 hours of exposure. This can have effects on healthy brain development and cognitive function, suggesting that PFHxS may act as a developmental neurotoxicant.

Lee et al. (2014) found that PFHxS causes apoptosis of neuronal cells in rats. They identified a possible pathway for this toxic effect, which indicates that a mechanism other than ROS may play an important role. Beesoon et al. (2012) detected exceptionally high serum concentrations of PFHxS (27.5–423 ng mL^{-1}) in a Canadian family, with the highest concentration in the youngest children. The exposure was linked to dust ingestion and/or inhalation in the home where the household carpets were repeatedly treated with possible PFAS-containing formulations. These studies suggest that continuous exposure to PFHxS in humans should be a matter of concern, and children could be more vulnerable to the toxic effects of PFHxS.

The effects of PFBS on the growth and sexual development of amphibians were evaluated by Lou et al. (2013). They exposed *Xenopus laevis* tadpoles at a series of concentrations of PFBS (0.1–1000 µg L^{-1}). PFBS did not show a significant effect on the survival and growth. PFBS had no effect on the sex ratio and gonadal histology. However, it exhibited adverse effects on the hepatic histology and sexual development of the amphibian *Xenopus laevis* at environmentally relevant concentrations.

A study on partition of environmental chemicals between maternal and fetal blood and tissues was performed by Needham et al. (2011). PFHxS was found to cross the placental wall, but there was a low correlation between maternal and cord serum concentrations. The observed PFAS concentrations were higher in the maternal serum than in cord serum, and the ratio suggested that the length of the PFAS chain, as well as the active group, affected the ability to pass the placenta. PFAS with a shorter chain length showed relatively higher cord serum concentrations than those with a longer chain length. PFAS with sulfonic acid as the active group appeared to pass more easily into the fetal circulation than those with carboxylic acid as the active group.

Newsted et al. (2008) evaluated the effects of PFBS on mallard and northern bobwhite quail. There were no acute or chronic toxic effects associated with PFBS, and any differences between the control group and the exposed group were only temporary. PFBS was observed to be 150 times less toxic than PFOS and did not bioaccumulate.

Viberg et al. (2013) studied the effects of PFHxS on neonatal rat development. PFHxS was found to have adverse effects on the behavior of these rats once they matured to adults. A single dose of PFHxS caused permanent changes to their spontaneous behavior and habituation to new environments. They identified the cholinergic system as a potentially affected part of the brain.

In summary, toxic effects have been observed for short-chain PFAS; however, they appear to be less toxic than long-chain compounds. In general, toxicity appears to decrease with decreasing chain length. PFSAs seem to have a higher potential for toxicity than carboxylic counterparts, the PFCAs.

20.5 ENVIRONMENTAL FATE

20.5.1 Sediment

Three studies were performed by Ahrens et al. on PFAS partitioning in water, sediment, and suspended particulate matter. The first study, performed on pore water and sediment from Tokyo Bay, Japan (Ahrens et al., 2009a), found that PFSAs bind more strongly to sediment than PFCAs, and longer-chain PFAS bind more strongly than short-chain congeners. PFBA, PFPeA, PFHxA, and PFHpA were found exclusively in pore water, and PFHxS was found almost exclusively in sediment.

The second study, performed on water and suspended particulate matter from the Elbe River in Germany (Ahrens et al., 2009b), found PFBS, perfluoropentanesulfonic acid (PFPeS), PFHxS, PFHpS, PFBA, PFPeA, PFHxA, and PFHpA in the dissolved phase in water and only PFBS and PFHxS in the particulate phase.

PFAS had a much higher frequency of detection and higher concentrations in the dissolved phase, suggesting that they tended to remain dissolved instead of adsorbing onto particulate matter.

The third study was performed on water, suspended particulate matter, and sediment from Tokyo Bay, Japan (Ahrens et al., 2010), and found short-chain PFCAs almost exclusively in the dissolved phase, with only PFHxS being in both the dissolved phase and the suspended particulate matter. PFHxS, PFBA, PFPeA, PFHxA, and PFHpA were all detected, with PFBA accounting for 52% of all PFAS detected. The partitioning coefficients for the PFAS were calculated. Increasing the carbon chain length was found to increase the partitioning coefficient, increasing the favorability of adsorbing onto solids in the water. Between water and suspended particulate matter, with each additional CF_2 moiety, the log K_{OC} (organic carbon normalized partitioning coefficient) increased by 0.52–0.75 log units for the PFCAs. The functional group also had a crucial influence on the partitioning coefficient. The log K_{OC} for PFSAs was 0.71–0.76 log units higher than that for the PFCAs.

All three of these studies show that short-chain PFAS will favor remaining dissolved in water over adsorbing onto suspended particulate matter and sediment, while long-chain PFAS are more likely to adsorb onto solids (Ahrens et al., 2009a, 2009b, 2010). Increased adsorption onto sediment decreases the aqueous long-range transport potential of the compound, so short-chain PFAS have higher long-range transport potential than long-chain congeners due to their lower partitioning coefficients (Ahrens et al., 2009a, 2010). This was also found to be the case in a study by Wang P. et al. (2015), although only specifically for PFBA and PFBS, and a study by Li et al. (2011). PFCAs should also have higher transport potential than PFSAs, due to their lower partitioning coefficients (Ahrens et al., 2009a, 2010).

A study by Campo et al. (2016) examined short- and long-chain PFAS levels in river water and sediment collected from the Jucar River in Spain. PFPeA was detected as one of the most predominant congeners. Clear PFAS accumulations were found in the sediments downstream, suggesting that regulation dams in the basin might have acted as sinks allowing such accumulations.

Sun et al. (2012) investigated PFAS in municipal WWTPs in Tianjin, China. PFAS were found to partition into sludge. Longer-chain PFAS were found to partition into sludge more than shorter-chain congeners, and PFSAs partitioned more than PFCAs. PFAS levels in effluent water were lower than those in influent water, and this was attributed to partitioning into the sludge.

Zhou et al. (2013) examined the spatial distribution of PFAS contamination in Tangxun Lake, China. As they sampled farther away from the contamination source, they found that the concentrations of PFBS and PFBA had decreased linearly, while the concentrations of long-chain PFAS had decreased exponentially. This was related to partitioning into the sediment. Longer-chain PFAS were found to be less water soluble than shorter-chain congeners and were more likely to be in sediment than shorter-chain compounds. Short-chain PFAS were found in plants and fish in the area; however, the levels were lower than those in the surrounding water. Bioaccumulation potential was calculated and found to be relatively low (<1), suggesting that short-chain PFAS do not bioaccumulate.

20.5.2 SOIL

Gellrich et al. (2012) examined the retention of PFAS in soil. They found that the retention of PFAS in soil increases with increasing chain length. PFOS did not leach, while PFOA and shorter-chain PFAS did. The total concentration of short-chain PFAS in groundwater was much higher than that of long-chain PFAS due to higher soil retention of the long-chain congeners. The results also indicated that larger and more lipophilic molecules can displace shorter congeners from their binding sites in the soil.

Li et al. (2010) investigated PFAS in soil, sediment, and sludge samples collected in China. Perfluoropropionic acid (PFPrA), PFBA, PFPeA, PFHxA, PFHpA, and PFBS were detected in all three sample types. PFHxS was detected in most sediment samples and a few soil and sludge samples. All the solid matrices were directly air-dried before analysis, and the total PFAS (both short- and long-chain) concentrations in sludge were found to be the highest compared with those in soil and sediment.

Sepulvado et al. (2011) examined PFAS in typical municipal biosolids applied to agricultural soil. PFAS levels in soils treated with affected biosolids were found to depend on their levels in the biosolids. The results suggested that significant quantities of PFBS were formed through precursor degradation. The leaching potential of PFAS from soil was found to be dependent on chain length, with short-chain PFAS leaching more than long-chain congeners.

Tan et al. (2014) investigated PFAS in surface soils along the Koshi River in Nepal, a typical agricultural country with little industrialization and urbanization. PFBA, PFPeA, PFHxA, PFHpA, and PFBS were detected from the soils. PFAS were found mainly in the areas of human activity. PFAS were not produced locally, so all of the PFAS came from long-range atmospheric transport and/or usage and disposal of PFAS-containing products.

Venkatesan and Halden (2014) examined the fate of PFAS in soils treated with biosolids. They found that concentrations of PFCAs with a carbon chain length of eight or fewer decreased in soils over time, while those with chain length greater than eight increased. Based on the first-order loss curve, the mean half-lives were calculated to be 385, 403, 417, and 866 days, respectively, for PFBA, PFPeA, PFHxA, and PFHpA. The loss of these compounds may be related to potential leaching, plant uptake, volatilization from soil, or a combination of these.

20.5.3 HUMANS AND BIOTA

PFHxS was found in the blood of Swedish women (Axmon et al., 2014). Samples taken between 1987 and 2007 were tested. PFHxS serum levels peaked during the period 1990–2000. A dependence on season was found, with serum PFHxS levels being higher during the winter and lower during the summer. There was no statistically relevant relationship found with the women's age.

Rotander et al. (2015) found PFHxS in the blood of Australian firefighters. PFHxS levels were found to be correlated with exposure to 3M AFFF, even 10 years after it was phased out. Blood donation was also linked to reduced levels of PFAS. Possible associations between serum PFAA concentrations and five biochemical outcomes (i.e., serum cholesterol, triglycerides, high-density lipoproteins, low-density lipoproteins, and uric acid) were assessed. No statistical associations between any of these end points and serum PFAA concentrations were observed.

A study by Gebbink et al. (2016) found PFBS, PFHxS, and PFHpA in polar bear and killer whale livers from Greenland. Only PFHxS and PFHxA were found in ringed seal livers. No PFAS precursors were detected. Since this is a remote Arctic location, the findings suggested that there is potential for long-range transport for short-chain PFAS.

Uptake of PFAS by market-size rainbow trout was examined by Goeritz et al. (2013). They found that liver, blood, kidney, and skin were the main target tissues. Despite the relatively low PFAS contamination found in muscle and skin, the edible parts of the fish can significantly contribute to the whole-body burden due to their high weight proportion. Biomagnification factors were calculated for both short- and long-chain PFAS and found to be below the biomagnification threshold of 1.

Chu et al. (2015) studied PFAS buildup in black-footed albatross from Midway Island, North Pacific Ocean. PFHxS was detectable in liver, muscle, and adipose tissue. PFHxA and PFHpA were only found at low concentrations in the liver. Long-chain PFAS dominated the total amount of PFAS buildup.

20.5.4 Precipitation

Kwok et al. (2010) investigated 20 PFAS in precipitation samples from several countries. PFPrA was detected in all of the samples. PFHxA was found in 85% of samples from Japan, PFHpA was found in all samples from the United States, and PFBA was found in all samples from Hong Kong. They also found that a second rain event has a lower PFAS concentration than the first rain event, suggesting that precipitation scavenges PFAS from the atmosphere. This also demonstrates that short-chain PFAS are present in the atmosphere.

PFBA, PFPeA, PFHxA, PFHpA, and PFHxS were detected in rainwater samples collected at a fluorine chemical industrial park at Changshu and its neighboring regions in eastern China (Chen et al., 2016). PFOA was dominant in the samples collected near the industrial park, but other PFAS started to appear farther away from the area, suggesting some sources other than the park, such as sewage from rural areas or atmospheric deposition of those compounds.

Meyer et al. (2011) examined PFAS during snowmelt in Toronto, Canada. PFHxA, PFHpA, PFBS, and PFHxS were all detected. PFHxA had a very high abundance, and this may be due to the fact that it is the replacement of long-chain PFAS in various consumer products. Concentrations in snow were lower than those in meltwater for short-chain PFAS by one order of magnitude. Concentrations of PFAS in snow increased over the sampling period, and this may be due to collection of the particle-bound PFAS. Snow contributed less than 10% of the short-chain PFAS in the nearby river water.

20.5.5 Arctic Ice and Water

Kwok et al. (2013) analyzed ice cores, snow, and water from Svalbard, a Norwegian island north of the Arctic Circle. In the ice cores, PFBA accounted for 39% of the

PFAS detected, and PFPeA and PFHxA were also detected. PFBA and PFPeA were the dominant PFAS in surface water, and PFHxS, PFHxA, and PFHpA were also detected in surface water. Two PFAS sources were found, glacier contamination from long-range atmospheric transport and downstream contamination caused by industrial and municipal activities of a nearby settlement. Of these, the long-range deposition is more interesting, as it demonstrates that PFBA and PFPeA, which were found in the glacial region, have airborne long-range transport potential.

PFAS, including short-chain compounds (i.e., PFBS, PFHxS, and PFHpA) and FTOHs, were found in lakes in the Canadian Arctic (Stock et al., 2007). A possible source for one of the lakes was waste from a nearby airport base. Possible use of AFFFs at the base may have contributed. Like long-chain PFAS, PFHxS was also detected in atmospheric samples, suggesting that atmospheric transport and deposition was a route for short-chain PFAS contamination in the Canadian Arctic.

20.6 TREATMENT

Methods that have been found to be effective for short-chain PFAS include reverse osmosis (RO) (Olsen and Paulson, 2008; Quinones and Snyder, 2009; Thompson et al., 2011; Appleman et al., 2013, 2014; Tabtong et al., 2015), nanofiltration (NF) (Appleman et al., 2013), and adsorption with granular activated carbon (GAC) (Olsen and Paulson, 2008; Quinones and Snyder, 2009; Brede et al., 2010; Carter et al., 2010; Hansen et al., 2010; Eschauzier et al., 2012; Appleman et al., 2013, 2014; Zaggia et al., 2016).

RO has shown significant removal for all PFAS, including PFBA in US full-scale water treatment systems (Appleman et al., 2014). NF demonstrated high PFAS removal in simulated real-world conditions (Appleman et al., 2013); however, further studies are recommended to corroborate these findings. Compared with RO and NF, adsorption with GAC is a relatively inefficient method for short-chain PFAS removal, requiring frequent carbon changing due to low sorption capacities for short-chain PFAS and, consequently, fast breakthrough times. Breakthrough times can be as short as 2 months for PFBA (Appleman et al., 2014). Because of this, GAC is an effective method only with constant monitoring of short-chain PFAS levels in the effluent and regular replacement of the carbon.

Resin has also been tested (Carter et al., 2010; Zaggia et al., 2016); however, removals only seem to be comparable to GAC removals at best. Regeneration is only possible with resins with lower removals. Resins with removals comparable to those of GAC could only be regenerated under harsh conditions (Zaggia et al., 2016). However, since there are a large variety of resins with a large variety of characteristics, it seems imprudent to discount them yet without further study. On that note, Chapters 14 and 21 provide more discussion on ion exchange resins and interesting results from a case study comparing GAC performance to resin performance for the removal of both long- and short-chain PFAS.

Appleman et al. (2014) performed bench-scale tests to investigate different PFAS treatment technologies, including RO, NF, GAC, and ion exchange methods. RO was the most effective method tested, managing to reduce all PFAS to below the detection limits. NF was the next most effective, reducing PFBA, PFPeA, and PFHxA to below the detection limits (Appleman et al., 2013). PFBS and PFHxS also had greater than 95% removal by NF in deionized water but were reduced to below the limits of detection when artificial groundwater and a fouling layer were present on the filtration membrane, which more accurately represents a real-world scenario (Appleman et al., 2013). GAC was initially effective at removing all PFAS, but there was rapid breakthrough for short-chain PFAS (Appleman et al., 2013, 2014). Without frequent replacement, GAC was found to be not very effective except when it was fresh, especially for very short-chain PFAS such as PFBA. In one water treatment utility, PFBA had a breakthrough time of only 2 months, so carbon replacement would have to be very frequent in order to achieve high removals (Appleman et al., 2014). Ion exchange resin was less effective at removing short-chain PFAS. It had >97% removal for PFHxS but only 81% and 46% removals for PFBS and PFHpA, respectively (Appleman et al., 2014).

Arnsberg, Germany, had elevated levels of PFAS in drinking water. Charcoal filters were installed to reduce the PFAS exposure of the residents. In 2010, Brede et al. (2010) conducted a follow-up biomonitoring study to measure the impact of the filters on the levels of PFAS in residents' plasma samples. The PFHxS plasma concentration had decreased since the filters had been instituted.

Carter et al. (2010) studied ion exchange resin and the GAC uptake of PFBS. PFBS reached equilibrium uptake by resin (i.e., Amberlite IRA-458) after 4 hours, and GAC (Filtrasorb 400) after 15 hours. PFBS had a lower GAC uptake than PFOS. PFBS uptake was irreversible for both resin and GAC using conventional regeneration methods.

Eschauzier et al. (2012) examined the treatment of drinking water by several conventional treatment approaches. They found that PFAS concentrations were not significantly affected by coagulation, sand filtration, or ozonation. GAC was observed to effectively remove certain PFAS. GAC had no significant effect on PFBA and PFHxA, while it increased the concentration of PFBS and decreased PFHxS. Short-chain PFAS concentrations dominated the total PFAS concentration in finished drinking water. GAC removal decreased with decreasing chain length and was lower for PFCAs than PFSAs.

Hansen et al. (2010) investigated the adsorption of PFAS from contaminated well water to activated carbon (AC) at environmentally relevant nanogram per liter concentrations. Powdered activated carbon (PAC) had higher adsorption capacity than GAC. PFSAs had stronger adsorption than PFCAs. Adsorption was found to increase with increasing chain length. Teflon adsorption was ineffective at removing PFAS.

Olsen and Paulson (2008) prepared a report for the State of Minnesota on the viability of point-of-use (POU) water treatment devices. POU devices with AC filters had difficulty removing PFBA from test solutions. The removal abilities of AC filters also decreased over time, resulting in breakthroughs after they had been used for too long, especially for PFBA. POU devices incorporating RO technology reduced all PFAS to below the detection limits in all but one device.

Quinones and Snyder (2009) looked into various drinking water treatment facilities across the United States. Successful removal of all target PFAS, including PFHxA and PFHxS, was only observed at a facility using microfiltration and RO in Orange County, California. Deep bed filtration, coagulation or filtration, ultraviolet (UV), and chlorination were all ineffective at removing PFHxA and PFHxS.

Tabtong et al. (2015) examined water treatment plants in Bangkok, Thailand. They found that many of the conventional water purification methods, including coagulation, sedimentation, chlorination, and sand filtration, were not effective and sometimes raised PFAS concentrations. Particularly, the PFBA level significantly increased during the conventional treatment process. They found that an advanced treatment process using GAC and RO was more effective than a conventional treatment process for removing PFAS (>86% removal); the RO step was especially effective.

Thompson et al. (2011) studied PFAS removal at water reclamation plants in South East Queensland, Australia. One plant using de-nitrification, ozonation, coagulation or flocculation, dissolved air flotation or sand filtration, and biologically AC generally failed to remove PFAS shorter than nine carbons. Another plant using ultrafiltration and RO was found to reduce all PFAS, including short-chain compounds (PFHxA, PFHpA, PFBS, and PFHxS), to below their limits of reporting.

Zaggia et al. (2016) tested anion exchange resins and compared their removal efficiencies with that of GAC. They found that GAC removal was inefficient and had an extremely fast breakthrough of PFBA. All the GAC filters tested reached the saturation point (0% removal) for PFBA within 3 months and then showed significant leaching of the PFBA previously sorbed in the following months. Early breakthrough of PFBA and PFSA was also observed for resins. Highly hydrophobic resin had higher removal capacity than non- and fairly hydrophobic resins, and it could only be regenerated under harsh conditions, and was deemed nonregenerable *in situ*. Longer-chain PFAS had higher sorption capacities than shorter-chain compounds. PFCAs had lower sorption capacities than PFSAs.

20.7 CONCLUSIONS

Due to recent and increasing regulatory attention being given to long-chain PFAS, short-chain PFAS are becoming more common in both production and the environment. Short-chain PFAS appear to be introduced into the environment mainly from industrial and municipal wastes; however, consumer products are also a source. Firefighting foams are also moving to short-chain PFAS and therefore make a contribution to the environment. Short-chain PFAS can also be generated through degradation of precursor compounds that are also released in wastes. This can lead to treated water having higher levels of short-chain PFAS than untreated water, due to the degradation of precursor compounds during wastewater treatment.

Short-chain PFAS have higher water solubility than longer-chain PFAS. Their sorption onto sediment and soil is dependent on chain length, with the shorter-chain PFAS having lower sorption than longer-chain compounds. As a result, short-chain PFAS have a greater potential for long-range transport in water. Short-chain PFAS have been found in precipitation and even in Arctic environmental samples, indicating their long-range atmospheric transport. Short-chain PFAS applied to soil through affected biosolids have also been found to leach from the soil into groundwater.

The toxicity and bioaccumulation potential of short-chain PFAS are also dependent on carbon chain length. The shorter the carbon chain, the more rapidly they are eliminated from organisms, the lower their observed toxicity, and the lower their bioaccumulation potential. Further research is required to determine at what chain length, if any, the toxic effects of short-chain PFAS are minimized.

Water treatment and removal of short-chain PFAS from the environment are very difficult. Three methods have been demonstrated to be effective: RO, NF, and GAC. Of these, RO is the only one that has demonstrated full removal in a real-world situation. NF has been demonstrated by one study to be effective under conditions mimicking real-world conditions; however, further study to corroborate these findings is recommended. GAC can be effective only for short periods of time, after which it loses its effectiveness and starts releasing the contaminants that it has adsorbed. This means that GAC would require constant monitoring and frequent replacement of carbon in order to remove short-chain PFAS.

ACKNOWLEDGMENTS

This study was funded by Environment and Climate Change Canada (ECCC)'s Compliance Promotion and Contaminated Sites Division (CPCSD) under the Federal Contaminated Sites Action Plan (FCSAP).

REFERENCES

Ahrens, L., N. Yamashita, L.W.Y. Yeung, S. Taniyasu, Y. Horii, P.K.S. Lam, and R. Ebinghaus. Partitioning behavior of per- and polyfluoroalkyl compounds between pore water and sediment in two sediment cores from Tokyo Bay, Japan. *Environmental Science and Technology*, 43:6969–6975, 2009a.

Ahrens, L., M. Plasmann, Z. Xie, and R. Ebinghaus. Determination of polyfluoroalkyl compounds in water and suspended particulate matter in the river Elbe and North Sea, Germany. *Frontiers of Environmental Science and Engineering China*, 3(2):152–170, 2009b.

Ahrens, L., S. Taniyasu, L.W.Y Yeung, N. Yamashita, P.K.S. Lam, and R. Ebinghaus. Distribution of polyfluoroalkyl compounds in water, suspended particulate matter and sediment from Tokyo Bay, Japan. *Chemosphere*, 79:266–272, 2010.

Anumol, T., S. Dagnino, D.R. Vandervort, and S.A. Snyder. Transformation of polyfluorinated compounds in natural waters by advanced oxidation processes. *Chemosphere*, 144:1780–1787, 2016.

Appleman, T.D., E.R.V. Dickenson, C. Bellona, and C.P. Higgins. Nanofiltration and granular activated carbon treatment of perfluoroalkyl acids. *Journal of Hazardous Materials*, 260:740–746, 2013.

Appleman, T.D., C.P. Higgins, O. Quinones, B.J. Vanderford, C. Kolstad, J.C. Zeigler-Holady, and E.R.V. Dickenson. Treatment of poly- and perfluoroalkyl substances in U.S. full-scale water treatment systems. *Water Research*, 51:246–255, 2014.

Axmon, A., J. Axelsson, K. Jakobsson, C.H. Lindh, and B.A.G. Jonsson. Time trends between 1987 and 2007 for perfluoroalkyl acids in plasma from Swedish women. *Chemosphere*, 102:61–67, 2014.

Beesoon, S., S.J. Genuis, J.P. Benskin, and J.W. Martin. Exceptionally high serum concentrations of perfluorohexanesulfonate in a Canadian family are linked to home carpet treatment applications. *Environmental Science and Technology*, 46:12960–12967, 2012.

Bogdanska, J., M. Sundstrom, U. Bergstrom, D. Borg, M. Abedi-Valugerdi, A. Bergman, J. DePierre, and S. Nobel. Tissue distribution of ^{35}S-labelled perfluorobutanesulfonic acid in adult mice following dietary exposure for 1–5 days. *Chemosphere*, 98:28–36, 2014.

Borg, D., B. Lund, N. Lindquist, and H. Hakansson. Cumulative health risk assessment of 17 perfluoroalkylated and polyfluoroalkylated substances (PFASs) in the Swedish population. *Environment International*, 59:112–123, 2013.

Brede, E., M. Wilhelm, T. Goen, J. Muller, K. Rauchfuss, M. Kraft, and J. Holzer. Two-year follow-up biomonitoring pilot study of residents' and controls' PFC plasma levels after PFOA reduction in public water system in Arnsberg, Germany. *International Journal of Hygiene and Environmental Health*, 213:217–223, 2010.

Campo, J., M. Lorenzo, F. Perez, Y. Pico, M. Farre, and D. Barcelo. Analysis of the presence of perfluoroalkyl substances in water, sediment and biota of the Jucar River (E Spain). Sources, partitioning and relationships with water physical characteristics. *Environmental Research*, 147:503–512, 2016.

Carter, K.E., and J. Farrell. Removal of perfluorooctane and perfluorobutane sulfonate from water via carbon adsorption and ion exchange. *Separation Science and Technology*, 45:762–767, 2010.

Castiglioni, S., S. Valseccho, S. Polesello, M. Ruscon, M. Melis, M. Palmiotto, A. Manenti, E. Davoli, and E. Zuccato. Sources and fate of perfluorinated compounds in the aqueous environment and in drinking water of a highly urbanized and industrialized area in Italy. *Journal of Hazardous Materials*, 282:51–60, 2015.

Chen, S., X. Jiao, N. Gai, X. Li, X. Wang, G. Lu, H. Piao, Z. Rao, and Y. Yang. Perfluorinated compounds in soil, surface water, and groundwater from rural areas in eastern China. *Environmental Pollution*, 211:124–131, 2016.

Chengelis, C.P., J.B. Kirkpatrick, N.R. Myers, M. Shinohara, P.L. Stetson, and D.W. Sved. Comparison of the toxicokinetic behavior of perfluorohexanoic acid (PFHxA) and nonafluorobutane-1-sulfonic acid (PFBS) in cynomolgus monkeys and rats. *Reproductive Toxicology*, 27:400–406, 2009.

Chu, S., J. Wang, G. Leong, L.A. Woodward, R.J. Letcher, and Q.X. Li. Perfluoroalkyl sulfonates and carboxylic acids in liver, muscle and adipose tissues of black-footed albatross (*Phoebastria nigripes*) from Midway Island, North Pacific Ocean. *Chemosphere*, 138:60–66, 2015.

Ding, G., M. Wouterse, R. Baerselman, and W.J.G.M. Peijnenburg. Toxicity of polyfluorinated and perfluorinated compounds to lettuce (*Lactuca sativa*) and green algae (*Pseudokirchneriella subcapitata*). *Archives of Environmental Contamination and Toxicology*, 62:49–55, 2012.

ENVIRON International. Assessment of POP criteria for specific short-chain perfluorinated alkyl substances. Prepared for FluoroCouncil, 2014.

Eriksen, K.T., O. Raaschou-Nielsen, M. Sorensen, M. Roursgaard, S. Loft, and P. Moller. Genotoxic potential of the perfluorinated chemicals PFOA, PFOS, PFBS, PFNA and PFHxA in human HepG2 cells. *Mutation Research/Genetic Toxicology and Environmental Mutagenesis*, 700:39–43, 2010.

Eschauzier, C., E. Beerendonk, P. Scholte-Veenendaal, and P. De Voogt. Impact of treatment processes on the removal of perfluoroalkyl acids from the drinking water production chain. *Environmental Science and Technology*, 46:1708–1715, 2012.

Gebbink, W.A., R. Bossi, F.F. Riget, A. Rosing-Asvid, C. Sonne, and R. Dietz. Observation of emerging per- and polyfluoroalkyl substances (PFASs) in Greenland marine mammals. *Chemosphere*, 144:2384–2391, 2016.

Gellrich, V., T. Stahl, and T.P. Knepper. Behavior of perfluorinated compounds in soils during leaching experiments. *Chemosphere*, 87:1052–1056, 2012.

Goeritz, I., S. Falk, T. Stahl, C. Schafers, and C. Schlechtriem. Biomagnification and tissue distribution of perfluoroalkyl substances (PFASs) in market-size rainbow trout (*Onchorhynchus mykiss*). *Environmental Toxicology and Chemistry*, 32:2078–2088, 2013.

Gorrochategui, E., E. Perez-Albaladejo, J. Casas, S. Lacorte, and C. Porte. Perfluorinated chemicals: Differential toxicity, inhibition of aromatase activity and alteration of cellular lipids in human placental cells. *Toxicology and Applied Pharmacology*, 277:124–130, 2014.

Hagenaars, A., L. Vergauwen, W. De Coen, and D. Knapen. Structure–activity relationship assessment of four perfluorinated chemicals using a prolonged zebrafish early life stage test. *Chemosphere*, 82:764–772, 2011.

Hansen, M.C., M.H. Borresen, M. Schlabach, and G. Cornelissen. Sorption of perfluorinated compounds from contaminated water to activated carbon. *Journal of Soils and Sediments*, 10:179–185, 2010.

Hong, S., J.S. Khim, T. Wang, J.E. Naile, J. Park, B. Kwon, S.J. Song et al. Bioaccumulation characteristics of perfluoroalkyl acids (PFAAs) in coastal organisms from the west coast of South Korea. *Chemosphere*, 129:157–163, 2015.

Houtz, E.F., R. Sutton, J. Park, and M. Sedlak. Poly- and perfluoroalkyl substances in wastewater: Significance of unknown precursors, manufacturing shifts, and likely AFFF impacts. *Water Research*, 95:142–149, 2016.

Ikeda, T., K. Aiba, K. Fukuda, and M. Tanaka. The induction of peroxisome proliferation in rat liver by perfluorinated fatty acids, metabolically inert derivatives of fatty acids. *Journal of Biochemistry*, 98:475–482, 1985.

Kim, J., N.M. Tue, T. Isobe, K. Misaki, S. Takahashi, P.H. Viet, and S. Tanabe. Contamination by perfluorinated compounds in water near waste recycling and disposal sites in Vietnam. *Environmental Monitoring and Assessment*, 185:2909–2919, 2013.

Kotthoff, M., J. Muller, H. Jurling, M. Schlummer, and D. Fiedler. Perfluoroalkyl and polyfluoroalkyl substances in consumer products. *Environmental Science and Pollution Research*, 22:14546–14559, 2015.

Kowalczyk, J., S. Ehlers, A. Oberhausen, M. Tischer, P. Furst, H. Schafft, and M. Lahrssen-Wiederholt. Absorption, distribution, and milk secretion of the perfluoroalkyl acids PFBS, PFHxS, PFOS, and PFOA by dairy cows fed naturally contaminated feed. *Journal of Agricultural and Food Chemistry*, 61:2903–2912, 2013.

Kwok, K.Y., S. Taniyasu, L.W.Y. Yeung, M.B. Murphy, P.K.S. Lam, Y. Horii, K. Kannan, G. Petrick, R.K. Sinha, and N. Yamashita. Flux of perfluorinated chemicals through wet deposition in Japan, the United States, and several other countries. *Environmental Science and Technology*, 44:7043–7049, 2010.

Kwok, K.Y., E. Yamazaki, N. Yamashita, S. Taniyasu, M.B. Murphy, Y. Horii, G. Petrick, R. Kallerborn, K. Kannan, K. Murano, and P.K.S. Lam. Transport of perfluoroalkyl substances (PFAS) from an Arctic glacier to downstream locations: Implications for sources. *Science of the Total Environment*, 447:46–55, 2013.

Lasier, P.J., J.W. Washington, S.M. Hassan, and T.M. Jenkins. Perfluorinated chemicals in surface waters and sediments from northwest Georgia, USA, and their bioaccumulation in *Lumbriculus variegatus*. *Environmental Toxicology and Chemistry*, 30:2194–2201, 2011.

Lee, I. and H. Viberg. A single neonatal exposure to perfluorohexane sulfonate (PFHxS) affects the levels of important neuroproteins in the developing mouse brain. *NeuroToxicology*, 37:190–196, 2013.

Lee, Y.J., S.Y. Choi, and J.H. Yang. PFHxS induces apoptosis of neuronal cells via ERK1/2-mediated pathway. *Chemosphere*, 94:121–127, 2014.

Li, F., C. Zhang, Y. Qu, J. Chen, L. Chen, Y. Liu, and Q. Zhou. Quantitative characterization of short- and long-chain perfluorinated acids in solid matrices in Shanghai, China. *Science of the Total Environment*, 408:617–623, 2010.

Li, F., H. Sun, Z. Hao, N. He, L. Zhao, T. Zhang, and T. Sun. Perfluorinated compounds in Haihe River and Dagu Drainage Canal in Tianjin, China. *Chemosphere*, 84:265–271, 2011.

Lou, Q., Y. Zhang, Z. Zhou, Y. Shi, Y. Ge, D. Ren, H. Xu, Y. Zhao, W. Wei, and Z. Qin. Effects of perfluorooctanesulfonate and perfluorobutanesulfonate on the growth and sexual development of *Xenopus laevis*. *Ecotoxicology*, 22:1133–1144, 2013.

Meyer, T., A.O. De Silva, C. Spencer, and F. Wania. Fate of perfluorinated carboxylates and sulfonates during snowmelt within an urban watershed. *Environmental Science and Technology*, 45:8113–8119, 2011.

Needham, L.L., P. Grandjean, B. Heinzow, P.J. Jorgensen, F. Nielsen, D.G. Patterson Jr., A. Sjodin, W.E. Turner, and P. Weihe. Partition of environmental chemicals between maternal and fetal blood and tissues. *Environmental Science and Technology*, 45:1121–1126, 2011.

Newsted, J.L., S.A. Beach, S.P. Gallagher, and J.P. Giesy. Acute and chronic effects of perfluorobutane sulfonate (PFBS) on the mallard and northern bobwhite quail. *Archives of Environmental Contamination and Toxicology*, 54:535–545, 2008.

Olsen, G.W., S. Chang, P.E. Noker, G.S. Gorman, D.J. Ehresman, P.H. Lieder, and J.L. Butenhoff. A comparison of the pharmacokinetics of perfluorobutanesulfonate (PFBS) in rats, monkeys, and humans. *Toxicology*, 256:65–74, 2009.

Olsen, P.C., and D.J. Paulson. Removal of perfluorochemicals (PFC's) with point-of-use (POU) water treatment devices. Performance Evaluation for Water Science and Marketing, LLC, 2008.

Quinones, O., and S.A. Snyder. Occurrence of perfluoroalkyl carboxylates and sulfonates in drinking water utilities and related waters from the United States. *Environmental Science and Technology*, 43:9089–9095, 2009.

Rayne, S. and K. Forest. Perfluoroalkyl sulfonic and carboxylic acids: A critical review of physicochemical properties, levels and patterns in waters and wastewaters, and treatment methods. *Journal of Environmental Science and Health*, 44:1145–1199, 2009.

Rotander, A., L.L. Toms, L. Aylward, M. Kay, and J.F. Mueller. Elevated levels of PFOS and PFHxS in firefighters exposed to aqueous film forming foam (AFFF). *Environment International*, 82:28–34, 2015.

Sepulvado, J.G., A.C. Blaine, L.S. Hundal, and C.P. Higgins. Occurrence and fate of perfluorochemicals in soil following the land application of municipal biosolids. *Environmental Science and Technology*, 45:8106–8112, 2011.

Stock, N., V.I. Furdui, D.C.G. Muir, and S.A. Mabury. Perfluoroalkyl contaminants in the Canadian Arctic: Evidence of atmospheric transport and local contamination. *Environmental Science and Technology*, 41:3529–3536, 2007.

Sun, H., X. Zhang, L. Wang, T. Zhang, F. Li, N. He, and A.C. Alder. Perfluoroalkyl compounds in municipal WWTPs in Tianjin, China—Concentrations, distribution and mass flow. *Environmental Science and Pollution Research*, 19:1405–1415, 2012.

Sundstrom, M., J. Bogdanska, H.V. Pham, V. Athanasios, S. Nobel, A. McAlees, J. Eriksson, J.W. DePierre, and A. Bergman. Radiosynthesis of perfluorooctanesulfonate (PFOS) and perfluorobutanesulfonate (PFBS), including solubility, partition and adhesion studies. *Chemosphere*, 87:865–871, 2012a.

Sundstrom, M., S. Chang, P.E. Noker, G.S. Gorman, J.A. Hart, D.J. Ehresman, A. Bergman, and J.L. Butenhoff. Comparative pharmacokinetics of perfluorohexanesulfonate (PFHxS) in rats, mice, and monkeys. *Reproductive Toxicology*, 33:441–451, 2012b.

Tabtong, W., S.K. Boontanon, and N. Boontanon. Fate and risk assessment of perfluoroalkyl substances (PFASs) in water treatment plants and tap water in Bangkok, Thailand. *Procedia Environmental Sciences*, 28:750–757, 2015.

Tan, B., T. Wang, P. Wang, W. Luo, Y. Lu, K.Y. Romesh, and J.P. Giesy. Perfluoroalkyl substances in soils around the Nepali Koshi River: Levels, distribution, and mass balance. *Environmental Science and Pollution Research*, 21:9201–9211, 2014.

Thompson, J., G. Eaglesham, J. Reungoat, Y. Poussade, M. Bartkow, M. Lawrence, and J.F. Mueller. Removal of PFOS, PFOA and other perfluoroalkyl acids at water reclamation plants in South East Queensland Australia. *Chemosphere*, 82:9–17, 2011.

Venkatesan, A.J. and R.U. Halden. Loss and in situ production of perfluoroalkyl chemicals in outdoor biosolids-soil mesocosms. *Environmental Research*, 132:321–327, 2014.

Viberg, H., I. Lee, and P. Eriksson. Adult dose-dependent behavioral and cognitive disturbances after a single neonatal PFHxS dose. *Toxicology*, 304:185–191, 2013.

Wang, F., K. Shih, and J.O. Leckie. Effect of humic acid on the sorption of perfluorooctane sulfonate (PFOS) and perfluorobutane sulfonate (PFBS) on boehmite. *Chemosphere*, 118:213–215, 2015.

Wang, P., Y. Lu, T. Wang, Z. Zhu, Q. Li, Y. Zhang, Y. Fu, Y. Xiao, and J.P. Giesy. Transport of short-chain perfluoroalkyl acids from concentrated fluoropolymer facilities to the Daling River estuary, China. *Environmental Science and Pollution Research*, 22:9626–9636, 2015.

Wilhelm, M., S. Bergmann, and H.H. Dieter. Occurrence of perfluorinated compounds (PFCs) in drinking water of North Rhine-Westphalia, Germany and new approach to assess drinking water contamination by shorter-chained C4-C7 PFCs. *International Journal of Hygiene and Environmental Health*, 213:224–232, 2010.

Zaggia, A., L. Conte, L. Falletti, M. Fant, and A. Chiorboli. Use of strong anion exchange resins for the removal of perfluoroalkylated substances from contaminated drinking water in batch and continuous pilot plants. *Water Research*, 91:136–146, 2016.

Zhao, L., Y. Zhang, S. Fang, L. Zhu, and Z. Liu. Comparative sorption and desorption behaviors of PFHxS and PFOS on sequentially extracted humic substances. *Journal of Environmental Sciences*, 26:2517–2525, 2014.

Zhou, Z., Y. Liang, Y. Shi, L. Xu, and Y. Cai. Occurrence and transport of perfluoroalkyl acids (PFAAs), including short-chain PFAAs in Tangxun Lake, China. *Environmental Science and Technology*, 47:9249–9257, 2013.

21 Case Study
Pilot Testing Synthetic Media and Granular Activated Carbon for Treatment of Poly- and Perfluorinated Alkyl Substances in Groundwater

Brandon Newman and John Berry

CONTENTS

21.1 Pilot Test Background, Objectives, and Methodology 467
21.2 Pilot Test Equipment Setup and Operations 468
    21.2.1 GAC Operations Summary 472
    21.2.2 Resin Operations Summary: First Loading Cycle 472
21.3 Pilot Test Analytical Results for Gac and Virgin Resin 473
21.4 Pilot Test Operations and Analytical Results for Resin Regeneration 479
    21.4.1 Second Virgin Resin Loading Cycle Operations and Results 480
    21.4.2 Postregeneration Resin Loading Cycle Operations and Results 480
21.5 Regenerant Solution Recovery and Waste Stream Minimization 482
21.6 Conclusions 482

21.1 PILOT TEST BACKGROUND, OBJECTIVES, AND METHODOLOGY

From November 2015 to May 2016, Amec Foster Wheeler Environment & Infrastructure, Inc. (Amec Foster Wheeler) and Emerging Compounds Treatment Technologies, Inc. (ECT) tested pilot-scale *ex situ* treatment technologies for the treatment of poly- and perfluorinated alkyl substances (PFAS) in groundwater at the former Pease Air Force Base in Portsmouth, New Hampshire (Pease), in support of the US Air Force Civil Engineer Center's (AFCEC) PFAS response activities at Pease. The historical use of aqueous film-forming foam (AFFF) during firefighting training activities at the site contaminated groundwater with PFAS. The pilot test's primary goal was to determine the best available PFAS treatment technology for reducing groundwater concentrations at Pease below US Environmental Protection Agency (USEPA) health advisories (HAs) for perfluorooctanesulfonic acid (PFOS) and perfluorooctanoic acid (PFOA): 0.07 µg/L for both the individual constituents and the sum of PFOS and PFOA concentrations.

An evaluation of best available technologies for PFAS removal and treatment in groundwater reviewed technologies for pilot testing to support the design of a groundwater treatment plant at the site. Two technologies were selected for pilot testing: ECT's Sorbix A3F synthetic media (resin) ion exchange and adsorption technology and Calgon Carbon Corporation's (Calgon) Filtrasorb 400 (F400) granular activated carbon (GAC) technology.

Pilot test objectives included:

- Demonstrate that both Sorbix A3F resin and F400 GAC can remove PFAS below achievable laboratory quantification limits and therefore USEPA HAs for PFOS and PFOA.
- Determine the breakthrough curves and removal capacities for PFAS in both Sorbix A3F resin and F400 GAC.
- Demonstrate in-place regeneration of Sorbix A3F resin to near-virgin PFAS removal performance.

Calgon specified four GAC vessels in series, each containing 1.2 cubic feet (CF), or 9 gallons, of Filtrasorb 400 GAC. Each vessel provided a 5-minute empty bed contact time (EBCT), for an overall EBCT of 20 minutes. Samples were collected weekly for 8 weeks, with downstream vessel samples only analyzed if sample results from upstream vessels showed breakthrough.

ECT specified three resin vessels in series, each containing 1.2 CF (9 gallons) of Sorbix A3F resin, with each vessel providing a 2.5-minute EBCT, for an overall EBCT of 7.5 minutes. Samples were collected routinely, with downstream vessel samples only analyzed if sample results from upstream vessels showed breakthrough.

Sample collection and handling followed Amec Foster Wheeler standard operating procedures. PFAS samples were analyzed by a modified USEPA 537 method using liquid chromatography–tandem mass spectrometry.

21.2 PILOT TEST EQUIPMENT SETUP AND OPERATIONS

Pilot test equipment was installed in an existing groundwater treatment plant building. The plant was built to treat contaminants as required by USEPA's 1994 Record of Decision. Since USEPA's August 2015 administrative order for PFAS cleanup, the plant now treats PFAS contamination from select groundwater extraction wells as an interim measure using GAC.

The process flow diagram in Figure 21.1 shows the two parallel pilot test process units. Figures 21.2 through 21.5 show photographs of the fabrication and installation of pilot test equipment at the plant.

Case Study

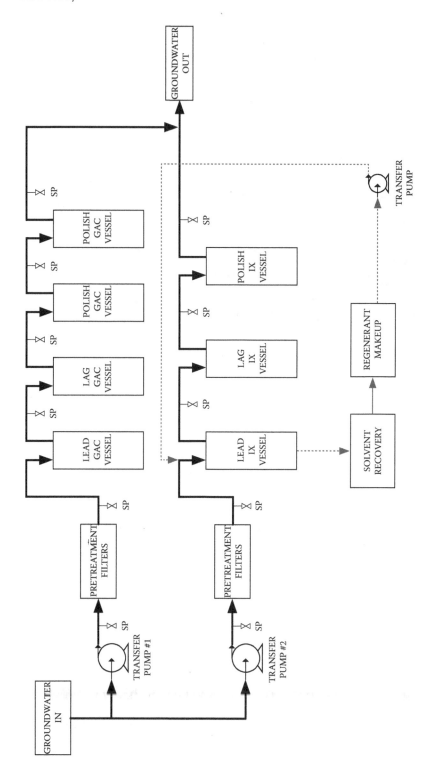

FIGURE 21.1 Pilot test process flow diagram.

FIGURE 21.2 Photograph of existing treatment plant equipment before connections.

FIGURE 21.3 Photograph of GAC prefilters, vessels, and piping fully installed.

Case Study

FIGURE 21.4 Photograph of resin vessels and piping fully installed.

FIGURE 21.5 Photograph of overall pilot test setup during forward flow operation.

21.2.1 GAC Operations Summary

The GAC system was loaded continuously during the pilot test to evaluate breakthrough with a few interruptions from plant shutdowns. The GAC process operated as follows:

1. GAC process water was pumped from the existing groundwater extraction network through two cartridge filters to prevent particulate matter from fouling the GAC.
2. The GAC system pump operated at approximately 1.8 gallons per minute (gpm) to provide the desired 5-minute EBCT per vessel.
3. PFAS samples were collected at the influent and each sample port twice during week 1 and weekly through week 8. Downstream samples were only analyzed if upstream samples showed breakthrough of PFOS or PFOA.
4. The GAC system was planned to operate in this way for 8 weeks unless otherwise recommended by Calgon, based on analytical results. The GAC process ended operation as planned.

The GAC process treated a total volume of 100,486 gallons of water, or 11,165 BVs, through the lead vessel. No changes to the GAC process equipment or operating parameters were made during its operation. No operational issues affected GAC performance aside from temporary and brief shutdowns of the existing plant.

The final sample collected showed that an approximately 69% breakthrough of total detected PFAS was achieved through the lead vessel (5-minute EBCT) effluent, compared with influent sample results. This coincided with an approximately 71% breakthrough of PFOA and an approximately 50% breakthrough of PFOS.

21.2.2 Resin Operations Summary: First Loading Cycle

ECT divided the resin pilot test activities into an initial loading cycle to evaluate virgin resin removal capacity for PFAS and subsequent alternating cycles of resin loading and regeneration to evaluate the effectiveness of regenerating spent resin. The resin process operated as follows during the first loading cycle:

1. The resin system pump operated at approximately 3.6 gpm to provide the desired 2.5-minute EBCT per vessel.
2. Resin process water was pumped from the existing groundwater extraction network through two cartridge filters to prevent particulate matter from fouling the resin.
3. The team operated the vessels to produce a 2.5-minute EBCT through each vessel. This allowed sampling at 2.5-, 5-, and 7.5-minute contact times through the series.
4. The team observed breakthrough of PFOA and PFOS in virgin media for the first loading cycle before moving on to the regeneration trials.

Case Study

The resin process treated 422,645 gallons of water, or 46,961 BVs, through the lead vessel during the first loading cycle. The first resin loading cycle operated longer than planned due to the resin's unanticipated performance at removing PFOS (i.e., higher than expected capacity). This cycle was extended to observe and better understand the breakthrough curve for PFOS.

The final sample of the first loading cycle showed an approximately 40% breakthrough of detected PFAS through the lead vessel (2.5-minute EBCT). This coincided with an approximately 50% breakthrough of PFOA and an approximately 3% breakthrough of PFOS.

21.3 PILOT TEST ANALYTICAL RESULTS FOR GAC AND VIRGIN RESIN

Table 21.1 summarizes the range of influent PFAS concentrations observed during the pilot test.

Figure 21.6 compares PFOA breakthrough curves for GAC and resin vessels during the first loading cycle. Figure 21.7 shows similar data for PFOS. Figures 21.8 through 21.19 compare breakthrough curves for observed PFAS compounds at the common 5-minute EBCT for each media.

As summarized in Table 21.2, the Sorbix A3F resin outperformed F400 GAC at PFOA and PFOS removal.

Compared with a 5-minute EBCT, the Sorbix A3F resin treated more than eight times as many BVs as F400 GAC before PFOS was observed at concentrations exceeding the USEPA HA and six times as many BVs before PFOA was observed at

TABLE 21.1
PFAS Influent Concentrations

Analyte	AnalyteAcronym	Influent Concentrations Observed during Pilot Test (µg/L)		
		Low	High	Average
6:2 fluorotelomer sulfonate	6:2 FS	15	22	18
8:2 fluorotelomer sulfonate	8:2 FS	0.055	0.3	0.23
Perfluorobutane sulfonate	PFBS	0.81	1.3	1.1
Perfluorobutanoic acid	PFBA	0.89	2.1	1.3
Perfluoroheptane sulfonate	PFHpS	0.85	1.4	1.1
Perfluoroheptanoic acid	PFHpA	1.6	2.2	1.9
Perfluorohexane sulfonate	PFHxS	18	25	22
Perfluorohexanoic acid	PFHxA	5.9	8.9	7.7
Perfluorooctanoic acid	PFOA	9.1	13	12
Perfluorooctane sulfonate	PFOS	4.2	32	26
Perfluoropentanoic acid	PFPeA	3.1	5.1	4.2
Sum of observed PFAS		65	112	94

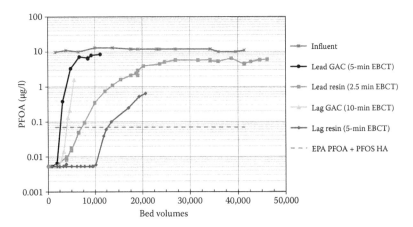

FIGURE 21.6 GAC and resin breakthrough curves—PFOA.

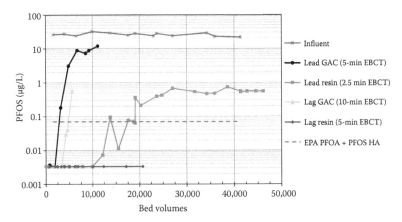

FIGURE 21.7 GAC and resin breakthrough curves—PFOS.

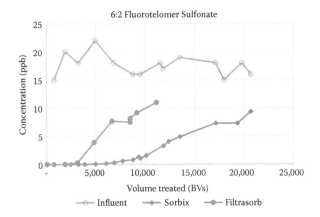

FIGURE 21.8 Loading cycle 1 breakthrough curves at 5-minute EBCT—6:2 fluorotelomer sulfonate.

Case Study

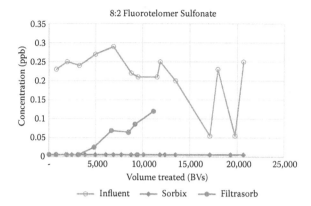

FIGURE 21.9 Loading cycle 1 breakthrough curves at 5-minute EBCT—8:2 fluorotelomer sulfonate.

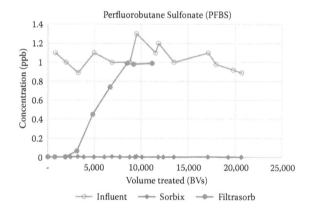

FIGURE 21.10 Loading cycle 1 breakthrough curves at 5-minute EBCT—perfluorobutane sulfonate (PFBS).

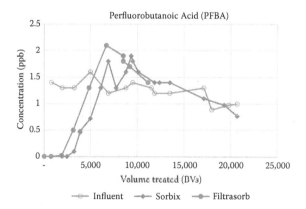

FIGURE 21.11 Loading cycle 1 breakthrough curves at 5-minute EBCT—perfluorobutanoic acid (PFBA).

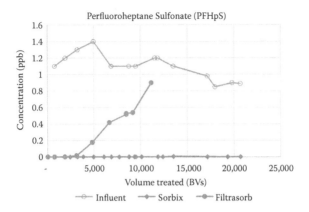

FIGURE 21.12 Loading cycle 1 breakthrough curves at 5-minute EBCT—perfluoroheptane sulfonate (PFHpS).

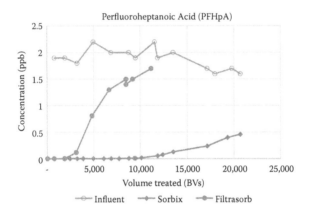

FIGURE 21.13 Loading cycle 1 breakthrough curves at 5-minute EBCT—perfluoroheptanoic acid (PFHpA).

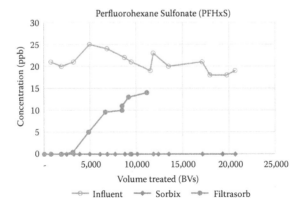

FIGURE 21.14 Loading cycle 1 breakthrough curves at 5-minute EBCT—perfluorohexane sulfonate (PFHxS).

Case Study

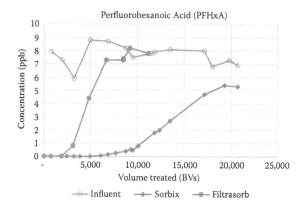

FIGURE 21.15 Loading cycle 1 breakthrough curves at 5-minute EBCT—perfluorohexanoic acid (PFHxA).

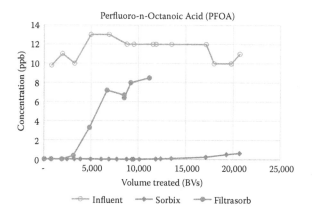

FIGURE 21.16 Loading cycle 1 breakthrough curves at 5-minute EBCT—perfluoro-n-octanoic acid (PFOA).

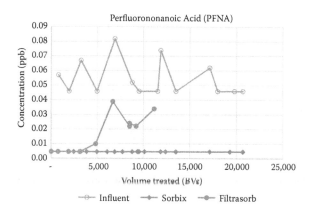

FIGURE 21.17 Loading cycle 1 breakthrough curves at 5-minute EBCT—perfluorononanoic acid (PFNA).

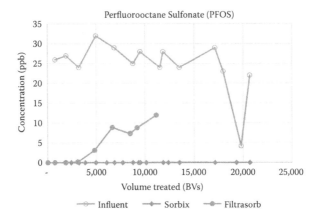

FIGURE 21.18 Loading cycle 1 breakthrough curves at 5-minute EBCT—perfluorooctane sulfonate (PFOS).

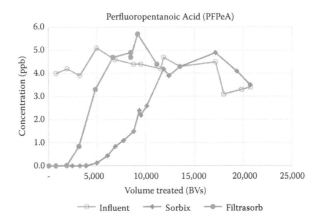

FIGURE 21.19 Loading cycle 1 breakthrough curves at 5-minute EBCT—perfluoropentanoic acid (PFPeA).

TABLE 21.2
GAC and Resin Breakthrough Comparisons: PFOS and PFOA

Vessel	Position	EBCT (minutes)	PFOA Breakthrough at HA (Bed Volumes Treated)	PFOS Breakthrough at HA (Bed Volumes Treated)
GAC 1	Lead	5	3147	3147
GAC 2	Lag	10	4254	5582
GAC 3	Polish	15	Not observed at 3721	Not observed at 3721
Resin 1	Lead	2.5	7479	17,592
Resin 2	Lag	5	13,515	Not observed at 20,675
Resin 3	Polish	7.5	Not observed at 13,783	Not observed at 13,783

TABLE 21.3
GAC and Resin Mass Removal Rates

Leakage Criteria	EBCT (minutes)	GAC	Resin
Initial detection of either PFOA or PFOS*	2.5		0.86 mg/g 0.037 lb/CF media
	5	0.40 mg/g 0.012 lb/CF media	1.58 mg/g 0.068 lb/CF media
	7.5		PFOA/PFOS not detected at end of cycle
	10	0.73 mg/g 0.021 lb/CF media	
	15	PFOA/PFOS not detected at end of cycle	
Combined PFOA/PFOS capacity at 0.07 ppb leakage	2.5		0.93 mg/g 0.040 lb/CF media
	5	0.40 mg/g 0.012 lb/CF media	1.66 mg/g 0.072 lb/CF media
	7.5		PFOA/PFOS not detected at end of cycle
	10	0.73 mg/g 0.021 lb/CF media	
	15	PFOA/PFOS not detected at end of cycle	

concentrations exceeding the USEPA HA. No HA exceedances were observed in the polish resin vessel or the third GAC vessel, and therefore no samplers were analyzed from the fourth GAC vessel.

A mass-to-mass comparison was calculated using totalized volumes treated in each train and mass loading calculations from influent and effluent analytical data. Directly compared at the 5-minute EBCT, the Sorbix A3F resin removed 1.66 mg total PFAS observed in the influent per gram of resin before breakthrough was observed at the USEPA HA. F400 GAC removed 0.40 mg total PFAS observed in the influent per gram of GAC before breakthrough was observed at the USEPA HA. Table 21.3 presents the mass removal rates of the media at different EBCTs.

21.4 PILOT TEST OPERATIONS AND ANALYTICAL RESULTS FOR RESIN REGENERATION

To evaluate the resin's ability to be regenerated, three regeneration trials were conducted throughout the pilot test using a proprietary regeneration procedure with a solution of solvent and brine. The objective for each regeneration was to restore the PFAS treatment capacity of the lead vessel's resin to the approximate capacity of virgin resin.

The first and second regenerations were experimental and did not restore the resin to virgin capacity based on postregeneration samples. Lessons learned from the first two regenerations were applied to develop the third regeneration procedure, summarized below, which effectively regenerated the resin column to near-virgin conditions:

1. The lead resin was changed out and filled with virgin resin, and then reloaded with PFAS. This loading cycle continued until total flow through the vessel reached the point of expected breakthrough at the HA limit.
2. Process water in the vessel was purged with one BV of 10% brine solution to prime the resin for desorption of PFAS compounds.
3. Ten BVs of regenerant were pumped through the resin in countercurrent flow (i.e., entering from the effluent port on the vessel and discharging through the influent port).
4. Once the 10 BVs of regenerant were pumped through, 10 BVs of potable water were used to rinse the bed in countercurrent flow.
5. Once the potable water rinse was complete, the resin vessel was returned to service for a final loading cycle to confirm if the regeneration was successful.
6. Samples were collected through the regeneration procedure to evaluate the desorption characteristics of the regeneration process.

21.4.1 Second Virgin Resin Loading Cycle Operations and Results

The resin in the lead resin vessel was replaced with virgin resin and continued to operate at a 2.5-minute EBCT. This second virgin loading cycle treated 124,293 gallons (10,951 BVs) of groundwater. This loading cycle was planned to run long enough to approach breakthrough of PFOA at the HA, based on breakthrough results from the first virgin loading cycle. It was assumed for this groundwater source that HA breakthrough for PFOA would be the trigger point for regenerations in a full-scale application. PFAS breakthrough curves were similar between the first and second virgin resin loading cycles.

21.4.2 Postregeneration Resin Loading Cycle Operations and Results

The lead resin vessel was returned to service after the regeneration procedures and a final loading cycle started. This cycle treated 77,455 gallons (6824 BVs) of groundwater.

The first sample collected in the regenerated loading cycle showed the sum of detected PFAS concentrations as 0.090 µg/L after 1367 BVs of treatment, compared with 0.052 µg/L after 1598 BVs from the first virgin loading cycle and 0.054 µg/L after 1104 BVs from the second loading cycle. The last sample collected in the regenerated loading cycle showed the sum of detected PFAS concentrations as 4.1 µg/L after 6824 BVs of treatment, compared with 4.6 µg/L after 6489 BVs from the first virgin loading cycle and 4.8 µg/L after 7608 BVs from the second virgin loading cycle.

PFOS was only detected between the reporting limit and detection limits (0.0077 µg/L in the final sample), and PFOA exceeded the HA (0.10 µg/L in the final sample) at 6823 BVs, as expected. This corresponded to the first detections of PFOA observed in the first virgin resin loading cycle 0.048 μg/L at 6489 BVs treated and the second virgin resin loading cycle at 0.053 μg/L at 7608 BVs treated.

Figure 21.20 shows the breakthrough curves of detected PFAS concentrations during the second virgin loading cycle and the regenerated resin loading cycle.

Figure 21.21 compares total PFAS effluent concentration for virgin resin to the regenerated resin.

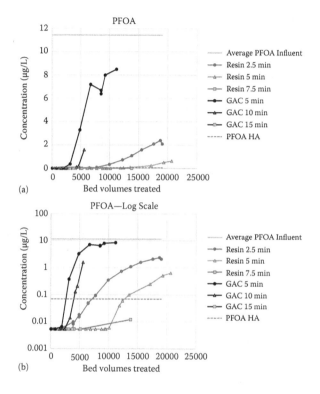

FIGURE 21.20 Virgin resin and regenerated resin breakthrough curves at a 2.5-minute EBCT—detected PFAS compounds in (a) linear and (b) logarithmic scales.

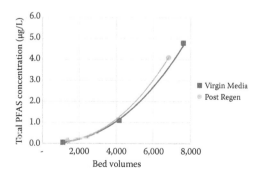

FIGURE 21.21 Resin breakthrough curves—virgin and regenerated resin.

Overall, compared with the two virgin resin loading cycles, PFAS removal results for the regenerated resin were consistent with those for virgin resin. The regeneration procedures developed during the third regeneration were successful in restoring the resin's PFAS adsorption capacity to near-virgin conditions.

21.5 REGENERANT SOLUTION RECOVERY AND WASTE STREAM MINIMIZATION

To minimize the consumption of raw materials and disposal of off-site wastes for potential full-scale applications, preliminary testing of distillation and superloading of the spent regenerant were performed during the pilot test.

1. *Distillation of spent regenerant*—Based on preliminary testing, recovered (distilled) solvent can be recycled to the regenerant supply tank for reuse. The spent regenerant contains solvent, brine, and PFAS. With the different boiling points of water, solvent, and PFAS, it is possible to distill the solvent fraction of the solution for recycling and reuse, while leaving behind the desorbed PFAS compounds in the brine solution fraction ("still bottoms") of the regenerant solution.
2. *Superloading the still bottoms*—Based on preliminary testing, it is possible to recycle the still bottoms from the distillation step by processing the concentrated PFAS and brine solution waste stream at low flow rates (long EBCT) through adsorption media (referred to as "superloading" in the case study). Recovered brine solution can be recycled to the regenerant supply tank for reuse and the superloaded media would be disposed off-site. The long ECBTs in the superloaders are used to maximize the adsorption capacities of the superloader media, thereby minimizing the amount of waste requiring disposal.

21.6 CONCLUSIONS

Pilot test activities achieved the test objectives as follows:

1. Both Sorbix A3F resin and F400 GAC demonstrated removal of PFAS below achievable laboratory quantification limits and the USEPA HAs for PFOS and PFOA.
2. Collected data allowed for development of breakthrough curves and calculation of removal capacities for PFAS in both Sorbix A3F resin and F400 GAC.
3. The Sorbix A3F resin was regenerated in place to near-virgin PFAS removal performance under simulated full-scale operating conditions.

Compared with a 5-minute EBCT, the Sorbix A3F resin treated more than eight times as many BVs as F400 GAC before PFOS was observed at concentrations exceeding the USEPA HA and six times as many BVs before PFOA was observed at concentrations exceeding the USEPA HA. A mass-to-mass comparison at a 5-minute EBCT showed that the Sorbix A3F resin removed 1.66 mg total PFAS per gram of resin before breakthrough was observed at the USEPA HA, while F400 GAC removed 0.40 mg total PFAS observed in the influent per gram of GAC before breakthrough was observed at the USEPA HA.

Based on the calculated loading capacities and the successful resin regeneration demonstrated during the pilot test, the full-scale regenerable resin system planned for the site (200 gpm of 90 ppb total PFAS) would treat more than 100 million gallons of groundwater per year and extract approximately 80 lb of PFAS compounds. Through the regeneration, distillation, and superloading processes, it is anticipated that the 80 lb of PFAS compounds could ultimately be superloaded onto approximately four or five 55-gallon drums of superloader media for off-site disposal via hazardous waste landfilling or incineration.

Author Index

A

Agrawal, Abinash, 417–446
Alden, David F., 405–416
Anderson, Janet K., 353–372
Anderson, R. Hunter, 113–122, 353–372
Archibald, John, 405–416

B

Berry, John, 467–484
Birk, Gary M., 405–416
Brannon, Miranda, 103–112
Braun, Joseph M., 171–202
Brown, Carl E., 383–404, 447–466
Buck, Robert C., 3–34
Burgess, Justin, 447–466

C

Cantu, Gabriel, 203–240
Carr, Stephanie, 303–324
Chandramouli, Bharat, 67–82
Crimi, Michelle, 255–302

D

Dasu, Kavitha, 35–66
Dickenson, Eric R. V., 241–254
Dyson, Sean M., 373–382

H

Holsen, Thomas M., 255–302

K

Kempisty, David M, 3–34, 35–66, 113–122, 203–240
Klein, David, 171–202
Korzeniowski, Stephen H., 3–34

L

Long, G. Cornell, 353–372

M

Matthis, John, 303–324
Mills, Marc A., 35–66, 417–446

N

Newman, Brandon, 467–484
Nickelsen, Michael G., 325–352
Nzeribe, Blossom N., 255–302

P

Pabon, Martial, 3–34
Porter, Ronald C, 353–372

R

Rice, Penelope, 123–170

S

Schmidt, Christopher, 373–382
Sinnett, Marilyn M., 325–352
Stewart, Richard J., 405–416
Stratton, Gunnar, 255–302
Stubbs, John E., 373–382

T

Thagard, Selma M., 255–302

V

Verdugo, Edgard M., 241–254
Via, Steve, 83–90
Volchek, Konstantin, 383–404, 447–466

W

Wagner, Andrew J., 91–102
Wang, Bo., 417–446
Woodard, Steven, 325–352

X

Xing, Yun, 203–240

Y

Yao, Yuan, 383–404, 467–484

Subject Index

0-series ions, 206, 208
1,1,2,2,3,3,4,4,5,5,6,6-dodecafluorohexyl)oxy]-
    1,1,2,2-tetrafluoroethanesulfonate
    (F-53B), 246
^{19}F-NMR, 204, 206, 209, 212
1m-PFOS, 204, 213, 216–217, 220–221, 223–225
2-(N-ethylperfluorooctane sulfonamido) ethanol
    (N-Et-FOSE), 215, 220–221
2-(N-methylperfluorooctane sulfonamido)
    ethanol (N-Me-FOSE), 215
3M, 6, 9, 17, 18
3M test for soil PFAs removal, 389
3m-PFOA, 204, 227, 232
40/40/20 rule for a GAC bed, 307
4m-PFOS, 204, 225, 232
5 Whys method, 97
5m-PFOA, 226, 232
5m-PFOS, 204, 206, 212, 220, 222, 225, 228, 232
6:2 FTAC, 450
6:2 FTMAC, 450
6:2 FTOH, 39, 41, 424, 448
6:2 FTUCA, 242
8:2 fluorotelomer acid (FTCA), 39, 44, 51, 248
8:2 fluorotelomer alcohol (8:2 FTOH), 248, 418,
    424–425, 430
8:2 fluorotelomer aldehyde (8:2 FTAL), 248, 430
8:2 fluorotelomer unsaturated acid
    (FTUCA), 242, 248
8:2 FTUA, 430
9-series ions, 206, 208

A

Abiotic degradation, 222
Accumulation and elimination, 125–127,
    218–219, 226
    In fish and wildlife, 218–219
    In occupational workers, 226
Acrylic gel anion exchange (AIX), 244–246, 250
Activated carbon, 336, 373–374, 379, 384,
    389–396, 405–410
Activated persulfate, 257, 432
Acyl coenzyme A (CoA), 229
Adelaide hot fire training ground (HFTG) test,
    410
Adipocyte differentiation, 183–184
Adiposity in children, 183
Adiposity, 174, 183–185, 187–188
Administrative Procedures Act (APA), 108
Adsorbable organic fluorine (AoF), 69
Adsorption pores, GAC, 308–310, 317, 319–320

Adsorption sites & mechanisms, 328–329
Adsorption to activated sludge, 243
Adsorption to AIX resin, 244–246
Adsorption to GAC and PAC, 243–244, 430–431
Adulthood obesity, 174
Advanced oxidation process (AOP), 375,
    431–433, 248
Advanced reduction processes (ARPs), 248, 456
Aeration packed towers, 242
AFFF, 9–12, 14, 25, 16, 116–118, 355–369, 418,
    429, also see firefighting foam
    AFFF release volume, 356
    ECF-based, 9–11
    Fluorotelomer-based, 11–12
    Formulations, 429
    History, 16; *see also* QPL
    Impacted sites, 355, 357, 359, 361, 363, 367, 369
    Impacted soil and groundwater & treatment,
        417–429
    Ingredients & composition, 14
    Use, best practice, 25–27
Air Force Civil Engineer Center's (AFCEC), 104
Air National Guard (ANG), 118
Airborn PFAs, 428
Air-sparged hydrocyclone (ASH), 435
AIX resin: pH dependency, 246
AIX resins, 245
Alcohol resistant (AR)-AFFF, 13
Alcohol resistant foam-forming fluoroprotein
    (AR-FFFP), 13, 26
Amec Foster Wheeler standard operating
    procedures, 467–468
American Water Works Association (AWWA), 97
Ammonium perfluorooctanoate (APFO), 229–230
Amphoteric surfactants, 11
Analytical methods, 40
Animal studies vs. epidemiology data, 172–173
Animal studies, 135–137
Anion exchange resin, 327, 396
Anion exchange, 244–246
Anionic PFAs, 40, 77, 409, 429
Anionic surfactants, 11, 384, 392
AOP pretreatment, 378
APFO degradation, 275
Apical effects, 125, 152, 157
Apoptosis, 453
Apoptotic, 140, 142, 147, 150–151
Apparent density (AD) (ASTM D2854), 317
Aquatic toxicity, 21–22
ARP process: Adsorption and reaction on the
    surface, 279

ARP process: Adsorption onto the bubble-gas-interface, 274
ARP, byproducts desorption and mass transfer, 279
Atmospheric contributions, 428
Atmospheric transport and deposition, 458
Atochem Forafac™, 12
Attention-deficit/hyperactivity disorder (ADHD), 185
Australian government airport authority case study, 411
Aviation rescue firefighter (ARFF), 411

B

Base Realignment and Closure (BRAC), 104, 118
Bench-scale tests, 242–243, 246–249, 285, 291, 332, 398, 458
Binding affinity to HSA, 232–233
Binding PFAs in soil, 410–411
Bioaccumulation factor (BAF), 218–219, 387–388
Bioaccumulation in Coastal organisms, 452
Bioaccumulation in earth worm, 388
Bioaccumulation in green algae, 452
Bioaccumulation in plants, 387–388
Bioaccumulation, 411–412, 451–452
Bioaccumulation: In aquatic worm, 388
Biochars, 244
Biomagnification, 457
Biosolids, 74, 77, 368, 385–387, 456, 460
Biota, 211, 218, 220–221, 383, 387, 418, 447, 452, 456
Biota-sediment accumulation factors, 452
Biotransformation, 219–222, 355, 361, 364, 367–370
Body mass index (BMI), 183–184
Boron-doped diamond (BDD), 249, 263
Branched APFO, 229–230
Branched fluorosurfactants, 11
Branched isomers, 48, 57, 59–60, 203–206, 209–210, 212, 214, 217–218, 220–222, 225–230, 420, 431, 450
Breakthrough curve, GAC, 243–245
Breakthrough, 243–245, 341–342, 378–379, 407, 431, 458–460, 468, 472, 473–483
Breakthrough, IEX resin, 473–483
Breastfeeding duration, 181–183

C

C4, C6, C8, C10, 6, 366, 409, 411, 448
C8 isomers, 205
Ca homeostasis, 151
Calgon Filtrasorb 400 (F400), 391, 431, 459, 467–468
Canonical structure, 365
Canonical variable, 364

Carbon reactivation, 304
Carboxymethyl cellulose (CMC), 436
Carcinogenicity & genotoxicity, 148–149
Case study: Loading cycle 1 breakthrough curves, 474–478
Categorical association, 368
Categorical data analysis, 361
Cationic PFAs, 408–409
Cationic resin, 327
Cationic surfactants, 11
Cavitation bubbles, 269, 433
Cell membrane alterations, 151
Cell viability, 231
Cellular growth and proliferation, 231
Cetyltrimethylammonium bromide (CTAB), 392
Chain length effect, 242, 244–246, 249, 268, 366, 384, 388, 409, 452, 460
Challenges, dealing with PFOA and PFOA HALs, 106
Challenges, Public water systems (PWS) addressing drinking water health advisories (HALs), 103–109
Chelating agents, 259, 392
Chemical activation of persulfate, 258
Chemical oxidation/Reduction, 248–249
Chemical reduction processes, 276
Chemical scrubber, 313
Chemisorption, 329
Chicken embryo hepatocytes (CEH), 229–230
Chicken embryos, 230
Childhood obesity, 183–185, 187
Chiral molecules, 204, 212
Chiral stationary phases (CSPs), 213
Chloramination, 242
Chlorination, 242, 244, 436, 459–460
Chlorine dioxide, 242
Chromatography effect, 334
Ciba-Geigy Lodyne™, 12
Class A fires, 12, 25
Class B fires, 9, 12, 26, 116
Class C fires, 25
Clean water act (CWA), 91
Clinical signs, 230
Coagulation, 242, 459–460
Coagulation-enhanced sorption, 395
Co-contaminants, 257, 262, 287, 292, 344, 368, 375, 386, 390, 393, 419, 429, 435, 437
Commons PFAs, names and molecular weight, 257
Commonwealth Scientific and Industrial Research Organization (CSIRP), 408
Comparison of Analytical instrument, 42
Comprehensive Environmental Response, Compensation, and Liability Act (CERCLA), 117–118
Contaminant candidate list (CCL), 85

Subject Index

Contaminant immobilization agent, 412
Contaminated soil excavation, 393
Conventional wastewater treatment, 243
Conventional water treatment, 242
Corse soil fraction, 393
Cost effectiveness, 259
Covalent triazine-based framework (CTF), 246
Critical separate phase concentration (CSPC), 429
Cross-contamination issues, 51–54
Cylindrical pressure vessels, 338
CYP1A4/5, 229
CYP4B1, 229
Cytotoxicity, 137, 453

D

DC plasma, 285–286
Decarboxylation reaction, 260–261, 285
Dechlorination, 436
Decreased neonatal survival, 176
Delayed ossification, 176–177
Department of Defense (DoD), 115–116
Desulfonation, 261, 266, 292
Detection and quantitation limits, 56
Detection frequency (DF), 355, 363
Developmental delay, 176–179
Devolatilization, 308
Digested sludge, 385–386
Direct anodic oxidation, 263
Direct electron-transfer, 266
Direct push technology (DPT), 356
Direct sources of PFAs, 215
Direct thermal decomposition, 285
Dissociation constant (Kd), 232
Dissolved air flotation, 242–243
Dissolved organic carbon (DOC), 429, 244, 268
Dissolved organic matter (DOM), 375, 431
Distillation and superloading process, 343–344
Divinylbenzene (DVB), 326
DOD's Environmental Restoration Program (DERP), 116
Drinking water advisories, 106
Drinking water guideline (DWG), 152–153
Drinking water sanitary surveys, 93–94
Drying of spent carbon, 307–308

E

Early life exposure: animal studies, 173–179
Early life exposure: epidemiological data, 179
Early life/ Prenatal/gestational exposure, 174–175
ECF-products, 2, 204, 215, 217, 233
Economic Co-operation and Development (OECD), 7, 386, 420
Ecotoxicological evaluation of PFOS, 388, 152

ECT Sorbix A3F, 467
Effects of AOP on GAC adsorption capacity, 378
Electrical discharge plasma-based AOP, 285
Electrical discharge, 285
Electrical process, 397
Electrochemical degradation of PFAS, Rate constants, 268
Electrochemical oxidation, 263–269
Electrochemical oxidation/reduction, 249
Electrostatic interaction, 246, 331, 408, 409, 412, 450
Elimination half-lives, 220, 451
Emerging Compounds Treatment Technologies, Inc. (ECT), 467
Emerging contaminant monitoring as a host nation guest, 91–100
Emerging contaminants (ECs), 92, 104, 113–118
Emerging technologies for PFAS analysis, 43–48
    Fluorine Nuclear Magnetic Resonance (^{19}F NMR), 48
    Particle-induced Gamma-Ray Emission (PIGE), 48
    Total fluorine analysis or Combustion ion chromatography (CIC), 43
    Total oxidizable precursor (TOP) assay, 48
Emerging technologies for PFAS treatment, 434–436
    Air-sparged hydrocyclone (ASH), 435
    Electrochemical treatment, 434–436
    Nanomaterial technology, 435
Empty bed contact time (EBCT), 336–338, 346, 375, 378, 468, 472–483
Enantiomer analysis, 212–213
Enantiomer fraction (EF), 212
Enantiomers, 204, 212, 233
Enantioselective biotransformation, 216
Enantioseparation, 213
Endocrine disruption, 186
Enriched accumulation of L-PFOS, 219
Environmental friendliness, 259
Environmental risks, 266
Environmental site remediation, 383
Environmental transformation, 449
Environmentally relevant PFAs, 421–426
Epidemiological, 127, 138, 143, 145–146, 149
Equipment blank, 357
Estrogen receptor signaling, 145
European Chemicals Agency (ECHA), 151
European Food Safety Authority (EFSA), 124
Ex situ treatment, 393
Excavation and specialized landfills and encapsulation, 389
Extractable organic fluorine (EPF), 427
Extraction completeness, 77
Extraction methods, 38–40

F

F600 GAC, 375, 378–379
Female reproductive toxicity, 143
Fenton's agent, 432
Ferrous iron, 258–259, 276
Fertility and reproductive organs, 141, 143–144
Fiberglass-reinforced plastic (FRP), 338
Field blank, 357
Fine soil fraction, 393
Fire training areas (FTAs), 116
Fire-fighting foam coalition (FFFC), 23
Firefighting foam, AFFF, 14
Firefighting foam, Aquatic toxicity, 22
Firefighting foam, Environmental properties, 21
Firefighting foam, Field test performance & efficiency, 23
Firefighting foam, fluorinated, 13
Firefighting foam, fluorine-free (F3), 13
Firefighting foam, Selection & use, 20–23
Firefighting foam: Expansion ratio, 13
Firefighting foam: Fuel shedding, 21, 23
Firefighting foam: Spreading coefficient (SC), 15
Firefighting foam: Fuel repellency, 21–22
Firefighting training ground (FTGs), 428
Fire Protection Association of Australia (FPAA), 8
Fixed bed vs. backwashable, 339
Flocculation, 242, 460
Flow distribution and distributor, 338
Fluoreosurfactants chemistry, 4
Fluoreosurfactants ECF process, 6
Fluoreosurfactants fluorotelomer, 6
Fluoreosurfactants long chain & short chain, 6–8
Fluorine nuclear magnetic resonance (^{19}F NMR), 48
Fluorotelomer acrylates (FTAs), 256
Fluorotelomer alcohols (FTOHs), 256
Fluorotelomer aldehyde (FTAL), 248
Fluorotelomer process – or telomerization, 5, 9
Fluorotelomer sulfonic acids (FTSAs), 127
Fluorotelomer unsaturated carboxylic acids (FTUCAs), 449, 458
Fluorotelomers (FTs), 4–29, 124, 242, 357
Foam concentrate, 8, 9, 11, 14
Food-grade carbon, 316
Former firefighting training area (FFTA), 384, 391
FOSA, 248
FOSAA, 220, 248
FOSE, 220
Frank Lautenberg Chemical Safety for the 21st Century Act, The 114
Free radicals, 257–259, 261, 269, 274, 276, 279, 433–434
FTIs, 387
Full-scale application, 247, 249–250, 292, 318, 342
Full-scale PerfluorAd system, 406
Functional groups & exchange sites, 327–328
Functional observational battery (FOB), 147

G

GAC adsorption, 458–459
GAC capacities, 378–379
GAC polishing, 407
Gas chromatography-mass spectrometry (GC-MS), 42–43, 74, 211
Geminal diperfluoromethyl isomers, 206
Geometric mean half-life, 126
Gestational development, 174
Gestational exposure, 175–176; see also early life exposure
Glomerular filtration rate (GFR), 180–181, 242
Granular activated carbon (GAC), 243, 304, 336, 374, 430–431
Grass-soil accumulation factors (GSAFs), 387
Groundwater aquifers, 374
Groundwater pump and treat, 391
Groundwater samples, 356
Groundwater treatment technologies, 467
Growth and sexual development of amphibians, 454
Growth trajectories, 184
Growth-restrictive effect, 184
Guideline and non-guideline studies, 127–128, 137, 142, 144

H

Half-lives, 450
Harmful algal bloom (HAB), 109
Health advisories (HAs), 83–86, 98, 468
Health advisory levels (HALs), 105–106
Health effect guideline values, 153–156
Health effects & toxicology, 127–150
Health risk reduction and cost analysis (HRRCA), 106
Hematology, 230
Hepatic hypertrophy, 174–175, 188
Hepatic isomer profiles, 230
Hepatic peroximal β-oxidation, 230
Hepatocellular adenomas, 148
Hepatocellular hypertrophy, 128–129, 131–135, 138, 147, 152, 160, 175, 452
Hepatotoxicity, 136–137, 139, 452
HepG2 cells, 453
Hexadecyl trimethyl ammonium bromide (CTAB), 275
HF elimination, 261, 266
High density polyethylene (HDPE) containers, 357
High volume of AFFF release, 356
High-throughput analysis, 210–211
Historical AFFF release, 356
Historical PFAS in the USAF, 113–116
HMG-CoA, 229
HOME study cohort, 182

Subject Index

Homeostasis of hormones, 177
Host nation: ways to connect, 99–100
Host nations: policy gaps, 98–99
HPLC-MS/MS, 205
Human Equivalent Dose (HED), 155
Human health risk assessment of PFAs, 123–170
Human serum albumin (HAS), 226, 232
Hydrocarbon fuel, 9, 12, 25, 418
Hydrocarbon-based surfactants, 418
Hydrochlorofluorocarbon (HCFC), 449
Hydrofluorocarbon (HFC), 449
Hydrolysis, 260–261, 266, 286
Hydrophobic interaction, 245–246, 331, 408–409, 428, 450
Hydrophobicity, 4, 243, 460
Hydropyrolysis combustion ion chromatography, 69
Hydroxy radical reaction, 269

I

IC50 value, 138, 143, 146
Idaho National Laboratory (INL), 375
IEX cycles, 341–343
IEX efficiency, effect of water quality, 332–334
IEX kinetics, 329
IEX resin structure model, 327
IEX resin: Selectivity coefficients, 329–330
IEX resin: Sensitivity to inorganic content, 332–334
IEX resin: Structure & porosity, 327
IEX resin: Thermal stability, 330
IEX resins for softening, TOC removal, 336
IEX swelling, 330
IEX system: Design parameters, 336–341
IEX: Change-out, 341–343
IEX: PFAS removal mechanisms, 331
IEX: Influent equalization, 336
IEX: Iron fouling, 335
IEX: Lead-lag configuration, 340–342
IEX: Lead-lag-polish configuration, 340–342
IEX: Loading cycle, 341
IEX: Operating capacity, 329
IEX: Regeneration cycle, 304, 342
IEX: Regeneration trigger, 341
IEX: Single-use vs. regenerable systems, 341
IEX: Total capacity, 329
Immobilization, 394–396
Immune cells responses, 140
Immunosurpression in humans, 141
Immunotoxicity, 139–141
Impaired gap junction communication, 151
Impaired glucose homeostasis, 183–184
In silico isomer study, 221
In situ chemical oxidation (ISCO), 74, 257, 434
In situ groundwater remediation, 276
In utero exposure, 172, 177, 179

Indirect anodic oxidation, 263
Indirect electrolysis, 263
Indirect sources of PFAs/precursors, 215
Industrial and municipal waste, 448
Industrial reactivation site, 315
Inflammatory responses, 140–141
Inhalation studies, 125, 453
Instrument detection limit (IDL), 56
Interaction plot, 368–369
Interfacial adsorption sites, 269
Interlaboratory comparisons, 49
Interlaboratory study (ILS), 49
Intracellular distribution, 126–127
Iodine number (I_2#) (ASTM D4607), 316
Ion exchange (IEX) resins, 326, 431
Ion exchange for PFAs removal, 325–351
IRA 67, 246
IRA67, 246
Isomer and serum biochemistry, 225
Isomer biotransformation: diSPA, 220
Isomer characterization & profiling, 205–213
Isomer composition, 204, 209, 212, 223, 226, 232
Isomer partitioning in water, particulate and sediment, 214–215, 218
Isomer pattern in Humans, 222–229
Isomer profiles in various matrices, 223
Isomer profiling of snow samples, 215
Isomer-specific interactions with biomolecules, 231–232
Isomeric degradation *in vitro*, 220–221
Isomer-profiling of water samples, 214
Isomer-specific biotransformation & bioaccumulation, 218–222
Isomer-specific toxicity, 229–231

J

Japan Environmental Governing Standards (JEGS 2016), 94
JEG3 cells, 231, 453
Joint Public Epidemiological Action Center for Health (JPEACH), 96

K

Kidney and testicular cancer, 148
Kidney weight, 147
Korea Environmental Governing Standards (KEGS 2012), 94

L

Laboratory water, 54, 245
Lactation, 128, 175, 181, 183
Lactational exposure, 175, 188
LC-PFAs, 123

Leachability data of RemBind treated soils, 413
Leachate, 393, 411, 413, 418, 433, 210, 318–319
Leaching potential of PFAS, 458, 450, 455
Lead dioxide (PbO2) electrode, 263
Legacy MilSpec AFFF, 373
Leydig cell adenoma, 142, 148–149
Limit of detection (LOD), 210
Limit of quantification (LOQ), 386, 268
Linear & branched isomers, 203–204
Linear APFO, 230
Linear discriminant analysis, 360
Linear fluorosurfactants, 11
Lipid catabolism, 174
Lipid homeostasis, 174–175
Lipid metabolism, 135–136, 231
Liquid chromatography, 40–42
Liquid chromatography-tandem mass spectrometry (LC-MS/MS), 69
Liquid chromatography-time of flight spectrometry (LC-TOF), 74
Liquid extraction, 357
Liquid phase activated carbon pools, 315
Liquid-liquid extraction (LLE), 40
Liver fatty acid-binding protein (L-FABP), 229
Liver toxicity, 128–139; see also hepatic toxicity
Logical regression, 368
Long-range transport potential, 450, 455, 458
Low volume of AFFF release, 356
Lower explosive limit (LEL), 348
Lowest observed effect levels (LOELs), 128–134
Lowest-concentration minimum reporting level (LCMRL), 57
L-PFOS, 218–219, 229–230

M

m/z 499→80 transition
m/z 499→99 transition, 206
Magnetic mesoporous carbon nitride (MMCN) adsorbent, 246
Mag-PMCAs, 246
Male reproductive toxicity, 141–143
Mammary gland development & differentiation, 144, 178–179
Margin of exposure (MOE), 156–157
Mass removal rates, 479
Mass transfer of contaminants to the surface, 279
Materials of Evolving Regulatory Interest Team (MERIT), 114–116
Maternal blood, serum and cord blood, serum samples, 227
Maternal serum PFAS concentration & birth weight, 180
Maximum contaminant level goal (MCLG), 84, 106
Maximum contaminant levels (MCLs), 84, 105
Maximum tolerable dose (MTD), 147

Mean particle diameter (MPD), 319
Membrane filtration, 247–248
Membrane filtration: Fouling, 247
Mesh size (ASTM D2852), 317
Metals content, 322
Method detection limit (MDL), 56, 210
Method reporting limit (MRLs), 247
Methylperfluorooctane sulfonamidoaceditc acid (MeFOSAA), 225, 47, 39, 71, 72
Mg-aminoclay (MgAC) coated nZVI, 279, 281, 436
Microfiltration (MF), 247
Migration potential, 429; see also mobility
MIL-F-24385, 15–17
Military Specification (MilSpec), 16, 354, 373
Milk production, 179, 181
Mineralization of PFCAs, 399, 433–434, 131, 142, 268, 274, 279, 289–290
Minimum reporting limit (MRL), 57
Minnesota Pollution Control Agency (MPCA), 389
Mitochondrial bioenergetics, 151
Mobile phases,HPLC, 53, 57, 206, 210–211, 213
Modified clay, MatCARE, 396
Molecular ion fractionation, 206, 218
Monomethylbranched isomers, 212
Morris water maze, 147
mRNA expression, 229
Multiple extraction procedure (MEP), 412
Multiple hearth furnace, 306, 312
Multiple-point high-voltage electrode reactor, 285
Multi-stage gas-liquid electrical discharge column reactor, 285

N

Nanofiltration (NF), 247, 458
Nanomaterial technology, 435; see also emerging technologies
Nanoscale ZVI (nZVI), 259, 276, 279, 281, 283, 430, 435–436
National Health and Nutrition Examination Surveys (NHANES), 124, 182
National Primary Drinking Water Regulations (NPDWRs), 105
Natural AFFF impacted groundwater, 375
Natural organic matter, 217, 245–246, 257–259, 332–333
Natural streams/water, 357, 245
Necrotic, 150
Neoplastic effects, 148–149
N-ethyl perfluorooctane sulfonamidoacetic acid (N-EtFOSAA), 220, 248
N-ethyl perfouorooctane sulfonamide (N-EtFOSA), 206, 220–221, 248
Neurodevelopment, 185–186

Subject Index 493

Neurotoxicity, 147–148
No observed effect levels (NOELs), 128–134
No-aqueous phase liquids (NAPLs), 429
Non-FTA AFFF impacted sites, 177, 353–372
Nonneoplastic effects on organ systems, 127–148
Non-thermal plasma, 282
NSF/ANSI standard 61, 316
Nuremberg field pilot test, 406–407

O

Obesity & hepatic hypertrophy, 174–176
Obesogens, 183–184, 187
Occupational Environmental Health Site Assessment (OEHSA), 98
OECD 416 studies, 128
OECD 422 design, 127
Off-site reactivation, 304, 315
Offspring weight, 175
Oleophobic, 4
Onset of puberty, 177
On-site carbon reactivation, 304, 316
Oral exposure to PFAS, 125, 127
Oral gavage, 230
Organic content of soil, 384
Organic matter, 244, 408
Organic solvent extraction, 394
Organophilic clays, 410
Orthogonal chromatography, 41
Oversea military installations and host nations (HNs), 92
Overseas Environmental Baseline Guidance Document (OEBGD), 94
Overweight, 183–184
Oxidants, 257
Oxidation and reduction approaches for PFAs removal, 255–302
Oxidation, 330, 336
Oxidative degradation, 433
Oxidative stress response, 231
Ozonation, 242, 248, 432, 459–460
Ozone/Hydrogen peroxide (O3/H2O2), 242

P

Pancreatic tumors, 149
Parenteral administration, 125
Particle-induced gamma-ray emission (PIGE), 48, 69
Particulate matter, 454–455
Partition between maternal and fetal blood & tissues, 454
Partitioning coefficients, 455
Partitioning in sediment, 454–455
PerfluoAd, 405–407
Perfluocarbon intermediates, 274

Perfluorinated carboxylic acids (PFCA), 123–127, 138, 141, 143–147, 354–355, 366–368, 384–384, 387–388, 397–398, 418, 420–421, 456
Perfluorinated compounds (PFCs), 41, 49, 104
Perfluorinated sulfonic acids (PFSA), 4, 7, 37–38, 41, 67–68, 74, 123–127, 138, 141, 143–147, 453–455, 459–460
Perfluoroalkane sulfonamide ethanols (FOSEs), 220–221, 243, 248, 427
Per-fluoroalkyl sulfonamids (FASA), 256, 425, 427
Perfluorodecanoic acid (PFDA), 124–127, 135–156
Perfluorododecanoic acid (PFDoDA), 127–128, 132, 135, 139, 141, 145–147, 151–152
Perfluoroheptanoic acid (PFHpA), 124, 126, 136
Perfluorohexane sulfonate (PFHxS), 7, 10, 38, 44–47, 71–74, 117, 125–130, 135, 138–139, 141, 146–152, 387, 393, 412, 421, 427, 429, 448–451, 453–460, 473, 477
Perfluorohexanoic acid (PFHxA), 7, 38, 44–47, 71–74, 117, 125–130, 135, 138–139, 141, 146–152, 387, 393, 412, 421, 427, 429, 448–451, 453–460, 473, 477
Perfluoromethyl isomers, 206
Perfluorononanoic acid (PFNA), 124–127, 135–156, 176, 179, 180, 182
Perfluorooctadecanoic acid (PFOcDA), 127, 132, 135, 139, 141, 147, 151
Perfluorooctane sulfonamide acetic acid (FOSAA), 248
Perfluorooctanesulfonyl fluoride (POSF), 215
Perfluoropentane sulfonate (PFPeS), 124–125
Perfluoropropanoic acid (PFPrA), 47, 257, 263–265, 270–273, 278, 284, 456–457
Perfluorosulfamide amino carboxylate (C4 – C8 PFSaAm), 409
Periodate, 274
Permanently confined micelle arrays (PMCAs), 246
Permanganate, 257
Permeable reactive barrier, 436
Peroxisomal β-oxidation, 135
Peroxisome proliferation, 133–135, 139, 175, 453
Peroxy radical species, 266
Peroxydisulfate (persulfate), 397–398
Persistent organic pollutant (POP), 456
Persistent, bioaccumulative and toxic (PBT), 7
Persulfate activation, peroxide, 398
Persulfate activation: heat, 260–261, 398
Persulfate activation: hydrothermal, 258
Persulfate activation: iron activation, 259–260
Persulfate activation:microwave heating, 258
Persulfate alkaline pH-activation, 260
Persulfate oxidation, 390–391, 397–399, 432
Petroleum hydrocarbons (TPHs), 396, 406, 408, 435, 304

PFAS analysis methods for water matrices, 44–47
PFAs bioavailability, 411–412
PFAS contaminated soil remediation: Nondestructive methods, 389–396
PFAS contaminated soil: removal, 389–390
PFAS contaminated soil: Destructive methods, 396–398
PFAs contributing source, 364, 369, 214
PFAs decomposition by activated persulfate, 260–261
PFAs decomposition by electrochemical oxidation, 263–269
PFAS decomposition by sonolytic degradation, 269–276
PFAs decomposition by ZVI, 279
PFAS defluorination efficiency, 249, 261–263, 268–269, 275, 280–281
PFAS degradation efficiency, 262, 268–269
PFAS fate & transport, 27, 353, 429, 437
PFAs fate in soil, 384–387, 428
PFAS HALs: Regulatory process, 108
PFAS HALs: Response actions, 106–107
PFAs in ground water, 427–430
PFAS in human blood samples, 225, 227, 427
PFAS in human plasma samples, 212, 222, 223
PFAs in human serum samples, 443, 206, 210–211, 214–215, 227–228
PFAS in precipitation, 457
PFAS in snowmelt sample, 457
PFAS isomer: in Japanese rice fish, 220
PFAS isomer: in rats, 220
PFAS isomers, characterization, profiling and toxicity, 203–238
PFAS pharmacokinetics: excretion, 125–127, 451
PFAS potency analyses, 152
PFAs precursors, 386–387
PFAs precursors, bioavailability, 387
PFAs precursors, biodegradation, 243, 430
PFAS radicals, 266
PFAS removal from soil: Supercritical fluid extraction, 394
PFAS sorption & mobility in soil, 384
PFAS sorption to humic acids, 450
PFAS sorption to lab equipment, 450
PFAS sorption: electrostatic interaction, 246, 408–409, 412, 450
PFAS sorption: Hydrogen bonding, 450
PFAS source tracking, ECF-based products vs Telomerized products, 214–215
PFAS source tracking: direct vs indirect exposure, 215–218
PFAs sources, 418
PFAS temporal trends in humans, 224–225
PFAs toxicity from early life exposure, 171–187
PFAS, nomenclature, 419
PFAS-contaminated FTAs, 116

PFASs monitored in UCMR3, 87
PFBA, 124, 245, 262, 268, 448, 452, 454–455, 458–459, 385
PFBS, 124, 245–246, 268, 385, 393, 448, 450–451, 453–455, 456, 459–460
PFCAs & PFSAs, 123
PFDA, 124, 386, 393, 418
PFDeA, 268, 432
PFDoDA, 139, 141
PFECHS, PFPCPeS, 219
PFHpA, 124, 245, 262, 393, 448, 451, 454, 456–458, 460
PFHpS, 274, 448, 451
PFHxA, 262, 268, 393, 448, 451, 453, 454, 456–458, 460
PFHxS, 124, 246, 268, 274, 450–451, 454, 457–458, 460
PFHxS, linear and branched isomers, 231
PFNA, 124, 176, 393, 453
PFOA, 245, 248, 261–262, 268, 279, 286–287, 289, 386, 407, 432, 433, 467–482, 472
PFOcDA, 141
PFOS, 6–29, 246, 261, 274, 286, 288–289, 378–379, 386, 407, 433, 456, 467–482
PFOSA, 206, 357
PFPeA, 393, 418, 448–449, 454, 456–458
PFPeS, 454
PFPnA, 262
PFPrA, 456
PFUnDA, 141
PFUnDA, 418
Pharmacokinetics, 125–127, 450–451
Phase I impact assessment, 115
Phase II assessment, 115
Photochemical oxidation, 432
Photodegradation, 248
Photo-induced electron, 280
Photolytic oxidation, (Photolysis), 248, 432
Photoreductive defluorination, 433
Physical carbon loss, 321
Physiochemical properties, 427
Pilot test: GAC and resin operations, 467–487
Pilot testing for groundwater PFAS removal, 467–482
Placental samples, 178, 182, 228
Plasma based water treatment (PWT), 249
Plasma degradation: Bubble gas-liquid interface, 286
Plasma process, 397
Plasma technology, 282–291
Plasma-reactive species, 285
Point-of-use (POU) devices, 459
Polycyclic aromatic hydrocarbons (PAH), 406
PolyPFAs, 419, 427
Polytetrafluoroethylene (PTFE), 52
Pooled reactivation, 315

Subject Index

Portable water activated carbon, 316
Potassium permanganate, 242
Potential AFFF-impacted sites, 118
Powdered activated carbon (PAC), 244
Powdered activated carbon (PAC), 431, 459, 395
PPAR activation, 150
PPAR- and non-PPAR-mediated signaling, 140
PPAR β/δ, 150
PPAR-α, 231
PPAR-α activation, 137, 140–143, 148, 150–151
PPAR-γ, 136, 140, 150–151
Precursor bioaccumulation, 368
Precursor biotransformation, 355, 361, 369–370
Precursor compounds, 449
Precursor degradation, 216
Precursor isomers, 206
Precursor volatility, 76
Precursors, 256
Precursors: Perflurooctane sulfonamide-based side-chain polymers, 387
Preferential clearance, 230
Preliminary assessments (PAs), 117
Prenatal exposure, 174–175, 178
PrePFAs, 419, 427; *see also* PFAs precursor
PrePFOS, 216
Pressurized solvent extraction, 394
Prevalent PFOS isomers, 204
Primary standards, 84; *see also* MCLs
Prior history of breastfeeding, 183
Product ions of PFOS isomers, 207–208
Proper *vs* pool reactivation, 320
Property change during reactivation, 318
Prostate cancer, 148
Provisional health advisory (PHA), 456
Provisional health-related indication value(HRIV), 448
Provisional tolerable daily intake (pTDI), 152
Psychomotor development, 185
Purolite A520E, 245
Purolite A532E, 245
Purolite A600E, 245
Purolite Ferrl X A33e media, 245
Pyrolysis, 269
Pyrolyzation, 309

Q

QPL Database comparison, 18–19
QPL list MIL Spec MIL-F-24385, 20
Qualified products list (QPL), 15–20
Quality assurance and quality control, 54–55
Quality control (QC) considerations for PFAS analysis, 58–59
Quality of underlying analytical procedure, 75
Quantitation challenges, 57
Quantitation methods, 55–56

R

Racemic *vs* non racemic, 216
Radical scavengers, 259, 262
Rainwater sample, 457
Rapid small-scale column tests (RSSCT), 374
RCRA hazardous, 321
Reaction completeness, 76
Reaction efficiency, 259
Reactivated carbon quality control, 316–317
Reactivated carbon: Ash content, 320
Reactivated carbon: Particle shape, 319
Reactivated carbon: Particle size distribution, 319
Reactivated carbon: Transport pores, 307–308, 319–320
Reactivation chemistry/process, 306–311
Reactivation equipment & furnaces, 312
Reactivation furnaces, 306
Reactivation on PFAs destruction, 318
Reactivation options, 313–316
Reactivation process, 305
Reactive oxygen species (ROS) production, 140, 150, 231, 453
Read-Across Assessment Framework (RAAF), 151
Recovery and reuse, 343–344
Red blood cell (RBC) count, 147
Reduced birth weight, 176
Reductant, 257, 276
Reductive defluorination, 279
Reductive degradation, 433
Reductive dehalogenation, 434
Reference doses (RfDs), 153
Regulated contaminant, 105
Regulatory agency RMs for PFAS, 152–157
Regulatory background for PFAS, 114–116
Rejection capacities, 247
RemBind ingredients, 408
RemBind mechanisms, 409
RemBind plus, 396
RemBind, 396, 408–413
Rembind: Aluminum hydroxide (Al(OH)3), 408
RemBind: Kaolinite, 396, 408–409
Remedial Investigations (RIs), 118
Remediation of PFAS-contaminated Soil-
Renal tubular hypertrophy, 147
Reproductive toxicity, 452
Resin (Sorbix A 3F) operation summary, first loading cycle, 472
Resin breakthrough curves, 481
Resin exchange, 458–459
Resin regeneration, 245, 331, 458, 479–482
Resource conservation and recovery act (RCRA), 315
Reverse osmosis (RO), 458
Rinse cycle, 343
Risk communication, 107

Risk Management (RM), 35, 83–87, 109, 114–115
Risk management of PFAS in drinking water, 83–88
Risk mitigation, 410
Risk of autism spectrum disorders, 185
River bank filtration (RBF), 249
Rodents *vs* humans studies, 143
Root cause analysis (RCA), 96
Rotary kiln, 306, 313

S

Safe Drinking Water Act (SDWA), 84, 91, 105–106
Sand filtration, 459
Sanitary survey trends, 94–95
Scan-watch-act process, 115
SDWA management tools, 84–85
Second virgin resin loading cycle, 480
Secondary maximum contaminant levels (SMCL), 84
Sediment, 118, 214, 216–219, 356
Sequestering agents, 336
Serum clearance, 451
Serum corticosterone levels, 140
Serum lipid parameters, 230
Serum PFAS levels & lipids levels, 137
Serum PFOA concentration, 230
Serum testosterone level, 142
Serum thyroid hormone levels, 146
Sex organ weights, 142
Short chain PFAS source: Waste disposal sites, 449
Short chain PFAS source: Waste water effluents, 73, 428, 449
Short chain PFAS sources, 448–449
Short chain PFAS: In humans & biota, 456–457
Short-chain PFAS toxicity: In JEG-3 cells, 453
Short-chain PFAS toxicity, 451–454
Short-chain PFAS toxicokinetics, 451
Short-chain PFAs, 447–461
Shorter-chain and long-chain PFAS, 227
Simens A, 244, 245
Simultaneous characterization of multiple PFAS, 210
Site inspection (SI), 118
Sludge samples, 456
Sludge-amended soil, 386
Smart combination *in situ* oxidation/reduction (SISOR), 398
Sodium dithionite, 276
Sodium dodecylbenzene sulfonate (SDBS), 390, 392–393
Sodium dodecylbenzene sulfonate (SDBS), 392
Soil and groundwater PFAs remediation, 405–413
Soil and groundwater remediation, 383

Soil aquifer treatment (SAT), 249
Soil flushing additives, 392
Soil flushing, 392–393
Soil reference values (SRVs), 389
Soil remediation, 383–399
Soil retention, 456
Soil stabilization, 410, 412
Soil treatment, 411
Soil washing, 393–394
Solid phase extraction (SPE), 38–39, 357
Solids suspended & dissolved, 333
Solvent extraction, 394
Solvent recovery unit (Distiller), 349
Sonochemical degradation: Bubble collapse intensity, 269
Sonochemical degradation: Bubble oscillation, 269
Sonochemical degradation: Bubble-gas-liquid interface, 269
Sonochemical oxidation (sonolysis), 249, 258, 433–434
Sonochemical process, 269
Sonolysis: effects of surfactants, 275
Sorbix A3F: In-place regeneration, 468
Sorption barriers, 389
Source water assessment protection programs (SWAPPs), 94–97
Source water assessment protection programs, 97–98
Source water protection and root cause analysis (RCA), 96
Spatial and vertical distribution of PFASs, 428
Spatial distribution, 455
Spent carbon or exhausted carbon, 304
Spent carbon profiling & reactivation, 321
Spent carbon reactivation devolatilization, 308
Spent carbon reactivation 40/40/20 rule, 307
Spent carbon reactivation afterburner, 313
Spent carbon reactivation baghouse, 313
Spent carbon reactivation chemical scrubber, 313
Spent carbon reactivation chemistry/process, 306–311
Spent carbon reactivation drying of spent carbon, 307–308
Spent carbon reactivation equipment & furnaces, 312
Spent carbon reactivation gasification, 309–310
Spent carbon reactivation multiple hearth furnace, 312
Spent carbon reactivation pyrolyzation, 309
Spent carbon reactivation rotary kiln, 313
Spent carbon reactivation, 303–323
Spent carbon reactivation: Hardness and abrasion drop, 320

Subject Index

Spent carbon reactivation: rules, 311
Spent carbon reactivation: Segregated reactivation, 315–316
Spent carbon, dewatered, 307
Spent carbon, wet, 307
Spent carbon: ignitability, 322
Spent regenerant, 342–348, 468
Spermogenesis, 141–142
Splenic & thymic atrophy, 139
SREBP2, 229
Standardized anthropometric measurements, 184
Stationary phases, 206–213
Still-bottoms handling, 349
Stockholm Convention on Persistent Organic Pollutants, 256
Strategic Environmental Research and Development Program (SERDP), 117
Strongly acidic cation (SAC), 327
Strongly basic anionic (SBA), 327
Styrene, 326
Sulfate radicals, 258
Sulfonamide ethanols, 256
Sulfonamidoethanol (NMeFBSE), 418
Surface soil samples, 356, 456
Surface water, 38, 42–48, 50–53, 84, 94, 97, 219, 242, 304, 355–364, 374, 418–420, 448–450
Surface-mediated electron transfer process, 279
Symbios plasma reactor, 285
Synthetic precipitation leaching procedure, 412
Systemic half-life, 125–127

T

Technical standards, 210–211, 213
Terminal functionalities, 4
Tetrachlorethylene (PCE), 258
Tetrafluoroethylene (TFE), 6
Thermal cleavage of C-C bond, 285
Thermal plasma, 282
Thermal reactivation of spent carbon, 305
Thermodynamic index of irreversibility (TII), 395
Thyroid follicular cell adenoma, 146, 148
Thyroid follicular cell hypertrophy, 145
Thyroid hormone deficiency, 175
Thyroid toxicity, 145–147
Tin oxide (SnO2), 263
Tissue distribution, 451
Tissue partitioning, 126–127
Titanium oxide (TiO2), 263
TOC removal, 377
TOP assay, 48, 70–77; *see also* in "Emerging technologies"
TOP assay: Molar yields, 71–72

TOP assay: Regulatory applications, 74–75
Total administered dose (TADs), 137
Total fluorine analysis or combustion ion chromatography (CIC), 43
Total mineralization, 274
Total organic content (TOC), 304, 374, 377
Total PFAs method ("ΣPFAs"), 205
Toxic equivalency factor, 152
Toxic substances control ACT, 114, 452
Toxicity characteristic leaching procedure (TCLP), 412
T-PFOS, 219, 229–230
Transcriptional responses, 231
Transformation products, 267
Transition metals, 258
Transplacental transfer efficiencies (TTE), 227
Transplacental transfer, 226
Treatment technique (TT), 105
Treatment technologies, 430, 458–460
Treatment technology effectiveness & cost, 390–391
Treatment train of AOP and GAC, 373–381
Treatment train, 375
Treatment walls, 389
Trichloroethylene (TCE), 258
Trifloroacetic acid (TFA), 449
Two way ANOVA, 361
Type 1 SBA IEX resin, 328
Type 2 SBA IEX resin, 328

U

UCMR3 (Third Unregulated Contaminant Monitoring Rule), 87–88
Ultrafiltration (UF), 247
Ultra-high-performance liquid chromatography (UPLC), 376
Ultra-short chain products, 76–77
Ultrasonication (sonolysis), 269–282
Ultraviolet irradiation, 258
UMSICHT, 405
Unintended consequences for UCMRs, 108
United States Forces Japan Public Health Working Group (JPHWG), 95–96
University of Queensland study, 411–412
Unregulated Contaminant Monitoring Rule (UCMR), 108
Unregulated contaminants, 105–107
USAF Civil Engineer Center (AFCEC), 116–117, 467
USAF installations, 355
USAF PWS, 104
USEPA 537 REV 1, 49, 537
UV irradiation pathway, 433
UV photolysis, 242, 248
UV/H2O2 AOP, 242, 375, 432

V

Vacuum ultraviolet (VUV), 432
Vacuum-enhanced multiphase extractor (VEMPE), 431, 391
Validation tests (Rembind), 412
Van der Waals, 409
Vapor phase carbon pools, 315
Virgin carbon makeup, 321
Virgin carbon, 304
Vitrification & incineration, 396–397
Volatile halides, 322
Volatile organic compounds (VOCs), 258, 304, 433
Volatilization of HF, 267
Volatilization of hypofluorous acid (HOF), 267
Volatilization, 243
Volume percent of char, 319

W

Wastewater treatment plant (WWTPs), 242–243, 448
Water Research Foundation (WRF), 97
Water treatment technologies for PFAs removal, 241–254
Weakly acidic cation (WAC), 327
Weakly basic anionic (WBA), 327

X

Xenobiotic metabolic enzyme induction, 136

Z

Zerovalent iron (ZVI), 258–259, 276, 434
ZVI aggregation, 276